THE ROUTLEDGE HANDBOOK OF TOURISM GEOGRAPHIES

The Routledge Handbook of Tourism Geographies, 2nd Edition, offers a comprehensive re-evaluation of the recent developments; conceptual, theoretical and empirical debates; and critical issues in this field of study.

Reflecting on and building from its original aim of rethinking geographical approaches to tourism, the volume explores contemporary tourism contexts and concepts, as marked by the present era of polycrises, setting out renewed and reoriented perspectives on tourism geographies into the mid-2020s. Across its diverse range of contributions, the Handbook navigates the complexities of tourism as a shifting construct, situating tourism geographies within the socio-spatial, economic and environmental implications of tourism, leisure and mobilities in the new contexts of global change, ecological transition and digital transformation. The volume aims to provide a nuanced and detailed analysis of established and emerging discourses and debates within tourism geographies, underscoring the field's inherent criticality and ideal positioning for understanding and catalysing complex global and local scenarios in contemporary tourism, leisure and mobilities.

Written by leading scholars in the tourism geographies field, this text is an invaluable resource for students, researchers and scholars working in the areas of tourism, geography and related disciplines, encouraging dialogue across areas of study.

Julie Wilson is Associate Dean for Research and Associate Professor in the Faculty of Economics and Business Studies at the Open University of Catalonia (UOC), and the current Chair of the Tourism, Leisure and Global Change Commission of the International Geographical Union (IGU). Her research interests focus on the analysis of tourism impacts and the socio-spatial transformation of urban/rural landscapes, the role of culture and creativity in the generation of new forms of sustainability in tourism, geographies of the platform economy and evolutionary economic geography as interpretative frameworks for sustainable tourism topics.

Dieter K. Müller is Professor of Human Geography, Umeå University Sweden and a former Chair of the IGU Commission of Tourism, Leisure and Global Change. His research addresses the geographies of second homes and the relationship between tourism and regional change in northern peripheries. Furthermore he has an interest in the institutional development of tourism geographies.

"Twelve years after the first Handbook, leading researchers from around the world have compiled an impressive update on research trends in tourism geographies. Key geographical concepts such as space, place, environment, sustainability and mobility provide a powerful framework for examining a wide range of issues related to tourism. The innovative and comprehensive approach will help us to imagine a more diverse, socially and environmentally just world of tourism."

Professor Carolin Funck, *Hiroshima University, Japan*

"Global travel is radically affecting places, people and ecosystems. The new Handbook is not just an update of the original text from 2012, but a comprehensive and much-needed sequel which provides critically evocative perspectives on tourism-related transformations in the Anthropocene. The editors have curated a compelling collection of original and theoretically novel contributions to uncover the contemporary geographies of tourism and mobilities. This is a much welcome volume pushing the boundaries of tourism social science."

Dr. Szilvia Gyimóthy, *Copenhagen Business School, Denmark*

"This new *Handbook of Tourism Geographies* delivers a rich, authoritative collection of cutting-edge chapters by leading authors in the field. It provides a contemporary perspective on the development and future of the field and is a must have reference book for researchers, teachers and students of tourism geographies. It will become a classic in its field."

Professor Chris Cooper, *School of Events, Tourism and Hospitality Management, Leeds Beckett University, UK*

THE ROUTLEDGE HANDBOOK OF TOURISM GEOGRAPHIES

2nd Edition

Edited by Julie Wilson and Dieter K. Müller

Routledge
Taylor & Francis Group

LONDON AND NEW YORK

Designed cover image: Getty Images / Tatsiana Volskaya

Second edition published 2025
by Routledge
4 Park Square, Milton Park, Abingdon, Oxon, OX14 4RN

and by Routledge
605 Third Avenue, New York, NY 10158

Routledge is an imprint of the Taylor & Francis Group, an informa business

First edition published by Routledge 2011

British Library Cataloguing-in-Publication Data
A catalogue record for this book is available from the British Library

ISBN: 978-1-032-26051-8 (hbk)
ISBN: 978-1-032-26054-9 (pbk)
ISBN: 978-1-003-28630-1 (ebk)

DOI: 10.4324/9781003286301

Typeset in Times New Roman
by Newgen Publishing UK

For Eva and Biel (JW)

CONTENTS

Contents

Contents

FIGURES

TABLES

ABOUT THE EDITORS

Julie Wilson is Associate Dean for Research and Associate Professor in the Faculty of Economics and Business Studies at the Open University of Catalonia (UOC), where she teaches in the field of tourism geographies and sustainability. She began her academic career in the United Kingdom, where in 2002 she obtained her PhD in Human Geography from the University of West of England, Bristol. She has since held four competitive international postdoctoral fellowships (Batista i Roca Anglo-Catalan Programme, AGAUR/regional government of Catalonia; Marie Curie EU Intra-European postdoctoral fellowship; Beatriu de Pinós postdoctoral fellowship AGAUR/regional government of Catalonia and Fulbright Schuman Advanced Research Award, undertaking a Fulbright postdoctoral research stay based between Columbia University and City University of New York CUNY.

In 2016 she took up an associate professor post with the Faculty of Economics and Business at the Open University of Catalonia (UOC), while continuing her active long-term collaboration with the GRATET group of the URV until the present day. In 2017, together with the tourism and sustainability team at the UOC, she became a founding member of the New Perspectives on Tourism and Leisure (NOUTUR) Research Group.

In recognition of her outstanding international contribution to the field of tourism geographies, in 2019 she was awarded the Roy Wolfe International Prize and in 2022, the John Rooney Prize from the American Association of Geographers (AAG), followed in 2020 by an appointment as Chair of the Tourism, Leisure and Global Change Commission of the International Geographical Union, having been a board member for two previous periods (www.igutourism.org).

Her research interests focus on the analysis of tourism impacts and the socio-spatial transformation of urban/rural landscapes, the role of culture and creativity in the generation of new forms of sustainability in tourism, geographies of the platform economy and evolutionary economic geography as interpretative frameworks for sustainable tourism topics. She has published 10 edited books and more than 40 refereed journal articles and book chapters and is a member of the editorial boards of the JCR/SJR Q1 journals *Tourism Geographies* and *Current Issues in Tourism.*

In recent years she also has accumulated extensive experience in multilevel governance of sustainable tourism, as a member of the advisory group of the Barcelona Strategic Tourism Plan 2020 (2016), and as an external expert for the 'BleuTourMed' European Sustainable Tourism Community Horizontal Project (Interreg MED, 2018–19).

Dieter K. Müller holds a PhD from Umeå University and is now employed as Professor. Currently, he is Deputy Vice-chancellor, Umeå University, with special responsibility for research and research education within the Social Sciences and the Humanities. Müller has research interests with respect to tourism and regional development, mobility and tourism in peripheral areas. His research interests specifically include almost all aspects of second homes and second home related mobility, Sami tourism, nature-based tourism, tourism labour markets, regional development and rural change particularly in Northern peripheries and Polar areas.

Dieter is an elected member of the Royal Swedish Academy of Sciences. He is the past chair of International Geographical Union's Commission on Tourism, Leisure and Global Change. Moreover, Müller is one of the editors for *Geographies of Tourism* and *Global Change*, a book series published by Springer, and Resource editor for the *Scandinavian Journal of Hospitality and Tourism*. Also, he serves on the editorial board of *Current Issues in Tourism, Tourism Geographies, Matkailututkimus – The Finnish Journal of Tourism Research, Tourism and Hospitality Management*, the *Croatian Geographical Bulletin, Frontiers in Sustainable Tourism*, the *Journal of Arctic Tourism*, and *Zeitschrift für Tourismuswissenschaft*.

CONTRIBUTORS

Loretta Bellato is a settler woman from Melbourne, and an early career researcher. Loretta is a Regenesis Regenerative Practitioner Series graduate, Master of International Development, RMIT and Master of International Sustainable Tourism Management, Monash University. She is an adjunct researcher at the Centre for Urban Transitions, Swinburne University of Technology.

Asunción Blanco-Romero is Associate Professor in the Department of Geography at the Universitat Autònoma de Barcelona. She is a member of the TUDISTAR-UAB research group. Her research focuses on tourism and local development, territorial planning, tourism and degrowth, proximity tourism, and geography and gender issues in regional development.

Macià Blázquez-Salom is a Full Professor in the Department of Geography at the University of the Balearic Islands. His research interests include tourism, sustainability and spatial planning. As a way to link activism and research, he collaborates with social movements in Spain and Latin America.

Helen Briassoulis holds a PhD in Regional Planning, University of Illinois at Urbana-Champaign. She has taught at the University of Cincinnati. Since 1994, she has been with the Department of Geography, University of the Aegean. Her research interests include socio-ecological systems analysis, land use change, desertification, planning, policy integration and tourism.

Patrick Brouder is an Associate Professor at Thompson Rivers University, situated at the confluence of the North and South Thompson Rivers on the traditional and unceded Secwépemc territory in western Canada. Patrick leads research projects in the areas of Indigenous tourism, creative economies, and regional evolution. He is an editor of the journal *Tourism Geographies*.

Franz Buhr is a researcher at the Centre of Geographical Studies, University of Lisbon. His work lies at the intersection between migration and urban studies. He is interested in how migrants have shaped city spaces, their everyday mobilities and place-consumption practices. He is the coordinator of PriMob, the IMISCOE research initiative on privileged mobilities.

Ernest Cañada, PhD (Geography), is a Postdoctoral Margarita Salas Fellow at the University of the Balearic Islands. He was the founder and coordinator of Alba Sud, an independent research centre specialised in critical tourism studies. His research has focused on both Spain and Latin America.

Rita de Cássia Ariza da Cruz is graduate, Master and PhD in Human Geography, Department of Geography, University of São Paulo-USP, where she works as an Associate Professor. She is a board member of the International Geographical Union Tourism Commission and leader of the International Research Network "Tourism and contemporary socio-territorial dynamics". Her research and bibliographic production mainly involves Regional Geography and Tourism, political economy and the production of space.

T.C. Chang is Associate Professor of Geography at the National University of Singapore. His research interests include urban arts and culture, and Asian tourism development. He is co-editor of *Interconnected Worlds: Tourism in Southeast Asia* (Routledge, 2001) and *Asia On Tour: Exploring the Rise of Asian Tourism* (Routledge, 2009).

Joseph Cheer is Professor of Sustainable Tourism and Heritage at Western Sydney University. He is Co-Editor in Chief of *Tourism Geographies* and presently Co-Chair of the World Economic Forum (WEF) Global Future Council on the Future of Sustainable Tourism, Co-Chair of the American Association of Geographers (AAG) Recreation, Tourism and Sport Group, and board member International Geographical Union (IGU) Tourism Commission on Tourism Leisure and Global Change, and is a co-founder and current board member of Atlas Critical Tourism Studies – Asia Pacific (CTS-AP).

Harry Coccossis, PhD (Cornell) is Emeritus Professor of Spatial and Environmental Planning, University of Thessaly, Greece. Key public administration positions and international consultant on tourism planning, spatial and environmental planning, coastal zone management, island development. Scientific-research work and publications on tourism development planning, sustainable development and environmental planning.

Agustín Cocola-Gant has a PhD in human geography from Cardiff University, and is an assistant researcher at IGOT, the Institute of Geography and Spatial Planning at the University of Lisbon. His research lies at the intersection of urban and tourism studies, with a focus on the touristification of cities and the short-term rental market.

Tim Coles is Professor in the Department of Management at the University of Exeter Business School. His research interests are in the sustainable development of tourism, and he has a longstanding interest in innovation, business models and the environmental management of travel and tourism enterprises.

Katie D. Dudley is in the Department of Recreation and Leisure Studies at California State University, Long Beach. Her research is centred around the racialisation of labour, organisational conspiracies, and social harms within tourism systems with an emphasis on the sociocultural, political and economic factors shaping the tourism workforce.

Tara Duncan is a Professor at Dalarna University, Sweden, and Chair of ATLAS (Association for Tourism and Leisure Education and Research). Tara's main area of research focuses on the intersections between tourism, work, and mobilities with an emphasis on sustainability, decent work, and dignity within tourism and hospitality careers.

David Timothy Duval is a Professor at the University of Winnipeg. His interests are in the economic and legal regulation of commercial air transport in Canada.

Phoebe Everingham is a lecturer in the department of Geography and Planning at Macquarie University, Sydney. She is interested in embodied intercultural encounters within tourism spaces and the possibilities for emotions and affect to unsettle neo-colonial and neoliberal power dynamics in tourism development.

Robert Fletcher is an Associate Professor in the Sociology of Development and Change group at Wageningen University in the Netherlands. He is the author of *Failing Forward: The Rise and Fall of Neoliberal Conservation* (U of California, 2023) and *Romancing the Wild: Cultural Dimensions of Ecotourism* (Duke University, 2014).

Niki Frantzeskaki is a Chair Professor in Regional and Metropolitan Governance and Planning at the Department of Human Geography and Spatial Planning, Faculty of Geosciences, Utrecht University in the Netherlands. Her research interests focus on urban sustainability transitions and transformations and their governance.

Valérian Geffroy holds a PhD in Geography, and is currently a fixed-term lecturer and research associate at the École Normale Supérieure de Lyon. His research interests include tourism and mobility, outdoor sport, theories of practice and of spatiality, and critical studies of quantification.

C. Michael Hall is a Professor in the Department of Management, Marketing and Tourism, University of Canterbury, New Zealand. He also holds visiting positions at Kyung Hee, Linnaeus, Lund and Oulu universities. Co-editor of Current Issues in Tourism and Field Editor of Frontiers in Sustainable Tourism, he has published widely on tourism, regional development, global environmental change, sustainability and food.

Olga Hannonen is a researcher at the Business School, University of Eastern Finland, and an Honorary Research Associate at the Department of Geography, Royal Holloway, University of London. She specialises in human-nature interactions, lifestyle and trans-border mobility, including second-home tourism, residential mobilities and digital nomadism.

Edward H. Huijbens is an Icelandic Geographer and graduate of Durham University. He chairs Wageningen University's research group in cultural geography. Edward works on spatial theory, issues of regional development, landscape perceptions, the role of transport in tourism and polar tourism.

Dimitri Ioannides is Chaired Professor of Human Geography in the Institution of Economics, Geography, Law and Tourism and researcher in the European Tourism Research Institute (ETOUR) at Mid Sweden University. He is interested in matters relating to the social-equity dimension of sustainability, including the geographies of tourism workers and work. His latest co-authored books are the *Handbook of Tourism Impacts* and *Peer to Peer Accommodation and Community Resilience.*

Josep Antoni Ivars Baidal is Professor in Regional Geography and Director of the Institute of Tourism Research in the University of Alicante. He has worked in public administrations involved in tourism planning and management. His current research interests focus on smart tourism, smart destinations, tourism planning and innovation.

Gunnar Thór Jóhannesson (PhD), is Professor in the Department of Geography and Tourism at the University of Iceland. His recent research has been on destination dynamics, mobilities and placemaking with a focus on the entanglement of nature and culture in Arctic and sub-Arctic regions.

Natalie L. B. Knowles is a Climate Scientist, wilderness conservationist, and outdoor athlete. Her research focuses on sport, outdoor recreation and tourism. Within these sectors, Nat uses participatory research and creative writing to understand the root drivers of biodiversity loss and climate change, and aims to re-imagine our social-ecological systems to align with environmental justice and a sustainable future.

Dominic Lapointe is a Professor in the Department of Urban and Tourism Studies at Université du Québec à Montréal. He holds the *Chaire de recherche sur les dynamiques touristiques et les relations socioterritoriales* and leads the *Groupe de recherche et d'intervention tourisme territoire et société* (GRITTS) at UQAM, while being the head editor of Téoros, the oldest French language tourism studies journal. His work explores the production of tourism space and its role in the capitalist system expansion and its biopolitical dimensions. Its latest research looks at climate change, social innovations and critical perspective in tourism studies.

John Macilree worked as a Policy and Communications Adviser for the Government of New Zealand, including as an international air rights negotiator and airline economic regulator. He is Past President of the New Zealand Division of the Royal Aeronautical Society and is a licensed pilot.

Paolo Mura is a Professor at Zayed University, Abu Dhabi. He holds a PhD in Tourism from the University of Otago, New Zealand. His research areas explore tourist experiences and behaviour, including gendered experiences and representations in tourism, travelling subcultures, expressions of art, and critical and qualitative approaches to research.

Ivan Murray is an Associate Professor in the Department of Geography at the University of the Balearic Islands. His research merges political ecology, political economy and ecological economics of touristification.

Velvet Nelson is a Professor in the Department of Environmental and Geosciences at Sam Houston State University in Huntsville, Texas. She teaches undergraduate courses in human geography and environmental studies and has written the textbook *An Introduction to the Geography of Tourism* to support tourism geographies education.

Sanjay K. Nepal is a Professor in the Department of Geography and Environmental Management at the University of Waterloo, Canada. His research is focused on sustainable tourism and wildlife-human conflicts in protected areas. His most recent research is focused on the intersections between tourism, natural disaster and social capital in Indonesia, Nepal and the Philippines.

Spyros Niavis, PhD, is an economist and Assistant Professor of Regional Economics at the Department of Planning and Regional Development, University of Thessaly. His research interests include efficiency and productivity analysis, regional development, and sustainability assessment. He has worked extensively in transport, tourism, energy, and agriculture, especially in the Mediterranean.

Christian A. Nygaard is an applied social and urban Economist and Director of the Centre for Urban Transitions (CUT) at Swinburne University of Technology. Andi's primary research interests are in the processes of urban change and institutional innovation.

Pau Obrador-Pons is an Assistant Professor in Tourism and Events Management at Northumbria University in Newcastle. His research focuses on the relationship between place, culture, tourism, and management. He explores various topics including dwelling-in-mobilities, mass tourism, embodiment, mobilities, the beach, tourism education, and festivals and placemaking.

Rannveig Ólafsdóttir is a Professor of Geography and Tourism studies at University of Iceland. Her research concentrates on exploring the nexus between tourism and the environment, encompassing diverse aspects, such as tourism environmental impacts, climate change, sustainability, geotourism, spatio-temporal modelling, wilderness mapping, land use management and public participation.

Theodora Papatheochari, PhD, is a spatial planner and Adjunct Lecturer at the Department of Planning and Regional Development, University of Thessaly. She has extensive experience in sustainability assessment, spatial planning, and tourism development. Her scientific interests focus mainly on sustainable tourism, integrated coastal zone management, and maritime spatial planning.

Meng Qu holds a PhD from Hiroshima University, serves as an Associate Professor and Vice Director at the Center for Advanced Tourism Studies, Hokkaido University. Grounded in interdisciplinary perspectives encompassing tourism geography and rural/island studies, his research underscores the importance of socially embedded art, creative geography, and regional revitalisation.

Bernadette Quinn, PhD, is a Human Geographer and Senior Lecturer in the Faculty of Arts & Humanities, Technological University Dublin. Much of her research focuses on the roles that arts festivals play in transforming space, reproducing place and shaping identities. Her work is widely published in journals and edited collections.

Maartje Roelofsen is a Postdoctoral Researcher in the Cultural Geography Group at Wageningen University and Research in the Netherlands. Within the context of tourism, Maartje's research has focused on the impact of digitalisation on hospitality and tourism work and understandings of home and (un)homeliness.

Antonio Paolo Russo is Professor of Urban Geography with the Department of Geography, Rovira i Virgili University, Tarragona, Spain. Dr. Russo's main research interests in his various publications and coordinated projects are the economic, cultural and social geographies of tourist cities and tourism development issues in cities and regions.

Jarkko Saarinen is a Professor of Human Geography at the University of Oulu, Finland. He serves as a Visiting Professor at the University of Johannesburg, and at the Uppsala University, Sweden. His research interests include sustainable tourism and localities, climate change adaptation, resilience, tourism-community relations, and nature conservation.

Anna Dóra Sæþórsdóttir is a Professor in Tourism Studies at the University of Iceland and has a PhD in human geography from the University of Oulu, Finland. Her main research interests are in nature-based tourism, tourist experiences, wilderness, tourism management, tourism carrying capacity, sustainable use of natural resources and rural development.

Daniel Scott is a Professor and Research Chair in the Department of Geography and Environmental Management at the University of Waterloo. He has worked extensively on the transition to a low carbon tourism economy and tourism sector adaptation to the complex impacts of a changing climate.

Filka Sekulova is a postdoctoral researcher in the Internet Interdisciplinary Institute at the Universitat Oberta de Catalunya, with a background in ecological economics, psychology, and urban environmental justice.

Noam Shoval is a Professor of Geography at the Hebrew University of Jerusalem. His main research interests are urban geography and planning, urban tourism, and the implementation of advanced tracking technologies in various areas of spatial research, such as tourism and urban studies and medicine.

Mathis Stock (BA in Geography (Bochum), MA in Geography (Paris), PhD in Geography (Paris)) is currently Full Professor at Lausanne University, Switzerland. He is editor-in-chief of the francophone journal *Mondes du tourisme* and heads the Swiss graduate school in digital studies. His research mainly relates to the modalities of inhabiting in contemporary societies in which mobility and digitality play a predominant role.

Theano S. Terkenli is a Professor and founding member of the first geography department in Greece, University of the Aegean, since 1994. She is faculty at various interdisciplinary and international graduate programs, researching, publishing and teaching in cultural geography, landscape geography and critical perspectives on tourism. She is also a board member of several international and Greek associations.

Maria Thulemark is a Senior Lecturer at Dalarna University, Sweden. Maria conducted her PhD in Human Geography (2015) and her research focuses on the links between tourism, hospitality, mobility and work. Her recent interest lies in decent work, intersectionality and hospitality careers.

Dallen J. Timothy is Professor of Community Resources and Development, and Senior Sustainability Scientist at Arizona State University. He also holds visiting professorships in Mexico, China, Spain, and South Africa. His current tourism research in Africa, Asia, North America, Europe and the Pacific islands includes perspectives on geopolitics and globalisation, international borders, pilgrimage, heritage management, retail consumption, human mobility, and sense of place.

Hazel Tucker is Professor of Tourism at the University of Otago, New Zealand. Publishing on tourism's influence on socio-cultural relationships and change, gender and tourism, colonialism / postcolonialism and emotional dimensions of tourism encounters, Hazel's most recent book is *Tourism and the Spectre of Unlimited Change* (Routledge, 2024).

Egbert van der Zee is working as a Tourism Policy Advisor at the municipality of Utrecht, the Netherlands. Previously, he held positions at Utrecht University and the University of Leuven. His main research interests are the changing impact of tourism on cities, spatial data analysis, platform economy and the interrelation of social media and tourism.

Gustav Visser is an urban geographer. He completed his undergraduate education at Stellenbosch University and received his doctorate from the London School of Economics and Political Science. He is a Professor of Geography at Stellenbosch University. His research concerns identity-based consumption and urban morphological change.

Sarah N.R. Wijesinghe is a lecturer at the Department of Marketing, Languages & Tourism, University of Lincoln, United Kingdom. Her research focuses on coloniality, neocolonialism, decolonisation, identity and representation, and critical theory.

Xiaojun Fan is an associate professor at the School of Business Administration, Guangdong University of Finance, Guangzhou, China. Her research interests include industrial heritage tourism and heritage interpretation.

Xu Honggang is Dean of the School of Tourism Management, Sun Yat Sen University, Guangdong. Her research interests are primarily in sustainable tourism. She has published over 100 journal papers, including in Tourism Management, Annals of Tourism Research and the *Journal of Sustainable Tourism*. In recent years, Professor Xu also initiated several special issues on gender and tourism. She is a member of the International Academy of Tourism Study. She is also the Chairperson of Chinese Tourism Geography Commission and vice-chair of the International Geographical Union Tourism Commission. Prof. Xu is a key expert within the Monitoring Centre for UNWTO Sustainable Tourism Observatories (MCSTO) and supports the monitoring of Huangshan world heritage site and Xidi and Hongcun world heritage site. She obtained the Roy Wolfe Award from the Recreation, Tourism and Sport Specialty Group of the American Association of Geographers for her contribution to tourism geography research.

Jonathan Yachin is a Senior Lecturer in Tourism Studies at the School of Culture and Society and a researcher at the Centre for Tourism and Leisure Research at Dalarna University, Sweden. Jonathan is interested in entrepreneurship, rural tourism, resilience thinking and the application of behavioural studies in tourism contexts.

FOREWORD

The world of tourism geographies continues to undergo enormous transformations as inflection points in its trajectory are enforced by wider global events, including tourism recovery, pandemic hangover, prolonged conflicts and wars, flaring geopolitical choke points, climate change and global warming, decarbonisation concerns, looming global economic recession, declining confidence in public institutions and governments, the rise of Artificial Intelligence and a social media environment that has become subject to disinformation and misinformation, among other dynamics. It is within this milieu that tourism geographies – tourism space, place and environment – as delineated by the journal *Tourism Geographies*, has gone on to develop its scholarly credentials.

In linking tourism geographies to wider geographical discourses, and in acknowledgement of the "2023 Geography Awareness Week," *National Geographic* made two statements that suggest that geography and tourism, make for an ideal coupling on account of the many crossovers:

> Geography isn't just about maps and coordinates; it's also about stories and offers different ways to learn, understand, appreciate and embark on a journey of reflection as we experience the world in our everyday lives. Geography is a way of looking at the world, a way of understanding why things are where they are and how people's movements and cultural, societal and political interactions and movements impact all of us.
>
> (www.nationalgeographic.org/society/education-resources/
> programs/geography-awareness-week/)

Such a positioning lends a convenient segue to commentary about the sub-field of tourism geographies, or the geographies of tourism – more or less the same thing. In his opening editorial remarks in the first issue of *Tourism Geographies*, Founder and Editor-in-Chief Alan A. Lew observed that there was "a realisation by many geographers that we are far better represented in the international academy of tourism scholars than one might expect for a discipline that is actually fairly modest in size" (1999b: 1).

The links between tourism and geography are argued by Lew to be intertwined in notions of place, "Place is such an encompassing concept that it requires a multidisciplinary and eclectic openness, a characteristic that has been both a hallmark and a challenge for geography. And place is also an intrinsic element of tourism, as all tourism involves some form of relationship between people and places that they call 'home' and 'not home'" (Lew, 1999b: 1). In the subsequent and second issue of *Tourism Geographies*, Lew referred to the emergence of "Tourism as a place and space phenomenon" (1999a: 141) as a hallmark of tourism geographies.

Prominent geographer, and arguably one of the earliest exponents of tourism geographies research, Richard W. Butler foreshadowed the evident links between tourism geographies and sustainability concerns, drawing on the Brundtland report which, at the time, had been in the public sphere for just over a decade. Butler quipped, "Geographers have long been interested in the relationships between tourism and the environments, both physical and human, in which it operates, and it is logical that they would be particularly interested in the discussion and application of sustainable development in the context of tourism" (1999: 8).

Together, Lew and Butler were arguably some of the earliest exponents ringing in tourism geographies and whose work and exhortations are fundamental to the foundation of the sub-discipline. Many others have subsequently grabbed the reins and taken up the momentum since. This second edition of *The Routledge Handbook of Tourism Geographies* led by Julie Wilson and Dieter Müller is evidence of the maturation of the sub-field, formerly at the periphery and now firmly established, in both tourism and geographies scholarly endeavours.

In the most recent Scopus Cite score Tracker monthly update (5 November 2023 at the time of writing), the journal *Tourism Geographies* was ranked 1st in the subject area, Tourism, Leisure and Hospitality Management, and 2nd in Geography, Planning and Development, respectively. While journal rankings are but one measure of overall esteem, this achievement signals that the sub-field has, for all intents and purposes, arrived as a serious undertaking in the scholarly sphere.

Coupled with the emergence of tourism geographies, is its presence in the activities of international institutions. This is evidenced in the popularity and following of the study area in the International Geographical Union (Commission on Tourism, Leisure and Global Change), American Association of Geographers (Recreation, Tourism and Sport), Royal Geographical Society (Geographies of Leisure and Tourism Research Group), Human Geography Commission of China Society of Geography, and the Spanish Asociación Española de Geografía, among others.

Notwithstanding, the need for rigour and robustness to underpin tourism geographies research, Dieter Müller (2019) has been at the forefront in tracing the trajectory of the sub-field and in *A Research Agenda for Tourism Geographies*, he highlights the travails of tourism geographies research, emphasising the blurring of boundaries, and raises questions about how exactly to define it, and how institutional shifts and research trends continues to shape and reshape it.

Müller's expansive bibliometric review mapping the development of the tourism geographies subfield is revealing in that it shows robust and steady growth in publications related to it. Moreover, Müller (2019: 20) expresses caution against assuming that the foundation, now established, could withstand wider institutional and disciplinary jostling: "institutional changes, with a dispersion of tourism geographers to departments outside geography, have created a situation in which tourism geographies may be challenged as a subdiscipline of geography, but at the same time have allowed the opportunity to break out of the sometimes rigid frameworks and practices of academic disciplines".

In the inaugural and first edition of *The Routledge Handbook of Tourism Geographies*, Julie Wilson (2012) called for consideration of post-disciplinarity, bringing attention to the way the progression of the sub-field was being developed by scholars who were multivalent in their disciplinary origins and predilections. Considering that this appeal was made just over a decade ago, the second edition of *The Routledge Handbook of Tourism Geographies* promises a comprehensive update of the sub-field.

Over the ensuing years, tourism geographies has consolidated and grown from its unassuming origins to now be positioned as a bona fide sub-field of its parent, geography. At the same time, its multi, cross and interdisciplinary nature has seen it morph into a legitimate pathway for questions around tourism and the evident intersections with space, place and environment, as well as the implications of all of this for the people and places in which this unfolds.

As Chris Gibson opined, "Geographers have been responsible for innovations in tourism research, some of which were substantial in an interdisciplinary sense", and "Geographers researching tourism have long sought audiences outside geography – perhaps even more so than within their discipline" (2008: 408). That said, Merigó et al.'s (2019) bibliometric analysis of twenty years of *Tourism Geographies* shows a vast network of internationally diverse research, with a rich disciplinary pedigree, and advancing both theoretical and empirical contributions.

Notwithstanding the advances evident in this volume, Müller (2018: 172) offers a reminder that hubris and complacency must be cautioned against, for the vicissitudes and fickleness of institutions and scholarly disciplines means that "breaking out from the cages of academic routine is not an easy endeavour, not least since academic disciplines come along with paradigms and strict set of traditions about what problems have to be researched and in what ways".

Joseph M. Cheer
Western Sydney University, Australia

References

Butler, R.W. (1999). Sustainable tourism: A state-of-the-art review. *Tourism Geographies* 1(1), 7–25.

Gibson, C. (2008). Locating geographies of tourism. *Progress in Human Geography* 32(3), 407–422.

Lew, A.A. (1999a). A foundation in place and space. *Tourism Geographies* 1(2), 141.

Lew, A.A. (1999b). A place called tourism geographies. *Tourism Geographies* 1(1), 1–2.

Merigó, J.M., Mulet-Forteza, C., Valencia, C., & Lew, A.A. (2019). Twenty years of Tourism Geographies: A bibliometric overview. *Tourism Geographies* 21(5), 881–910.

Müller, D.K. (2018). Time to reconsider tourism geographies? *Tourism Geographies* 20(1), 172–174.

Müller, D.K. (2019, Ed.). *A Research Agenda for Tourism Geographies*. Edward Elgar.

Wilson, J. (2012, Ed.). *The Routledge Handbook of Tourism Geographies*. Routledge.

ACKNOWLEDGEMENTS

JW:

I am writing these acknowledgements in December 2023 at an incredibly special place in the Osona region of Central Catalonia; the *Santuari de Bellmunt*. Standing proudly on top of a rocky outcrop at 1246m above sea level and built directly into the rock, this sanctuary/hostelry has incredible views over the Plain of Vic, the Pyrenees, the Ges Valley and the sub-county of the Bisaura. In fact, on clear days, the mythical, mysterious mountain of Montserrat can even be seen from here. This place was introduced to me by my wonderful friend and colleague, Maartje Roelofsen, and due to its stunning geographical aspect, it feels somewhat like being on top of the world, or 'your secret in the sky', as owners and self-proclaimed hermits Marina and Santi like to refer to it.

In kicking off these acknowledgements, I would like to start by thanking Maartje, not only for clueing me into this place as the most perfect writing retreat imaginable, but also for her great company and encouragement over our past three years as colleagues at the Universitat Oberta de Catalunya. This was definitely not an easy book to complete, what with a serious pandemic in between, numerous delays and both Dieter and I having significant institutional roles and responsibilities, so finishing the volume in this beautiful place left a warm and positive memory of the whole process. Dank-u-wel, my friend.

I am particularly grateful for the support and encouragement of my academic 'home' – the Faculty of Economics and Business of the Universitat Oberta de Catalunya, with very special thanks to Xavier Baraza Sánchez, Mª Jesús Martínez Argüelles and Àngels Fitó Bertran, all three of whom have, as successive Deans of the Faculty of Economics and Business, continually paved the way for my research, including this volume. Thanks also to all the people that form part of my two research 'families' – the NOUTUR Research Group (UOC) and the GRATET Research Group (Rovira i Virgili University).

To Segis: Thank you for absolutely everything and more.

And of course, I would like to thank my co-editor Dieter Müller most immensely, for being such a collegiate, generous and patient co-editor and truly an exceptional scholarly example to follow, with the added bonus of having a fantastic sense of humour. This has been a long process, though undertaken in the best possible company – många tack, min vän.

DM:

I am fortunate to be blessed with fine colleagues at an excellent geography department at Umeå University. It is one of the few places around the world where tourism geographies were allowed to develop into a strong and integrated part of geography. This makes it an attractive research environment and certainly, I would like to thank my colleagues and friends at the department for making this possible and thriving. The Umeå department is actually a lot of fun and indeed, while I am away for administrative assignments, I always look forward to returning.

Being a member of an international community of tourism geographers provided during all years a great deal of inspiration and fun as well. Hence, contributing this volume and hopefully adding to the development of a field of research gives actually great satisfaction. I am thus very grateful to Julie Wilson for inviting me to join her on this journey. And even though the project was conducted in difficult years and took longer time than anticipated, it was always a great pleasure to have online chats with Julie regarding the progress of the book project and a lot of other things. Hence, I hope for many more meetings and joint undertakings with Julie in the future and certainly, the result of this collaboration is not only a book but also a friendship – thank you Julie and welcome to Umeå whenever you want.

In terms of thanks that are due on both our parts, it is first of course very important to acknowledge all of the contributors to this Handbook and thank them for their efforts, consistency and professionalism. Many colleagues around the globe have had – and continue to have – a very tough time in the wake of the COVID-19 pandemic, current geopolitical conflicts, environmental crises and many other turbulent situations. This has meant that the project has been much more complex to complete than we could have imagined, but given the highly challenging contexts that many have experienced, managing to come through and deliver quality chapters in spite of these circumstances is highly admirable.

Supporting us throughout the past few years has been our editorial team at Routledge/Taylor & Francis, Emma Travis and Harriet Cunningham; always constructive and friendly, not to mention patient and polite! Many thanks to you both, as well as to the whole team behind the Handbook's production.

As with the first edition, this new Handbook was also developed very much under the auspices of the International Geographical Union's Commission on Tourism, Leisure and Global Change. As the current and former chairs of this Commission, we are extremely proud of its achievements and activities over the last few decades and we definitely consider the present volume as an endeavour 'made in IGU Tourism'. Many thanks to all the IGU Tourism Commission current board members, most of whom have contributed chapters to the volume: Joseph Cheer, Rita da Cruz, Edward Huijbens, Dominic Lapointe, Velvet Nelson, Bernadette Quinn, Theano Terkenli, Vuk Tvrtko Opačić, Gustav Visser and Xu Honggang, as well as former chairs C. Michael Hall, Jarkko Saarinen, Alan A Lew, all of the honorary board members and the IGU President, Mike Meadows.

Sincere thanks are due to the whole editorial team behind the landmark T&F journal *Tourism Geographies*, without a doubt trailblazing the future of our field and going from strength to strength in terms of its academic and social impact internationally, with particular thanks to co-editors-in-chief, Joseph Cheer and Mary Mostefanezhad and former editor-in-chief, Alan Lew.

We would also like to thank the various academic organisations in the tourism geographies domain for their ongoing dynamism and energies in maintaining our field year after year, many of whom have also contributed chapters to this volume: the ATLAS Special Interest Group (SIG) on Space, Place, Mobilities in Tourism (coordinated by Paolo Russo, Chiara Rabbiosi and Wilbert den Hoed); the Association of Spanish Geographers (AGE) Working Group 10 on Geography of Tourism, Leisure and Recreation (coordinated by Josep Ivars, Carmen Hidalgo, Anna Torres and Paco Femenia); the Recreation, Tourism and Sport Specialty Group of the Association of American Geographers (AAG) (coordinated by Jennie Germann-Molz and Joseph Cheer); and the Geography of Leisure and Tourism Research Group (GLTRG) of the Royal Geographical Society (with the Institute of British Geographers) (coordinated by Tijana Rakić).

Finally, we are also very grateful to the following collaborators and friends for all the conversations, cheer and goodwill over the years: Alberto Amore, Jordi Andreu, Salvador Anton Clavé, Albert Arias, Bailey Ashton Adie, Andreas Back, Loretta Bellato, Macià Blázquez, Patrick Brouder, Harald Buijtendijk, Ernest Cañada, August Corrons, Raquel Camprubí, Doris Carson, *'the'* Maria Casado-Díaz, *'the'* Ana Casado-Díaz, Alba Colombo, Chris Cooper, David Cullen, Carolyn Daher, Javier Delgado, Cenk Demiroglu, Tara Duncan, Amal Elasri, Alberto Forte, Koen Frenken, Txell Fuguet, Carolin Funck, Christian Garaus, Lluís Garay, Maria Laura Gasparini, Joan Miquel Gomis, Francesc González, Sara González, Brynhild Granås, Antonio Guevara, Inés Gutiérrez Cueli, Aaron Gutiérrez, Szilvia Gyimóthy, C. Michael Hall, Henrik Halkier, Gijsbert Hoogendoorn, Ed Huijbens, Dimitri Ioannides, Laura James, Hubert Job, Linda Lundmark, Hug March, Roger Marjavaara, Alessia Mariotti, Marius Mayer, Carles Méndez, Claudio Minca, Soledad Morales, Kike Navarro, Vuk Tvrtko Opačić, Richard Pettersson, Robert Pettersson, María José Piñeira, Enrico Porfido, Solène Prince, Maite Puigdevall, Meng Qu, Bernadette Quinn, Ramon Ribera Fumaz, Trini Rovira, Paolo Russo, Jarkko Saarinen, Cinta Sanz Ibañez, Regina Scheyvens, Noam Shoval, Andrew Smith, Elsa Soro, Pere Suau-Sanchez, Dallen Timothy, Egbert van der Zee, Konstantina Zerva, Jazmine Zhang, Malin Zillinger.

Barcelona/Umeå, Spring 2024

PERMISSIONS ACKNOWLEDGEMENTS

Special thanks are due the academic publishers Taylor & Francis, Sage and Elsevier for their kind permission to include material previously published in their journal publications as Chapters 4 (Mura & Wijesinghe), 5 (Blázquez-Salom, Murray, Fletcher, Sekulova, Blanco-Romero & Cañada), 11 (Everingham, Obrador & Tucker), 23 (Bellato, Frantzeskaki & Nygaard) and 24 (Saarinen & Nepal), earlier or partial versions of which were previously published under the following auspices:

Mura, P. & Wijesinghe, S.N.R. (2023). Critical theories in tourism – a systematic literature review. *Tourism Geographies* 25(2–3), 487–507. DOI: 10.1080/14616688.2021.1925733

Murray, I., Fletcher, R., Blázquez-Salom, M., Blanco-Romero, A., Cañada, E., & Sekulova, F. (2023). Tourism and degrowth. *Tourism Geographies* 1–11.

DOI: 10.1080/14616688.2023.2293956

Everingham, P., Obrador, P. & Tucker, H. (2021). Trajectories of embodiment in tourist studies, *Tourist Studies* 21(1), 70–83.

DOI: 10.1177/1468797621990300

Bellato, L., Frantzeskaki, N. & Nygaard, C.A. (2022) Regenerative tourism: a conceptual framework leveraging theory and practice, *Tourism Geographies* 25(4), 1026–1046.

DOI: 10.1080/14616688.2022.2044376

Jarkko Saarinen & Sanjay K. Nepal (2016) Conclusions: Towards a political ecology of tourism – Key issues and research prospects. In: Sanjay Nepal and Jarkko Saarinen (eds.) *Political Ecology and Tourism* (pp. 253–261). Routledge.

Additionally, Figure 21.1 from Chapter 21 (Bauer Knowles & Scott) was previously published as part of the following article and is reused in this volume with kind permission from Elsevier:

Scott, D., Hall, C. & Gössling, S. (2019). Global tourism vulnerability to climate change. *Annals of Tourism Research* 77(C), 49–61.

PART I

Introduction

1

KEEPING PACE WITH THE EVOLUTION OF TOURISM GEOGRAPHIES WITHIN SPACE, PLACE, SOCIETY AND ENVIRONMENT

Julie Wilson and Dieter K. Müller

Introduction

This volume represents a major reconsideration of the nuanced landscape of tourism geographies concepts and contexts as examined over a decade ago in the first edition of the *Routledge Handbook of Tourism Geographies*. The introduction to the first edition argued that in a time of post-disciplinarity, and identity crises in geography more generally, there was "a need to take stock of what has gone before within the sub-discipline of tourism geographies, prior to reorienting our thinking towards the future" (Wilson, 2012: 1). Looking back at this assertion in early 2024, firmly situated within that very future context and navigating the present era of complexity in the form of polycrisis ("a marker of our age, capturing overlapping and interconnected crises beyond cause and effect", Henig & Knight, 2023:3; Tooze, 2022) or permacrisis (Turnbull, 2022), the time has now come to once again reorient that 'reoriented' thinking from 2012 towards the next ten years and more of tourism geographies research and scholarship.

The previous introductory chapter (Wilson, 2012) outlined the need to debate which new (or renewed) research perspectives may be the most adequate for advancing our knowledge and emergent ideas on tourism activity in all its complexity and dynamism. It is not an easy task to assert which research perspectives are best adapted to making sense of a highly diverse series of complex and often polemical phenomena – tourism is after all a slippery, shifting construct; a mobile subject in itself. Understanding it has never been more challenging than in the critical underlying contexts of the 2020s (Bianchi & Milano, 2024) and, in parallel to these shifting contexts, it is important to analyse just how and why the past decade has reshaped the contours of the field of tourism geographies.

In taking on this somewhat major challenge, this chapter introduces a new and fully revised edition of the *Routledge Handbook of Tourism Geographies* and situates the field of tourism geographies as of 2024 in the context of academic, societal, economic, environmental and cultural change in relation to tourism over the past decade since the first edition in 2012. This Handbook

DOI: 10.4324/9781003286301-2

aims to comment on the intervening years since the first edition, marked by major societal, technological and environmental transformations. Meeting this aim necessitated a complete rethink and overhaul of the Handbook from a decade earlier. As such, readers will find that the contributions to this Handbook are for the most part entirely new (in fact, only two chapters have been reconfigured from the first edition), in response to the incredible transformation of the tourism geographies landscape and the concepts and contexts that form it.

In this sense, the decade or so that passed between the two editions of this Handbook has been one of considerable evolution within the conceptual, methodological and analytical landscape of tourism geographies, as well as a constantly changing socio-spatial, political, cultural and environmental reality. If the tourism geographies field had already shifted immensely before the COVID-19 pandemic, it is now once again undergoing a major shift, as new approaches emerge (including for example the 'regenerative turn', mobility justice, sustainability transitions and new post-pandemic politics) and the socio-spatial, economic and environmental implications of tourism, leisure and mobilities in the new context of global change, ecological transition and digital transformation start to become clearer.

The first edition (Wilson, 2012) aimed to respond to a clear increase in academic attention to lifestyle-related mobility and the use, creation and (re)presentation of space and place in tourism, as a state of the art in tourism geographies internationally. While this overall aim remains the same, the state of the art we are referring to has shifted enormously in the decade since the first edition was published. In recent years, some landmark new texts have been published, capturing the essence of just how much tourism geographies as a field and discipline have evolved (see, for example, Müller, 2019, which explored research approaches in tourism geographies from different perspectives).

Importantly, to this evolutionary trajectory we might add the considerable socio-spatial and geopolitical implications for the future of tourism and travel mobilities that the COVID-19 pandemic presented. As such, the main rationale of this new volume remains to provide a fresh examination of existing and emerging debates in tourism geographies in the context of disciplinary changes visible in geography and society more generally, continuing to highlight tourism geography's inherent 'criticality, pervasiveness and scope to catalyse cutting-edge research' as argued a decade and a half ago by Chris Gibson (2008: 15).

We aim to provide tourism geographies scholars with comparisons and contrasts between numerous conceptual frameworks, theoretical and methodological approaches that might be adopted when researching issues and topics related to tourism geographies and related fields. In terms of intended readership in teaching/learning context, the volume is aimed not only at researchers but also at advanced undergraduates and taught and research postgraduates in those fields where human geography and tourism studies intersect. However, the very nature of geography as an (in)discipline means that the volume will be salient for researchers from a diverse range of disciplinary backgrounds (Cresswell, 2024). In addition, in a climate where tourism geographers are increasingly to be found working outside of geography departments, the volume represents an opportunity for mainstream geographers to get an idea about what is on the agenda of their tourism-focused contemporaries.

The new Handbook aims to provide a bridge between the state of the art definitional territory of Wiley Blackwell's *Companion to Tourism* (Lew, Williams & Hall, 2014) and Müller's (2019) volume that was more focused on research perspectives on tourism geographies (Elgar), as well as Julie Wilson and Salvador Antón Clavé's 2013 volume, *Geographies of Tourism: European Research Perspectives* (Emerald), which also focused more on new and old research traditions, and national/linguistic diversity in the field.

Along similar lines to these earlier volumes, but at a broader scale and in more diversity of detail, this second edition aims to cover the state of the art in terms of what has been done over the past decade or so, how it has been done, and what needs to be done/could be done in the future.

Considerations When Using the New *Routledge Handbook of Tourism Geographies* as a Research Companion

This second edition includes new contributions from various fields and disciplines (sociologists, political scientists, economic geographers, cultural geographers, trained geographers now working outside of geography, and anthropologists, urban and regional studies scholars). As with the first edition, we hope to have left enough space for due consideration of post-disciplinarity questions within an essentially discipline-based field while at the same time not abandoning important disciplinary traits and traditions (see Müller & Hall, Chapter 2). Again, as with the first edition, contributing to this Handbook represented a significant challenge for authors (not to mention a somewhat complex editing task). A limited word count and a strict brief to map out each topic and set a research agenda within a piece around half the length of the average journal article is tough by anyone's standards. But we believe these contributions have more than managed this, representing a further strength of this volume: concise, yet reflexive, and comprehensive, yet clearly targeted for the task of mapping and advancing research in tourism geographies.

In terms of the (inevitable) limitations of this new edition, we would like to emphasise that despite our best intentions and efforts, the coverage and scope are certainly far from exhaustive. We hope that in spite of these circumstances, there is enough wider applicability for researching related tourism contexts and enough attention paid to certain topics within the final lineup of chapters, even if not the subject of a dedicated chapter in themselves. It was our original intention, for example, to include specific chapters on the geopolitical turn in tourism geographies, more-than-physical geographies, cultural political economy approaches, indigeneity, race and ethnicity perspectives, the tourism-migration nexus and labour geographies of tourism. However, over the course of the volume's development, some of these ideas were unable to be mobilised as chapters (especially given the impacts of the COVID-19 pandemic on many potential contributors' capacities). That having been said, these ideas have often been referred to more implicitly within the volume in proximity to other, related topics.

Structure

In the introductory Part I of this Handbook, Chapters 1–3 provide a comprehensive trajectory of recent developments in the field. Following this first introduction, in Chapter 2 Dieter Müller and C. Michael Hall set out the current state of the art in the field of tourism geographies, analysing how the relationship between tourism geographies and their parent discipline of geography has long been and still remains complex and often fraught. In Chapter 3, Rita de Cássia Ariza da Cruz reflects on the inherent challenges facing the geographies of tourism and development in the twenty-first century, contributing to the state of the art in research and knowledge on the intersection between these two domains.

Part II, 'Critical Geographies of Tourism', includes eight chapters with strong theoretical and epistemological orientations. Paolo Mura and Sarah Wijesinghe in Chapter 4 examine the diverse opportunities offered by critical theory in defining optimal scenarios for sustainable tourism, calling for a strengthening in the nexus between critical tourism scholarship and activism beyond academia. Chapter 5 examines the notion of degrowth in the context of tourism geographies. Inspired by a longstanding body of research problematising the imperative and consequences of economic growth more generally, initiated by social scientists, Macià Blázquez-Salom, Ivan Murray, Robert Fletcher, Filka Sekulova, Asunción Blanco-Romero and Ernest Cañada, who provide a state-of-the-art analysis of degrowth perspectives applied to debates on (sustainable) tourism development.

In Chapter 6, Dominic Lapointe conceptualises how tourism space is constituted, as framed via the theoretical triad of alterity, mobility, and territory, as well as the continuous territorialisation/deterritorialisation movement of tourism, drawing upon Deleuze and Guattari's philosophy. Taking a cue from the contributions to geography of actor-network theory (ANT) and assemblage thinking (AT), in Chapter 7 Helen Briassoulis sets out a critical agenda for incorporating assemblage thinking into research on tourism geographies, proposing 'assemblage tourism geographies' as a promising terrain for tourism studies.

Honggang Xu and Xiaojun Fan analyse the state of the art in critical studies of gender equality and sustainable development in tourism geographies in Chapter 8, arguing that tourism constitutes and is constituted by gendered scapes and that any socio-spatial understanding of tourism cannot be separated from gender perspectives. In Chapter 9, Gustav Visser provides a critical analysis of tourism, place, space and queer sexuality, setting out how the connections between place, space, sexuality and tourism that have been acknowledged over the past few decades are changing in both form and function.

Following this are two chapters that address the performative, sensory and embodied dimensions of tourism as a geographical research subject. Gunnar Þór Jóhannesson centres on performativity, space and tourism in Chapter 10, examining how the concepts of performance and space can be valuable for understanding tourism dynamics and their repercussions, while Phoebe Everingham, Pau Obrador and Hazel Tucker discuss the multisensorial notion of embodied encounters in tourism geographies research in Chapter 11, analysing how an 'embodied turn' has allowed consideration of a wider range of bodies in tourism research, beyond its traditional focus on male, white and middle-class figurations.

In Part III, 'Place Perspectives on Tourism Geographies', six chapters set out an agenda for a diversity of place-based approaches to tourism geographies research. In Chapter 12, Theano Terkenli considers landscape perspectives on tourism geographies research, exploring state-of-the-art inroads at the interdisciplinary interface of tourism and landscape, highlighting their range and significance, in terms of theory, empirical practice, approach, policy, ethics, and future prospects. TC Chang examines place making and unmaking in tourism in Chapter 13, arguing that just as places are 'made', they may also be contested, resisted and 'unmade', necessitating engagement with the politics of tourism place making.

Meng Qu's Chapter 14 on rural creative tourism geography for community revitalisation argues for new interdisciplinary research frameworks that avoid partial interpretations of the role of rural art interventions and underpin debates between art in society, society in art, art in tourism, tourism in art, tourism in rural areas, and rurality in tourism. Geographies of festival and event spaces and places are the focus of Chapter 15 by Bernadette Quinn, who points out that while festivals and events did not merit a dedicated chapter in the 2012 first edition of the Handbook, specifically geographical literature on festivals and events has burgeoned since then.

Mathis Stock and Valérian Geffroy ask some critical and salient questions in Chapter 16 as to what tourism geographies might gain from a more direct conversation with urban theories and they provide an analysis of geographical debates on tourism-driven urbanisation processes. In

Chapter 17, on historical geographies of tourism and heritage, Dallen Timothy, examines three geo-heritage settings of interest to historical geographies: tourism landscapes, human mobilities and religious tourism.

Part IV, 'Sustainability Transitions in Tourism Geographies', contains seven chapters and takes a major cue from Piotr Niewiadomski and Patrick Brouder's recent (2024) special issue of the journal Tourism Geographies on the same topic.

Chapter 18, by Jarkko Saarinen traces the origins and background of sustainability thinking in relation to tourism, navigating through concepts such as wise use, carrying capacity, crowding and the limits to growth. He also examines current and future connections between the tourism industry and the 2030 Sustainable Development Goals. Complementarily, Edward Huijbens' contribution on tourism and the anthropocene (Chapter 19) centres on the anthropocene's utility as a conceptual lens through which to view and understand tourism. Countering the growth paradigm, Huijbens debates 'earthly tourism', which advocates for tourism that is deeply aware of and responsive to the climate crisis, emphasising the challenges associated with the planet's ecological limits and variances. Also along complementary lines, in Chapter 20, Rannveig Ólafsdóttir and Anna Dóra Sæþórsdóttir offer a fresh perspective on resilience thinking and tourism by framing tourism as an ecosystem service within socio-ecological systems (SES). They set out to unravel tourism's potential for sustainable management, advocating for a blend of holistic pluralism and geographical methodologies, exploring how tourism can both harmoniously integrate into and benefit socio-ecological frameworks.

On the theme of tourism's role in the climate crisis, Chapter 21 by Natalie L. B. Knowles and Daniel Scott take a forward-looking stance on the intersection of tourism and global climate dynamics, scrutinising the future trajectory of tourism amid the dual challenges of decarbonisation and escalating climate disruptions. They examine how the tourism sector might navigate and adapt to evolving environmental scenarios, emphasising the critical need for sustainable development practices in response to climate change. The role of governance in approaching tourism more responsibly forms the focus of Chapter 22, by Dora Papatheochari, Spiros Niavis and Harry Cocossis. Their contribution examines different governance-based responsibility actions adopted by destinations, as well as trends towards responsibility generated by tourist behaviour.

Loretta Bellato, Niki Frantzeskaki and Andi Nygaard introduce the concept of regenerative tourism in Chapter 23, as a paradigm shift from traditional sustainable approaches, arguing for tourism activities to be viewed as positive interventions that enhance the resilience and capacity of places, communities, and their guests to thrive within interconnected social-ecological systems. On related terms, Chapter 24, by Jarkko Saarinen and Sanjay Nepal, focuses on political ecologies of tourism. Their chapter underscores the potential of political ecology frameworks in comprehensively examining the intricate dynamics among tourism, local communities, and natural resources, as well as the consequential environmental and social transformations that ensue.

'Digital Transformation, Platform Economy and Tourism Geographies' is the title of **Part V** of this volume, encompassing four chapters that focus on the state of the art of digital transformation and tourism geographies.

In Chapter 25, Maartje Roelofsen explores the intricate relationship between digital technologies and the tourism industry, analysing how digitalisation has become deeply integrated into the organisation, production and consumption of tourism, profoundly shaping places, mobility patterns, social interactions, and daily experiences. Importantly, she also draws attention to the digital inequalities that often stem from the particular social, cultural, political, and economic contexts. Antonio Paolo Russo and Josep Ivars, in Chapter 26, examine 'smart' tourism destinations through the lens of social inclusion, analysing the complexities of how 'smart' technologies intersect with tourism and the implications for social equity within these spaces.

A nuanced exploration of the relationship between tourism, surveillance and capitalist configurations is set out by Katie Dudley in Chapter 27. Dudley's chapter examines how capitalist ideologies permeate and adapt within tourism contexts, shedding light on the ways in which capitalist logic shapes and influences broader societal norms and patterns within tourism. Chapter 28, by Egbert van der Zee, zones in on the significant impacts of the platform economy on tourism places and spaces. He advocates for a geographical lens to understand the development and consequences of the platform economy and platform capitalism in tourism, emphasising the importance of considering the different scales and contexts that influence and are influenced by these platforms.

In Part VI, 'Geographies of Tourism Mobilities', five chapters explore different dimensions of the contemporary tourism-mobilities nexus, with a particular emphasis in some cases on its most recent manifestations in a post-COVID-19 context.

Chapter 29 transcends conventional discourse on the relationship between tourism and mobility, as Maria Thulemark and Tara Duncan examine the effects of changing patterns of (im)mobility induced by the COVID-19 pandemic on the tourism system. Continuing the theme of tourism mobilities, digital nomadism forms the focus of Chapter 30 by Olga Hannonen, engaging with the emerging body of literature on digital nomad mobilities, situating digital nomadism within conceptual and empirical developments in tourism geographies, including the mobilities and experience turn, (re)construction of the tourist gaze as well as place attachment and belonging. Franz Buhr and Agustín Cocola-Gant, in Chapter 31, also follow up the mobilities theme by analysing the nexus between increasingly mobile lifestyles and urban transformation in focusing on issues such as the relationship between short-term rentals and gentrification processes, who (usually) are transnational gentrifiers and how their place consumption practices are reshaping urban territories.

In Chapter 32, centred on transportation geographies and tourism, David Timothy Duvall and John MaCilree evaluate the current state of transportation networks and flows and their implications for global tourism geographies, with a particular focus on air transport and the implications of the COVID-19 pandemic for air transport geographies. Noam Shoval argues the importance of time geography for analysing tourism activity in Chapter 33, maintaining that a better understanding of the logic of visitor activities in time and space not only serves a number of practical purposes in tourism industries, planning and management, but also develops the existing concept of time geography, contributing to the theoretical foundations of tourism (geographies) research.

The three chapters in **Part VII, 'Economic, Entrepreneurship and Business Perspectives on Tourism Geographies',** are rooted in an in-depth exploration of the intricate interplay between economic geography, entrepreneurial activity and business innovation practices within the multifaceted landscape of tourism geographies.

In Chapter 34, Tim Coles offers a geographical examination of recent research on business innovation within the tourism sector. With a primary emphasis on innovation among privately owned enterprises, Coles navigates through the intricacies of innovation discourses, acknowledging the interchangeability in this context of terms such as *businesses, enterprises* and *firms*. Importantly, he also underlines that innovation is not confined solely to the private sector within tourism, as public- and voluntary-sector organisations, including destination management entities, can also engage in innovative practices and processes.

Chapter 35 addresses tourism entrepreneurship in relation to the nexus between entrepreneurs and their spatial environment. Dimitri Ioannides and Jonathan Yachin focus on the attributes and motivations of tourism entrepreneurs and propose the notion of 'spatial bricolage'; a framework for investigating tourism entrepreneurship as centred on the interplay between owner-managers of tourism businesses and their geographical context.

Patrick Brouder and Dimitri Ioannides, in Chapter 36, build on the extant studies in the economic geography of tourism, highlighting four critical perspectives for the future: political economies; sustainability transitions; work and workers and policy development. They argue that tourism studies can benefit from deeper engagement with emerging and established research in economic geography whereby evolutionary economic geography (EEG) becomes one clear framework to coalesce these four focal areas.

In **Part VIII, 'Challenges for the Future of Tourism Geographies Education',** Velvet Nelson sets out an agenda for strategy development in tourism geographies education in Chapter 37. Adopting a Threats, Opportunities, Weaknesses, Strengths (TOWS) matrix approach, Nelson argues tourism geographies education is well-positioned to tackle the wide-ranging changes in the tourism system as well as the significant impacts that tourism has on peoples and places. She maintains that the need for tourism geographies education has never been stronger, despite coming under threat in recent years.

In Part IX, the last in the volume, we seek to synthesise and contrast the complex, shifting perspectives on tourism geographies for the 2020s – perspectives that emerge from the previous parts of the volume (Chapter 38), focusing on the new realms and parametres of the tourism geographies field (now more than a decade into its 'post-disciplinary age' and bracing for the rapidly consolidating era of the polycrisis) as well as sketching out key topics of tourism geographies scholarship for the coming decade and beyond.

References

Bianchi, R.V., & Milano, C. (2024). Polycrisis and the metamorphosis of tourism capitalism. *Annals of Tourism Research* 104, 103731. https://doi.org/10.1016/j.annals.2024.103731

Cresswell, T. (2024). *Geographical Thought; A Critical Introduction* (2nd edition). Wiley.

Gibson, C. (2008). Locating geographies of tourism. *Progress in Human Geography* 32(3), 407–422.

Henig, D., & Knight, D.M. (2023). Polycrisis: Prompts for an emerging worldview. *Anthropology Today* 39(2), 3–6. DOI: 10.1111/1467-8322.12793

Lew, A.A., Williams, A.M., & Hall, C.M. (2014, Eds.). *The Wiley Blackwell Companion to Tourism*. Wiley Blackwell.

Müller, D.K. (2019, Ed.). *A Research Agenda for Tourism Geographies*. Edward Elgar.

Niewiadomski, P., & Brouder, P. (2024). From 'sustainable tourism' to 'sustainability transitions in tourism'?, *Tourism Geographies* 26(2), 141–150. DOI: 10.1080/14616688.2023.2299832

Tooze, A. (2022). Welcome to the world of the polycrisis. *Financial Times,* 28 October.

Turnbull, N. (2022). Permacrisis: What it means and why it's word of the year for 2022. *The Conversation,* 11 November.

Wilson, J. (2012). Tourism – The view from space. In J. Wilson (Ed.), *The Routledge Handbook of Tourism Geographies* (1st Edition) (pp. 1–6). Routledge.

Wilson, J., & Anton Clavé, S. (2013, Eds.). *Geographies of Tourism: European Research Perspectives*. Emerald.

2
THE TOURISM–GEOGRAPHIES NEXUS

Dieter K. Müller and C. Michael Hall

Introduction

The relationship between tourism geography and its parent discipline of geography has been complex, sometimes fraught, and certainly debated within the tourism geographies literature, and the question of whether tourism should be considered a discipline in its own right has caught particular attention (e.g., Butler, 2004; Hall, 2013; Hall et al., 2014; Müller, 2014b, 2019b; Tribe 1997, 2006). However, tourism has become institutionalised within academia, and numerous tourism departments and dedicated journal outlets have been established as well as committed positions on all academic levels. It is in these tourism departments that tourism geographers are increasingly to be found, rather than in geography departments, though regional variations occur, perhaps reflecting the higher education and employment policies of government as well as the initiatives of individual institutions (Hall, 2005; Hall et al., 2014; Müller, 2014b). In addition to the establishment of tourism departments, the detachment of tourism from geography is also manifested in academic positions, scientific publications, as well as conferences, that are all dedicated to the topic of tourism. Even core publications such as *Tourism Geographies*, *The Routledge Handbook of Tourism Geographies*, and *The Companion to Tourism* mirror this; most authors are affiliated with tourism departments, business schools or other social science institutions outside geography departments (Gill, 2012; Mérigo et al., 2019; Müller, 2014b).

Institutional change triggers, of course, also intellectual change since old ties to mainstream geography are often weakening, at the same time as new ties to researchers with other academic backgrounds emerge. Over time, several areas of mainstream geography have found themselves adopted by other fields. For example, retail geography, place promotion and much economic geography have been adopted by business schools and regional development departments to the point that their origins in geography are often forgotten. Nevertheless, content from traditional disciplines such as geography is also enriched with influences from other disciplines. Such development may therefore open new avenues and provide opportunities for post-disciplinary research (Coles et al., 2005, 2006, 2016; Coles & Hall, 2006; Hall, 2013; Müller, 2019a).

An apocalyptic fear for the disappearance of tourism geographies is thus far from justified. Instead, as the development of the journal *Tourism Geographies* as well as publications such as *The Routledge Handbook for Tourism Geographies* (Wilson, 2012a) or the *Companion to Tourism*

DOI: 10.4324/9781003286301-3"

(Lew et al., 2014) indicate – and ultimately of course this Routledge handbook – the subdiscipline is alive and kicking. Furthermore, various compilations reviewing global research within tourism geographies (Butler, 2004; Gibson, 2008, 2009, 2010; Hall & Page, 2009; Nepal, 2009) or providing regional accounts (Bao & Ma, 2011; Huijbens & Müller, 2022; Rogerson & Visser, 2020: Stock, 2021; Wilson & Anton Clavé, 2013; Xu & Wu, 2019), and add to this the impression of a vivid and innovative field of investigation. And, based on a bibliometric analysis, tourism geographers are highly influential in the wider field of tourism, especially in more specialised areas such as sustainable tourism, mountain tourism and recent research on COVID-19 (McKercher, 2008; Niñerola et al., 2019; Vishwakarma & Mukherjee, 2019; Bhatt et al., 2022; Ng, 2022).

Academic contributions to tourism geographies can be traced back to the early twentieth century and, thus, it can no longer be seriously claimed that tourism geographies are a new field of investigation (Hall & Page, 2014). In contrast, tourism geographies have created a substantial body of knowledge mirrored in an impressive number of publications. Hence, today tourism geographies are home to thematic subcategories, distinct scholarly traditions as well as specific schools of thought highlighting particular theories, topics, or applying specific methodologies (Lew et al. 2004, 2014; Hall & Page 2009; Wilson 2012; Wilson & Anton Clavé 2013). Still, as Ioannides (2006) and Hall (2013) note, major reference works in geography usually lack references to this research. Seemingly, tourism research is developing without any great interconnection to other subfields within geography (Gibson 2008; Hall & Page 2009) and, when it does happen, influences seem to be from the mother discipline to tourism geographies but not vice versa.

In a digital era, publications from tourism geographies are easily findable and often accessible. Hence, the detachment of tourism geographies from the overall geography discipline cannot be caused by the availability of research in tourism geographies. Instead, the institutional divorce may have been caused by an 'out of sight, out of mind' mentality among geographers housed in geography departments. Alternatively, the neglect of tourism geographies could be caused because it lacks relevance for wider geographical research.

In this context, tourism geographies have by no means resembled a stable practice but rather been transformed by multiple 'turns' and fashions (Hall, 2019). This has itself rendered critique for shortcomings and marginalisation (Bianchi, 2009; Debbage & Ioannides, 2012; Gibson, 2009; Hall & Page, 2009). Still, as several contributors note, this mirrors a constant struggle for being assessed as relevant by universities, funding organisations, the industry, academic peers, and students (Hall, 2019; Saarinen, 2019).

While such struggle may explain an almost obsessive orientation towards satisfying expectations on tourism geographies and following various kinds of theoretical trends and fashions in other disciplines, Hall (2019) reminds the community to remember its own past as well:

> In order to suggest where we should go in the future, we need to have a much better understanding of the past and the various works that have been significant and influential over time, not just those that are easily locatable online or fashionable. This also means that the "sullied" strategy of tourism geography should be to ensure that before or while borrowing from elsewhere tourism geographers need to look at their own work as well.
>
> (Hall, 2019, p. 30)

Against such a background, it is reasonable to bring together contributions on current thought within tourism geographies, but also to assess how these relate to development in the wider field of geography. Or as Hall (2013, p. 614) puts it, "There is value in asking what tourism geographers do".

Research in tourism geographies has been reviewed frequently and, hence, this chapter does not attempt to redo such efforts (e.g., Hall & Page, 2014; Lew et al., 2014; Wilson, 2012a). Instead, the chapter focuses on the relationship between tourism geographies and geography. The objective is to map and analyse the intersections of the two fields. As a background, however, a summary of recent findings on the standings of tourism geographies is provided.

Tourism Geographies

Doubtless, tourism is a spatial phenomenon creating linkages between places and people and, hence, applying a geographical perspective is self-evident and mandatory to understand how these processes unfold within and between different localities (Hall & Page, 2014). However, neither a distinct definition nor a delimitation of the discipline is agreed upon and, moreover, even the practice of tourism geographical research is not limited to geography departments or researchers trained as geographers or publications labelled accordingly (Hall & Page, 2014; Lew, 2001; Müller, 2014b), a situation perhaps even enhanced changing editorial positions in journals such as *Tourism Geographies*. Lew (2001) also noted that introductory textbooks in tourism geography had significant overlap with textbooks in tourism applying more multidisciplinary approaches. Still, in acknowledging the legacy of previous research Hall (2013) notes the following:

> Tourism geography is the study of tourism within the concepts, frames, orientations, and venues of the discipline of geography and accompanying fields of geographical knowledge. The notion of tourism geographies describes the multiple, and sometimes contested, theoretical, philosophical and personal orientations of those who undertake tourism research from geographical perspectives.
>
> (Hall, 2013, p. 602)

While an ambivalence in relation to definition is principally sound – disciplines should be dynamic and adapt to changes and new requirements – editors and authors will always need to select what should be included in compilations of the discipline. This particularly applies to introductory texts targeting students. Usually, these aim at providing a comprehensive picture of the field. As indicated by a review of four popular textbooks – all published in several editions and representing different periods in time – there are certain commonalities that lasted over time (Table 2.1). For example, an engagement with the tourism system is a common denominator for all books as is the awareness of spatial variation. Still, while Pearce (1995) offers predominantly a descriptive approach inspired by attempts to model tourism flows, which were popular at that time, the other books no longer engage with patterns of tourism flows to the same degree. Instead, a commitment towards analysing the relationship between tourism and various aspects of place becomes prevalent.

The volume by William and Lew (2015) addresses this by focusing the reciprocal relations of tourism to economy, community, and the environment. Furthermore, temporal and cultural dimensions of tourism receive greater attention than in Pearce's work. Hence, both history and heritage are addressed, as are issues of identity and cultural production of tourism. A different approach towards grasping the geographies of tourism is taken in Nelson's textbook (2021). Here an ambition seems to be to indicate the overlap with other areas of the mother discipline and to clearly position tourism geographies as an integrated part of the geography discipline. Besides addressing the supply, demand, and impacts of tourism, Hall and Page (2014) provide a territorial approach, focusing tourism within different geographical contexts. This follows to a certain extent traditional segmentations of the wider geography discipline and resembles not least urban

Table 2.1 Contents of four popular textbooks in tourism geographies

Tourism Today: A Geographical Analysis (Pearce, 1995)	*Tourism Geography: Critical Understandings of Place, Space and Experience (Williams & Lew, 2015)*	*An Introduction to the Geography of Tourism (Nelson, 2021)*	*The Geography of Tourism and Recreation: Environment, place and space (Hall & Page, 2014)*
• Tourism models • The demand for tourist travel • Patterns of international tourism • Intra-national travel patterns • Domestic tourist flows • Measuring spatial variations in tourism • The national and regional structure of tourism • The spatial structure of tourism on islands • Coastal resorts and urban areas • Implications, applications, conclusions	Part I: Introduction: Tourism and geography • Tourism, geography and geographies of tourism Part II: The emergence of global tourism • The birth of modern tourism • International patterns of travel and tourism Part III: Tourism's economic, environmental and social relations • Cost s and benefits: the local economic landscape of tourism • Tourism, sustainability and environmental change • Socio-cultural relations and experiences in tourism Part IV: Understanding tourism places and spaces • Cultural constructions and invented places • Theming the urban landscape • The past as a foreign country: heritage as tourism • Nature, risk and geographic exploration in tourism • Consumption, identity and specialty tourism Part V: Applied and future tourism geographies • Planning and managing tourism development • Emerging and future tourism geographies	Part I: The Geography of Tourism • Geography and tourism • Basic concepts in tourism • Overview of tourism products Part II: The Geographic foundation of tourism • The historical geography of tourism • The transport geography of tourism • The physical geography of tourism • The Human geography of tourism Part III: The geography of tourism effects • The economic geography of tourism • The social geography of tourism • The environmental geography of t ourism • Sustainable tourism development Part IV: Tourism and place Tourism, representations of place • Tourism, representations of place, and social media • Experiences o place in tourism.	• Introduction: Tourism matters! • The demand for recreation and tourism • The supply of recreation and tourism • The impacts of tourism and recreation • Urban recreation and tourism • Rural recreation and tourism • Tourism and recreation in wilderness and protected areas • Coastal and marine recreation and tourism • Tourism, recreation planning, policy and governance • The future

geography and rural geography. Moreover, the richness of references in the book clearly indicates the maturity of the field and illustrates the variegated relations to the wider field of geography as well.

A common denominator for all textbooks is the understanding of tourism geographies as an applied science. Hence, sections and chapters are committed to the application of tourism knowledge in spatial planning. In this context it is worthwhile noting that two of the textbooks explicitly mentioned topics related to physical geography. This direct engagement with the physical world and the environment distinguishes tourism geographies from the wider field of tourism in which concerns with the physical environment are extremely limited and, when considered, it tends to be in relation to tourism planning, ecotourism, and environmental change, fields that are usually the domain of the geographer (Ruhanen et al., 2015; Singh et al., 2022). Indeed, in a review of Nordic tourism geographies, it was shown that research on the physical impacts of tourism, wilderness, and climate change in relation to tourism accounted for a significant share of the overall academic production (Huijbens & Müller, 2022). Thus, tourism geographies also played a role as frontrunner when it comes to questions of sustainable development (Hall & Lew, 1998; Saarinen, 2006).

Other reviews of tourism geographies provide variegated kinds of groupings. Hall and Page's (2009) review, *Progress in Tourism Management: From the Geography of Tourism to Geographies of Tourism*, which was also re-printed in the first edition of this book, discerns the following topics of tourism geographical research: explaining spatialities, tourism planning and places, development and its discontents, and tourism as an applied area of research. In a bibliometric approach ten years later, research on tourism impacts, protected areas and sustainability, the relationship to primary industries and land use, rural areas and accessibility, industry and economic development, and heritage, image and identity were identified as prominent topics for research (Müller, 2019b).

In summary, it can be noted that tourism geographies form a dynamic field of research that in large has followed in the footsteps of the mother discipline, independent of institutional belongings of the contributors. In this context, particularly, an engagement with place, or destinations has become the predominant topic for researchers applying a geographical perspective. Furthermore, the multifaceted human experiences of tourism have entered the research agendas at the same time as an interest in explaining actual tourism flows has declined. The latter is somewhat surprising, recognising the growing availability of spatial data and the predicted potential for tourism research (Bauder, 2019; Shoval et al., 2014; Shoval & Ahas, 2016).

Acknowledging Tourism Geographies

As mentioned earlier, bibliometric research has long revealed the core position of tourism geographers in the wider field of tourism (McKercher, 2008). While this analysis is relatively dated, it can be noted that such research is increasingly difficult to reproduce because of the institutional changes mentioned above. Still, the inherent spatial dimensions of tourism guarantee a place for geography in tourism research. Tourism shapes places and, in fact, Hall (2005) identified tourism research as the social science of mobility.

Hence, while tourism geographies are acknowledged within tourism research, more concern has targeted the role of tourism geographies for geographical research. For example, Ioannides (2006) pointed out the marginal role of tourism in economic geography, but ten years later saw a new opportunity for bridging the gap between tourism and economic geographies through the application of evolutionary economic geography (Ioannides & Brouder, 2016). Even Gibson (2008, p. 407) noted, "Although not taken seriously by some, and still considered marginal by many, tourism constitutes an important point of intersection".

Similar accounts and considerations have been given by Butler (2015), Hall (2013), Müller (2014b, 2019b), and Saarinen (2019). Shaw (2010) provided evidence that even within geography departments, the position of tourism research has been contested. Butler (2015) sees a reason for this development in the increasing specialisation within research and the consequent loss of common ground. He noted thus in respect to the geography community:

> As a geographer, this author long ago lost the battle to keep up to date in most aspects of geographical research, and if truth be told, also lost interest in doing so, as much of the research became peripheral to his interests (tourism and leisure related topics) and even banal in terms of some of the dogma and theoretical viewpoints being proposed.
>
> (Butler, 2015, p. 20)

Hence, Butler realises the divorce of tourism geographies and geographies but does not lament the development. And, indeed, even Müller (2014b) does not see the diverging pathways as a problem for the single researcher. Rather, he notes, students of geography were the ones suffering from being deprived of learning about an important force shaping places and networks with influence far beyond the immediate realm of tourism. Even Nepal (2009) mentions the repercussions of a disciplinary divorce for geography; the implications of tourism are so significant for place, space, and environment that ignoring them truncates the knowledge creation on these categories.

While these accounts address the relationship between tourism and geography on a more conceptual level, the impact of one discipline on another can today be measured in bibliometric terms, as well. A recent review of publications in the journal *Tourism Geographies* reveals that many scholars citing contributions published in the journal are themselves trained as geographers (Merigó et al., 2019). However, the list of the most citing institutions includes only a few geography departments with prominent tourism research. Other listings of tourism geographies research institutions mirror this as well (Müller, 2019b). Furthermore, a predominance of English-speaking countries among the affiliations of the citing authors indicates the Anglo-American hegemony in global academic publishing, which has only been reinforced by government imposed metrics on the quality of publications, even in those countries where English is not the mother tongue (Hall, 2013).

Merigó and colleagues (2019) analyse which journals include citations to *Tourism Geographies*. Both *Progress in Human Geography* and the *Annals of the American Association of Geographers* are among the top 10 citing journals. Moreover, *Environment and Planning, Economic Geography, Geojournal* and *Geoforum* are other geography journals to be found among the 30 listed journals. Their links to *Tourism Geographies* are, however, relatively weak. An analysis of co-citations, which refers to the co-presence in reference lists of a third journal paper, shows that *Tourism Geographies* is indeed more closely related to other geography journals compared to *Annals of Tourism Research* and *Tourism Management, Tourist Studies* and *Journal of Sustainable Tourism*, all of which are more oriented toward other tourism journals (Merigó et al., 2019). Still, most citations are shared with *Annals of Tourism Research* and *Tourism Management*. The analysis illustrates as well that *Tourism Geographies* is indeed referred to in many variegated social and environmental sciences.

While *Tourism Geographies* certainly has been the pivotal publication channel for research in the field, books on progress and state-of-the-art within tourism geographies have been available, too. The most prominent volumes are certainly *The Companion to Tourism* published in two editions (Lew, et al. 2004, 2014) and the *Routledge Handbook of Tourism Geographies* (Wilson, 2012a). The latter presents not least theoretical and conceptual approaches to tourism geographies, ordered in five sections and 32 chapters covering various theoretical perspectives, topical approaches, and situated experiences. The second edition of the *Companion* does not explicitly limit itself to portraying tourism geographies only. Indeed, the relationship of tourism and geography is addressed in the volume's introduction (Hall, et al., 2014). Still, the fact that all editors are geographers is doubtless reflected in the nine sections and 50 chapters of the book, addressing theoretical perspectives, the production and consumption of tourism, globalisation, the relation to the environment, as well as planning and governance.

Both compilations had a significant impact on scientific literature and were cited frequently. However, more important, they also offer a good entry point and quick orientation for young scholars or researchers from other disciplines looking to familiarise themselves with major topics within tourism geographies. Hence, a mapping exercise scrutinising all citations to all chapters of the two volumes, registered in Scholar Google in early 2023, revealed that many PhD students acknowledged the volumes in their thesis work. More than 210 citations can be found in PhD theses, though far from all PhD theses, are indexed and traceable in the database. Furthermore, the exercise showed that both volumes and their chapters are also cited in books and journals in languages other than English. Not least, Spanish-and German-speaking publications pick up the volumes.

It is notable that the previous edition of the *Routledge Handbook of Tourism Geographies* as an entire volume was widely acknowledged within a couple of books committed to reviews of tourism geographies (Wilson & Anton Clavé, 2013; Müller, 2019b). In contrast, references in academic journals to the entire volume can be predominantly found in geography journals. This indicates that the book has been visible to other than tourism geographers, too, and even here the volume yielded notable interest including outside the English-speaking literature.

A thorough look at the different *Routledge Handbook* chapters reveals a great variety regarding their citation patterns. While this can be expected, certain chapters were more often cited in geography and other more thematically oriented social science publications, while others mainly were acknowledged in tourism journals and books. Indeed, for the *Routledge Handbook of Tourism Geographies*, of the 600 recorded citations, 61.8 per cent are to be found in other tourism publications (Table 2.2).

A common denominator for the chapters that are frequently cited in non-tourism publications is that they address topics that are popular and pivotal within geography as well. This becomes obvious in chapters discussing, for example, lifestyle mobility (Casado-Diaz, 2012), space and place (Chang, 2012; Gibson, 2012), the city (Selby, 2012) and gentrification (Wilson & Tallon, 2012), where a clear majority of references can be found in non-tourism publications. Also, the most frequently cited chapters (Casado-Diaz, 2012; Anton Clavé, 2012; Edensor & Falconer, 2012; Hall, 2012) have more than average numbers of citations in non-tourism publications. Chapters assessing the position of tourism geographies are most frequently cited with a more than average presence in tourism publications (Butler, 2012; Hall & Page, 2012). Hence, the discussion of the relationship between tourism geographies and the mother discipline seems to be no undertaking for the left-behind mainstream geographers. Otherwise, even chapters addressing various kinds of spatial perspectives and theoretical lenses appear to be contributions that are acknowledged mainly in the tourism literature.

Citations patterns for the *The Wiley-Blackwell Companion to Tourism* (Lew et al., 2014) resemble largely what could be seen for *The Routledge Handbook of Tourism Geographies*. Being more voluminous, the *Companion* yields all together more than 1250 citations. Almost identical even here, 60.3 per cent of all citations can be found in tourism publications (Table 2.3).

Table 2.2 Citations to a selection of frequently cited chapters of *The Routledge Handbook of Tourism Geographies* as recorded in the Google Scholar Database, Spring 2023[a]

Chapters (Selection)	Citations Journals			Books			TOTAL		
	Tourism	Geography	Other	Tourism	Geography	Other	Sum	Tourism %	Non-tour. %
Spatial analysis: a critical tool for tourism geographies (Hall, 2012)	18	10	11	24	3	5	71	59,2	40,8
Exploring the geographies of lifestyle mobilities: current and future fields of inquiry (Casado-Diaz, 2012)	5	9	11	5	2	9	41	24,4	75,6
Sensuous geographies of tourism (Edensor & Falconer, 2012)	12	6	3	8	1	8	38	52,6	47,4
Rethinking mass tourism (Anton Clavé, 2012)	9	6	6	10	3	1	35	54,3	45,7
From the geography of tourism to geographies of tourism (Hall & Page, 2012)	9	3	3	18	0	1	34	79,4	20,6
Tourism geographies or geographies of tourism: where the bloody hell are we? (Butler, 2012)	7	2	0	16	5	0	30	76,7	23,3
Performance, space and tourism (Larsen, 2012)	4	5	2	13	0	5	29	58,6	41,4
Geographies of gentrification and tourism (Wilson & Tallon, 2012)	5	5	8	7	1	1	27	44,4	55,6
A radical departure: a critique of the critical turn in tourism studies (Bianchi, 2012)	9	0	2	8	1	3	23	73,9	26,1
The "mobilities turn" and the geographies of tourism (Duncan, 2012)	6	1	3	8	0	3	21	66,7	33,3
Geographies of tourism: space, ethics ad encounter (Gibson, 2012)	5	3	4	5	2	2	21	47,6	52,4
Time geography and tourism (Shoval, 2012)	9	3	3	3	1	1	20	60,0	40,0

(Continued)

Table 2.2 (Continued)

Chapters (Selection)	Citations	Journals			Books			TOTAL		
		Tourism	*Geography*	*Other*	*Tourism*	*Geography*	*Other*	*Sum*	*Tourism %*	*Non-tour. %*
The economy of tourism spaces: a multiplicity of critical turns (Debbage & Ioannides, 2012)		8	1	2	5	1	0	17	76,5	23,5
Future spaces ad postcolonialism in tourism (Keen & Tucker, 2012)		10	0	0	5	1	1	17	88,2	11,8
Tourism: the view from space (Wilson, 2012b)		0	1	1	11	1	0	14	78,6	21,4
Tourism geographies and post-structuralism (Gale, 2012)		6	2	0	4	1	1	14	71,4	28,6
Geographies of tourism and the city (Selby, 2012)		4	4	1	1	1	2	13	38,5	61,5
Geography and the marketing of tourism destinations (Lew, 2012)		8	0	0	2	0	2	12	83,3	16,7
Landscape perspectives on tourism geographies (Knudsen et al., 2012)		7	1	0	0	1	2	11	63,6	36,4
Making and unmaking places in tourism geographies (Chang, 2012)		2	1	1	1	3	2	10	30,0	70,0
Total		172	73	74	199	32	50	600	61,8	38,2

Note: The classification was done manually by the author. Tourism publications usually contained the term tourism and analogous geography publications contained the term geography. Citations in the journal *Tourism Geographies* were classified as tourism.

[a] Chapters with more than average citations in non-tourism publications are shown with grey background.

Table 2.3 Citations to a selection of frequently cited chapters of *The Wiley-Blackwell Companion to Tourism* as recorded in the Google Scholar Database, Spring 2023[a]

	Citations						TOTAL		
	Journals			Books					
Chapters (Selection)	*Tourism*	*Geography*	*Other*	*Tourism*	*Geography*	*Other*	*Total*	*Tourism %*	*Non-tour. %*
Cultural geographies of tourism (Crang, 2014)	45	5	24	52	17	27	170	57,1	42,9
Entrepreneurial cultures and small business enterprises in tourism (Shaw, 2014)	33	2	17	21	1	13	87	62,1	37,9
Gender and feminist perspectives in tourism research (Pritchard, 2014)	25	1	12	37	1	6	82	75,6	24,4
The tourist gaze 1.0, 2.0, and 3.0 (Larsen, 2014)	25	2	16	6	2	14	65	47,7	52,3
Transport and tourism (Page & Connell, 2014)	24	4	12	8	2	4	54	59,3	40,7
Tourism and public policy: contemporary debates and future directions (Jenkins et al., 2014)	25	1	5	6	0	3	40	77,5	22,5
Tourism in the development of regional and sectoral innovation systems (Weidenfeld & Hall, 2014)	20	1	7	9	1	1	39	74,4	25,6
Political economy of tourism: regulation theory, institutions, and governance networks (Mosedale, 2014)	8	3	5	14	4	4	38	57,9	42,1
Coastal and marine tourism: emerging issues, future trends, and research priorities (Orams & Lück, 2014)	15	0	11	6	3	3	38	55,3	44,7
Tourist flows and spatial behavior (McKercher & Zoltan, 2014)	11	6	3	9	1	4	34	58,8	41,2
Speaking heritage: language, identity, and tourism (Hall-Lew & Lew, 2014)	9	1	8	8	0	8	34	50,0	50,0
Tourism innovation: products, processes, and people (Williams, 2014)	12	1	9	6	2	2	32	56,3	43,8

(*Continued*)

Table 2.3 (Continued)

Chapters (Selection)	Citations Journals			Books			TOTAL		
	Tourism	Geography	Other	Tourism	Geography	Other	Total	Tourism %	Non-tour. %
GPS, smartphones, and the future of tourism research (Shoval et al., 2014)	15	2	5	6	1	1	30	70,0	30,0
Medical tourism (Ormond, 2014)	2	1	12	2	5	7	29	13,8	86,2
Trends in tourism, shopping, and retailing (Timothy, 2014)	10	0	8	6	2	2	28	57,1	42,9
Tourism after the postmodern turn (Minca & Oakes, 2014)	11	1	3	7	1	4	27	66,7	33,3
Religion and spirituality in tourism (Stausberg, 2014)	10	1	4	10	0	2	27	74,1	25,9
The sustainable development of tourism: a state-of-the-art perspective (Weaver, 2014)	6	1	5	3	1	10	26	34,6	65,4
Landscapes of tourism (Terkenli, 2014)	6	2	12	0	0	5	25	24,0	76,0
Progress in second-home tourism research (Müller, 2014a)	7	2	5	7	2	0	23	60,9	39,1
Tourism motivations and decision making (Pearce, 2014)	9	0	3	7	0	2	21	76,2	23,8
Tourism and the visual (Scarles, 2014)	12	0	3	5	0	1	21	81,0	19,0
Taking stock of sport tourism research (Hinch et al., 2014)	10	0	4	4	0	2	20	70,0	30,0
Total	444	45	260	321	52	146	1268	60,3	39,7

Notes: The classification was done manually by the author. Tourism publications usually contained the term tourism and analogous geography publications contained the term geography. Citations in the journal *Tourism Geographies* were classified as tourism.

[a] Chapters with more than average citations in non-tourism publications are shown with grey background.

A chapter on medical tourism (Ormond, 2014) is almost exclusively cited in geography and other non-tourism publications. Even Weaver's review of sustainable development of tourism (2014) is acknowledged in the geography and social science literature. Overall, it can be registered that theoretically inclined reviews highlighting various kinds of schools of thinking and their approaches on tourism tend to yield more attention outside the tourism literature. In contrast, chapters addressing certain forms of tourism or other more tourism-specific topics are predominantly cited within the tourism literature.

Conclusion

This chapter had the ambition to throw some light on the intersections of tourism geographies and the mother discipline of geography. Indeed, it shows that geographers do not totally neglect tourism geographies, which have firmly developed side by side and in relation with other geographies. In fact, the assessment of the citation patterns to *The Routledge Handbook of Tourism Geographies* and *The Wiley Blackwell Companion to Tourism* reveal that roughly 40 per cent of all citations can be found in publications from geography and other social sciences. Hence, this analysis does not mirror the more lamenting accounts on the position of tourism geographies, which seem to refer rather to institutional issues and a neglect of tourism when core topics of the geography discipline are defined and discussed. Furthermore, it is obvious that tourism geographers, despite some doubts (Butler, 2012), continue to identify themselves as geographers, which is also mirrored in their presence at major geography summits such as the American Association of Geographers' (AAG) annual meetings or the International Geographical Union's (IGU) events. Even textbooks clearly reproduce geographical approaches and topics.

Hence, an institutional divide (Hall, 2013; Müller, 2014b) does not imply a loss of geographical thinking and, indeed, it may have added to the vibrant development of tourism geographies by exposing tourism geographers to thinking from outside the discipline. Indeed, the existence outside geography departments often implies post-disciplinary positions (Coles et al., 2005, 2006, 2016) and is in line with the overall stated need for multi-disciplinary solutions for today's pressing complex societal challenges.

As Coles and Hall (2006) somewhat provocatively suggested almost two decades ago, "The geography of tourism is dead. Long live geographies of tourism." Tourism geographies are thus very much alive and kicking. They are strong enough to not only apply models and ideas from other fields of geography within a tourism context, but also to develop their own strongholds, as for example in relation to issues of climate change and sustainable development (Huijbens & Müller, 2022) or to mobility research (Hall, 2005), or as shown here, in relation to cultural, economic and environmental geographies. This is good practice. It reinforces theoretical renewal and contributes to continuing an applied orientation addressing real-world problems.

References

Anton Clavé, S.A. (2012). Rethinking mass tourism, space and place. In J. Wilson (Ed.), *The Routledge Handbook of Tourism Geographies* (pp. 230–237). Routledge.

Bao, J., & Ma, L.J.C. (2011). Tourism geography in China, 1978–2008: Whence, what and whither? *Progress in Human Geography 35*(1), 3–20.

Bauder, M. (2019). Engage! A research agenda for big data in tourism geography. In D.K. Müller (Ed.), *A Research Agenda for Tourism Geographies* (pp. 149–157). Edward Elgar.

Bhatt, K., Seabra, C., Kabia, S.K., Ashutosh, K., & Gangotia, A. (2022). COVID crisis and tourism sustainability: An insightful bibliometric analysis. *Sustainability 14*(19), Art. 12151.

Bianchi, R.V. (2009). The 'critical turn' in tourism studies: A radical critique. *Tourism Geographies 11*(4), 484–504.

Bianchi, R.V. (2012). A radical departure: A critique of the critical turn in tourism studies. In J. Wilson (Ed.), *The Routledge Handbook of Tourism Geographies* (pp. 46–54). Routledge.

Butler, R. (2004). Geographical research on tourism, recreation and leisure: Origins, eras and directions. *Tourism Geographies 6*(2), 143–162.

Butler, R. (2015). The evolution of tourism and tourism research. *Tourism Recreation Research 40*(1), 16–27.

Butler, R.W. (2012). Tourism geographies or geographies of tourism: Where the bloody hell are we? In J. Wilson (Ed.), *The Routledge Handbook of Tourism Geographies* (pp. 26–34). Routledge.

Casado-Diaz, M. (2012). Exploring the geographies of lifestyle mobility: Current and future fields of enquiry. In J. Wilson (Ed.), *The Routledge Handbook of Tourism Geographies* (pp. 120–125). Routledge.

Chang, T.C. (2012). Making and unmaking places in tourism geographies. In J. Wilson (Ed.), *The Routledge Handbook of Tourism Geographies* (pp. 133–138). Routledge.

Coles, T., & Hall, C.M. (2006). The geography of tourism is dead: Long live geographies of tourism and mobility. *Current Issues in Tourism 9*(4–5), 289–292.

Coles, T., Hall, C.M., & Duval, D.T. (2005). Mobilizing tourism: A post-disciplinary critique. *Tourism Recreation Research 30*(2), 31–41.

Coles, T., Hall, C.M., & Duval, D.T. (2006). Tourism and post-disciplinary enquiry. *Current Issues in Tourism 9*(4–5), 293–319.

Coles, T., Hall, C.M., & Duval, D.T. (2016). Tourism and postdisciplinarity: Back to the future? *Tourism Analysis 21*(4), 373–387.

Crang, M. (2014). Cultural geographies of tourism. In A.A. Lew, C.M. Hall, & A.M. Williams (Eds.), *The Wiley Blackwell Companion to Tourism* (pp. 66–77). Wiley Blackwell.

Debbage, K.G., & Ioannides, D. (2012). A multiplicity of 'critical turns'? In J. Wilson (Ed.), *The Routledge Handbook of Tourism Geographies* (pp. 149–156). Routledge.

Duncan, T. (2012). The 'mobilities turn' and the geography of tourism. In J. Wilson (Ed.), *The Routledge Handbook of Tourism Geographies* (pp. 113–119). Routledge.

Edensor, T., & Falconer, E. (2012). Sensuous geographies of tourism. In J. Wilson (Ed.), *The Routledge Handbook of Tourism Geographies* (pp. 74–81). Routledge.

Gale, T. (2012). Tourism geographies and post-structuralism. In J. Wilson (Ed.), *The Routledge Handbook of Tourism Geographies* (pp. 37–45). Routledge.

Gibson, C. (2008). Locating geographies of tourism. *Progress in Human Geography 32*(3), 407–422.

Gibson, C. (2009). Geographies of tourism: Critical research on capitalism and local livelihoods. *Progress in Human Geography 33*(4), 527–534.

Gibson, C. (2010). Geographies of tourism: (Un)ethical encounters. *Progress in Human Geography 34*(4), 521–527.

Gibson, C. (2012). Geographies of tourism: Space, ethics and encounter. In J. Wilson (Ed.), *The Routledge Handbook of Tourism Geographies* (pp. 55–60). Routledge.

Gill, A. (2012). Travelling down the road to postdisciplinarity? Reflections of a tourism geographer. *The Canadian Geographer 56*(1), 3–17.

Hall, C.M. (2005). *Tourism: Rethinking the Social Science of Mobility*. Prentice Hall.

Hall, C.M. (2012). Spatial analysis: A critical tool for tourism geographies. In J. Wilson (Ed.), *The Routledge Handbook of Tourism Geographies* (pp. 178–188). Routledge.

Hall, C.M. (2013). Framing tourism geographies: Notes from the underground. *Annals of Tourism Research 43*, 601–623.

Hall, C.M. (2019). Tourism/geography/mobilities are dead, long live tourism/geography/mobilities: Or returning to yet another turn: The inexorable search for relevant research agendas in tourism geographies. In D.K. Müller (Ed.), *A Research Agenda for Tourism Geographies* (pp. 23–32). Edward Elgar.

Hall, C.M., & Lew, A.A. (Eds.) (1998). *Sustainable Tourism: A Geographical Perspective*. Longman.

Hall, C.M., & Page, S.J. (2009). Progress in tourism management: From the geography of tourism to geographies of tourism – a review. *Tourism Management 30*(1), 3–16.

Hall, C.M., & Page, S.J. (2012). From the geography of tourism to geographies of tourism. In J. Wilson (Ed.), *The Routledge Handbook of Tourism Geographies* (pp. 27–43). Routledge.

Hall, C.M., & Page, S.J. (2014). *The Geography of Tourism and Recreation: Environment, Place and Space*. 4th ed. Routledge.

Hall, C.M., Williams, A.M., & Lew, A.A. (2014). Tourism: Conceptualizations, disciplinarity, institutions, and issues. In A.A. Lew, C.M. Hall, & A.M. Williams (Eds.), *The Wiley-Blackwell Companion to Tourism* (pp. 3–24). Wiley.

Hall-Lew, L.A., & Lew, A.A. (2014). Speaking heritage: Language, identity, and tourism. In A.A. Lew, C.M. Hall, & A.M. Williams (Eds.), *The Wiley Blackwell Companion to Tourism* (pp. 336–348). Wiley Blackwell.

Hinch, T., Higham, J., & Sant, S.L. (2014). Taking stock of sport tourism research. In A.A. Lew, C.M. Hall, & A.M. Williams (Eds.), *The Wiley Blackwell Companion to Tourism* (pp. 413–424). Wiley Blackwell.

Huijbens, E.H., & Müller, D.K. (2022). The socio-spatial articulations of tourism studies in nordic geography. In P. Jakobsen, E. Jönsson, & H. Gutzon Larsen. (Eds.), *Socio-Spatial Theory in Nordic Geography* (pp. 169–190). Springer.

Ioannides, D. (2006). Commentary: The economic geography of the tourist industry: Ten years of progress in research and an agenda for the future. *Tourism Geographies 8*(1), 76–86.

Ioannides, D., & Brouder, P. (2016). Tourism and economic geography redux: Evolutionary economic geography's role in scholarship bridge construction. In P. Brouder, S. Anton Clavé, A. Gill, & D. Ioannides (Eds.), *Tourism Destination Evolution* (pp. 183–194). Routledge.

Jenkins, J.M., Hall, C.M., & Mkono, M. (2014). Tourism and public policy: Contemporary debates and future directions. In A.A. Lew, C.M. Hall, & A.M. Williams (Eds.), *The Wiley Blackwell Companion to Tourism* (pp. 542–555). Wiley Blackwell.

Keen, D., & Tucker, H. (2012). Future spaces of postcolonialism in tourism. In J. Wilson (Ed.), *The Routledge Handbook of Tourism Geographies* (pp. 97–102). Routledge.

Knudsen, D.C., Rickly-Boyd, J.M., & Metro-Roland, M.M. (2012). Landscape perspectives on tourism geographies. In J. Wilson (Ed.), *The Routledge Handbook of Tourism Geographies* (pp. 201–206). Routledge.

Larsen, J. (2012). Performance, space and tourism. In J. Wilson (Ed.), *The Routledge Handbook of Tourism Geographies* (pp. 84–90). Routledge.

Larsen, J. (2014). The tourist gaze 1.0, 2.0, and 3.0. In A.A. Lew, C.M. Hall, & A.M. Williams (Eds.), *The Wiley Blackwell Companion to Tourism* (pp. 304–313). Wiley Blackwell.

Lew, A.A. (2001). Literature review: Defining a geography of tourism. *Tourism Geographies, 3*(1), 105–114.

Lew, A.A. (2012). Geography and the marketing of tourism destinations. In J. Wilson (Ed.), *The Routledge Handbook of Tourism Geographies* (pp. 181–186). Routledge.

Lew, A.A., Hall, C.M., & Williams, A.M. (Eds.) (2004). *The Companion to Tourism*. Wiley.

Lew, A.A., Hall, C.M., & Williams, A.M. (Eds.) (2014). *The Wiley-Blackwell Companion to Tourism*. 2nd ed. Wiley.

McKercher, B. (2008). A citation analysis of tourism scholars. *Tourism Management 29*(6), 1226–1232.

McKercher, B., & Zoltan, J. (2014). Tourist flows and spatial behavior. In A.A. Lew, C.M. Hall, & A.M. Williams (Eds.), *The Wiley Blackwell Companion to Tourism* (pp. 33–44). Wiley Blackwell.

Merigó, J.M., Mulet-Forteza, C., Valencia, C., & Lew, A.A. (2019). Twenty years of Tourism Geographies: A bibliometric overview. *Tourism Geographies 21*(5), 881–910.

Minca, C., & Oakes, T. (2014). Tourism after the postmodern turn. In A.A. Lew, C.M. Hall, & A.M. Williams (Eds.), *The Wiley Blackwell Companion to Tourism* (pp. 294–303). Wiley Blackwell.

Mosedale, J. (2014). Political economy of tourism. In A.A. Lew, C.M. Hall, & A.M. Williams (Eds.), *The Wiley Blackwell Companion to Tourism* (pp. 55–65). Wiley Blackwell.

Müller, D.K. (2014a). Progress in second-home tourism research. In A.A. Lew, C.M. Hall, & A.M. Williams (Eds.), *The Wiley Blackwell Companion to Tourism* (pp. 389–400). Wiley Blackwell.

Müller, D.K. (2014b). 'Tourism geographies are moving out' – A comment on the current state of institutional geographies of tourism geographies. *Geographia Polonica 87*(3), 299–312.

Müller, D.K. (2019a). Infusing tourism geographies. In D.K. Müller (Ed.), *A Research Agenda for Tourism Geographies* (pp. 60–70). Edward Elgar.

Müller, D.K. (2019b). Tourism geographies: A bibliometric review. In D.K. Müller (Ed.), *A Research Agenda for Tourism Geographies* (pp. 7–22). Edward Elgar.

Nelson, V. (2021). *An Introduction to the Geography of Tourism*. 3rd ed. Rowman Littlefield.

Nepal, S. (2009). Traditions and trends: A review of geographical scholarship in tourism. *Tourism Geographies 11*(1), 2–22.

Ng, S.L. (2022). Bibliometric analysis of literature on mountain tourism in Scopus. *Journal of Outdoor Recreation and Tourism 40*, Art. 100587.

Niñerola, A., Sánchez-Rebull, M.V., & Hernández-Lara, A.B. (2019). Tourism research on sustainability: A bibliometric analysis. *Sustainability 11*(5), Art. 1377.

Orams, M.B., & Lück, M. (2014). Coastal and marine tourism. In A.A. Lew, C.M. Hall, & A.M. Williams (Eds.), *The Wiley Blackwell Companion to Tourism* (pp. 479–489). Wiley Blackwell.

Ormond, M. (2014). Medical tourism. In A.A. Lew, C.M. Hall, & A.M. Williams (Eds.), *The Wiley Blackwell Companion to Tourism* (pp. 425–434). Wiley Blackwell.

Page, S., & Connell, J. (2014). Transport and tourism. In A.A. Lew, C.M. Hall, & A.M. Williams (Eds.), *The Wiley Blackwell Companion to Tourism* (pp. 155–167). Wiley Blackwell.

Pearce, D. (1995). *Tourism Today: A Geographical Analysis*. 2nd ed. Longman.

Pearce, P.L. (2014). Tourism motivations and decision making. In A.A. Lew, C.M. Hall, & A.M. Williams (Eds.), *The Wiley Blackwell Companion to Tourism* (pp. 45–54). Wiley Blackwell.

Pritchard, A. (2014). Gender and feminist perspectives in tourism research. In A.A. Lew, C.M. Hall, & A.M. Williams (Eds.), *The Wiley Blackwell Companion to Tourism* (pp. 314–324). Wiley Blackwell.

Rogerson, J.M., & Visser, G. (Eds.) (2020). *New Directions in South African Tourism Geographies*. Springer.

Ruhanen, L., Weiler, B., Moyle, B.D., & McLennan, C.L.J. (2015). Trends and patterns in sustainable tourism research: A 25-year bibliometric analysis. *Journal of Sustainable Tourism 23*(4), 517–535.

Saarinen, J. (2006). Traditions of sustainability in tourism studies. *Annals of Tourism Research 33*(4), 1121–1140.

Saarinen, J. (2019). Not a serious subject?! Academic relevancy and critical tourism geographies. In D.K. Müller (Ed.), *A Research Agenda for Tourism Geographies* (pp. 33–41). Edward Elgar.

Scarles, C. (2014). Tourism and the visual. In A.A. Lew, C.M. Hall, & A.M. Williams (Eds.), *The Wiley Blackwell Companion to Tourism* (pp. 325–335). Wiley Blackwell.

Selby, M. (2012). Geographies of tourism and the city. In J. Wilson (Ed.), *The Routledge Handbook of Tourism Geographies* (pp. 232–239). Routledge.

Shaw, G. (2010). A long and winding road: Developing tourism geographies. In S.J.L. Smith (Ed.), *The Discovery of Tourism* (pp. 79–92). Emerald.

Shaw, G. (2014). Entrepreneurial cultures and small business enterprises in tourism. In A.A. Lew, C.M. Hall, & A.M. Williams (Eds.), *The Wiley Blackwell Companion to Tourism* (pp. 120–131). Wiley Blackwell.

Shoval, N. (2012). Time geography and tourism. In J. Wilson (Ed.), *The Routledge Handbook of Tourism Geographies* (pp. 174–180). Routledge.

Shoval, N., & Ahas, R. (2016). The use of tracking technologies in tourism research: the first decade. *Tourism Geographies 18*(5), 587–606.

Shoval, N., Isaacson, M., & Chhetri, P. (2014). GPS, smartphones, and the future of tourism research. In A.A. Lew, C.M. Hall, & A.M. Williams (Eds.), *The Wiley Blackwell Companion to Tourism* (pp. 251–261). Wiley Blackwell.

Shoval, N., Kwan, M.P., Reinau, K.H., & Harder, H. (2014). The shoemaker's son always goes bare-foot: Implementations of GPS and other tracking technologies for geographic research. *Geoforum 51*, 1–5.

Singh, R., Sibi, P.S., & Sharma, P. (2022). Journal of ecotourism: A bibliometric analysis. *Journal of Ecotourism 21*(1), 37–53.

Stausberg, M. (2014). Religion and spirituality in tourism. In A.A. Lew, C.M. Hall, & A.M. Williams (Eds.), *The Wiley Blackwell Companion to Tourism* (pp. 349–360). Wiley Blackwell.

Stock, M. (Ed.). (2021). *Progress in French Tourism Geographies: Inhabiting Touristic Worlds*. Springer.

Terkenli, T.S. (2014). Landscapes of tourism. In A.A. Lew, C.M. Hall, & A.M. Williams (Eds.), *The Wiley Blackwell Companion to Tourism* (pp. 282–293). Wiley Blackwell.

Timothy, D.J. (2014). Trends in tourism, shopping, and retailing. In A.A. Lew, C.M. Hall, & A.M. Williams (Eds.), *The Wiley Blackwell Companion to Tourism* (pp. 378–388). Wiley Blackwell.

Tribe, J. (1997). The indiscipline of tourism. *Annals of Tourism Research 24*(3), 638–657.

Tribe, J. (2006). The truth about tourism. *Annals of Tourism Research 33*(2), 360–381.

Vishwakarma, P., & Mukherjee, S. (2019). Forty-three years journey of tourism recreation research: A bibliometric analysis. *Tourism Recreation Research 44*(4), 403–418.

Weaver, D. (2014). The sustainable development of tourism: A State-of-the-art perspective. In A.A. Lew, C.M. Hall, & A.M. Williams (Eds.), *The Wiley Blackwell Companion to Tourism* (pp. 524–533). Wiley Blackwell.

Weidenfeld, A., & Hall, C.M. (2014). Tourism in the development of regional and sectoral innovation systems. In A.A. Lew, C.M. Hall, & A.M. Williams (Eds.), *The Wiley Blackwell Companion to Tourism* (pp. 578–588). Wiley Blackwell.

Williams, A.M. (2014). Tourism innovation: Products, processes, and people. In A.A. Lew, C.M. Hall, & A.M. Williams (Eds.), *The Wiley Blackwell Companion to Tourism* (pp. 168–178). Wiley Blackwell.

Williams, S., & Lew, A.A. (2015). *Tourism Geography: Critical Understandings of Place, Space and Experience*. 3rd. ed. Routledge.

Wilson, J. (Ed.) (2012a). *The Routledge Handbook of Tourism Geographies*. Routledge.

Wilson, J. (2012b). Tourism: The view from space. In J. Wilson (Ed.), *The Routledge Handbook of Tourism Geographies* (pp. 1–5). Routledge.

Wilson, J., & Anton Clavé, S.A. (Eds.) (2013). *Geographies of Tourism: European Research Perspectives*. Emerald.

Wilson, J., & Tallon, A. (2012). Geographies of gentrification and tourism. In J. Wilson (Ed.), *The Routledge Handbook of Tourism Geographies* (pp. 103–112). Routledge.

Xu, H., & Wu, Y. (2019). The contribution of tourism geography to the field of geography overall. In D.K. Müller (Ed.), *A Research Agenda for Tourism Geographies* (pp. 50–59). Edward Elgar.

3

GEOGRAPHIES OF TOURISM AND DEVELOPMENT

Facing Twenty-First Century Challenges

Rita de Cássia Ariza da Cruz

Introduction

According to the United Nations Refugee Agency[1] 82.4 million people were forcibly displaced within their own countries or to another in search of peace, work, food or dwelling in 2020. These are refugees, anonymous faces of a humanitarian tragedy emerging from wars, famine, power struggles and the persecution of ordinary citizens for diverse reasons, such as race, beliefs, political opinion and nationality. To this contingent we may add the approximately 272 million international migrants, the sum total reached in 2019, according to the International Organization for Migration (IOM),[2] which highlights the search for work as the main reason for international migration. The migratory crisis of the early twenty-first century laid bare a longstanding reality related to forced population movement, much of which results from deep social and economic differences between nations and regions, which in turn have been theorised for over a century as the outcome of combined unequal development (Trotsky, 2016 [1909]; Mandel, 1970; Smith, 2008 [1984]; Lowy, 1995; Harvey, 2007; Peck, 2016).

This brief allusion to the theme of international refugees and migrants aims to draw attention to the contradictions involved in the issue of social, economic and spatial development reproduced in the twenty-first century, in which tourism figures as a dramatic expression of the idiosyncrasies of a globalised world, contradictorily defined by hypermobility (Hall, 2010; Dumont, 2020), on the one hand, and by unequal mobility or even (im)mobility, on the other (Mostafanezhad, Cheer & Sin, 2020).

Besides and beyond the refugee crises, the hardest challenge in the early 21st century has undoubtedly been the COVID-19 pandemic, which shook the foundations of the world economy, forcing nations, regions and their respective governments to re-evaluate their projects, strategies and development expectations. The pandemic also exposed, in a very clear way, the deep gap between rich and poor countries, or between the so-called Global North and South, concerning their capacity to react to the crisis and all its repercussions. With regards to tourism, one of the sectors most harshly affected by the pandemic, researchers the world over have spent much time analysing ongoing processes as well as the post-crisis period, adopting a critical and propositional approach to the development of the activity. In this regard, social and spatial differences between

DOI: 10.4324/9781003286301-4

places, regions and countries soon proved to be fundamental for understanding the impacts of the pandemic on tourism development.

Another challenge for scholars of tourism and development in the beginning of this century is related to the development of the 'sharing economy' and its spatial, social and economic impacts, which are still to be fully understood. The Airbnb phenomenon, for example, has garnered increasing attention from geography researchers focusing on tourism, who have analysed its impact, frequently associated with processes of *overtourism* and gentrification.

Starting from a dialectical perspective, this chapter aims to contribute to and analyse the state of the art in research and knowledge regarding the interfaces between geographies of tourism and development, highlighting the challenges specific to the beginning of this century and also giving a certain prominence to the Global South.

Geographies of Tourism and Development: Recent Contributions

As Gascón and Cañada (2007) warn, the relationship between tourism and development is more complex than is usually assumed. The production of geographic thought regarding the relationship between development and tourism harkens back to the mid twentieth century and, throughout the last couple of decades, given the increased importance of the tourism phenomenon and the deep transformations affecting the world, and academic and scientific work around these themes have not only multiplied but formed tight relations with each other.

When analysing the role of development theories for the understanding of the tourism phenomenon, Sharpley (2022: 2) points out that David Harrison would have been the first to "apply development theory more broadly to tourism", notwithstanding the earlier and important contributions of authors like Stephen G. Britton, John M. Bryden and Louis A. Pérez. Indeed, according to Palomino-Schalscha (2012: 416), "over time research in both development and the study of tourism for development have evolved in interconnected trajectories".

Taking into account the different theories of development, tourism has gone through different stages, according to Wallingre (2017): the pre-Fordist model of production, from which early industrial tourism would have been derived; the Fordist model of production, during which mature industrial tourism emerged; and the post-Fordist, recognised as a flexible and sustainable production model, which would have fostered the emergence of post-tourism or post-industrial tourism.

Palomino-Schalscha (2012) highlights the roles of modernisation, dependence, neoliberalism, post-structuralism and post-colonialism theories as important paths in the search for understanding the development of tourism in all its complexity. Based on a critical revision of the literature on the "political economy of tourism development" published by Bianchi (2018: 92), we would add to this set the theory of regulation which, according to the author, "in contrast to dependency and underdevelopment theories, underlines the importance of the state in the production and reproduction of capitalist development".

We also find in Palomino-Schalscha (2012) a notable mention of 'alternative theoretical approaches', which emerged in the 1980s and acquired strength regarding the top-down model of development, calling for more social participation and focus on local scales, among its particularities. It is from this moment on, according to the author, that practice and study on alternative forms of tourism flourish, such as "ecotourism, sustainable tourism, pro-poor tourism and volunteer tourism, among many others" (Palomino-Schalscha, 2012: 423). This affirmation is corroborated by Wallingre (2017) in her theoretical review on development and tourism in Latin America. According to the author, "since the last stage of the last century, broader, more inclusive debates have been held regarding development involving environmental issues, conservation (of nature and culture) and the general improvement of living standards" (Wallingre, 2017: 4)[3]. In this sense, we can highlight the works of Butler (1974), Pigram (1980), Murphy (1983), Pearce (1989), Wahab and Pigram (1997) and Pearce and Butler (1999), among others.

Concerning the relationship between tourism, development and planning, Saarinen, Rogerson and Hall (2017) identify the post-WW2 period as the moment in which research began to emphasise themes such as "the modelling of tourism development, related questions of tourist supply and demand, and enquiries concerning locations and flows of tourists" (p. 310). Saarinen, Rogerson and Hall (2017) identify the 1980s as a moment of paradigm shift, from which point on, according to the authors, studies on tourism planning and development began to evolve more intensely in urban areas. Flourishing practice and studies in alternative forms of tourism are directly associated with positions adopted by international agencies, such as the United Nations (UN) and, more directly, with the Millennium Development Goals (MDGs), which have led to studies relating the UN MDGs to tourism development, as alluded to by Saarinen, Rogerson and Manwa (2011), Saarinen and Rogerson (2014) and Hall (2019).

In the last decade or so (2010–2021), theoretical approaches to neoliberalism, post-colonialism and post-development contributed, to some extent, to new approaches on the relationship between tourism and development, as found in the works of Mosedale (2016) and Onofre and Marin (2022). In turn, approaches focusing on issues such as participation, social empowerment and sustainability (Milano, 2016; Scheyvens & Van der Watt, 2021) remain relevant in academic debates on tourism geographies and development (Palomino-Schalscha, 2012). And, finally, the centrality that certain notions acquired – such as the sharing economy, overtourism, tourismphobia, tourism degrowth, gentrification, crises and the pandemic – allowed the emergence of new debates and theoretical and methodological contributions which are crucial for reaching a better understanding of the challenges posed by the twenty-first century.

Current Issues in the Geographies of Tourism and Development

Tourism and development in the Global South

Sarmento and Brito-Henriques (2013) claim that the Global South is not a "strict geographical categorisation of the world, but one which is based on economic inequalities and power imbalances having a certain cartographic continuity" (p. 2). Taking into account that the Global South includes "mainly regions of the world where poverty, environmental crisis, human and civil rights abuses, and ethnic intolerance are dramatic issues", the authors consider issues such as identity friction, cultural change and "the responsibility of tourism on poverty reduction and sustainable development" (p. 6) as paramount subjects for analysing tourism development in these countries.

The specificities and particularities of the Global South indicate that tourism explanatory models developed in the Global North cannot be automatically transposed to such distinct countries and places, despite the hegemonising structural processes deriving from economic and financial globalisation and neoliberalism. On the other hand, 'hegemonic tourism' has much to do with the development and modernisation of peripheral countries, as pointed out by Onofre and Marín (2022) in analysing the case of Latin America, based on a theoretical discussion of 'accumulation by dispossession', who consider tourism development in the region a kind of 'extractive activity'.

For Gascón and Cañada (2007: 12), tourism is presented in the Global South "as a fast track to economic growth and development within the current context of the predominance of the neo-liberal model".[4] Whatever the case, the adoption of development models 'from the North'

(Wallingre, 2017), including tourism development models, was and continues to be common to many countries in the Global South. Another side of the coin is that much of the academic production on tourism and development in the Global South – mainly from Latin America, the Caribbean and African Lusophone countries, such as Cape Verde, Mozambique and Angola – remain unknown by Global North academics, as most of them are published exclusively in Portuguese or Spanish. This suggests that comprehension of the geographies of development and tourism, at least in part of the Global South, still reproduces a hegemonic view from the Global North.

The sharing or collaborative economy and tourism development

According to Schor (2016), it is practically impossible to define the sharing economy in such a way that the concept reflects all the complexity found in the universe of shared or collaborative experiences, both for profit and non-profit, which has emerged since the end of the twentieth century, involving "recirculation of goods, increased utilisation of durable assets, exchange of services, and sharing of productive assets" (p. 2). Concerning shared or collaborative economies and their relationship to tourism, Dredge and Gyimóthy (2015) carried out a wide-ranging literature review and identified five pervasive claims on the theme, from which to seek a more reflective agenda for research. The authors found significant differences between the analysed approaches and, among other findings, arrived at a concluding statement: "Whether the collaborative economy is good or bad, what are its advantages and disadvantages, who wins and who loses, have not been fully revealed, much less researched" (p. 18).

Airbnb, one of the most successful experiences in the shared economy, together with Uber, are a recurring focus for studies on the relationship between shared/collaborative economies and tourism (Oskam, 2020; Wearing, MacDonald & Taylor, 2019). Apart from strongly impacting on the development of tourism, Airbnb has become an economic, social and geographic phenomenon frequently associated with aggravating conflicts related to the use and appropriation of space by tourism, as studied by Richards, Brown and Dilettuso (2020), Morales-Pérez, Garay and Wilson (2022) and González-Pérez (2020), to processes of gentrification, as analysed by Mermet (2017), Wachsmuth and Weisler (2018), Milano (2018), Jover and Dias-Parra (2019), Robertson, Oliver and Nost (2022); González-Pérez (2020) and Cocola-Gant et al. (2021), and overtourism, analysed by Milano (2018), Jover and Dias-Parra (2019), Oskam (2020) and Celata and Romano (2022).

Different approaches have demonstrated the various spatial implications of Airbnb in the development of tourism in different regions and locations, generally inserted into the mass tourism market before the emergence of 'platform tourism' (Söderström & Mermet, 2020). Domènech et al. (2019), in applying Local Moran's spatial autocorrelation indicator and the geographic information system to the analysis of Airbnb in Switzerland, arrive at a result that, based on Clivaz (2013), points toward a possible transition from a 'construction tourism' model, to one of 'service-tourism'. Their conclusion highlights the importance of a geography-based analysis model, given the centrality of a territory's characteristics in the spatialisation and growth of Airbnb (Domènech et al., 2019: 256). In addition, based on a statistical and geospatial analysis of AirDNA data applied to the bohemian neighbourhoods of Utrecht, Ioannides, Röslmaier and van der Zee (2019) warn of the fact that Airbnb may take on the role of an 'urban tourism bubble' expansion (see also van der Zee, Chapter 28 of this volume).

Starting from a relational approach, which situates the shared economy in the conflicting and contradictory process of the production of space, these studies provide important theoretical–methodological contributions for the geographies of development and tourism, which range from approaches to neoliberalism to those of financialisation and 'platform capitalism', passing through issues such as regulatory aspects, perceptions of tourists and inhabitants and socio-spatial segregation. Generally, a critical approach to the meanings and directions of shared/collaborative economies for the development of tourism prevails among these authors, considering most of all their socio-spatial impacts on different places.

Overtourism, tourismphobia and tourism degrowth

The debate on what has come to be called, in the twenty-first century, 'overtourism' harkens back at least four decades with studies on mass tourism, according to Milano, Novelli and Cheer (2019a). However, it is only since the last decade, thanks to sensationalist media narratives, that overtourism and the term 'tourismphobia' have become objects of increasing academic interest, according to Milano, Novelli and Cheer (2019a). For Dodds and Butler (2019), overtourism is 'a new term for an old problem' and, according to Koens, Postma and Papp (2018), it is an imprecise concept surrounded by myths.

Milano, Cheer and Novelli (2018: 2) point toward an oversimplification in analysis of the phenomenon. Furthermore, they indicate a possible causal relation between "the present era of unprecedented affluence and hyper mobility" and overtourism, affirming it is an issue of shared responsibility between local governments and destination managers.

Concerning theories and methods applied to tourism development in its relations with overtourism, Weber et al. (2020) highlight the use of approaches based on the tourism irritation index, the tourism carrying capacity and the tourism area life cycle (TALC), as well as on measuring systems proposed by McKinsey & Company and the World Travel & Tourism Council in "Nine Core Metrics Developed by McKinsey & Company and WTTC", both published in 2017 (Weber et al., 2020: 24–25).

Overtourism and tourismphobia have been related almost exclusively to countries of the Global North, especially European countries, where mass tourism has undoubtedly reached critical levels. This does not mean, however, that overtourism is not present in the development of tourism in the Global South, considering the region's particularities (Maingi, 2019; Pecout & Ricaurt, 2019).

Critiques of predatory economic development of sites, oblivious to social demands and interests, are not recent and have developed throughout time in debates that, in antagonising growth and development, serve to emphasise the issue of degrowth (Higgins-Desbiolles et al., 2019).

In recent years, an increase in interest in the degrowth problem can be easily felt in geographic approaches to tourism, specially concerning the Global North (Higgins-Desbiolles et al., 2019; Fletcher et al., 2019; Hall, Lundmark & Zhang, 2021; Cañada, 2021), and also associated with other themes, such as sustainability, community-based tourism, tourism planning and policy, and the COVID-19 pandemic (Butcher, 2021; Butcher, 2023; Seyfi & Hall, 2021; Fletcher et al., 2021; Cheer, 2020).

Much like other emerging themes, degrowth opens a window of possibilities for critically evaluating contemporary developments in tourism (see also Blàzquez et al., Chapter 5 of this volume).

Gentrification and tourism development

Although the debate over the processes of 'bourgeoisification' of spaces previously inhabited by the working class goes back to the 1960s, when sociologist Ruth Glass coined the concept of gentrification, approaches that correlate it to tourism development only gained visibility in the 1980s. From that moment, the international dissemination of a neoliberal urbanism, based on entrepreneurship as an urban governance strategy (Harvey, 1989) clearly expresses the strategic importance assumed by tourism in the refunctionalisation and revaluation of urban areas (Hall, 2014 [1988]).

Since then, this theme has been highlighted in approaches to tourism and development, as mentioned earlier in this text in relation to the phenomenon of Airbnb, but also with regard to cultural heritage (Hiernaux & Gonzáles, 2014; Paes, 2017; Gravari-Barbas & Guinand, 2017; Jover & Diaz-Parra, 2019), contemporary mobilities (López-Gay, Russo & Cocola-Gant, 2021; Sánchez-Aguilera & Gonzáles-Pérez, 2021), transnational migration (Jover & Diaz-Parra, 2019; Cocola-Gant & López-Gay, 2020), and overtourism and tourismphobia (Milano, 2018; Milano, Novelli & Cheer, 2019b; Celata & Romano, 2022), among other aspects related to the development of contemporary tourism (for a more in-depth exploration, see Buhr & Cocola-Gant, Chapter 31 of this volume).

Crisis events, COVID-19 and tourism development

In an analysis of the effect of crisis on the tourism sector by Hall (2010), the author draws attention to two central aspects: first, that many of the crises affecting tourism are "crisis events", given that they occur in an "identifiable place and time"; second, the growing potential of these events for having a global impact, given the hypermobility characteristic of a globalised world, in increasingly interconnected economies (Hall, 2010: 401).

This framework, outlined by the author a decade before the onset of the contemporary crisis, acquired dramatic overtones from 2020 onward. Undoubtedly the COVID-19 pandemic, the effects of which on tourism were devastating, is the most important and most overwhelming crisis of the 21st century, considering its social and 'unprecedented economic impacts' (UNWTO) as well as its geographic scope. In a few days we were led from a condition of hypermobility to a drastic interruption (Dumont, 2020) and a context of forced immobility that lasted for months (Mostafanezad, Cheer & Sin, 2020; Adey et al., 2021). This was something that the other crises of this century, such as the 2008 financial crash and 2009 H1N1 pandemic, were unable to do, despite their notoriously negative effects felt on a global scale.

Geographic approaches to tourism and development during pandemic times have provided a major contribution to understanding the impact of the current economic crisis on the sector in different locations, where innumerable case studies published in various languages and on the future of tourism (Seyfi & Hall, 2021; Renaud, 2020; Cheer, 2020; Gossling, Scott & Hall, 2021; Lew et al., 2020; Falcón, Martínez, Sánchez & Martínez, 2021) point to a challenge that still instigates researchers all over the world.

Whether from a perspective focused on thematic crises or on the issue of contemporary mobility, from a more geopolitical approach (Seyfi, Hall & Shabani, 2023), or from a critical perspective on the unsustainable nature of global tourism and its challenges (Ioannides & Gyimóthy, 2020), or adopting a more hopeful approach on the possibilities of tourism helping to build a more environmentally and socially just society (Mostafanazhed, 2020), the state of the art in research on geographies of tourism and development in the pandemic reveals the engagement of researchers with the theme and the wide range of possible theoretical, methodological and conceptual approaches. Given the easing of the pandemic, we may observe the gradual resumption of international travel, associated with progress in vaccination programmes worldwide and the reopening of borders. Accompanying this process closely is a challenge and an opportunity for researchers on geographies of tourism and development.

Concluding Considerations

The first two decades of this century clearly mirror what the geographer Milton Santos (2000) called 'contemporary acceleration', at times leaving us under the impression that we have lived a hundred years in twenty! The advancing financialisation process, expansion of *fintechs*, increasing technological changes, emergence of space tourism, a major financial crisis, a migration crisis, two pandemics – in short, a period of history marked by important economic, political and social instabilities.

This is the context in which tourism develops today, as a product of the contradictions of the contemporary world, but also as a reproducer of inherited contradictions and as a vehicle for new ones. Thus, tourism can be taken as an interpretive key, considering different scales of analysis – from local to global – to the idiosyncrasies of the contemporary world, as pointed out at the beginning of this chapter.

A research agenda for the future must seek to overcome gaps that still remain in the approaches of tourism and development geographies, such as the effects on the activity of different forms of neoliberalism and neoliberalisations, for which political economy stands as a theoretical field to be more fully explored, as Mosedale (2016) maintains. An analytical perspective based on political economy could, for example, contribute to a better understanding of the relationship between tourism and uneven geographic development, as well as to critical approaches to the profound changes that the world of work, and work in tourism specifically, are going through today (Ioannides & Zampoukos, 2018).

Another perspective that could be better explored by future studies concerns the so-called 'geographies of marginalisation' (Trudeau & Mcmorran, 2011), located in the sub-fields of social and cultural geographies, concerning transformations in economic and political relations, "social justice movements, gender empowerment initiatives and intense focus on rectifying development concerns, especially poverty alleviation and under development" (Cheer, 2018: 729). The same applies to creative tourism, an emerging theme, understudied from a developmental perspective (Pimenta; Ribeiro; Remoaldo, 2021).

From a look towards Europe, Scheyvens and Biddulph (2018) point to the challenge of promoting the development of a more inclusive tourism, which could provide a tool for greater integration in the Continent, 'not least in relation to making refugees welcome', but also with a greater commitment to the UN SDGs. Faced with crises, contradictions and even unwanted impacts related to mass tourism at the beginning of the twenty-first century, as well as difficulties in promoting an economic, social and environmentally sustainable development of the activity, Knafou (2017: 7) proposes *"un tourisme qui s'interrogue constamment sur lui-même, sa démarche, ses objectifs, ses finalités, ses moyens tout en se développant"* – "tourism which constantly questions itself, its approach, its objectives, its ends and its means, as it develops", editors' translation).

We do not know how tourism will develop in the coming years, whether the so-called 'new normal' will actually exist or what types of crises will occur, but we do know that there are many open doors and a world of theoretical and methodological possibilities to explore. Certainly, tourism and development geographies will remain a fertile ground for future research.

Notes

1 www.unhcr.org/figures-at-a-glance.html. Accessed on 2 October 2021.
2 https://publications.iom.int/system/files/pdf/wmr_2020.pdf. Accessed on October 2nd 2021.

3 Free translation. In the original: "[…] desde la última etapa del siglo pasado se sostienen debates ampliados y más integradores respecto del desarrollo los cuales involucran los temas ambientales, la conservación - naturaleza y cultura - y la mejora general de los estándares de vida […]" (Wallingre, 2017, p. 4).

4 Free translation. In the original: "[…] el turismo se presenta como una vía rápida para el desarrollo y el crecimiento económico, dentro del actual contexto de predominio del modelo neoliberal". (Gascón; Cañada, 2007, p. 12).

References

Adey, P., Hannam, K., Sheller, M., & Tyfield, D. (2021). Pandemic (Im)mobilities. *Mobilities* 16, 1–19.

Allis, T., Moraes, C.M.S., & Sheller, M. (2020). Revisitando as mobilidades turísticas. *Revista Turismo em Análise* 31(2), 271–295.

Almeida-García, F., Cortés-Macías, R., & Parzych, K. (2021). Tourism impacts, Tourism-Phobia and gentrification in historic centers: The cases of Málaga (Spain) and Gdansk (Poland). *Sustainability* 13, 408, 1–25.

Bianchi, R. (2018). The political economy of tourism development: A critical review. *Annals of Tourism Research* 70, 88–102.

Butcher, J. (2021). Debating tourism degrowth post COVID-19. *Annals of Tourism Research* 89, art. 103250.

Butcher, J. (2023). Covid-19, tourism and the advocacy of degrowth. *Tourism Recreation Research* 48(5), 633–642.

Butler, R.W. (1974). The social implications of tourist developments. *Annals of Tourism Research* 2(2), 100–111.

Cañada, E. (2021). Community-based tourism in a degrowth perspective. In K. Andriotis (Ed.), *Issues and Cases of Degrowth in Tourism* (pp. 42–63). CABI.

Celata, F., & Romano, A. (2022). Overtourism and online short-term rental platforms in Italian cities. *Journal of Sustainable Tourism* 30(5), 1020–1039.

Cheer, J.M. (2018). Geographies of marginalization: Encountering modern slavery in tourism. *Tourism Geographies* 20(4), 728–732.

Cheer, J.M. (2020). Human flourishing, tourism transformation and COVID-19: A conceptual touchstone. *Tourism Geographies* 22(3), 514–524.

Clivaz, C. (2013). Acceptance of the initiative on second homes: Emergence of a new development model for Swiss winter sports resorts? *Journal of Alpine Research | Revue de Géographie Alpine*. DOI:10.4000/rga.2274

Cocola-Gant, A., Hof, A., Smigiel, C., & Yrigoy, I. (2021). Short-term rentals as a new urban frontier: Evidence from European cities. *Environment and Planning A: Economy and Space* 53(7), 1601–1608.

Cocola-Gant, A., & Lopez-Gay, A. (2020). Transnational gentrification, tourism and the formation of 'foreign only' enclaves in Barcelona. *Urban Studies* 57(15), 3025–3043.

Dodds, R., & Butler, R. (2019). The phenomena of overtourism: A review. *International Journal of Tourism Cities* 5(4), 519–528.

Domènech, A., Schegg, R., Larpin, B., & Sclaglioni, M. (2019). Disentangling the geographical logic of Airbnb in Switzerland. *Erdkunde* 73(4), 245–258.

Dredge, D., & Gyimóthy, S. (2015). The collaborative economy and tourism: Critical perspectives, questionable claims and silenced voices. *Tourism Recreation Research* 40(3), 286–302.

Dumont, J-F. (2020). Le covid-19: La fin de la géographie de l'hypermobilité? *Les Analyses de Population Avenir* 29(11), 1–13.

Falcón, V.V., Martínez, B.S., Sánchez, F.C., & Martínez, N.G. (2021). Impacto de la Covid-19 en el turismo de Latinoamérica y el Caribe. *Revista Universidad y Sociedad* 13(3), 460–466.

Fletcher, R., Blanco-Romero, A., Salom, M.B., Cañada, E., Murray, I., & Sekulova, F. (2021). Socialisation at scale: Post-capitalist tourism in a Post-COVID-19 world. In F. Higgins-Desbiollles, A. Doering, & B. Bigby (Eds.), *Socializing Tourism: Rethinking Tourism for Social and Ecological Justice* (pp 229–242). Routledge.

Fletcher, R., Murray, I, Blanco-Romero, A., & Salom, M.B. (2019). Tourism and degrowth: An emerging agenda for research and praxis. *Journal of Sustainable Tourism* 27(12), 1745–1763.

Gascón, J., & Cañada, E. (2007). *Turismo y desarrollo. Herramientas para una Mirada crítica*. Enlace.

Giampiccoli, A., & Saayman, M. (2018). Community-based tourism development model and community participation. *African Journal of Hospitality, Tourism and Leisure* 7(4), 1–27.

González-Pérez, J.M. (2020). The dispute over tourist cities: Tourism gentrification in the historic Centre of Palma (Majorca, Spain). *Tourism Geographies* 22(1), 171–191.

Gössling, S., Scott, D., & Hall, C.M. (2021). Pandemics, tourism and global change: A rapid assessment of COVID-19. *Journal of Sustainable Tourism* 29(1), 1–20.

Gravari-Barbas, M., & Guinand, S. (2017). *Tourism and Gentrification in Contemporary Metropolises.* Routledge.

Hall, C.M. (2010). Crisis events in tourism: Subjects of crisis in tourism. *Current Issues in Tourism* 13(5), 401–417.

Hall, C.M. (2019). Constructing sustainable tourism development: The 2030 agenda and the managerial ecology of sustainable tourism. *Journal of Sustainable Tourism* 27(7), 1044–1060.

Hall, C.M., Lundmark, L., & Zhang, J.J. (Eds.) (2021). *Degrowth and Tourism: New Perspectives on Tourism Entrepreneurship, Destinations and Policy.* Routledge.

Hall, P. (2014). *Cities of Tomorrow: An Intellectual History of Urban Planning and Design since 1880.* 4th ed. Wiley-Blackwell.

Harvey, D. (1989). From managerialism to entrepreneurialism: The transformation in urban governance in late capitalism. *Geographisca Annaler* 71(1), 3–17.

Harvey, D. (2007). *The Limits to Capital.* Verso.

Hiernaux, D., & Gonzáles, C.I. (2014). Turismo y gentrificación: Pistas teóricas sobre una articulación. *Revista de Geografía Norte Grande* 58, 55–70.

Hiernaux-Nicolas, D. (2020). Nuevas encrucijadas para el turismo. *Estúdio y perspectivas en turismo* 29(3), 996–1011.

Higgins-Desbiolles, F., Carnicelli-Filho, S., Krolikowski, C., Wijesinghe, G., & Boluk, K. (2019). Degrowing tourism: Rethinking tourism. *Journal of Sustainable Tourism* 27(12), 1926–1944.

Ioannides, D., & Gyimóthy, S. (2020). The COVID-19 crisis as an opportunity for escaping the unsustainable global tourism path. *Tourism Geographies* 22(3), 624–632.

Ioannides, D., Röslmaier, M., & van der Zee, E. (2019). Airbnb as an instigator of 'tourism bubble' expansion in Utrecht's Lombok neighbourhood. *Tourism Geographies* 21(5), 822–840.

Ioannides, D., & Zampoukos, K. (2018). Tourism's labour geographies: Bringing tourism into work and work into tourism. *Tourism Geographies* 20(1), 1–10.

Jover, J., & Díaz-Parra, I. (2019). Gentrification, transnational gentrification and touristification in Seville, Spain. *Urban Studies* 57(15), 3044–3059.

Knafou, R. (2017). Le tourisme réflexive: un nuveau fondement d'un tourisme durable. *Arbor* 193(785), art. 395.

Koens, K., Postma, A., & Papp, B. (2018). Is overtourism overused? Understanding the impact of tourism in a city context. *Sustainability* 10(12), art. 4384.

Lew, A.A., Cheer, J.M., Haywood, M., Brouder, P., & Salazar, N.B. (2020). Visions of travel and tourism after the global COVID-19 transformation of 2020. *Tourism Geographies* 22(3), 455–466.

López-Gay, A., Cocola-Gant, A., & Russo, A.P. (2021). Urban tourism and population change: Gentrification in the age of mobilities. *Population, Space and Place* 27, 1–17.

Lowy, M. (1995). La théorie de le développement inegal. *Revue Actuel Marx* 18, 111–120.

Maingi, S.W. (2019). Sustainable tourism certification, local governance and management in dealing with overtourism in East Africa. *Worldwide Hospitality and Tourism Themes* 11(5), 532–551.

Mandel, E. (1970). The law of uneven development. *New Left Review* 59, art. 19.

Merigó, J.M., Mulet-Forteza, C., Valencia, C., & Lew, A.A. (2019). Twenty years of Tourism geographies: A bibliometric overview. *Tourism Geographies* 21(5), 881–910.

Mermet, A.-C. (2017). Airbnb and tourism gentrification: Critical insights from the exploratory analysis of the 'Airbnb syndrome' in Reykjavík. In M. Gravari-Barbas, & S. Guinand (Eds.), *Tourism and Gentrification in Contemporary Metropolises* (pp. 52–74). Routledge.

Milano, C. (2016). Antropologia, turismo y desarrollo en custión: el turismo comunitário a debate. *Quaderns de linstitute Català d'Antropologia* 32, 145–167.

Milano, C. (2017). *Overtourism and Tourismphobia: Global Trends and Local Contexts.* Ostelea School of Tourism & Hospitality.

Milano, C. (2018). Overtourism, malestar social y turismofobia. Un debate controvertido. *Pasos* 16(3), 551–564.

Milano, C., Novelli, M., & Cheer, J.M. (2018). Overtourism: A growing global problem. *The Conversation*, July 18, 2018.

Milano, C., Novelli, M., & Cheer, J.M. (2019a). Overtourism and Touritsmphobia: A journey through four decades of tourism development, planning and local concerns. *Tourism Planning & Development* 16(4), 353–357.

Milano, C., Cheer, J.M., & Novelli, M. (2019b). Overtourism: An evolving phenomenon. In C. Milano, J.M. Cheer, & M. Novelli (Eds.), *Overtourism: Excesses, Discontents and Measures in Travel and Tourism* (pp. 1–17). CABI.

Milano, C., Novelli, M., & Cheer, J.M. (2019c). Overtourism and degrowth: A social movements perspective. *Journal of Sustainable Tourism* 27(12), 1857–1875.

Minassian, H. (2012). Patrimonialisation et gentrification: Le cas de Barcelone. *Cahier Construction Politique et Sociale des Territoires* 12(1), 49–58.

Morales-Pérez, S., Garay, L., & Wilson, J. (2022). Airbnb's contribution to socio-spatial inequalities and geographies of resistance in Barcelona. *Tourism Geographies* 24(6–7), 978–1001.

Mosedale, J. (Ed.) (2016). *Neoliberalism and the Political Economy of Tourism*. Routledge.

Mostafanezhad, M. (2020). Covid-19 is an unnatural disaster: Hope in revelatory moments of crisis. *Tourism Geographies* 22(3), 639–645.

Mostafanezhad, M., Cheer, J.M., & Sin, H.L. (2020). Geopolitical anxieties of tourism: (Im)mobilities of the COVID-19 pandemic. *Dialogues in Human Geography* 10(2), 182–186.

Murphy, P.E. (1983). Tourism as a community industry: An ecological model of tourism development. *Tourism Management* 4(3), 180–193.

Onofre, A.A.V., & Marín, A.I.M. (2022). Trazos teóricos para el análisis del turismo como actividade extractiva. In A.A. Briceño, & M.A. Villa (Eds.), *Abordajes Críticos del Turismo: Conceptualizaciones y Estudios de Caso* (pp. 31–58). Edicionaes Navarra.

Oskam, J. (Ed.) (2020). *The Overtourism Debate: NIMBY, Nuisance, Commodification*. Emerald Publishing.

Paes, M.T.D. (2017). Gentrificação, preservação patrimonial e turismo: Os novos sentidos da paisagem urbana ma renovação das cidades. *Geousp – Espaço e Tempo* 21(3), 667–684.

Palomino-Shalcha, M. (2012). Geographies of tourism and development. In J. Wilson (Ed.), *The Routledge Handbook of Tourism Geographies* (pp. 414–432). Routledge.

Pearce, D. (1989). *Tourist Development*. Longman.

Pearce, D., & Butler, R.W. (Eds.) (1999). *Contemporary Issues in Tourism Development*. Routledge.

Peck, J. (2016). Uneven regional development. In D. Richardson, N. Castree, M. Goodchild, W. Liu, A. Kobayashi, & R. Marston (Eds), *The International Encyclopedia of Geography*. Wiley-Blackwell.

Pecot, M., & Ricaurte-Quijano, C. (2019). '¿Todos a Galápagos?' Overtourism in wilderness areas of the global South. In C. Milano, J.M. Cheer, & M. Novelli (Eds.), *Overtourism: Excesses, Discontents and Measures in Travel and Tourism* (pp. 70–85). CABI.

Pigram, J.J. (1980). Environmental implications of tourism development. *Annals of Tourism Research* 7(4), 554–583.

Pigram, J.J., & Wahab, S. (1997). *Tourism, Development and Growth*. Routledge.

Pimenta, C.A.M., Ribeiro, J.C., & Remoaldo, P.C. (2021). The relationship between creative tourism and local development: A bibliometric approach for the period 2009–2019. *Tourism & Management Studies*, 17(1), 5–18.

Renaud, L. (2020). Reconsidering global mobility: Distancing from mass cruise tourism in the aftermath of COVID-19. *Tourism Geographies* 22(3), 679–689.

Richards, S., Brown, L., & Dilettuso, A. (2020). The Airbnb phenomenon: The resident's perspective. *International Journal of Tourism Cities* 6(1), 8–26.

Robertson, D., Oliver, C., & Nost, E. (2022). Short-term rentals as digitally-mediated tourism gentrification: Impacts on housing in New Orleans. *Tourism Geographies* 24(6–7), 954–977.

Saarinen, J., Rogerson, C., & Manwa, H. (2011). Tourism and millennium development goals: Tourism for global development? *Current Issues in Tourism* 14(3), 201–203.

Saarinen, J., & Rogerson, C.M. (2014). Tourism and the millennium development goals: Perspectives beyond 2015. *Tourism Geographies* 16(1), 23–30.

Saarinen, J., Rogerson, C.M., & Hall, C.M. (2017). Geographies of tourism development and planning. *Tourism Geographies* 19(3), 307–317.

Sánchez-Aguilera, D., & Gonzáles-Pérez, J.M. (2021). Geographies of gentrification in Barcelona: Tourism as a driver of social change. In J. Dominguez-Mujica, J. McGarrigle, & J.M. Parreño-Castellano (Eds.), *International Residential Mobilities: From Lifestyle Migrations to Tourism Gentrification* (pp. 243–268.). Springer.

Santos, M. (2000). *Por Uma Outra Globalização*. Record.

Sarmento, J., & Brito-Henriques, E. (Eds.) (2013). *Tourism in the Global South*. Lisbon: Center for Geographic Studies.

Scheyvens, R., & Biddulph, R. (2018). Inclusive tourism development. *Tourism Geographies* 20(4), 589–609.

Scheyvens, R., & Van der Watt, H. (2021). Tourism, empowerment and sustainale development: A new framework for analysis. *Sustainability* 13(22), art. 12606.

Schor, J. (2016). Debating the sharing economy. *Journal of Self-governance and Management Economics* 4(3), 7–22.

Seyfi, S., & Hall, C.M. (2021). COVID-19 pandemic, tourism and degrowth. In C.M. Hall, L. Lundmark, & J. Zhang (Eds.), *Degrowth and Tourism: New Perspectives on Tourism Entrepreneurship, Destinations and Policy* (pp. 220–238). Routledge.

Seyfi, S., Hall, C.M., & Shabani, B. (2023). COVID-19 and international travel restrictions: The geopolitics of health and tourism. *Tourism Geographies* 25(1), 357–373.

Sharpley, R. (2022). Tourism and development theory: Which way now? *Tourism Planning & Development* 19(1), 1–12.

Smith, N. (2008). *Uneven Development: Nature, Capital and the Production of Space*. University of Georgia Press.

Söderström, O., & Mermet, A.-C. (2020). When Airbnb sits in the control room: Platform urbanism as actually existing smart urbanism in Reykjavík. Frontiers in Sustainable Cities 2. DOI: 10.3389/frsc.2020.00015

Trotsky, L. (2016). *1905*. 2nd ed. Haymarket Books.

Trudeau, D., & McMorran, C. (2011). The geographies of marginalization. In V.J. Del Casino Jr, M. Thomas, P. Cloke, & R. Panelli (Eds.), *A Companion to Social Geography* (pp. 437–453). Blackwell.

Veríssimo, M., Moraes, M., Breda, Z., Guizi, A. & Costa, C. (2020). Overtourism and tourismphobia: A systematic literature review. *Tourism: An International Interdisciplinary Journal* 68(2), 156–169.

Wachsmuth, D., & Weisler, A. (2018). Airbnb and the rent gap: Gentrification through the sharing economy. *Environment and Planning A: Economy and Space* 50(6), 1147–1170.

Walingre, N. (2017). Enfoques del desarrollo y el turismo en América Latina. *Divulgatio* 1(3), 27–43.

Weber, F., Eggli, F., Meier-Crameri, U., & Stettler, J. (2020). *Measuring Overtourism: Indicators for Overtourism: Challenges and Opportunities*. Hochshule Luzern.

Wearing, S.L., Macdonald, M., & Taylor, G. (2019). Neoliberalism and global tourism. In D.J. Timothy (Ed.), *Handbook of Globalisation and Tourism* (pp. 27–43). Edward Elgar.

Zeng, Z., Chen, P.-J., & Lew, A.A. (2020). From high-touch to high-tech: COVID-19 drives robotics adoption. *Tourism Geographies* 22(3), 724–734.

PART II

Critical Geographies of Tourism

4

CRITICAL THEORIES AND SUSTAINABLE TOURISM FUTURES

Paolo Mura and Sarah N. R. Wijesinghe

Introduction

Debates concerning the role of critical approaches to tourism (as a field of inquiry, industry, and socio-cultural phenomenon) have proliferated in the tourism literature in the last 20 years (Ateljevic et al., 2007; Bianchi, 2012; Higgins-Desbiolles & Whyte, 2014; Morgan et al., 2018; Pritchard et al., 2011). Drawing upon – and going beyond – the tenets of critical approaches to inquiry advocated by the Frankfurt School (Horkheimer, 1937/2002; Horkheimer & Adorno, 1947/2002), critical tourism scholars have embraced a multiplicity of diverse perspectives and theoretical stances (e.g., feminism, critical race theory, queer theories) to unveil the unequal power structures underpinning and shaping tourism. While discussions about 'the critical' have intensified in the aftermath of the First Critical Tourism Studies (CTS) Conference in 2005 (Ateljevic et al., 2007), critical approaches have become even more relevant in envisioning sustainable tourism futures in the post–COVID-19 era. Indeed, the temporary halt of the tourism industry during the pandemic has contributed to the reiteration of previous critical debates concerning the need to 'rethink' both the industry and the academy to pursue more democratic, just and equal future scenarios (Higgins-Desbiolles et al., 2019).

Drawing upon the themes emerging from a recent systematic review on criticality in tourism (Mura & Wijesinghe, 2023), this chapter discusses the nexus between critical theory and sustainable tourism futures. More specifically, after presenting an overview of the main topics and issues underpinning criticality in tourism, this work highlights the opportunities offered by critical approaches to achieve the sustainable 'new normal' that has been called out in post–COVID-19 tourism practices. Importantly, the chapter points to the need to strengthen the nexus between tourism critical scholarship and activism outside academia in order to engage with – and achieve – critical praxis.

Critical Theories in Tourism: An Overview

The body of knowledge mobilising aspects of critical theory in tourism (Frankfurt School and beyond) has grown since the 1980s, with a particular emphasis on contextualising the role and meanings of tourism in contemporary capitalist societies. For instance, early work by Britton

DOI: 10.4324/9781003286301-6

(1991), which refers to the notion of 'culture industry' to contend that in a capitalist system 'sameness' is produced through the massification and commoditisation of leisure experiences, tourism represents one of the leisure experiences reproducing existing power structures and ideologies. Higgins-Desbiolles and Whyte (2014) also point out that, despite being overlooked by tourism scholars, early critical perspectives concerning the development and impacts of tourism have appeared since the 1970s in material published outside academic circles, mostly by non-governmental organisations (NGOs). Generally, critical approaches to tourism knowledge production and representation (Adams, 2020; Aitchison, 2001; Botterill, 2003) alongside critical approaches to tourism as a lived experience outside academic circles (its impacts on places, peoples and economies) (see Britton, 1991; Higgins-Desbiolles & Whyte, 2014) are not new in tourism. However, propelled further by a 'critical turn' (Ateljevic et al., 2007; Chambers, 2007; Pritchard et al., 2011) and an overall critique of positivist research, debates concerning the meanings and implications of 'the critical' in tourism have also gained momentum and more visibility in the last 20 years.

Recent work has also moved beyond critical discussions concerning contemporary capitalist production and consumption patterns to highlight the limitations of the epistemological and ontological assumptions of tourism theory (i.e., Hollinshead, Suleman & Nair, 2021; Hollinshead, Suleman & Vellah, 2021). In doing so, the 'critical' has taken centre stage to scrutinise ways of producing, knowing and representing tourism (as both socio-economic phenomenon and field of inquiry). This includes important criticism on the supposed 'critical perspectives' of the critical turn agenda in tourism as well (Bianchi, 2012). As such, Chambers and Buzinde (2015) and Higgins-Desbiolles and Whyte (2013) have highlighted issues concerning the co-creation of knowledge with the marginalised, participatory aspects, the Eurocentric nature of general knowledge production, and lack of reflexivity by researchers throughout the process.

Recently, Morgan et al. (2018) have referred to embracing post–disciplinary approaches, strengthening the dialogue with business approaches, developing research agendas that contemplate multiple worldviews and engaging in reflexivity as the main challenges faced by critical tourism scholars in the future. Despite these challenges and the various points of criticism raised about the critical turn, it is important to reiterate that 'as we approach the second decade of the new millennium, the role of critical tourism studies scholarship is arguably more vital than ever' (Morgan et al., 2018; p. 185). In this light, in a previous study we analysed the current state of critical theory research in tourism scholarly work to understand how critical theories and approaches have been employed by tourism scholars in the last 40 years. By scrutinising publications in tourism journals, books and book chapters in four languages (English, Spanish, Portuguese and Italian) in the last 43 years (1977–2020), we explored the main critical approaches mobilised, the areas / topics of interest, authorship and the emancipatory outcomes pursued (see Mura & Wijesinghe, 2023).

Overall, the work of critical theorists in tourism presented several insights that are important to consider in envisioning sustainable tourism futures. Indeed, aligned with the existing body of knowledge (Bramwell & Lane, 2011; Liburd & Edwards, 2010), many papers analysed reiterate an important nexus between critical theory and sustainable tourism (i.e., Dwyer, 2017; Hanna, et al. 2015; Grimwood, 2015; Jamal, Carmago, & Wilson, 2013; Sowards, 2012). Notably, a main point emerging from the articles analysed is that employing a critical lens is necessary to scrutinise and question the current status quo and develop forms of tourism that promote social justice and the inclusion of multiple perspectives, which can be attained by giving voice to all the stakeholders involved in the development of tourism alongside the need to respect local cultures and the environment (i.e., Barrios, Prowse & Vargas, 2020; Bramwell, 2010; Dredge, 2006). It is evident that being 'critical' and developing critical perspectives to tourism are articulated as fundamental steps towards achieving sustainability.

However, it was also apparent that while an assessment of the authorship of critical work in tourism studies pointed to balanced figures in terms of gendered distribution, the affiliation of the authors portrayed the dominant role of Western universities in propelling critical scholarship. Also, a significant lack of reflexivity in critical work emerged in aspects of both knowledge production as well as in so-called 'emancipatory' initiatives. The earlier critiques of the agenda of the 'critical turn' and 'hopeful tourism' (see Bianchi, 2012; Chambers & Buzinde, 2015; Higgings-Desbiolles & Whyte, 2013), are aligned with the work highlighting the dire need to reflect on authors' own positionality and Eurocentric values in tourism epistemology, which hinder a path of inclusive criticality (Chambers & Buzinde, 2015; Everingham, Peters & Higgins-Desbiolles, 2021). The lack of reflexive thought on our role and rights in advocating more emancipatory and inclusive inquiry may lead to conceal – and/or also to reproduce – existing power structures. For example, the fact that most of the published articles are produced by authors based in Western universities (with little to no collaboration with scholars/or epistemologies outside of the so-called non-West), should lead us to reflect upon the power structures (e.g., postcolonial, linguistic, among others; see Tucker, 2012) that underpin tourism critical scholarship, especially if we consider that most critical work involves communities in the 'Global South' considered 'marginalised' or 'minority' (see Wijesinghe, Mura & Bouchon, 2017). This certainly has implications for the discourses of criticality itself.

It remains evident that critical and emancipatory stances in tourism are compromised by an enduring Eurocentric gaze (see Hobson & Sajed, 2017). Although calls for decolonising the tourism academy have been prominent in the last decade (Chambers & Buzinde, 2015) and post-colonial/decolonial perspectives have gained momentum, embracing 'alternative' forms of agency and forms of thinking that reflect a plurality of paradigms (i.e., indigenous) remains imperative. The need to globalise the discipline in a more pluralistic manner remains a path requiring attention. Consequently, perhaps, a significant increasing number of studies has brought into context the importance of critical pedagogical aspects and academic leadership (Boluk, Cavaliere & Duffy, 2019; Boyle, Wilson & Dimmock, 2015; Schweinsberga, 2018).

Another important aspect emerging from our assessment of criticality in tourism refers to the lack of research attempting to bridge academia, industry and communities. In this regard, while only a few studies approached criticality by employing non-academic literature (e.g., work published by NGOs, industry, governments), most critical work in tourism is grounded in schol-arly academic literature. It is important to re-emphasise the crucial role of critical academics in connecting and establishing a dialogue with the non-academic world as a first step toward collab-orative and inclusive research that enables sustainable and inclusive practices elsewhere, essen-tially producing a ripple effect (Ainley & Kline, 2013; Dredge & Gyimóthy, 2015). This is even more important in the current global reality the tourism industry has to face.

Critical Theories in Envisioning 'Sustainable Realities' in the Post–COVID-19

In *The Critical Turn in Tourism Studies – Innovative Research Methods*, published in the aftermath of the first conference on criticality in tourism as the manifesto of the Critical Tourism Studies (CTS) community, Ateljevic et al. (2007, p. 3) define critical tourism scholarship as 'more than simply a way of knowing, an ontology, it is a way of being, a commitment to tourism inquiry which is pro-social justice and equality and anti-oppression: it is an academy of hope'. In this line of thought, critical perspectives also play a crucial role in realising future sustainable realities that have been called out in the post–COVID-19 era. As such, dialogues that fundamentally question tourism and remind us of the need for sustainable, just and equitable practices in tourism have risen in the last two years with the advent of the pandemic and subsequent lockdowns.

Indeed, calls for transformation and new forms of capitalism that would move beyond the business-as-usual ethos in the tourism industry have been louder than ever (Cave & Dredge, 2020). Across the world, COVID-19 has restricted physical mobilities with closed national borders, suspended international transportation and restrictive lockdowns, leading to a phenomenon of 'temporary de-globalisation' (Niewiadomski, 2020). The scale of impact on the tourism industry is wide economically but, simultaneously, the situation has fundamentally brought to the fore the often neglected and 'thrown under the carpet' issues that drove mass international tourism (i.e., overtourism, labour conditions, economic-orientation, social and environmental degradation, overdependence, tourist-focused development/enclavity) and further exposed the 'limits of the neoliberal orthodoxy' (Bianchi, 2021; Lapointe, 2020 and Chapter 6, this volume; Niewiadomski, 2020).

The opportunities to reflect, rethink and rebuild via unlearning the dominant exploitative economic-oriented discourses that gave rise to systemic inequalities within tourism have been profound (Benjamin, Dillette & Alderman, 2020). Many have voiced the rare opportunity the pandemic has presented to the industry to reflect critically on its 'old entrenched ways of doing business' and transform into a 'new normal' that is more equitable and just (Ateljevic, 2020; Crossley, 2020; Haywood, 2020; Higgins- Desbiolles, Doering & Bigby, 2021). Against this background of temporary 'de-globalisation' (Niewiadomski, 2020), the need for a 'new relationship with capitalism and new measures of success in tourism' has gained importance.

Immediately after, there has been a significant proportion of critical work that has reiterated the need for a restructuring towards a 'new normal' in tourism, capable of challenging neoliberalist production and consumption. Several special issues (for example 'Vision of Travel & Tourism after Global COVID-19 Transformation of 2020' in *Tourism Geographies*) have encouraged critical perspectives from scholars across the world and opened a dialogue to visualise tourism beyond its neoliberal economic practices. These include, for instance, the need to diffuse holistic principles of the degrowth movement (see Büscher & Fletcher, 2017; Higgins-Desbiolles et al., 2019), consider sustainability beyond economic interests (Lapointe et al., 2018), responsible tourism, localisation, and moral capitalism (Haywood, 2020), pro-equity (Benjamin et al., 2020), and inclusion of indigenous epistemologies (Cheer, 2020; Everingham & Chassagne, 2020; Niewiadomski, 2020). Scholars have also called for collective efforts by policymakers, business sector, experts, academics, tourism students, and tourists themselves, at global and local levels (Romagosa, 2020), to undertake deep critical assessments of 'what is happening, what isn't; what's flourishing, what isn't; what's possible, what isn't ... why and why not' (Haywood, 2020, p. 605).

While there is an upsurge of initiatives across the industrial world to move 'outside the logic of growth and expansion' or 'prosperity without growth' (i.e., ecological civilisation in China, sufficiency economy in Thailand, Earth democracy in India), such perspectives are still in their infancy in tourism research with a minority of scholars reiterating the importance of engaging with sustainable discourses beyond their usual growth/capitalist values (Chassagne & Everingham, 2019; Crossley, 2020; Higgins-Desbiolles, et al., 2021; Rastegar, Higgins-Desbiolles & Ruhanen, 2021). Indeed, discourses of sustainability in tourism remain largely influenced by old development narratives where environmentalism and culture are subjugated to the worldwide expansion of capitalism (i.e., the global environmental–economic paradigm) (Fitchett et al., 2021). Paradoxically, sustainable development is erected as the 'conceptual roof for both violating and healing the environment' (Sachs, 2019, p. 28). In this worldview, sustainability and the economy are intertwined, where the economy is presented as the healing agent for transforming social and environmental ills (see Wijesinghe & Higgins-Desbiolles, 2024). As such, in global sustainability agendas, starting from the Rio+20 declaration to the 2030 Agenda for Sustainable Development, development (traditionally embraced as economic growth) is promoted, even after decades of evidence that existing forms of growth strategies, regardless of 'green economy', 'circular economy', or 'eco-efficiency', have not led to the healing of the environment.

This has further been presented in recent studies where tourism recovery policies and projections are analysed (i.e., Wijesinghe, 2022). Hence, further engagements with critical perspectives are needed to highlight the capitalist contradictions that exist in tourism (for instance, the idea that nature and all its aspects can be saved and protected by privatising and exploiting them; see Nepal, 2020).

In this sense, some have also noted that the global pandemic has enabled not only a moment to 'reconsider' and 'transform', but also an important step towards unlearning hegemonic notions of prosperity and relearning alternative ways forward (Benjamin et al., 2020). This is particularly true for the so-called 'developing nations', which remain 'captive' within neoliberalist development discourses (Wijesinghe, 2022). To do so, much like what has previously been observed in relation to critical work in tourism, an important element that is further pushed is the importance of education. Education and dialogue play an essential role in inducing Pernecky's (2020) 'hope-as-utopia', a space to critique and imagine new ways of thinking and being in tourism that are resistive, just and centred within the local. Such induction and orientation are also required for nation-state decision-makers, especially in former colonies that continue to function within limited (unchallenged) views of neoliberalism (and coloniality). In imagining new ways, as reiterated by previous critical works in tourism (see Mura & Wijesinghe, 2023), reflexivity must remain an ongoing process in challenging positionality of knowledge itself, and hence criticality in itself. Since tourism moves across cultures, it remains imperative to imagine sustainable and alternative realities in a multiplicity of 'local unique conditions' (Tomassini & Cavagnaro, 2020).

The Way Forward: Reconsidering the Dialectical Relations between Critical Theory, Sustainable Tourism and Political Activism in Tourism

As argued above, critical tourism scholarship can play a crucial role in shaping sustainable tourism futures in the post–COVID-19 era. However, how can the 'critical' propel change outside tourism academic circles and promote forms of tourism sustainability beyond tourism academia? One of the most important tenets and legacies of critical theory concerns its call for action, namely forms of political academic activism aimed at propelling democratic pluralism and societal change. In this respect, critical theorists belonging to different disciplines have often been involved in forms of political activism and have taken an active role in translating/transmitting academic knowledge to social movements and for political causes (see Davis, 1994 as an emblematic example of a critical theorist engaged in activist scholarship).

While academic activism can assume different meanings and forms (see Blomley, 2008, on its various aspects and manifestations, including direct and indirect participation in social movements and other forms of activism leading to social change), Piven (2010, p. 806) points out that 'the motivating idea is that academic work can be useful in ameliorating the big problems of our society, problems such as inequality and insecurity, or militarism and imperial overreach, or the corruption of democratic procedures, or ecological degradation'. At the same time, critical scholarship has also recognised that social change is a process informed by different types of knowledge, not only that which is produced within academic circles but also that constructed and circulated within activist groups and social movements outside academia (Choudry, 2020). From this perspective, critical academic work (in general and in tourism) and political activism/knowledge can be framed in a dialectical and circular relationship in which each informs and influences the other.

It has been argued that such dialectical nexus is imperative in the current political scenario – a scenario characterised by an overall critique of the role of intellectuals and scientists in society, the rise of white supremacism and right-wing populism, and by more-pronounced socio-economic inequalities among communities and social groups (Macdonald & Young, 2018). Despite this, the ambiguity concerning the role of academia and universities in contemporary societies has often limited attempts of 'activist scholarship' and 'scholar activism'. Indeed, universities have become ambiguous contexts as locales representing both traditional sites for struggle, contestation and social change through campus activism and neoliberalist and market-driven forces in which elitist approaches to knowledge and market-oriented dynamics are often produced and reiterated (Choudry, 2020). From this perspective, while still-remaining sites for academic freedom and free thinking, academic spaces have also become 'spaces of tension' between neoliberal/capitalist ideologies and criticality. As such, activist scholars often face obstacles and attacks from both peers and other stakeholders outside academia, for example, politicians and policy makers (Flood et al., 2013). As Macdonald and Young (2018, p. 530) have pointed out, within the social science disciplines, discourses concerning pluralism and multiplicity of ideologies are tolerated 'so long as they do not threaten the hegemonic discourse of science and liberalism. [...] In this contemporary context, activism can be safely studied at a distance as just another subject in the scholastic marketplace'.

Within this scenario, it is not surprising that, outside academia, examples are limited of tourism critical scholarship being engaged in forms of political activism striving to promote new approaches to sustainable tourism. Drawing on the idea that social change is a process driven by collective – rather than individual – actions (Choudry, 2020), there have been various calls, before and after the pandemic, for a more cooperative and politically active tourism academy, mostly propelled by a 'critical turn' in tourism (Ateljevic et al., 2007; Pritchard et al., 2011) and various demands to 'rethink' tourism as both an industry and a subject of academic inquiry (Cave & Dredge, 2020; Higgins-Desbiolles et al., 2019; Hollinshead et al., 2021). Yet, whether and how these calls (mostly from scholars based in Western institutions) have initiated or contributed to debates or forms of political change outside tourism academia is a subject of debate. Exceptions to this status quo do exist. Hales, Dredge, Higgins-Desbiolles and Jamal (2018) represent noteworthy examples of tourism scholars actively engaged in forms of activism with communities outside academic circles. Furthermore, by linking critical race theory (CRT) to social movement theory (SMT), work by Dillette and Benjamin (2022) and McGhee, Kline and Knollenberg (2014) has brought to the fore the possibilities for tourism critical scholarship and forms of travel (e.g., the Black Travel Movement) to propel social activism and change.

Among the various forms of scholarly activism, Hales et al. (2018) emphasise the importance of 'embedded situated methodologies' – namely, forms of action inquiry in which researchers become entangled with 'Other' communities by establishing deep and long-lasting relationships of trust – in contributing to politically activist critical research and sustainable futures for tourism (as both an academic field and industry). As methodological pluralism and qualitative participatory research approaches informed by moral stances to inquiry have been conceived as political acts capable of producing change in academia and society (Denzin & Lincoln, 2011), they could pave the way for more sustainable forms of tourism. From this perspective, sustainable tourism is conceived as both an industry and a social force that should incorporate and value the contribution and input of communities and other stakeholders in the research, planning and implementation processes. This aspect, which should be regarded as pivotal in any research agenda concerning sustainable tourism, has often been neglected by tourism scholars (including critical tourism scholars), who have tended to approach communities and people affected by tourism with the main aim to 'collect data' and publish in academic journals and books. Here, it needs to be emphasised that members of local communities or stakeholders outside academia do not have access to traditional academic publications or, if they do, may not read them (Mura, 2020). As such, although valuable within academic circles, the body of knowledge produced by critical scholars rarely establishes forms of dialogue with the participants before, during and after the research process. Hence, there seems to exist a gap between critical tourism scholars researching sustainability and its potential non-academic audience.

Non-traditional ways to engage with 'Other' groups and non-academic communities (e.g., migrants and children), which have the potential to lead to transformative experiences and realities for both researchers and participants, have begun to emerge in the tourism literature. For example, approaches to inquiry that mobilise forms of arts-based methods and embodied knowledge (see Rydzic et al., 2013; Mura et al., 2021) have reasserted the importance of performance ethnography and embodied knowledge in establishing deeper forms of engagement and collaboration with non-academic Others. Yet, the participatory and emancipatory possibilities (for both communities and scholars) of non-traditional qualitative research (e.g., qualitative approaches that go beyond interviews) has not yet fully explored within tourism academic circles (Wilson et al., 2020), especially within discourses of tourism sustainability. Recently, Bertella (2023) has proposed a care-based academic activism model as a frame to conduct research leading to sustainable transformations. Importantly, the model recognises the central role of scholars' reflexivity in contrasting the current hegemonic anthropocentric approach to sustainability research and considers critical thinking, attentiveness, imagination and responsiveness as pivotal components. Overall, the opportunities behind a deeper dialogue between critical approaches to inquiry and political activism for sustainable tourism futures are multiple and promising. Yet, they deserve further consideration within tourism academic circles.

References

Adams, K.M. (2020). What western tourism concepts obscure: Intersections of migration and tourism in Indonesia. *Tourism Geographies*, 23(1), 1–26.

Ainley, S., & Kline, C. (2013). Moving beyond positivism: Reflexive collaboration in understanding agritourism across North American boundaries. *Current Issues in Tourism*, 17(5), 404–413.

Aitchison, C. (2001). Theorizing other discourses of tourism, gender and culture: Can the subaltern speak (in tourism)? *Tourist Studies*, 1(2), 133–147.

Ateljevic, I. (2020). Transforming the (tourism) world for good and (re)generating the potential 'new normal'. *Tourism Geographies*, 22(3), 467–475.

Ateljevic, I., Pritchard, A., & Morgan, N. (Eds.) (2007). *The Critical Turn in Tourism Studies: Innovative Research Methodologies*. Elsevier.

Barrios, L.M., Prowse, A., & Vargas, V.R. (2020). Sustainable development and women's leadership: A participatory exploration of capabilities in Colombian Caribbean fisher communities. *Journal of Cleaner Production*, 264, 121277.

Benjamin, S., Dillette, A., & Alderman, D. (2020). "We can't return to normal": Committing to tourism equity in the post-pandemic age. *Tourism Geographies*, 22(3), 476–483.

Bertella, G. (2023). Care-full academic activism for sustainable transformations in tourism. *Current Issues in Tourism*, 26(2), 212–223.

Bianchi, R. (2012). A Radical Departure: A Critique of the Critical Turn in Tourism Studies. In J. Wilson (Ed.), *The Routledge Handbook of Tourism Geographies* (pp. 63–71). Routledge.

Bianchi, R. (2021). Tourism, COVID-19 and Crisis: The Case for a Radical Turn. In F. Higgins-Desbiolles, A. Doering, & B.C. Bigby (Eds.), *Socialising Tourism: Rethinking Tourism for Social and Ecological Justice* (pp. 93–108). Routledge.

Blomley, N. (2008). The spaces of critical geography. *Progress in Human Geography*, 32(2), 285–293.

Boluk, K.A., Cavaliere, C.T., & Duffy, L.N. (2019). A pedagogical framework for the development of the critical tourism citizen. *Journal of Sustainable Tourism*, 27(7), 865–881.

Botterill, D. (2003). An autoethnographic narrative on tourism research epistemologies. *Loisir et Société / Society and Leisure*, 26(1), 97–110.

Boyle, A., Wilson, E., & Dimmock, K. (2015). Transformative education and sustainable tourism: The influence of a lecturer's worldview. *Journal of Teaching in Travel & Tourism*, 15(3), 252–263.

Bramwell, B. (2010). Participative planning and governance for sustainable tourism. *Tourism Recreation Research*, 35(3), 239–249.

Bramwell, B., & Lane, B. (2011). Critical research on the governance of tourism and sustainability. *Journal of Sustainable Tourism*, 19(4–5), 411–421.

Britton, S. (1991). Tourism, capital, and place: Towards a critical geography of tourism. *Environment and Planning D: Society and Space*, 9(4), 451–478.

Büscher, B., & Fletcher, R. (2017). Destructive creation: Capital accumulation and the structural violence of tourism. *Journal of Sustainable Tourism*, 25(5), 651–667.

Cave, J., & Dredge, D. (2020). Regenerative tourism needs diverse economic practices. *Tourism Geographies*, 22(3), 503–513.

Chambers, D. (2007). Interrogating the 'Critical' in Critical Approaches to Tourism Research. In I. Ateljevic, A. Pritchard, & N. Morgan (Eds.), *The Critical Turn in Tourism Studies: Innovative Research Methodologies* (pp. 105–120). Elsevier.

Chambers, D., & Buzinde, C. (2015). Tourism and decolonisation: Locating research and self. *Annals of Tourism Research*, 51, 1–16.

Chassagne, N., & Everingham, P. (2019). Buen Vivir: Degrowing extractivism and growing well-being through tourism. *Journal of Sustainable Tourism*, 27(12), 1909–1925.

Cheer, J.M. (2020). Human flourishing, tourism transformation and COVID-19: A conceptual touchstone. *Tourism Geographies*, 22(3), 514–524.

Choudry, A. (2020). Reflections on academia, activism, and the politics of knowledge and learning. *The International Journal of Human Rights*, 24(1), 28–45.

Crossley, É. (2020). Ecological grief generates desire for environmental healing in tourism after COVID-19. *Tourism Geographies*, 22(3), 536–546.

Davis, A.Y. (1994). Afro images: Politics, fashion, and nostalgia. *Critical Inquiry*, 21(1), 37–45.

Denzin, N.K., & Lincoln, Y.S. (Eds.) (2011). *The Sage Handbook of Qualitative Research*. Sage.

Dillette, A., & Benjamin, S. (2022). The Black travel movement: A catalyst for social change. *Journal of Travel Research*, 61(3), 463–476.

Dredge, D. (2006). Networks, conflict and collaborative communities. *Journal of Sustainable Tourism*, 14(6), 562–581.

Dredge, D., & Gyimóthy, S. (2015). The collaborative economy and tourism: Critical perspectives, questionable claims and silenced voices. *Tourism Recreation Research*, 40(3), 286–302.

Dwyer, L. (2017). Saluting while the ship sinks: The necessity for tourism paradigm change. *Journal of Sustainable Tourism*, 26(1), 29–48.

Everingham, P., & Chassagne, N. (2020). Post COVID-19 ecological and social reset: Moving away from capitalist growth models towards tourism as Buen Vivir. *Tourism Geographies*, 22(3), 555–566.

Everingham, P., Peters, A., & Higgins-Desbiolles, F. (2021). The (im)possibilities of doing tourism otherwise: The case of settler colonial Australia and the closure of the climb at Uluru. *Annals of Tourism Research*, 88, 103178.

Fitchett, J., Lindberg, F., & Martin, D. (2021). Accumulation by symbolic dispossession: Tourism development in advanced capitalism. *Annals of Tourism Research*, 86, 1–10.

Flood, M., Martin, B., & Dreher, T. (2013). Combining academia and activism: Common obstacles and useful tools. *Australian Universities Review*, 55(1), 17–26.

Grimwood, B.S. (2015). Advancing tourism's moral morphology: Relational metaphors for just and sustainable arctic tourism. *Tourist Studies*, 15(1), 3–26.

Hales, R., Dredge, D., Higgins-Desbiolles, F., & Jamal, T. (2018). Academic activism in tourism studies: Critical narratives from four researchers. *Tourism Analysis*, 23(2), 189–199.

Hanna, P., Johnson, K., Stenner, P., & Adams, M. (2015). Foucault, sustainable tourism, and relationships with the environment (human and nonhuman). *GeoJournal*, 80, 301–314.

Haywood, K.M. (2020). A post-COVID future: Tourism community re-imagined and enabled. *Tourism Geographies*, 22(3), 599–609.

Higgins-Desbiolles, F., Carnicelli, S., Krolikowski, C., Wijesinghe, G., & Boluk, K. (2019). Degrowing tourism: Rethinking tourism. *Journal of Sustainable Tourism*, 27(12), 1926–1944..

Higgins-Desbiolles, F., Doering, A., & Bigby, B.C. (2021). *Socialising Tourism: Rethinking Tourism for Social and Ecological Justice*. Routledge.

Higgins-Desbiolles, F., & Whyte, K.P. (2013). No high hopes for hopeful tourism: A critical comment. *Annals of Tourism Research*, 40, 428–433.

Higgins-Desbiolles, F., & Whyte, K.P. (2014). Critical Perspectives on Tourism. In A. Lew, M. Hall, & A. Williams (Eds.), *The Wiley Blackwell Companion to Tourism* (pp. 88–98). John Wiley & Sons.

Hobson, J.M., & Sajed, A. (2017). Navigating beyond the eurofetishist frontier of critical IR theory: Exploring the complex landscapes of non-Western agency. *International Studies Review*, 19(4), 547–572.

Hollinshead, K., Suleman, R., & Nair, B. (2021). The unsettlement of tourism studies: Positive decolonization, deep listening, and dethinking today. *Tourism Culture & Communication*, 21(2), 143–160.

Hollinshead, K., Suleman, R., & Vellah, A. (2021). The reimagination of tourism studies: Positive renewal, restoration, and revival today. *Tourism Culture & Communication*, 21(3), 259–276.

Horkheimer, M. (2002). Traditional and Critical Theory. In *Critical theory: Selected* essays (M. J.O'Connell, Trans.). Continuum (original work published 1937).

Horkheimer, M., & Adorno, T. (2002). *Dialectic of Enlightenment*. Stanford University Press. (original work published 1947).

Jamal, T., Camargo, B., & Wilson, E. (2013). Critical omissions and new directions for sustainable tourism: A situated macro–micro approach. *Sustainability*, 5(11), 4594–4613.

Keen, D., & Tucker, H. (2012). Future Spaces of Postcolonialism in Tourism. In J. Wilson (Ed.), *The Routledge Handbook of Tourism Geographies* (pp. 114–119). Routledge.

Lapointe, D. (2020). Reconnecting tourism after COVID-19: The paradox of alterity in tourism areas. *Tourism Geographies*, 22(3), 633–638.

Lapointe, D., Sarrasin, B., & Benjamin, C. (2018). Tourism in the sustained hegemonic neoliberal order. *Revista Latino Americana de Turismologia*, 4(1), 16–33.

Liburd, J., & Edwards, D. (Eds.) (2010). *Understanding the Sustainable Development of Tourism*. Good Fellow Publishers.

Macdonald, B.J., & Young, K.E. (2018). Adorno and Marcuse at the barricades? Critical theory, scholar-activism, and the neoliberal university. *New Political Science*, 40(3), 528–541.

McGehee, N.G., Kline, C., & Knollenberg, W. (2014). Social movements and tourism-related local action. *Annals of Tourism Research*, 48, 140–155.

Morgan, N., Pritchard, A., Causevic, S., & Minnaert, L. (2018). Ten years of critical tourism studies: Reflections on the road less travelled. *Tourism Analysis*, 23(2), 183–187.

Mura, P. (2020). Ethnodrama and ethnotheatre in tourism. *Current Issues in Tourism*, 23(24), 3042–3053.

Mura, P., & Wijesinghe, S.N. (2023). Critical theories in tourism: A systematic literature review. *Tourism Geographies*, 25(2–3), 487–507.

Mura, P., Wijesinghe, S.N.R. & Matar, M. (2021). 'Some glimpses of an Asian PhD journey in tourism' – An ethnodrama. *Tourism Management Perspectives*, 40, 100908.

Nepal, S. (2020). Travel and tourism after COVID-19: Business as usual or opportunity to reset? *Tourism Geographies*, 22(3), 646–650.

Niewiadomski, P. (2020). COVID-19: From temporary de-globalisation to a re-discovery of tourism? *Tourism Geographies*, 22(3), 651–656.

Pernecky, T. (2020). Critical tourism scholars: Brokers of hope. *Tourism Geographies*, 22(3), 657–666.

Piven, F.F. (2010). Reflections on scholarship and activism. *Antipode*, 42(4), 806–810.

Pritchard, A., Morgan, N., & Ateljevic, I. (2011). Hopeful tourism: A new transformative perspective. *Annals of Tourism Research*, 38(3), 941–963.

Rastegar, R., Higgins-Desbiolles, F., & Ruhanen, L. (2021). COVID-19 and a justice framework to guide tourism recovery. *Annals of Tourism Research*, 91, 103161.

Romagosa, F. (2020). The COVID-19 crisis: Opportunities for sustainable and proximity tourism. *Tourism Geographies*, 22(3), 690–694.

Rydzik, A., Pritchard, A., Morgan, N., & Sedgley, D. (2013). The potential of arts-based transformative research. *Annals of Tourism Research*, 40, 283–305.

Sachs, W. (2019). *The Development Dictionary: A Guide to Knowledge as Power* (3rd Ed). Zed Books.

Schweinsberg, S., Heizmann, H., Darcy, S., Wearing, S., & Djolic, M. (2018). Establishing academic leadership praxis in sustainable tourism: lessons from the past and bridges to the future. *Journal of Sustainable Tourism*, 26(9), 1577–1586.

Sowards, S. K. (2012). Expectations, experiences, and memories: Ecotourism and the possibilities for transformations. *Environmental Communication*, 6(2), 175–192.

Tomassini, L., & Cavagnaro, E. (2020). The novel spaces and power-geometries in tourism and hospitality after 2020 will belong to the 'local'. *Tourism Geographies*, 22(3), 713–719.

Wijesinghe, S. (2022). Neoliberalism, Covid-19 and hope for transformation in tourism: The case of Malaysia. *Current Issues in Tourism*, 25(7), 1106–1120.

Wijesinghe, S., & Higgins-Desbiolles, F. (2024). A Critical Analysis of the United Nations Sustainable Development Goals. In K. Boluk, F. Higgins-Desbiolles, & M. Akhoundoghli (Eds.), *The Elgar Companion to Tourism and the Sustainable Development Goals* (pp. 18–29). Edward Elgar.

Wijesinghe, S.N.R., Mura, P., & Bouchon, F. (2017). Tourism knowledge and neocolonialism: A systematic critical review of the literature. *Current Issues in Tourism*, 22(11), 1263–1279.

Wilson, E., Mura, P., Sharif, S. P., & Wijesinghe, S.N. (2020). Beyond the third moment? Mapping the state of qualitative tourism research. *Current Issues in Tourism*, 23(7), 795–810.

5

TOURISM AND DEGROWTH

Beyond the Capitalist Growth Imperative

Macià Blázquez-Salom, Ivan Murray, Robert Fletcher,
Filka Sekulova, Asunción Blanco-Romero and Ernest Cañada

Introduction

The necessity and consequences of the exponential growth in tourism activity experienced throughout the world over the past half-century have been increasingly questioned by an expanding body of activists and critical researchers. One of the emerging responses within this debate concerns calls for reversing the trend and considering, or pursuing, tourism 'degrowth'. This discussion is inspired by a longstanding body of research problematising the imperative and consequences of economic growth more generally, initiated by social scientists such as Georgescu-Roegen (1971), Meadows et al. (1972), Illich (1973) and Gorz (1972), and recently updated by authors including Mies (2007), Latouche (2009) and Kallis et al. (2018). This chapter aims to provide a state-of-the-art overview on the application of degrowth perspectives to discussions of (sustainable) tourism development and to outline a future agenda for research and praxis continuing this important line of inquiry.

The Capitalist Growth Imperative and Its Discontents

An expanding body of research has shown that economic growth has done a fairly poor job overall in contributing to happiness, equity or justice (Schmelzer et al. 2022). Within capitalist societies, growth and continuous accumulation are pursued as ends in themselves, in order to overcome the system's inherent contradictions and ensure its long-term survival (Harvey, 1989). To achieve this, capitalism must always increase industrial levels of extraction of natural resources, exploitation of labour, consumption of energy and materials and appropriation of time and space previously outside the market, in order to increase the rate of profit and hence of capital accumulation. In this way, capitalism continually flees from its own contradictions, chief among them being the biophysical limits of growth on a finite planet, a major cause of the current global socio-ecological crisis (O'Connor, 1988; Foster et al., 2010; Harvey, 2015; Fraser, 2022). While these limits have long been identified (Meadows et al., 1972), they are increasingly framed as the 'planetary boundaries' of a healthy and sustainable life (Rockström et al., 2009). In response, critical social scientists have asserted the importance of attending to 'societal boundaries' emerging from "contested societal

DOI: 10.4324/9781003286301-7

processes that lead to collectively defined thresholds that societies commit not to trespass" (Brand et al., 2021: 276).

As we already exceed the capacity for resilience of the global environment, capitalism and the perpetual growth it demands are revealed to be ecologically and socially unsustainable (Moore, 2015; Hickel et al., 2021; Brockway et al., 2021). In response to growing discussion of the Anthropocene and 'Great Acceleration' that frame the human species as a geological force, many consider it more appropriate to describe this long period of global transformation and destruction to the 'web of life' (Moore, 2015) as the 'Capitalocene', understood as "the geology of capital accumulation" (Malm, 2016: 391; see also Moore, 2016). From the analytical perspective of degrowth, capitalism is becoming increasingly unviable in both economic and social terms (Schmelzer et al., 2022; Streeck, 2017). In recent decades, after all, the growing difficulties of ensuring the expanding reproduction of capital have led to an increase in social inequality (Piketty, 2019) and an intensification of uneven global geographical development (Brenner, 2019).

The question of economic growth is also central for the field of tourism development. One of the world's largest industries, international tourism is a form of capital accumulation that exerts significant influence on a planetary level. Especially since the global financial crisis of 2008, tourism's relevance to capitalist accumulation logics has redoubled, since the industry's growth has been widely promoted as a spatiotemporal solution to the crisis (Fletcher, 2011). Likewise, (sustainable) tourism development played an important role in the Green Economy agenda advanced at the Rio +20 Earth Summit in 2012 (Hall, 2015) and retains a central place in discussions of the Sustainable Development Goals (SDGs) and the EU Green Deal. The adoption of sustainability discourse and jargon in tourism development and elsewhere in response to the ecological crisis is one of the signs that capitalist growth is increasingly confronting its socio-ecological limits. This response is, however, largely intended to sustain capitalism itself rather than resolve its multiple contradictions (Fletcher, 2011).

The upward path of tourism growth experienced in the years immediately prior to COVID-19 was cut short by the onset of the pandemic in 2020. The subsequent global lockdown halted the tourist circuit in its tracks; indeed, tourism was the capitalist activity most immediately and significantly impacted by the pandemic. In this way, the pandemic exposed and magnified the enormous contradictions of the accumulation strategy tourism exemplifies (Cañada & Murray, 2021; Lew et al., 2022). The COVID crisis has hit particularly hard the most touristified societies, which have proven extremely vulnerable due to their overreliance on a single economic activity. By way of example, Spain, and particularly its two archipelagos (Balearic and Canary Islands) have suffered a significant increase in poverty and inequality during the pandemic, due to their high reliance on tourism (Murray & Martínez-Caldentey, 2020).

Foundations of Degrowth

The term degrowth was originally coined by Andre Gorz, in 1972, the same year in which *The Limits to Growth* (Meadows et al., 1972) also appeared, projecting the potential limits of capitalist development. Over time the booming degrowth literature has advocated for a planned and democratically organised reduction of the throughput of energy and resources in order to bring the economy into balance within the biophysical environment while also contributing to justice, equity and human flourishing (Chertkovskaya et al., 2019; Kallis et al., 2018).

Since 2008, discussion of degrowth has experienced a leap in scale, with the introduction of the discussion into Anglophone literature (Martínez-Alier 2009; Schneider et al., 2010). Since these first English-language texts on degrowth were published, the rate of publications has been dizzying, making degrowth one of the most prominent topics of current debate within both academia and social movements (e.g., D'Alisa et al., 2015; Kallis, 2018; Kallis et al., 2020; Hickel, 2020; Schmelzer et al., 2022).

Despite the novelty of the term degrowth, it should be noted that many of its core concepts are borrowed from various strands of the literature, albeit not under the umbrella of degrowth per se. In this sense, the theoretical roots of degrowth are found in radical and eco-socialist economics and political ecology proposals (Peet et al., 2010; Spash, 2017; Brownhill et al., 2022). To a large extent, the degrowth proposal emerges as a critique of the hegemonic vision of sustainable development, such as that outlined in the famous Brundtland Report (WECD, 1987), which clearly links sustainable development to the imperative of economic growth. Hence, its characterisation by Hall as BAU (Brundtland-as-Usual) (Hall et al., 2021). This criticism has gained strength in relation to the intensification of neoliberal capitalism and its increasing translation into neoliberal environmental governance (Pellizzoni, 2011).

It is important to highlight that, despite the increasing popularity of degrowth, there are some objections to it as well as proposals that fundamentally question use of the term altogether. By contrast, other proposals prefer to mobilise concepts such as post-growth, post-development, post-capitalism, prosperity without growth, eco-socialism, environmental justice, and good living, among others (Gibson-Graham, 2006; Jackson, 2009; Martínez-Alier et al., 2016; Cassiers et al., 2018; Kothari et al., 2019; Brownhill et al., 2022).

One of the most robust theoretical foundations for the critique of growth from a biophysical point of view is the work of Nicholas Georgescu-Roegen (1971) on economics and the law of entropy. Georgescu-Roegen (1971) asserted that economic process could not be divorced from the thermodynamic limitations that govern the rest of nature. The second law of thermodynamics, or law of entropy, establishes that in its transformation, energy inevitably loses its quality and degrades, reducing its future possibilities for human use. This dissipation of energy is irrevocable, unidirectional and marks the physical limit of industrial societies (Valero & Valero, 2014). Also acknowledging such dynamics, one of the founders of ecological economics, Herman Daly (1974), advanced the theory of the steady-state economy, according to which a point would be reached wherein growth becomes 'uneconomic', hence, having higher socio-ecological costs than benefits. Before that point, a steady state should be established by maintaining "constant stocks of physical well-being (artefacts) and a constant population, each maintained at some chosen, desirable level by a low rate of throughput" (Daly, 1974: 15).

Based on these premises, one of the key proposals in degrowth scholarship revolves around the pursuit of equitable, democratic, voluntary and planned reduction of the current material throughput. Hickel (2020) clarifies that this decrease is not focused solely on cutting GDP, but rather the amount of material and energy metabolised within a given society . In capitalist societies, reduction in GDP means recession or crisis, and those who tend to suffer the consequences of capitalist crises the most are the subaltern classes, since vulnerability is eminently a function of unequal ownership of and access to resources (Malm, 2020). Furthermore, the capitalist class, through the adoption of the shock doctrine, has benefited from past crises and catastrophes, and will presumably do the same in future catastrophes (Klein, 2007). Acknowledging these issues, degrowth is increasingly posed as an explicitly anti-capitalist political project (Schmelzer et al., 2022). For some, indeed, degrowth is considered an 'impossible theorem' within capitalism given the imperative of continuous economic accumulation to ensure the system's survival (Foster, 2011).

Degrowth is thus defined by its contrast to the expansive and colonial pattern inherent to capitalism, "a process of elite accumulation, the commodification of commons, and the appropriation of human labour and natural resources" (Hickel, 2021: 1107). The capitalist growth imperative is rooted in individualism and competitiveness, qualities accentuated by neoliberal doctrine (Harvey 2005). A spatial expression of economic growth is the geographical expansion of commodity frontiers, to extract free or 'cheap' natural resources and dump waste freely. This globalisation of supply expands the boundaries of accumulation by appropriation or dispossession, especially of labour, energy, food and minerals (Moore, 2015).

This also allows for most of the flow of energy and materials, the cause of the ecological crisis, to be consumed by the Global North, while its source is mainly in the South. Given uneven global geographic development, "degrowth in the North represents a process of decolonisation in the South, to the extent that it releases communities in the South from the pressures of atmospheric colonisation and material extractivism" (Hickel, 2021: 1109). Escobar (2015) asserts that by deconstructing the very notion and praxis of development, degrowth in the Global North (Büscher, 2019) could and shall facilitate redistribution and decolonisation policies and practices in the context of planetary boundaries. This entails contracting consumption of resources by the elite social classes and expanding prosperity for the disadvantaged and disenfranchised classes.

Degrowth is furthermore a call to abandon the centrality of economic growth as an overarching societal objective, and instead place care at the centre (Perez Orozco 2014) as a way to make space for direct democracy, political engagement, conviviality and a meaningful life . In this sense, one of the central foci of degrowth discussions is the fundamental, though insufficiently acknowledged, role of reproductive work (Mies, 2007; Waring, 2003). Reproductive activities provide the conditions for the continuous regeneration of society, although their pace tends to be slower than that of capitalist working times (Salleh, 2012). Degrowth thinking recognises that both women and the environment are marginalised vis-a-vis their (usually undervalued) positions within the formal economy, while they in fact provide the fundamental basis for the economy's reproduction and growth.

The political agenda outlined in degrowth scholarship is built upon multiple pillars (Schmelzer et al., 2022, Hickel, 2020), some of which can be summarised as: (1) the planned reduction of environmental impact; (2) restriction of the least necessary economic activity, to instead expand the most important sectors, such as health, education, access to information and care; (3) improvement to working conditions, through their regulation and establishment of a universal basic income (UBI), thereby ensuring autonomy and collective self-organisation; (4) reduction of inequality; (5) expansion of public goods and services; and (6) an ecological transition to reverse the environmental crisis.

Despite the fact that most of the academic literature on degrowth remains largely abstract and theoretical, a growing body of research is focused on its practical applications, too. Without aiming to be exhaustive, some of the topics analysed from a degrowth approach are as follows: degrowth-based alternatives in cities like Stuttgart (Schmid, 2021); food systems (Nelson and Edwards, 2021); housing (Nelson and Schneider, 2018); urban planning (Xue, 2021); energy transitions (Kunze and Becker, 2015); blue degrowth (Ertor & Hadjimichael, 2020); technology (Keschner et al., 2018); labour relations (Barca, 2019); degrowth and feminism (Hanacek et al., 2020); and degrowth and the environmental justice movement (Rodríguez-Labajos et al., 2019). Yet some still contend that degrowth remains too focused on small-scale and grassroots movements rather than confronting state and global politics (Schwartzman, 2012). However, the question of the state and institutional transformation has long been a focus of some degrowth discussion (Kallis, 2013; D'Alisa and Kallis, 2020), in terms of the need for redistribution of wealth via setting maximum income thresholds (D'Alisa et al., 2015), public regulation of activity licences, or establishing progressive pricing for consumption and disposal (Lehtinen, 2018).

Approaches to Tourism Degrowth

Global tourism has become a key factor in the process of environmental degradation at a planetary scale (Hall et al., 2015) and a paramount expression of a 'fossil capitalism' that has been revealed as highly vulnerable to chronic emergencies (Malm, 2016: 2020). Critical tourism scholars have long analysed the socio-environmental costs and conflicts associated with touristification. The questions of limits and carrying capacity are introduced early in tourism studies (Mathieson & Wall, 1982) and critical analysis of the tourist industry as a capitalist mode of accumulation has grown since Britton's (1991) initial analysis. Some of these critical approaches emerged within discussion of the potential for sustainable tourism (ST) as a component of the global sustainable development agenda. In 1998, Mowforth and Munt (2016 [1998]) published a vibrant critique of the potential to pursue sustainable tourism under capitalism. Yet the hegemonic vision of ST was very much Brundtland-as-Usual inspired, and erected on fragile theoretical foundations (Sharpley, 2020). As an example, while the UNWTO theoretically promotes ST, at the same time its unwavering promotion of global tourism growth seems to be blind to all the scientific evidence concerning impact on climate change of such growth (Gössling & Peeters, 2015; Sun et al., 2022). Tourism-driven economic growth cannot be simply 'made sustainable' for a number of reasons, including the inherent impossibility of sufficiently 'decoupling' growth from environmental impacts (Hickel & Kallis, 2020). In this regard, Chakraborty (2021) argues that for tourism to contribute to sustainability it must be reconceptualised from the perspective of biophysical limits.

Critiques of the hegemonic ST vision have made important contributions to re-politicising the sustainability question. However, there is a growing divide between critical academic discourse and tourism practice, which has predominantly adhered to this conventional vision (Sharpley, 2020). Consequently, some scholars have advanced alternative theoretical proposals that more radically problematise the relationship between tourism and environment. In 2009, coinciding with the global financial meltdown, two initial works advocated a paradigm shift towards degrowth tourism (Hall, 2009) and post-development tourism (Sharpley, 2009). Such calls echoed the mounting chorus of voices from academia and social movements opposing capitalist globalisation and demanding degrowth more generally as a response to the systemic crisis. As advocacy of degrowth was expanding in academia, critical tourism scholars joined the debate to also problematise touristification from a degrowth perspective. It is in this context that a succession of works have been published since 2018, most significantly the following books: *Degrowth in Tourism* (Andriotis, 2018); *Tourism and Degrowth* (Fletcher et al., 2020); *Degrowth and Tourism* (Hall et al., 2021); and *Issues and Cases of Degrowth in Tourism* (Andriotis, 2021).

When analysing the current state of research on tourism degrowth, Lundmark et al. (2021: 8) point out that there is "a strong focus on Europe, and especially the Mediterranean region, concentrating on Barcelona, Costa del Sol, Malaga and Marbella, as well as coastal tourism". This emphasis is not random, but has resulted from a convergence of two key factors: first, the presence of a dynamic research collective on degrowth based around the Autonomous University of Barcelona (UAB); second, the intensification of touristification combined with austerity politics and increasing inequality have been particularly significant in that region. Social movements and activist scholars located in these spaces have denounced this process of profound tourism commodification, commonly labelled 'overtourism', and in response have introduced discussion of *decreixement turistic* (or tourism degrowth) together with claims for the right to the city (Blanco-Romero et al., 2019).

Thus far, the body of research on tourism degrowth has focused *inter alia* on the following topics. First, development of the theoretical basis and research agenda on tourism degrowth (e.g., Fletcher et al., 2019, Higgins-Desbiolles, 2019). Second, critical analysis of tourism capital accumulation, its contradictions and social contestation (e.g., Navarro-Jurado et al., 2019). Third, from a demand-side perspective, a focus on degrowth-inspired travel (e.g., Andriotis, 2018; Díaz-Soria, 2017). Fourth, tourism degrowth policies and planning and their own contradictions (e.g., Blázquez-Salom et al., 2019). Fifth, the relationship between energy, climate change and tourism degrowth (e.g., Torres & Moranta, 2020; Adedoyin et al., 2020; Balsalobre-Lorente, 2020). Sixth, reconceptualisation of tourism in the light of climate change, emphasising the role of domestic or proximity tourism as a potential degrowth strategy (e.g., Ballantine, 2021; Cañada & Izcara, 2021). Seventh, discourse analysis of tourism degrowth and appropriation of the term by the ruling class for greenwashing (e.g., Valdivielso & Moranta, 2019). Eighth, putting degrowth tourism to work, particularly in relation to community-based tourism (e.g., Cañada, 2021; Ruíz-Ballesteros, 2021).

However, it is important to recognise that while the term 'tourism degrowth' is relatively recent, many of the ideas and topics explored under its rubric have been researched for some time from different theoretical perspectives . In this sense, discussion of tourism degrowth shares common threads with proposals and case studies from the political ecology of tourism (Mostafanezhad et al., 2016), tourism and feminism (Devine & Ojeda, 2017), political economy of tourism (Bianchi, 2018), post-capitalist tourism (Fletcher et al., 2023), and convivial tourism (Büscher & Fletcher, 2020), among others. More important than the term of analysis itself, therefore, is the critical problematisation of touristification and building a political agenda for transforming tourism for the reproduction of life instead of the reproduction of capital.

Finally, based on existing research and proposals from grassroots degrowth movements, tourism degrowth can be understood as a multi-layer strategy based on the following principles:

(1) Resistance against dispossession by conflictive touristification (Büscher & Fletcher, 2020).
(2) Planned reduction of the resources used and waste produced by tourism activities.
(3) De-touristification as a downsizing of tourism, particularly within highly touristified spaces, combined with a process of degrowth-inspired economic diversification.
(4) Post-capitalist economic and social re-organisation of the tourism industry, which implies its collective appropriation and socialisation.
(5) Rethinking tourism, leisure and recreation in times of chronic emergencies for the reproduction of life and conviviality.
(6) De-commodification of tourism, leisure and recreation.

Into the Future

Looking to the future, we propose that the potential for tourism degrowth can be enhanced by pursuing the various lines of research outlined by Fletcher et al. (2019). We consider it important for academics to contribute to the public debate on tourism's socio-ecological transformation via research such as this, in line with the theoretical and methodological framework of participatory action research.

The future research programme we propose is thus based on harnessing the potential for tourism degrowth as a strategy to re-politicise questions of tourism development in general and sustainable tourism in particular. This point of departure helps to liberate the study of tourism from any preconceived bias that *a priori* favours the industry. From this vantage point, a critical analysis of the role of tourism as a form of uneven capitalist development can be undertaken, drawing on the perspectives of political economy and political ecology. This analysis should include diagnosis of the social and biophysical costs of tourism development under capitalism, to assess potential for transforming the industry's political and economic organisation. The culmination of this inquiry will entail contributing constructively to policy and practice, through the study of the existing examples of touristic degrowth (explicit or implicit). For practical purposes, the most saturated tourist destinations are the best and most urgent 'laboratories' within which to investigate the potential of degrowth to reduce tourism's intensity and its impacts. Finally, we highlight the need to recognise and elaborate the 'right to metabolism', defined as a radical political project that explores the potential for proportional reduction of the flows of materials and energy required by tourism, as well as political reorganisation of these flows, to facilitate socio-ecological transformation.

References

Adedoyin, F. F., & Bekun, F. V. (2020). Modelling the interaction between tourism, energy consumption, pollutant emissions and urbanization: Renewed evidence from panel VAR. *Environmental Science and Pollution Research*, *27*(31), 38881–38900.

Andriotis, K. (2018). *Degrowth in Tourism: Conceptual, Theoretical and Philosophical Issues*. CABI.

Andriotis, K. (2021). *Issues and Cases of Degrowth in Tourism*. CABI.

Ballantine, P. W. (2021). Don't leave town till you've seen the country: domestic tourism as a degrowth strategy. In C. M. Hall, L. Lundmark, & J. J. Zhang (Eds.), *Degrowth and Tourism: New Perspectives on Tourism Entrepreneurship, Destinations and Policy* (pp. 187–201). Routledge.

Balsalobre-Lorente, D., Driha, O. M., Shahbaz, M., & Sinha, A. (2020). The effects of tourism and globalization over environmental degradation in developed countries. *Environmental Science and Pollution Research*, *27*(7), 7130–7144.

Barca, S. (2019). The labor(s) of degrowth. *Capitalism Nature Socialism, 30*, 207–216.

Bianchi, R. (2018). The political economy of tourism development: A critical review. *Annals of Tourism Research, 70*, 88–102.

Blanco-Romero, A., Blàzquez-Salom, M., Morell, M., & Fletcher, R. (2019). Not tourism-phobia but urbanphilia: Understanding stakeholders' perceptions of urban touristification. *Boletín de la Asociación de Geógrafos Españoles, 83*. DOI: 10.21138/bage.2834

Blazquez-Salom, M., Blanco-Romero, A., Vera-Rebollo, J. F., & Ivars-Baidal, J. (2019). Territorial tourism planning in Spain: From boosterism to tourism degrowth? *Journal of Sustainable Tourism, 27*(12), 1764–1785.

Brand, U., Muraca, B., Pineault, E., Sahakian, M., Schaffartzik, A., Novy, A., Streissler, C., Haberl, H., Asara, V., Dietz, K., Lang, M., Kothari, A., Smith, T., Spash, C., Brad, A., Pichler, M., Plank, C., Velegrakis, G., Jahn, T., Carter, A., Huan, Q., Kallis, G., Martínez Alier, J., Riva, G., Satgar, V., Teran Mantovani, E., Williams, M., Wissen, M., & Görg, C. (2021). From planetary to societal boundaries: An argument for collectively defined self-limitation. *Sustainability: Science, Practice and Policy, 17*(1), 265–292.

Brenner, N. (2019). *New Urban Spaces: Urban Theory and the Scale Question*. Oxford University Press.

Britton, S. G. (1991). Tourism, capital and place: Towards a critical geography of tourism. *Environment and Planning D: Society and Space, 9*, 451–478.

Brockway, P., Sorrell, S., Semieniuk, G., Kuperus Heun, N., & Court, V. (2021). Energy efficiency and economy-wide rebound effects: A review of the evidence and its implications. *Renewable and Sustainable Energy Reviews, 141*, 110781.

Brownhill, L., Engel-Di-Mauro, S., Giacomini, T., Isla, A., Löwy, M., & Turner, T. E. (Eds.) (2022). *The Routledge Handbook on Ecosocialism*. Routledge.

Büscher, B. (2019). From 'global' to 'revolutionary' development. *Development and Change, 50*(2), 484–494.

Büscher, B., & Fletcher, R. (2020). *The Conservation Revolution: Radical Ideas for Saving Nature beyond the Anthropocene*. Verso.

Cañada, E. (2021). Community-based tourism in a Degrowth perspective. In K. Andriotis (Ed.). *Cases of Degrowth in Tourism* (pp. 42–63). CABI.

Cañada, E., & Izcara, C. (Ed.) (2021). *Turismos de proximidad, un plural en disputa*. Barcelona: Alba Sud.

Cassiers, I., Maréchal, K., & Méda, D. (Eds.) (2018). *Post-growth Economics and Society. Exploring the Paths of a Social and Ecological Transition*. Routledge.

Chakraborty, A. (2021). Can tourism contribute to environmentally sustainable development? Arguments from an ecological limits perspective. *Environment, Development and Sustainability, 23*, 8130–8146.

Chertkovskaya, E., Paulsson, A., & Barca, S. (Eds.) (2019). *Towards a Political Economy of Degrowth*. Rowman & Littlefield.

D'Alisa, G., Demaria, F., & Kallis, G. (Eds.) (2015). *Degrowth: A Vocabulary for a New Era*. Routledge.

D'Alisa, G., & Kallis, G. (2020). Degrowth and the State. *Ecological Economics, 169*, 106486.

Daly, H. (1974). The economics of the steady state. *The American Economic Review, 64*(2), 15–21.

Devine, J., & Ojeda, D. (2017). Violence and dispossession in tourism development: A critical geographical approach. *Journal of Sustainable Tourism, 25*(5), 605–617.

Díaz-Soria, I. (2017). Being a tourist as a chosen experience in a proximity destination. *Tourism Geographies, 19*(1), 96–117.

Ertör, I., & Hadjimichael, M. (2020). Blue degrowth and the politics of the sea: Rethinking the blue economy. *Sustainability Science, 15*, 1–10.

Escobar, A. (2015). Degrowth, postdevelopment, and transitions: A preliminary conversation. *Sustainability Science, 10*(3), 451–462.

Fletcher, R. (2011). Sustaining tourism, sustaining capitalism? The tourism industry's role in global capitalist expansion. *Tourism Geographies, 13*(3), 443–461.

Fletcher, R., Blanco-Romero, A., Blázquez-Salom, M., Cañada, E., Murray, I., & Sekulova, F. (2023). Pathways to post-capitalist tourism. *Tourism Geographies, 25*(2–3), 707–728.

Fletcher, R., Murray, I., Blanco-Romero, A., & Blázquez-Salom, M. (2019). Tourism and degrowth: An emerging agenda for research and praxis. *Journal of Sustainable Tourism, 27*(12), 1745–1763.

Fletcher, R., Murray, I., Blanco-Romero, A., & Blázquez-Salom, M. (2020). *Tourism and Degrowth: Towards a Truly Sustainable Tourism*. Routledge.

Foster, J. B. (2011). Capitalism and degrowth: An impossibility theorem. *Monthly Review, 62*(8), 26–33.

Foster, J. B., Clark, B., & York, R. (2010). *The Ecological Rift: Capitalism's War on the Earth*. Monthly Review Press.

Fraser, N. (2022). *Cannibal Capitalism: How Our System is Devouring Democracy, Care and the Planet—and What We Can Do About It*. Verso.

Georgescu-Roegen, N. (1971). *The Entropy Law and the Economic Process*. Harvard University Press.

Gibson-Graham, J. K. (2006). *Postcapitalist Politics*. Minnesota Press.

Gorz, A. (1972). Proceedings from a public debate organized by the Club du Nouvel Observateur. Paris: Nouvel Observateur, 397.

Gössling, S., & Peeters, P. (2015). Assessing tourism's global environmental impact 1900–2050. *Journal of Sustainable Tourism, 23*(5), 639–659.

Hall, C. M. (2009). Degrowing tourism: Décrissance, sustainable consumption and steady-state tourism. *Anatolia, 20*(1), 46–61.

Hall, C. M. (2015). Economic greenwash: On the absurdity of tourism and green growth. In V. Reddy, & K. Wilkes (Eds.), *Tourism in the Green Economy* (pp. 339–359). Earthscan.

Hall, C. M., Gössling, S., & Scott, D. (Eds.) (2015). *The Routledge Handbook of Tourism and Sustainability*. Routledge.

Hall, C. M., Lundmark, L., & Zhang, J. J. (Eds.) (2021). *Degrowth and Tourism. New Perspectives on Tourism Entrepreneurship, Destinations and Policy*. Routledge.

Hanacek, K., Roy, B., Avila, S., & Kallis, G. (2020). Ecological economics and degrowth: Proposing a future research agenda from the margins. *Ecological Economics, 169*, 106495.

Harvey, D. (1989). *The Condition of Postmodernity*. Blackwell.

Harvey, D. (2005). *A Brief History of Neoliberalism*. Oxford University Press.

Harvey, D. (2015). *Seventeen Contradictions and the End of Capitalism*. Profile Books.

Hickel, J. (2020). *Less is More: How Degrowth will Save the World*. Windmill.

Hickel, J. (2021). What does degrowth mean? A few points of clarification. *Globalizations, 18*(7), 1105–1111.

Hickel, J., & Kallis, G. (2020). Is green growth possible? *New Political Economy, 25*(4), 469–486.

Higgins-Desbiolles, F., Carnicelli, S., Krolikowski, C., Wijesinghe, G., & Boluk, K. (2019). Degrowing tourism: Rethinking tourism. *Journal of Sustainable Tourism, 27*(12), 1926–1944,

Illich, I. (1973). *Tools for Conviviality*. Harper & Row.

Jackson, T. (2009). *Prosperity without Growth: Economics for Finite Planet*. Earthscan.

Kallis, G. (2013). Societal metabolism, working hours and degrowth: A comment on Sorman and Giampietro. *Journal of Cleaner Production, 38*, 94–98.

Kallis, G. (2018). *Degrowth*. Agenda Publishing.

Kallis, G., Kostakis, V., Lange, S., Muraca, B., Paulson, S., & Schmelzer, M. (2018). Research on Degrowth. *Annual Review of Environment and Resources, 43*(1), 291–316.

Kallis, G., Paulson, S., D'Alisa, G., & Demaria, F. (2020*). The Case for Degrowth*. Polity Press.

Kerschner, C., Wächter, P., Nierling, L., & Ehlers, M. H. (2018). Degrowth and technology: Towards feasible, viable, appropriate and convivial imaginaries. *Journal of Cleaner Production, 197*, 1619–1636.

Klein, N. (2007). *The Shock Doctrine: The Rise of Disaster Capitalism*. Metropolitan Books/Henry Holt.

Kothari, A., Salleh, A., Escobar, A., Demaria, F., & Dacosta, A. (Eds.) (2019). *Pluriverse: A Post-Development Dictionary*. Tulika Books.

Kunze, C., & Becker, S. (2015). Collective ownership in renewable energy and opportunities for sustainable degrowth. *Sustainability Science, 10*, 425–437.

Latouche, S. (2009). *Farewell to Growth*. Polity.

Lehtinen, A. A. (2018). Degrowth in city planning. *Fennia, 196*(1), 43–57.

Lew, A. A., Cheer, J. M., Brouder, P., & Mostafanezhad, M. (Eds.) (2022). *Global Tourism and COVID-19 Implications for Theory and Practice*. Routledge.

Lundmark, L., Zhang, J. J., & Hall, C. M. (2021). Degrowth and tourism: Implications and challenges. In C. M. Hall, L. Lundmark, & J. J. Zhang (Eds.), *Degrowth and Tourism. New perspectives on tourism entrepreneurship, destinations and policy* (pp. 1–21). Routledge.

Malm, A. (2016). *Fossil Capital: The Rise of Steam Power and the Roots of Global Warming*. Verso.

Malm, A. (2020). *Corona, Climate, Chronic Emergency*. Verso.

Martínez-Alier, J. (2009). Socially sustainable economic de-growth. *Development and Change, 40*(6), 1099–1119.

Martínez-Alier, J., Temper, L., Del Bene, D., & Scheidel, A. (2016). Is there a global environmental justice movement? *Journal of Peasant Studies, 43*(3), 731–55.

Mathieson, A., & Wall, G. (1982). *Tourism: Economic, Physical and Social Impacts*. Longman.

Meadows, D. H., Meadows, D. L., Randers, J., & Behrens III, W. W. (1972). *The Limits to Growth: A Report to the Club of Rome*. Universe Books.

Mies, M. (2007). Patriarchy and accumulation on a world scale – Revisited. *International Journal Green Economics, 1*(3/4), 268–75.

Moore, J. W. (2015). *Capitalism in the Web of Life: Ecology and the Accumulation of Capital*. Verso.

Moore, J. W. (Ed.) (2016). *Anthropocene or Capitalocene? Nature, History, and the Crisis of Capitalism*. Kairos.

Mostafanezhad, M., Norum, R., Shelton, E. J., & Thompson-Carr, A. (Eds.) (2016). *Political Ecology of Tourism Community, Power and the Environment*. Routledge.

Mowforth, M., & Munt, I. (2016). *Tourism and Sustainability: Development, Globalisation and New Tourism in the Third World*. Routledge.

Murray, I., & Martínez-Caldentey, M. A. (2020). Tourism and inequality: A pending debate. In G. X. Pons, A. Blanco-Romero, R. Navalón-García, L. Troitiño-Torralba, & M. Blázquez-Salom (Eds.), *Sostenibilidad Turística: Overtourism vs Undertourism* (pp. 593–606). Mon. Soc. Hist. Nat. Balears.

Navarro-Jurado, E., Romero-Padilla, Y., Romero-Martínez, J. M., Serrano-Muñoz, E., Habegger, S., & Mora-Esteban, R. (2019). Growth machines and social movements in mature tourist destinations Costa del Sol-Málaga. *Journal of Sustainable Tourism, 27*(12), 1786–1803.

Nelson, A., & Edwards, F. (Eds.) (2021). *Food for Degrowth: Perspectives and Practices*. Routledge.

Nelson, A., & Schneider, F. (Eds.) (2018). *Housing for Degrowth: Principles, Models, Challenges and Opportunities*. Routledge.

O'Connor, J. (1988). Capitalism, nature, socialism: A theoretical introduction. *Capitalism Nature Socialism, 1*(1), 11–38.

Peet, R.; Robbins, P., & Watts, M. (Eds.) (2010). *Global Political Ecology*. Routledge.

Pellizzoni, L. (2011). Governing through disorder: Neoliberal environmental governance and social theory. *Global Environmental Change, 21*(3), 795–803.

Perez Orozco, A. (2014). *Subversion de la economia. Aportes para un debate sobre el conflicto capital-vida*. Tranficantes de sueños.

Piketty, T. (2019). *Capital and Ideology*. Harvard University Press.

Rockström, J., Steffen, W., Noone, K., Persson, Å., Chapin III, F. S., Lambin, E. T., Lenton, M., Scheffer, M., Folke, C., Schellnhuber, H., Nykvist, B., De Wit, C. A., Hughes, T., van der Leeuw, S., Rodhe, H., Sörlin, S., Snyder, P.K., Costanza, R., Svedin, U., Falkenmark, M., Karlberg, L., Corell, R. W., Fabry, V. J., Hansen, J., Walker, B., Liverman, D., Richardson, K., Crutzen, P., & Foley, J. (2009). Planetary boundaries: Exploring the safe operating space for humanity. *Ecology and Society, 14*(2), 32.

Rodríguez-Labajos, B., Yánez, I., Bond, P., Greyl, L., Munguti, S., Ojo, G. U. & Overbeek, W. (2019). Not so natural an alliance? Degrowth and environmental justice movements in the global south. *Ecological Economics, 157,* 175–184.

Ruíz-Ballesteros, E. (2021). Community-based tourism and degrowth. In C. M. Hall, L. Lundmark, & J. J. Zhang (Eds.), *Degrowth and Tourism: New Perspectives on Tourism Entrepreneurship, Destinations and Policy* (pp. 187–201). Routledge.

Salleh, A. (2012). *Eco-sufficiency and Global Justice: Women Write Political Ecology*. Pluto Press.

Schmelzer, M., Vetter, A., & Vansintjan, A. (2022). *The Future is Degrowth: A Guide to a World beyond Capitalism*. Verso.

Schmid, B. (2021). *Making Transformative Geographies: Lessons from Stuttgart's Community Economy*. transcript Verlag.

Schneider, F., Kallis, G., & Martinez-Alier, J. (2010). Crisis or opportunity? Economic degrowth for social equity and ecological sustainability. *Journal of Cleaner Production, 18*(6), 511–518.

Schwartzman, D. (2012). A critique of degrowth and its politics. *Capitalism Nature Socialism, 23*(1), 119–125.

Sharpley, R. (2009). *Tourism Development and the Environment: Beyond Sustainability?* Earthscan.

Sharpley, R. (2020). Tourism, sustainable development and the theoretical divide: 20 years on. *Journal of Sustainable Tourism, 28*(11), 1932–1946.

Spash, C. L. (Ed.) (2017). *Routledge Handbook of Ecological Economics*. London: Routledge.

Streeck, W. (2017). *How Will Capitalism End? Essays on a Failing System*. Verso.

Sun, Y.-Y., Gössling, S., & Zhou, W. (2022). Does tourism increase or decrease carbon emissions? A systematic review. *Annals of Tourism Research, 97,* 103502.

Torres, C., & Moranta, J. (2020). Climate emergency in touristified economies: The necessary economic, ecological and social transition as the basis for an effective mitigation strategy. *Revista de Economía Crítica, 30,* 120–135.

Valdivielso, J., & Moranta, J. (2019). The social construction of the tourism degrowth discourse in the Balearic Islands. *Journal of Sustainable Tourism, 27*(12), 1876–1892.

Valero, A., & Valero, A. (2014). *Thanatia: The Destiny of the Earth's Mineral Resources. A Thermodynamic Cradle-to-cradle Assessment*. World Scientific.

Waring, M. (2003). Counting for something! Recognising women's contribution to the global economy through alternative accounting systems. *Gender and Development, 11*(1), 35–43.

WECD (1987). *Report of the World Commission on Environment and Development: "Our common future.* Oxford University Press.

Xue, J. (2021). Urban planning and degrowth: A missing dialogue. *Local Environment, 27*(4), 404–422.

6

ALTERITY, MOBILITY AND TERRITORY

Conceptualising Tourism Space in a World in Flux

Dominic Lapointe

Introduction

It is a truism to state that tourism is constituted through the triad of mobility, alterity and territory. While these three dimensions of the tourism phenomenon have already underpinned many inquiries, the intersection between the three is exposing certain issues that need to be deconstructed and explored. Following previous work on territory (Lapointe, 2021) and alterity (Lapointe, 2020), this chapter will tackle what both papers pointed to in their conclusions: the continuous territorialisation/deterritorialisation movement of tourism, and the relevance of Deleuze and Guattari's philosophy, especially the general theory of stratification (Jacques, 2014) as explained in *Milles Plateaux* (Deleuze & Guattari, 1980). Indeed, there is a whole naturalist dimension to Deleuze and Guattari's works on the concept of strata and the process of territorialisation and deterritorialisation that could offer new perspectives for understanding tourism. This discussion will be explored by conceptualising each element of the triad, starting with alterity, which operates along two threads: visitors experiencing the alterity of the place visited and the residents that co-exist with an industry catering to non-locals. These two threads are fuelled by mobility – mobility that accentuates alterity or can decrease it through virtual movement and stratification of presence and absence. Mobility that accumulates in the territorialisation process, constitutive of a subjectivity in place that reproduces in time, and the deterritorialisation process that virtualises subjectivity and praxis.

Alterity and Otherness: The Tourist as Alterity-Seeking Subject

The concept of alterity is closely linked to that of Otherness. As Picard and Di Giovine (2014) state, alterity in tourism is not only seeking differences, as it is bidirectional: from the tourist to the Other and back. Alterity comes from a process of delineation in which differences are marked as compared to the self. This process of delineation can involve space (somewhere else), time (before and after), cultures (difference in practices and beliefs) and bodily boundaries. As alterity-seeking beings, tourists are not just experiencing alterity, they are also producing it. Indeed, tourists act as the Other, but they also constitute the Other in the Self (op cit.). This

DOI: 10.4324/9781003286301-8

preconception of the Other in the Self is part of the drive to experience alterity, but in the context of encounters, it is also part of the production of the experience of alterity by the Other for the Self, and the Self for the Other.

Therefore, tourism equates to an intersubjective encounter (Hockert, 2018; Gibson, 2012; Jóhannesson, Chapter 10 of this volume). Inspired by Levinas's intersubjective ethics, Hockert (2018) underlines that alterity is a negotiated act, between Self and Other in a moral encounter. This understanding of the encounter in the negotiation of the Self moves away from the individual being defining itself autonomously, to emphasise the relationship between Self and Other as interdependent in their welcoming of each other, and responsibility without mastering the Other (Hockert, 2018). For Levinas (1979), the alterity is created in a relational interaction, where the subject, the Self, is created through the welcome, the hospitality, of the Other. Nonetheless, this hospitality of the host can turn them into a hostage of the guest (Derrida, 1999; Levinas, 1979). As Raffoul (2012) underlines, for Levinas, while ethics is the total hospitality of the infinity of the Other (Levinas, 1979), there is also the 'renversement' (reversal) of the subject, whereby the host becomes a hostage of the guest, the host becoming the guest, while the guest become the host of the host-turned-guest. It is important to acknowledge here that host and guest are both *Hôte* in the French text of Levinas. This reversal brings Derrida, as Raffoul explains, to express in an open formula the sense of this subjectivity reversal as being "at home in the Other home", which could mean the subject is at home in the Other's home, or in its home, the subject is Other. This is particularly relevant to contemporary tourism, where negotiated alterity cultivates alterity figures for the Other in the home of the Self, which redefine themselves according to this Othering. This is also exemplified in the 'Live like the local of AirBnB', where the Other becomes at home in the host home, while the host becomes hostage to the needs and care of the guest as Other, in a uniformising trend where, under the name of alterity-seeking in daily life, Otherness merges into sameness while expropriating the meaning and home of the host (Bélanger & Lapointe, 2021; Roelofsen & Minca, 2018).

While alterity is discussed within encounter (Picard and Di Giovani, 2018; Hockert, 2018) in tourism it also influences place and space (see also Everingham et al, Chapter 11 of this volume), as the structure of Othering that involves the Other in us would shape tourism areas at a crossroads between what the in-place tourism stakeholders think would be of interest to the visiting Others, but also what the visiting Others expect to experience in place.

Place and Space in Tourism Alterity

The tourism industry can have various levels of linkage with the local economy and way of life. It ranges from a complementary activity to a total delinkage. Alterity is constitutive of tourism, as tourists venture out of their daily life space to visit other places (Équipe MIT, 2002). Alterity does not always equate to distance, as alterity can be found close by, while for the hypermobile class, the nodal point of their life space, the space of their Self, can be distant and not part of Otherness as experienced by them as familiar. This question of alterity was part of the work of the *Équipe Mobilité, itinéraire, tourisme in France* (Équipe MIT, 2002): looking at tourism as an alterity-seeking activity in space. One of the examples that they give is the relationship between urban and rural zones in an urbanising world in which the generation born in cities considers rural space as a space of Otherness. Paradoxically, their practices as tourists create pockets of urban values and practices in the rural areas. We may even say that tourism is the process of commodification of alterity, stressing the importance of Levinas's ethics to challenge the impacts and relations established by tourism development and activities.

The other form of alterity that tourism builds is in place, where the space of tourism activity is being Othered by the resident population. This stems from the often 'enclavic' tendencies of tourism initiatives (Saarinen, 2016). There is a long tradition of witnessing how tourism grows symbolically and physically away from its host community. The concept of tourism bubble as coined by Judd (1999) is particularly self-explanatory for the process of setting apart space for tourism activities. Space assigned for tourists emerges as it becomes the dominant industry in an area. The paradox here, as Judd (1999) underlines, is in setting such a space apart, which tends to normalise and uniformise what the areas where tourism happens are, and what they ought to be. This creates an alterity with the remaining parts of the city, while making it more similar to other tourism bubbles elsewhere. In this manner, contradictions inherent to the tourism phenomenon start to emerge; on one hand, a desire for Otherness, while on the other, a transformation of this Otherness, making it more similar without being totally the same. All this occurs while unlinking part of tourism's practices, and offer, from the in-place economy and culture that is the origin of the sought-after alterity. As argued in earlier texts (Bélanger & Lapointe, 2021; Bélanger, Lapointe, & Guillemard, 2020; Lapointe, 2020; Lapointe & Coulter, 2020) this is not a fixed linear process, but rather, a dynamic one, where daily life and tourism, Self and Other, negotiate their place. Place is negotiated from the touristification of daily life to the 'mundanisation' of tourism activities, and back through to the spectacularisation of tourism space and place.

Mobility

One of the alterity markers in a world of tourism is mobility. As the world in which we live is characterised by ever-increasing mobility (Bauman, 1998, 2000; Urry, 1995) of people, goods, information and other components of non-human and more-than-human subjects (Braidotti, 2019), the alterity between mobility and immobility becomes more important than ever. Tourism is constitutive of this mobile world, while not being the only form of mobility at play in alterity. Indeed, tourism is at the intersection of mobility and fixities performing in a particular place (Baerenholdt, Haldrup, Larsen, & Urry, 2004); a performance in which Self and Others interact.

Economic access to tourism is a long-known marker of social class (Veblen, 1899). At the turn of the nineteenth century, in their theory of the leisure class, Veblen (1899) underlined what he called "conspicuous consumption" as a privilege of the upper classes to assert their status via non-productive consumption. If the concept of non-productive consumption is no longer used within a service and experiential economy, access to mobility – or lack thereof – acts as a class marker in a mobile world. If, as first thought, the lockdown episodes of the COVID-19 pandemic might look like a return to a less mobile world, this does not necessarily mean that access to mobility as a class marker has decreased; rather, the opposite may be true. The return to mobility within the ongoing pandemic is not only conditioned by sanitary restrictions, but also by prior social stratification, where privileged teleworkers managed to move to second homes or remote destinations, competing to attract digital nomads, while other workers have lost their jobs, home and mobility altogether (Adey, Hannam, Sheller & Tyfield, 2021; Rose, 2020).

Tourism is thus at the intersection of alterity and (im)mobility: a subjective creation activity that blurs ideas of Self and Others in place. In this sense, tourism places are (re)produced in intersubjective relations between fixities and mobility; between fixed subjects and mobile subjects, between here and there, here and from there, between presence and absence. The complexity of this entanglement can be hard to grasp theoretically, as each one of its threads tends to be addressed autonomously in research, while it seems central to our future understanding of tourism, as it folds and unfolds in times of pandemic and, eventually, in a post-pandemic new normality that is still hard to foresee.

Deleuze and Guattari's Strata

Tourism is a finely woven fabric of mobility and immobility that creates alterity and uniformity. Such movements flow back and forth and, arguably, look like a tide of flux. Exploring these movements requires a lens that can account for their persistence, transformations and flows of the parts and the whole. To employ this lens, it is pertinent to draw upon Deleuze and Guattari's work. Much has been written on their work on capitalism and schizophrenia, *Anti-Oedipus* and *A Thousand Plateaus* – an open text full of digressions and wit, while at the same time a difficult philosophical read and a stimulating conceptual toolkit. In *A Thousand Plateaus*, Deleuze and Guattari break away from a structuralist and fixed understanding of society and philosophy. Through the metaphor of the rhizome, they describe a world in constant flux, where being is a moving state of becoming. Often considered as an important post-structuralist work, and a very abstract and theoretical one, there is an underplayed naturalist dimension to Deleuze and Guattari's work. This naturalist dimension – important in *A Thousand Plateaus* – revolves around the concepts of strata: the process of territorialisation (and its counterpart, deterritorialisation). It is through the assemblage of such concepts that tourism can be used to understand how alterity and mobility interact with tourism place and space.

Following DeLanda (2006) and Dovey (2009), *A Thousand Plateaus* is considered here as a conceptual toolbox to build theory that can interact with complexity while trying to avoid reductionism. As Dovey (2009) expresses it:

> My task here is to unravel some of these conceptual tools in a manner that may be useful for the interpretation of place and its intersection with practices of power. An understanding of this work cannot be a simple top-down grasp of a unitary whole; an understanding of Deleuzian thought requires that we enter into this system of concepts rather than contemplate from the outside or from above. We might treat this assemblage of concepts like a strange place – we visit, we explore, we use it; we may or may not get a feel for the game of inhabiting, and we may or may not feel at home.
>
> (Dovey, 2009)

While DeLanda (2006) interprets society through assemblage theory, Dovey (2009) zeroes in on place and power. The Deleuze and Guattari system of concepts as a framework for interpreting tourism will now be explored.

In an effort to capture the complexity of heterogeneous relationships that allow a whole to be more than the sum of its parts, but without being contingent on the sum of its parts, Deleuze and Guattari (1980) came up with the concept of assemblage (*agencement*, which literally means pieced together; see also Briassoulis, Chapter 7 of this volume). The concept of assemblage points to a focus on relations of exteriority where the constituent elements cannot be reduced to their function as a whole while being at the same time part of a multiplicity of assemblage. One of the strengths of this theory is that it makes it possible to reduce the tensions between the scales of analysis and intervention without denying their structuring force. For example, Dovey (2009) illustrates how national policies entangle with practices, materiality and affect to transform local practices in cities, but also materialities and affect. The same author stresses that the theory of assemblages makes it possible to dynamically analyse the relations between such heterogeneous elements as experiences, representations and materiality, both in the reproduction of these links and in their transformation. Readers may refer to Briassoulis (Chapter 7 of this volume) for a more in-depth discussion of the whole assemblage apparatus. This chapter will, however, enter into Deleuze and Guattari's work, in the middle of the tourism phenomenon or in the middle of movements, intensity, or matters. 'Enter' being an important word here; as for Deleuze and Guattari, one is always in the middle of the event.

Central to Deleuze and Guattari's work is the notion of movement, which arguably echoes this chapter's interest in mobility. Movement is conceptually what is happening between assemblages; the exchanges of information and particles; materialities. Importance is placed upon virtual movement, which is not the opposite of real, but rather a counter to the actual. Virtual movement is a movement which has blurred boundaries, and does not have a clear and definite contour; an unpredictable movement. It is the movement between assemblages: that which is transformed by them and transforms them. This might echo involuntary mobility, nomads and *flâneurs*. On the other side, we have actual movement, which is a predictable, well-defined movement that seals the line of flights between new assemblages. Actual movement might echo pendular mobility and packaged uniformised experience through mass tourism. On one side we have a mobility that transforms the assemblage into continuous becoming, while on the other we have a reinforcement of the assemblage in a crystallisation into strata. As is often the case in Deleuze and Guattari's work, such movements are continuously happening – being and becoming, folding and unfolding.

Strata are at the core of Deleuze and Guattari's naturalism. They refer to the accumulation of singularities, the fixing of intensities and the capture of virtual movement in a reproducible actual movement (Deleuze and Guattari, 1980; Jacques, 2014). Strata is a non-linear conceptualisation of the perpetual movement of becoming and being in their coherence, but also with their loose ends that could reverse the strata in a new assemblage. If strata can be a stable assemblage, components of the assemblage – movement, particles, energies – can be assembled into more than one stratum, as their assemblage capacities, through actual and virtual movements, can exist at different scales and layers.

Strata in tourism might refer to the transformation of the virtual movement of emergent practices into an actual movement. The virtual movement of tourism implies a high level of alterity in a less predictable interaction between Self and Other, neither stabilised nor institutionalised, where tourism and tourists are not yet the dominant world-making force (Hollinshead & Suleman, 2018) at play in place. It is a virtual movement that encounters the actual movement of daily life in this place; a micro-assemblage with the strata of living in such a place, not just of moving and stopping there. Over time, this virtual movement of tourism can become the actual movement, unfolding the previous stratum to refold it along the actual practices of tourism and the more institutionalised and predictable practices of tourists. This is not necessarily an antagonistic process. While alterity is negotiated between Self and Other, between the Other in the Self, and the Self in the Other, there is a reversal of the strata of tourism: world-making, decoding and re-coding what is and what ought to be in place and space. While this tourism strata stabilised as the actual, it virtualised the emergence of daily life, marginalising daily-life practices from tourism practices. It is through this movement that we can observe the territorialisation/deterritorialisation process, which, I argue, is central for addressing the contemporary tourism phenomenon.

First, the territorialisation/deterritorialisation process is at the centre of Deleuze and Guattari's work. They describe the continuous movement between earth (which I will refer to as *Terra* so as not to share the translation of the word *Terre* into 'earth', as it breaks the etymological coherence of the Deleuzean system and tends to restrict its meaning) and territory in their 1994 work, *What is Philosophy?*:

> Movements of deterritorialisation are inseparable from territories that open onto an else-where; and the process of reterritorialization is inseparable from the earth, which restores territories. Territory and earth are two components with two zones of indiscernibility, deterritorialisation (from territory to the earth) and reterritorialization (from earth to territory).
>
> (Deleuze & Guattari, 1994: 85–86)

For Deleuze and Guattari, Terra is the plane of immanence, the free and overarching surface of the earth, the whole, while the territory is the section, the corners, the bounded space. Those two concepts are essentially movements towards each other, territorialising and deterritorialising, from the territory to the Terra, and from the Terra to the territory; folding and unfolding with movement, practices, continuities and ruptures. This movement is two-fold, with transcendental and immanent movement, where transcendental ideals and ideas can become deterritorialised, like principles of free movement opening frontiers. However, deterritorialisation can also be imminent, freeing the individual, the Self from the Terra. The Self is then sent in movement, thus becoming the Other and reterritorialising via new practices. This reterritorialising movement sets new boundaries on the whole and operates through the inner properties of the deterritorialised movement. As the movement free of Terra, it keeps its components and stability and reorders its assemblage from its inner qualities, a reordering that can redefine boundaries and reproducible movements anchoring back to Terra, thus drawing a new territory: a combination of physical and symbolic elements that have stabilised together. This reterritorialising movement might also generate lines of flight on the plane of immanence that revert it to a deterritorialised state, line of flight being the virtual, freed from the previous assemblage that could be recombined in a new assemblage. As we saw during the COVID-19 pandemic, the mobility restrictions sent lines of flight that created a new assemblage along local needs through services that previously catered to tourism (Lapointe, 2020).

For tourism, this movement from territory to Terra and so forth is central to a new understanding of the continuous dynamics of tourism and its multiple lines around the Earth, the Terra and their assemblage at the destination level. In a rare effort to conceptualise the destination as assemblage, Briassoulis (2017; see also Chapter 7 of this volume) tackled the territorialisation/deterritorialisation movement, mostly via DeLanda's (2006) understanding of the concepts. Territorialisation refers to the stability of the assemblage, while deterritorialisation refers to the destabilising of the assemblage (DeLanda, 2006; Briassoulis 2017).

> Territorialisation and coding processes are tourism-related and other habitual, routine practices followed by tourists, tourism entrepreneurs, and intermediaries, the state and locals that provisionally assemble or reassemble heterogeneous components, assigning them material and expressive roles and bringing [tourism assemblages] into being. Tourists pursue various practices, ranging from ordinary and material to specialised and discursive, while undertaking daily personal and other activities at a destination.
>
> (Briassoulis, 2017)

While it is concurrent with parts of Deleuze and Guattari's concepts, this particular angle for analysing the notion of destination brings an underlying system, just as the socio-ecological system reference acknowledges that it seems counter to the open and generative movement of the concepts, keeping it grounded in its philosophical basis.

Discussion

While this triad I have introduced (alterity, mobility and territory) might seem quite stretched conceptually, I will use a short excerpt from the beginning of 1982 Don DeLillo novel, *The Names*, to show the generative capacity of such a theoretical apparatus. In the novel, DeLillo's characters are American expats living a deterritorialised life, not fleeing their home for a freer and hedonist lifestyle as in, for example, the nineteen fifties and sixties American expat novels, but rather as professionals floating on the expansion of Western capital and politics at the dawn of the Cold War. They mention at one point, "Errors and failings don't cling to you the way they do back home" (pp. 50–51), expressing a freedom from Terra brought by mobility and the relative anonymity of deterritorialised life. We will not proceed to a formal literary analysis but focus on the opening paragraph of the book, as it expresses the levels of alterity introduced earlier and the territorialisation/deterritorialisation process set in movement by mobile subjects, tourists, expats, and so forth.

[F]or a long TIME I stayed away from the Acropolis. It daunted me, that somber rock. I preferred to wander in the modern city, imperfect, blaring. The weight and moment of those worked stones promised to make the business of seeing them a complicated one. So much converges there. It's what we've rescued from the madness. Beauty, dignity, order, proportion. There are obligations attached to such a visit. Then there was the question of its renown. I saw myself climbing the rough streets of the Plaka, past the discos, the handbag shops, the rows of bamboo chairs. Slowly, out of every bending lane, in waves of color and sound, came tourists in striped sneakers, fanning themselves with postcards, the philhellenes, laboring uphill, vastly unhappy, mingling in one unbroken line up to the monumental gateway. What ambiguity there is in exalted things. We despise them a little. I kept putting off a visit. The ruins stood above the hissing traffic like some monument to doomed expectations. I'd turn a corner, adjusting my stride among jostling shoppers, there it was, the tanned marble riding its mass of limestone arid schist. I'd dodge a packed bus, there it was, at the edge of my field of vision. One night (as we enter narrative time) I was driving with friends back to Athens after a loud dinner in Piraeus and we were lost in some featureless zone when I made a sharp turn into a one-way street, the wrong way, and there it was again, directly ahead, the Parthenon, floodlit for an event, some holiday or just the summer sound-and light, floating in the dark, a white fire of such clarity and precision I was startled into braking too fast, sending people into the dashboard, the backs of seats.

(DeLillo, 1982:1)

In this excerpt, the narrator exposes three forms of alterity: the tourists, the invisible tourism workers who are not named and the cosmopolitan mobile subjects incarnate in the narrator. This is spatially expressed in the dichotomy between the Acropolis and the modern city, as having different inherent qualities and relationalities. These relationalities are expressed in the sentence, "There are obligations attached to such a visit", but also in the way in which the narrator is projecting himself going to the Acropolis, expressing the Other in the Self. There is considerable mobility in this excerpt: the tourists labouring uphill, the narrator and his friends as deterritorialised expats, the cars moving in the city and the lights, among others.

The narrator alludes to a deterritorialising process in describing the Acropolis as an *alterity space* in Athens, a space out there, a space floating above the modern city, in all its expressive qualities (in that so much converges there), deterritorialised from its environment. It also underlines a reterritorialising process: the Acropolis as a tourism space reterritorialised around its inner qualities – mainly its expressive qualities – to stabilise as a tourism assemblage with the disco, the shops, the tourists with postcards, and so forth. This process deterritorialises the non-tourist city, mentioned as a featureless zone – imperfect – blaring, from the stratification of the omnipresent tourist territory of the Acropolis (floating in the dark). The non-tourist city described as an alterity in this small excerpt is part of a different assemblage that reterritorialises and deterritorialises lines of flights with different assemblages, while still assembling lines of flights with the tourism assemblage through workers, goods, policies, and so forth. Crises like the financial crash of 2008 or the health crisis of the COVID-19 pandemic can also reorder the strata, sending expressive and material elements into movement that transforms the assemblage. In the case of the Acropolis and tourism, the financial crisis strengthens the tourism assemblage as a rare source of income and currency in a hard-hit deterritorialising economic assemblage that reterritorialises around tourism practices and values. In the case of the sanitary crisis, the assembling of a new actor – a virus – with its inherent mobile and moving qualities, still feeds the transformation of the assemblage; reterritorialising through micro mobile alterity while deterritorialising previous cosmopolitan assemblages.

Conclusion

Engaging with the theoretical apparatus of Deleuze and Guattari (1980) is always a perilous effort that bears the risk of turning into some kind of intellectual *virtuosi* exercise that may sound like some overly long old hard-rock electric guitar solo that only proves someone knows how to play guitar, without contributing much to the piece. Humbly, in line with Briassoulis (Chapter 7 of this volume), I consider that in terms of addressing the complexity and multiplicity of the tourism phenomenon, assemblage theory is one of the keys for moving beyond some of the shortcomings of tourism research, including tourism geographies.

The part of assemblage theory expressed here appears highly relevant for generating a different way of understanding tourism beyond dichotomies such as local/global, Self/Other, mobile/immobile, rural/urban, mundane/spectacular, and so forth. At the same time, assemblage theory moves beyond those dichotomies without abandoning them. Indeed, in the practice of tourism geographies research, assemblage theory helps describe the complexities of the moving qualities of tourism along space, place, identity, affects, Self and Others. Instead of trying to isolate them as subjects of research, it links them through movement, within a post-qualitative position that seeks to understand their constant and multiple becoming instead of what they are and ought to be. As in the example we built with the DeLillo excerpt, assemblage theory allows us to see the movement of the different components, these being territorialising tourism while deterritorialising daily life that is still reterritorialised through new lines of flight. As Briassoulis (Chapter 7) expresses the capacity of assemblage theory to expose the multiple geographies of tourism, we focus on the territorialisation/deterritorialisation process at the nexus of mobility, alterity and territory. This focus offers tools to reconceptualise them as fluid and blurred, moving in a rhizomatic continuum of being and becoming, instead of rigid categories.

For tourism geographies this means challenging the place-based localisation paradigm that focuses on where tourism happens and is performed (Franklin, 2014) without focusing solely on cosmopolitan global mobility. The territorialising deterritorialising process allows us to dissect how mobilities contribute to both movements at different interconnected scales: deterritorialising at the local scale for some actors while reterritorialising at a higher scale, all the while creating, and destroying, alterity at both scales. Places and scales remain important when using the assemblage lens, but they are not things, or boxes. Rather, they are a combination of lines of flight at different intensities of stratification and possibilities. They are not static facts to be analysed, but are moving active matter that constantly and repeatedly assembles in continuous being, while also having potential reassembling movements. Tourism can be a technology of the Self, and therefore also a technology of Others, that contributes to subjectivity formation in the contemporary mobile world (Lapointe & Coulter, 2020). The (im)mobilities of the COVID-19 pandemic (Adey et al., 2021) did not stop this subjectivity formation, but reterritorialised it through different practices in an assemblage that keeps becoming, calling for many perspectives to understand emerging tourism geographies (Cheer, et al., 2021). This is also a call to understand the roles of previous assemblages that were deterritorialised and reterritorialised in the third decade of the twenty-first century through the pandemic.

References

Adey, P., Hannam, K., Sheller, M., & Tyfield, D. (2021). Pandemic (Im)mobilities. *Mobilities* 16(1), 1–19. doi:10.1080/17450101.2021.1872871

Baerenholdt, J. O., Haldrup, M., Larsen, J., & Urry, J. (2004). *Performing tourist places*. Ashgate.

Bauman, Z. (1998). *Globalization: The human consequences*. Polity Press.

Bauman, Z. (2000). *Liquid modernity*. Polity Press.

Bélanger, H., & Lapointe, D. (2021). Revitalisation et «bulles touristiques»: Une gentrification instantanée par la touristification du quotidien? *Recherches Sociographiques* 62(1), 149–173.

Bélanger, H., Lapointe, D., & Guillemard, A. (2020). Central neighbourhoods revitalization and tourists bubble: from gentrification to touristification of daily life in montréal. In J. Bean (Ed.), *Critical Practices in Architecture: The Unexamined* (pp. 71–94). Cambridge Scholar Publishing.

Braidotti, R. (2019). A theoretical framework for the critical posthumanities. *Theory, Culture & Society*, 36(6), 31–61.

Briassoulis, H. (2017). Tourism destinations as multiplicities: The view from assemblage thinking. *International Journal of Tourism Research* 19(3), 304–317. doi:10.1002/jtr.2113

Cheer, J. M., Lapointe, D., Mostafanezhad, M., & Jamal, T. (2021). Global tourism in crisis: Conceptual frameworks for research and practice. *Journal of Tourism Futures* 7(3), 278–294.

DeLanda, M. (2006). *A new philosophy of society: Assemblage theory and social complexity*. Bloomsbury Publishing.

Deleuze, G., & Guattari, F. (1980). *Mille Plateaux*. Editions de Minuit.

Deleuze, G., & Guattari, F. (1994). *What is philosophy?* Columbia University Press.

DeLillo, D. (1982). *The names*. Knopf.

Derrida, J. (1999). *Responsabilité et hospitalité. Manifeste pour l'hospitalité, Paris: Paroles l'Aube*, 121–124.

Dovey, K. (2009). *Becoming Places: Urbanism/architecture/identity/power*. Routledge.

Everingham, P., Obrador, P., & Tucker, H. (2024). Embodied encounters in tourism geographies research. In J. Wilson and D. K. Müller (Eds.), *The Routledge Handbook for Tourism Geographies*. 2nd edition. Routledge.

Équipe MIT. (2002). *Tourisme 1: Lieux communs, coll.* Mappemonde. Belin.

Franklin, A. (2014). Tourist studies. In P. Adey, D. Bissell, K. Hannam, P. Merriman, & M. Sheller (Eds.), *The Routledge handbook of mobilities* (pp. 74–84). Routledge.

Gibson, C. (2012) Geographies of tourism: Space, ethics and encounter. In J. Wilson (Ed.), *The Routledge handbook of tourism geographies* (pp. 55–60). Routledge.

Hollinshead, K., & Suleman, R. (2018). The everyday instillations of worldmaking: New vistas of understanding on the declarative reach of tourism. *Tourism Analysis* 23(2), 201–213. https://doi.org/10.3727/108354218X15210313504553

Höckert, E. (2018). *Negotiating hospitality: Ethics of tourism development in the Nicaraguan highlands*. Routledge.

Jacques, V. (2014). Deleuze pas à pas. Paris, Ellipses.

Jóhannesson, G.T. (2024). Performativity, space and tourism. In J. Wilson and D.K. Müller (Eds.), *The Routledge Handbook for Tourism Geographies*. 2nd edition. Routledge.

Judd, D.R. (1999). *Construct*. R. Judd & S.S. Fainstein (Eds.), The Tourist City (pp. 35–53). Yale University Press.

Lapointe, D. (2020). Reconnecting tourism after COVID-19: The paradox of alterity in tourism areas. *Tourism Geographies* 22(3), 633–638.

Lapointe, D., & Coulter, M. (2020). Place, Labor, and (Im)mobilities: Tourism and Biopolitics. *Tourism Culture & Communication* 20(2–3), 95–105.

Lapointe, D. (2021). Tourism territory/territoire(s) touristique(s): When mobility challenges the concept. In M. Stock (Ed.), *Progress in French Tourism Geographies: Inhabiting Touristic Worlds*, (pp. 105–116). Springer Nature.

Levinas, E. (1979). *Totality and infinity: An essay on exteriority* (Vol. 1): Springer Science & Business Media.

Picard, D., & Di Giovine, M. A. (2014). Introduction: Through other worlds. In D. Picard, & M. A. Di Giovine, (Eds.), *Tourism and the power of otherness: Seductions of difference* (pp. 2–28). Channel View Publications.

Raffoul, F. (2012). Chez lui chez l'autre. *Les temps modernes*, (3), 133–156.

Roelofsen, M., & Minca, C. (2018). The superhost. Biopolitics, home and community in the Airbnb dream-world of global hospitality. *Geoforum* 91, 170–181. doi:10.1016/j.geoforum.2018.02.021

Rose, J. (2020). Biopolitics, essential labor, and the political-economic crises of COVID-19. *Leisure Sciences* 43(1–2), 1–7. doi:10.1080/01490400.2020.1774004

Saarinen, J. (2016). Enclavic tourism spaces: Territorialization and bordering in tourism destination development and planning. *Tourism Geographies* 19(3), 425–437. doi:10.1080/14616688.2016.1258433

Urry, J. (1995). *Consuming Places*. Routledge.

Veblen, T. (1899). *The Theory of the Leisure Class: An Economic Study of Institutions*. BW Huebsch.

7

ASSEMBLAGE TOURISM GEOGRAPHIES

Helen Briassoulis

Tourisms, Geographies, Tourism Geographies

Tourism studies, regardless of focus, spatial level or outlook, overtly and covertly implicate entanglements and flows of people, goods, services, animate and inanimate beings, culture, institutions and politics. This is so because tourism is multi-component, multi-activity, multi-level, multi-dimensional and dynamic; tourists, tourism destinations, firms and products do not pre-exist as autonomous, isolated entities; they are continuously being made, performed and changing (Wilson, 2012; Lew et al., 2014).

Tourism has been studied from several disciplinary perspectives, mainly economics/management/planning, sociology, anthropology and geography and, since the 2000s, from interdisciplinary, socio-ecological perspectives. Consequently, approaches to its study have been shaped by the dominant disciplinary outlook, epistemology and methodology together with changes in tourism phenomena and broader, coevolving philosophical, scientific and societal developments.

Tourism geography is a distinct (sub)field of geographic inquiry at the intersection of geography and tourism, originally developed within the broader intellectual context of the geographic studies of the 1960s and 1970s (Hall & Page, 2009; 2012). The principal premise was that the scientific analysis of tourism should be based on modernist, reductionist/essentialist approaches founded on positivism/postpositivism, while adopting universal, aggregate, top-down, linear, structuralist/functionalist theories and employing formal quantitative analytical techniques (Briassoulis, 2017).

In the 1980s and 1990s, the various turns in social sciences (in particular, relational, critical, cultural, material, performativity, practice and spatial) generated nonreductionist/relational approaches that spread into geography (Briassoulis, 2017). These are founded on relational, nonpositivist epistemologies, mainly of the post-structuralist genre (social constructivism and critical realism), adopt disaggregate social, cultural, nonrepresentational and multi/interdisciplinary theories, and employ qualitative and mixed analytical techniques.

In 2010, Robbins and Marks introduced to the study of geographic phenomena the term *assemblage geographies*, in order to denote the contribution of actor–network theory (ANT) and assemblage thinking (AT). These post-structuralist approaches, both employing the Deleuzean notion of assemblage, constitute fitting frames of geographic inquiry that maintain the spatiality, temporality and materiality of pertinent phenomena.

DOI: 10.4324/9781003286301-9

Within this changing and challenging intellectual context and the growing recognition that reductionist/essentialist approaches could neither satisfactorily address diverse and dynamic tourism issues nor negotiate the role of tourism in socio-spatial development, the shift from tourism geography to tourism geographies occurred, and non-reductionist/relational approaches complemented or replaced reductionist/essentialist ones (Hall & Page, 2012; Ioannides & Debbage, 2014). Pertinent studies reject binaries and taxonomies and foreground the multiplicity, heterogeneity, hybridity, complexity, materiality, reflexivity, performativity, processuality, contextuality, emergence and contingency of tourism phenomena (Edensor, 2000; Saarinen, 2004; Cohen & Cohen, 2012). Cultural geographies of tourism (Crang, 2014), born out of the cultural turn, the new mobilities paradigm (NMP) (Sheller & Urry, 2004; Duncan, 2012) and critical tourism studies (CTS), associated with the post-structuralist mobilities and critical turns (Ateljevic et al., 2007; Bianchi, 2009; Gale, 2012) have been particularly influential in shaping the focus, content and method of tourism geographies since the 2000s.

Post-structuralism also informs ANT and AT, which have spread into tourism studies. ANT, originating in science and technology studies, has been very influential in tourism studies since the 2000s (Ateljevic et al., 2007; Debbage & Ioannides, 2012; Ioannides & Debbage, 2014). Drawing from social constructivism, the actor–network ontology claims that "it is the relations – and their heterogeneity – that are important, and not the things in themselves. … Entities achieve their form as a consequence of the relations in which they are located and the processes of translation they go through". (Van der Duim et al., 2013: 9). Tourism places are "assembled … of heterogeneous elements such as people, money, data, images and brands … accomplished and stabilised from time to time through networked relationships between the actors involved … and can be understood as ordering effects of those agents" (Jóhannesson, 2005: 137; Franklin, 2004; Ren, 2010). 'Tourismscapes' are "actor-networks … consisting of relations between people and things dispersed in time–space-specific patterns" (Van der Duim, 2007: 967).

While ANT has made greater inroads in tourism studies thus far, AT, remaining closer to the Deleuzoguattarian philosophy and assemblage ontology, has been more reluctantly adopted. The rest of this chapter introduces AT and then offers a concise critical presentation of its diffusion in the study of tourism geographies. Lastly, arguing for the suitability of the assemblage ontology/ analytic and AT on theoretical, methodological, empirical, policy and governance grounds, it suggests 'assemblage tourism geographies' as a promising terrain for tourism studies and outlines future research directions.

Assemblage Thinking and the Study of Tourism Geographies

Assemblage thinking: A brief introduction

AT denotes the use of the Deleuzoguattarian assemblage ontology/analytic and associated conceptual lexicon to study socio-spatial and other issues arising within socio-ecological milieus (SEM). It encompasses the works of Deleuze and Guattari (DG), Manuel DeLanda, who has produced an intelligible rendering of Deleuzoguattarian assemblage theory, and secondary literature adopting DG and/or DeLanda. AT is primarily informed by critical realism, as it assumes the existence of a mind-independent reality, while acknowledging the social construction of socio-spatial phenomena (Delanda, 2006). It emphasises ontology over epistemology, a distinctive feature of Deleuzoguattarian philosophy (Briassoulis, 2017).

In contrast to reductionist/essentialist approaches that conceive SEMs as seamless, pre-given wholes, possessing "a clear and distinct nature" (DeLanda, 2002: 12), AT conceives SEMs as 'multiplicities' comprising assemblages. Assemblage, an ontology of becoming, denotes the coming or fitting together of diverse, heterogeneous, material, immaterial, human (socio-cultural, institutional, political) and nonhuman (biophysical) components into dynamic, provisional, decomposable, but irreducible wholes to serve a purpose, and creating agency (Delanda, 2006; Anderson & McFarlane, 2011; Anderson et al., 2012). *Desire*, the state of unconscious drives *per* Deleuze, moves the process, constantly coupling continuous flows and partial, fragmentary and fragmented objects and ensuring the organised rather than random unity of assemblages (Deleuze & Guattari, 1987).

The heterogeneous components are relatively autonomous, have multiple memberships, variable spatio-temporal reach and importance and play material and symbolic/expressive roles (DeLanda, 2006; Anderson et al., 2012). Critical/limiting components support biophysical and human functions that secure the survival, maintenance and functioning of a SEM (Briassoulis, 2017). These usually change more slowly (e.g., climate, capital, technology, culture) than other components that change more rapidly and fluctuate widely over time (e.g., weather, demand, preferences, prices).

The composition and duration of assemblages are not predetermined and constant; they remain deliberately open (Anderson & McFarlane, 2011; DeLanda, 2006, 2011). Contingently obligatory *relationships of exteriority* link their components in contrast to logically necessary *relationships of interiority* that characterise and define pre-given wholes (DeLanda, 2006). Deleuze & Guattari (1987) distinguish rhizomatic (nonhierarchical, a-centred and horizontal) from arborescent (centralised and vertical) connections among components (Bonta & Protevi, 2004). Rhizomatic, self-organising structures co-exist with arborescent, centrally organised structures in practice.

Assemblages have *properties* (extent, density, intensity, potential, connectedness, diversity, resilience, and so forth), which are not the sum of the properties of their components. They result from complex, contextual and contingent interactions among components, especially the critical/ limiting ones, which exercise their *capacities*; the powers they possess to affect and be affected ('affect' *per* Deleuze). Properties are actual and known/knowable. Capacities are open and unpredictable since the variable, contextual and contingent interactions among innumerable components cannot be presaged (DeLanda, 2006; Anderson et al., 2012).

The *possibility space* of an assemblage (phase/state space of nonlinear systems) comprises the set of capacities of its (critical/limiting) components (DeLanda, 2002, 2006) within which *basins of attraction* develop around *attractors*; that is, final (minimum) states towards which a system spontaneously tends in the long run in the absence of constraints. Attractors exemplify patterns of behaviour and indicate the long-term *tendencies* of an assemblage (DeLanda, 2002).

Territorialisation/coding processes of assembly hold components together, 'produce' assemblages, secure their internal coherence, resulting in stratification and striated space (Bonta & Protevi, 2004). Habitual, routine, ordinary and specialised, material and discursive practices actualize these processes, underlining the constant *labour* needed to (re)connect heterogeneous components (Li, 2007; Briassoulis, 2017).

Deterritorialisation/decoding processes, activated by endogenous and exogenous, biophysical and human forces from all levels (e.g., natural and technological disasters, political unrest, environmental, socio-cultural and institutional change), modify the capacities and thresholds of components, break down their relationships and disrupt the coherence of assemblages (DeLanda, 2002, 2006; Bonta & Protevi, 2004; Anderson & McFarlane, 2011). Reterritorialisation follows either as adaptation (movements within the same basin of attraction) or as transformation (movement to another basin) (see also Lapointe, Chapter 6).

The repetition of these processes generates *multiplicities*, populations of assemblages, of historically contingent individuals that "define ... and progressively specify the nature of a multiplicity as they unfold" (DeLanda, 2002: 12). A point (state) in a basin represents an assemblage, an *individual singularity* (DeLanda, 2006, 2011). SEMs have multidimensional possibility spaces, complex distributions of attractors and multidimensional basins of attraction populated by co-functioning and spatio-temporally overlapping assemblages. Their features are empirically identified only (DeLanda, 2011).

Once assemblages emerge, they are real, immanent and establish a territory (Anderson & McFarlane, 2011) linking the micro, disaggregate behaviour (molecular *per* Deleuze) with the macro, average behaviour (molar *per* Deleuze) (Bonta and Protevi, 2004; DeLanda, 2006). Their ontology is flat; that is, they are unique individuals, differing in spatiotemporal scale but not in ontological status (DeLanda, 2006). They possess agency, because they act back on their components, enabling or constraining their relationships and characteristic identity. Both are multiple, composite, distributive and emergent, shaped by the capacities of the heterogeneous components.

Power in assemblages is multiple and decentralised; there are many, interacting sources of power, not a single/central governing centre (Anderson & McFarlane, 2011). Similarly, causality is emergent and nonlinear, "located not in a pre-given sovereign agent, but in interactive processes of assembly", immanent and over-determined; different assemblages emerge under different conditions (Anderson et al., 2012: 180).

Assemblage thinking and tourism studies: State-of-the-art

To appraise the current uptake and value of AT in tourism studies, over forty papers and book chapters, published up to 2020, have been reviewed. They variously use AT and negotiate tourism issues from four broad viewpoints: (a) individuals (tourists, tourism firms, various tourism actors), (b) tourism places/destinations, (c) types of tourism (health, urban, agritourism) and (d) tourism as a component of wider socio-spatial assemblages. Based on the strength of their AT use, the most pertinent pieces were used to support the following discussion.

The strength of AT use was loosely distinguished into 'shadow', weak, moderate and strong. Shadow use encompasses miscellaneous works employing post-structuralist, relational approaches using the term 'assemblage' without reference to DG, DeLanda or the secondary AT literature (Holloway, 2010; Holliday et al., 2014; Sommer & Kip, 2019; Haanpää et al., 2021). Weak use employs selected AT notions relying on the secondary literature without reference to DG and/ or DeLanda. Moderate use employs selected AT notions relying on DG and/or DeLanda and the secondary literature. Lastly, strong use encompasses works that rely on DG, DeLanda and the secondary literature and employ several AT concepts. Although shadow and weak use are still practised, the application of AT in tourism studies – moderate and strong use – slowly and selectively began in the early 2000s (Obrador, 2003) and has accelerated since 2010 signalling the growing recognition of its value and contribution to the analysis of tourism-related issues.

Conception of 'assemblage'. Anderson & McFarlane (2011) have penetratingly discussed three uses of assemblage – descriptor, concept and ethos. Shadow use studies employ it as a descriptor, a working description of collections/entanglements of material and immaterial components and flows that possess agency, signifying the dynamic formation of tourism geographies and avoiding their representation as static gatherings.

Most studies, however, employ assemblage as a concept, to a greater or lesser extent, following the Deleuzoguattarian thesis (and pertinent conceptual lexicon) which maintains that concepts, like assemblages, comprise inseparable finite, heterogeneous components, securing their (endo) consistency, and perform certain functions (Deleuze & Guattari, 1984); hence "a concept is what the assemblage determines it to be" (Deleuze & Guattari, 1987: 229). In the studies reviewed, assemblage, its components, and their relationships, processes of assembly and flat ontology are most commonly used.

Assemblage, components, relationships. The literature covers a broad and diverse array of issues: tourist experience (Obrador, 2003), national tourism assemblages (Crofts Wiley et al., 2010), popular humanitarian gaze (Mostafanezhad, 2014), everyday tourist practices (Barry, 2017), tourist well-being (Li & Chan, 2020), small tourism firms (Saxena, 2015), ghost tours (Holloway, 2010), rural tourism assemblages (Rosin et al., 2013), new farms/agritourism assemblages (Sutherland and Calo, 2020), rave tourism (Saldanha, 2006), volunteer tourism (Burrai et al., 2017), film-induced tourism (Mostafanezhad & Promburom, 2018), animal-based tourism (Haanpää et al., 2021), medical/therapeutic assemblages (Foley, 2014), medical tourism facilitators (Chee et al., 2017), international mountain leaders (Cousquer & Beames, 2013), tourism destinations (Briassoulis, 2017; 2018), small fishing community waterfront (Kadfak & Knutsson, 2017), festival attendees and places (Waitt & Duffy, 2010), urban hangouts (Sommer & Kip, 2019), urban tourist places (Sommer, 2019), urban tourism protest (Bruttomesso, 2018) and geopolitical island assemblages (Davis et al., 2020).

These objects of study are not treated as pre-existing totalities represented by reified, distinct categories and binaries; they are conceptualised as *socio-material assemblages*, emergent provisional gatherings of heterogeneous human and nonhuman, material and immaterial, mobile and immobile components coming together to serve a purpose. Briassoulis (2017) defines a *tourism assemblage* (TA) as a unique, place- and time-specific provisional whole emerging from spontaneous and planned processes that assemble diverse existing and new components associated with tourism-related activities (travel, accommodation, services, sightseeing), originating in a SEM, other SEMs and spatial levels, wedding tourism supply to tourism demand over a given period.

The heterogeneous components (tourists, entrepreneurs, hotels, trees, objects, feelings) are relatively autonomous and, participate in more than one tourism or non-tourism assemblage Some are absently present (McCann, 2011), for example, national policy, global discourses, past actors, ideas, and so forth, have variable spatio-temporal reach and play material and expressive roles (Kadfak & Knutsson, 2017; Mostafanezhad, 2020; Sutherland & Calo, 2020).

Contingently obligatory relationships of exteriority develop among components (Chee et al., 2017; Sutherland & Calo, 2020; Davis et al., 2020). Rhizomatic, horizontal relationships are implicitly or explicitly underlined hinting at the pivotal role of desire and affect in producing relational effects such as tourist experience (Obrador, 2003), protest against tourism (Bruttomesso, 2018), imagined relational capital (Saxena, 2015), communal experience (Waitt & Duffy 2010), geopolitical experience of places (Mostafanezhad & Promburom, 2018), place- and frontier-making (Mostafanezhad, 2020).

Particularly interesting are studies that negotiate Deleuzean machinic assemblages foregrounding the agentive role of nonhuman components, their reciprocal relationships with human components, the co-constitution of components with the emergent assemblage that defines a, however ephemeral, place/territory. This is exemplified in Anderson's (2012) surfed wave (surfers-boards-waves), Waitt & Duffy's (2010) festival places (bodies, music, affect, emotion), Rhoden & Kaaristo's (2020) water-boat-human assemblage, Wilson et al.'s (2019) campervan human/machine assemblage.

A distinct group of studies treats tourism as a component of wider socio-spatial assemblages. Tourism is formally or informally employed as a force that opens new lines of flight, transforming urban (King & Dovey, 2013; Hammelman & Montoya, 2020) or rural areas (Rosin et al., 2013).

In particular, Mostafanezhad (2014; 2020) has delved more deeply into the deployment of tourism in geopolitical projects playing material and symbolic roles in frontier-making practices. Tourism frontiers, enrolling tourists' affective experiences, state and tourism narratives, historical representations, promotions, land and infrastructure development projects, participate as components in territorialising processes facilitating the enclosure of new (overseas) land frontiers (Mostafanezhad & Promburom, 2018; Mostafanezhad, 2020).

Processes of assembly. Several studies implicitly or explicitly address the territorialisation processes underlying the coming together of components to provisionally compose (or de/re-compose) tourism assemblages. The material, corporeal, incorporeal and discursive practices that tourists, tourism entrepreneurs, intermediaries, the state and locals apply/perform while engaging in tourism supply- or demand-related activities, through which territorialisation transpires, are particularly foregrounded. These include daily routine, embodied, specialised, socio-cultural and spatial practices (Obrador, 2003; Saldanha, 2006; Crofts Wiley et al., 2010; Anderson, 2012; Bruttomesso, 2018; Rhoden & Kaaristo, 2020), professional accreditation practices (Cousquer & Beames 2013), management practices (Saxena, 2015), land use practices (Rosin et al., 2013), tourism infrastructure construction, promotion/advertising, geopolitical imaginaries and discourses (Burrai et al., 2017; Mostafazenhad 2020). Sutherland & Calo (2020) also consider coding processes in (agritourism) assemblage formation and the particular role legislation plays.

Hammelman & Montoya (2020) address the use of tourism as a territorialisation strategy in urban development and the resultant construction of urban-rural borders generating inequalities and social exclusion. Rosin et al. (2013) discuss the introduction of tourism into declining rural economies that initially deterritorialises (offering lines of flight), unsettling the old economic order, and then reterritorialises (lines of articulation), transforming rural into agritourism economies. Other deterritorialising forces are natural and technological disasters, disease outbreaks, political unrest, terrorism, environmental, socio-cultural and institutional change, immigration/outmigration, population and infrastructure ageing, tourism saturation and differentiation (Briassoulis, 2017).

Flat ontology. AT applications pay high attention to the flat ontology of assemblages owing to the nature of tourism that materialises through the articulation of past, present and future components from all spatial levels. The micro(molecular) level behaviour of tourists, entrepreneurs and locals is linked to the macro(molar) level behaviour of the state, tour operators, NGOs and others in the space of tourism assemblages, underlining their distinct spatiality. Mostafanezhad (2014) conceptualises the popular humanitarian gaze in volunteer tourism as a multi-level geopolitical assemblage of institutions, cultural practices and actors (international celebrities). Mostafanezhad & Promburom (2018), studying film-induced tourism, show how geopolitical assemblages of tourism, co-constituted by geopolitical and tourism imaginaries, everyday tourism encounters and practices, media engagements and political-economic relations contribute to collectively constructing tourism imaginaries of places. Davis et al. (2020) explain the flat ontology of emergent geopolitical assemblages in the Pacific Islands as they simultaneously engage with multiple powers and their political, economic and social influences realised by material infrastructures. Heikkinen et al. (2016) discuss the ontological politics of global discourses (climate change) at ecotourism destinations. Kadfak & Knutsson (2017) analyse the blurred entanglements of local and global relations driving socio-material waterfront transformation in a small fishing community (India).

A limited number of studies explicitly discuss other important AT concepts, although these are implicit in several AT applications. These include the immanence of emergent assemblages

(Saldanha, 2006), the role of 'more-than-human' actants in (agritourism assemblages) identity construction (Sutherland & Calo 2020), and the composite and distributed agency of assemblage properties (Haanpää et al., 2021). Briassoulis (2017) argues that the identity of tourism assemblages reflects the situated balance between tourism demand and supply.

Lastly, in most studies assemblage is used as a "certain ethos of engagement with the world, one that experiments with methodological and presentational practices in order to attend to a lively world of differences" (Anderson & McFarlane, 2011: 126). The appeal of thinking with assemblages and multiplicities lies in their accord with the dynamic nature of tourism phenomena. The multi-sectoral and multiscalar dynamics of the industry, geopolitical interests, socio-economic, technological, lifestyle and environmental changes continuously open new lines of flight, transform socio-ecological milieus, create 'flat' territories, enunciating the co-production of tourism, tourists and places.

Orientation of AT applications. AT has mostly been used to conceptually underpin the study of a wide-ranging repertoire of tourist geographic issues. In certain cases, resourceful combinations of AT with existing theories – Heidegger and nonrepresentational theory (Obrador, 2003), Bourdieu (Sutherland & Calo, 2020), NMP (Rhoden & Kaaristo, 2020) – have produced rigorous and meaningful theoretical schemata to negotiate emergent tourism geographic formations and processes.

Certain studies explicitly use Deleuzoguattarian analytical tools to develop their methodology. Crofts Wiley et al. (2010) apply Deleuzean hydrological analysis to study how a subject's articulation to multiple assemblages produces non-Euclidean social space. Li & Chan (2020) employ the DG biographical approach to study diaspora tourism. Barry (2017) employs DG's diagram to study the formation of everyday tourist practices.

The policy-relevant line of analysis drawing on DeLanda's works is yet to be followed – namely, the identification of critical/limiting components, possibility space, attractors, basins of attraction and tendencies of assemblages and assemblage properties. The capacities and thresholds of (critical/limiting) components, such as water, climate, capital, technology, tourism demand, attractions, institutions/governance and culture, are highly uncertain and variable. They can be irretrievably diminished (due to heavy tourist usage and for other reasons, such as resource exhaustion, congestion, destruction of historical artefacts, coastal erosion, bankruptcies) or augmented (desalinisation augments drinking water supply, infrastructure works improve accessibility, favourable development laws augment tourism infrastructure, changing local attitudes open areas to tourism, etc.). Component capacities determine the possibility space of assemblages and reveal their tendencies towards particular types of tourism, development patterns, and so forth, thus, signalling on-going or impending changes (adaptation or transformation) of SEMs (Briassoulis, 2017).

The relationship of AT with ANT is occasionally unclear, owing perhaps to the use of DG concepts and the greater diffusion of ANT in tourism studies. Bell et al. (2011), Oakes (2012) and Bell et al. (2015) apply the concept of assemblage, but in the context of ANT. Podoshen (2017) relies on DeLanda, but uses 'assemblage theory' to refer incorrectly to Latour and ANT. Sutherland & Calo (2020) follow DeLanda but report ANT assumptions. Ormond & Kaspar (2018) refer to Latour and ANT to negotiate health care assemblages and cities as healing machines, entanglements of urban and transnational medical travel/tourism care elements and relations. Ivanova & Buda (2020) combine the Deleuzoguattarian 'rhizome' with ANT to study heritage tourism. Rhoden & Kaaristo (2020) and Coffin (2021) exhibit the same tendency.

These cases raise two issues. The first concerns the understanding of the differences between AT and ANT. Both are commonly grounded in post-structuralism and draw on Deleuzean philosophy but they differ on epistemological, ontological and methodological grounds. ANT espouses social constructionism, prioritises the network (relationships) over the components, starts from description to arrive at explanation, and provides direct practical guidance for tracing the actor-network that seems to be valued in empirical applications (Müller & Schurr, 2016). AT is closer to critical realism, attends to both components and their relationships, employs a rich conceptual/analytical repertoire for description and explanation, but does not offer direct practical guidance for applications. DeLanda's contribution has been instrumental in bringing the Deleuzoguattarian ideas to bear on real-world phenomena and guiding empirical applications in several contexts, including tourism studies (Briassoulis, 2017). Since 1999, however, ANT, represented by Latour and other scholars (e.g., A. Mol & J. Law), has undergone changes, incorporating Deleuzean concepts and relaxing rigid assumptions concerning the actor-network (Müller & Schurr, 2016). This may have confounded their differences and favoured their confusion in applications.

The second issue concerns the prospects of systematically, instead of *ad hoc* and selectively, combining and synthesising AT and ANT in tourism studies, an under-researched issue deserving further exploration. This is a broader and challenging task as it requires deep knowledge of both the Deulezean philosophy and ANT as well as resolution of their deeper epistemological differences. The exploration needs first to inquire whether it is possible to establish a unified conceptual/analytical repertoire fusing ANT with Deleuzean concepts that are central in the study of emergent socio-material formations, such as affect (capacities) and desire. The second question is whether it is possible to produce concomitant harmonised methodological approaches capitalising on and refining ANT's empirical analysis tools as well as potentially introducing nonlinear analysis tools (cf. Müller & Schurr, 2016; Briassoulis, 2017).

Assemblage Tourism Geographies

The post-2010s literature frequently calls for the adoption of a post-disciplinary, problem-focused, flexible outlook to studying tourism and tourism geographies (Wilson, 2012; Gale, 2012), because changes in tourism, and society at large, necessitate more open and less restrictive approaches to sensibly support its context-sensitive analysis and governance. Tourism is now ubiquitous, as the right to leisure/tourism is widely accepted. It is considered the world's largest industry, and it offers a versatile mechanism to promote socio-political goals at all levels. The definition of 'tourist' and 'tourism' keep on changing following the digitisation and hybridisation of technology, society and the economy that have blurred the boundaries between activities, mobilities and places, among others.

Assemblage tourism geographies, denoting the application of AT in tourism studies, is a response to these calls. AT prioritises the non-reductionist/relational ontology of assemblage that renders typologies, dichotomies, boundaries and hierarchies associated with reductionist, essentialist 'tourism-only' totalities pointless, focusing on the particular and the unique that better suits the study of dynamic and novel issues in tourism geographies as its growing usage reveals.

Conceptualising the objects of tourism analysis as multiplicities of assemblages urges researchers to renegotiate taken-for-granted conceptual, theoretical, methodological and policy/planning issues. It offers an alternative lexicon to build all-encompassing theoretical frameworks and elaborate non-deterministic, dynamic causal explanations to unpack their multiplicity, materiality, hybridity, complexity, processuality, contextuality, co-constitution, co-evolution, performativity, spatiality and immanence, thus broadening the scope of tourism studies. By implication, AT directs analysis away from tourism-only totalities towards the study of the pertinent tourism and non-tourism assemblages formed over time as vibrant and meaningful objects of reference and the judicious and masterful synthesis of quantitative and qualitative techniques to address their characteristics.

The AT approach has far-reaching management, planning and policy implications. One-size-fits-all solutions as well as spatial, tourism and sustainability fixes are dismissed given the uncertain and situated outcomes of planned interventions in contemporary socio-ecological milieus. Assemblage-based adaptive management and governance offer greater promise to guide the emergence of tourism and non-tourism assemblages co-existing in concert under conditions of spatial and temporal variability of tourism resources, inherently uncertain tourism demand and inevitable trade-offs among multiple conflicting interests (Briassoulis, 2017; 2018).

To date, most studies of tourism geographies rely on the Deleuzoguattarian and DeLanda version of assemblage theory, the latter being more accessible and understandable to the less initiated. The works of other students of DG (see, for example, the Deleuze Connections series (Buchanan, n.d.), might be usefully exploited to enrich the understanding and application of AT concepts in tourism analysis.

Although the post-2010s applications make greater use of AT concepts, several others remain little-used or not used at all despite their importance in theoretically informing tourism geographic studies: Desire, rhizome, affect (capacities), body-without-organs, molar/molecular, smooth/striated space and coding processes, to mention a few. Additionally, the methodological rigour and policy relevance of studies will improve if the analysis follows DeLanda's methodological guidance, as previously discussed.

AT provides an all-encompassing conceptual framework and apposite integrative constructs for substantive theory-building through the synthesis of multidisciplinary scientific and traditional/lay knowledge. Extant efforts can be extended to include other social, economic and cultural theories to offer complete explanations of the continuously emerging types of tourism, tourists, firms, destinations and experiences. Moreover, recasting Butler's (1980) "Tourism Area Life Cycle" and related models in an assemblage-based framework may help to more meaningfully trace the evolutionary path of destinations through the sequence of multiplicities of tourism and other assemblages formed over time disposing of the use of reductionist theories of pre-defined stages of development. AT applications need to cover many more geographic areas than is currently the case, and the continuously emerging forms of tourism. Tourism geopolitics transpires as a critical research niche given tourism's place in both geopolitical affairs and everyday experiences as well as its leading role and deployment in the pursuit of overt and covert development goals (Mostafanezhad et al., 2021).

Assemblage tourism geographies, by dint of the analytic power of assemblage, impart a distinctive perspective to the study of tourism geographies that goes beyond conceptual, spatio-temporal, sectoral and other boundaries, categories and fixes. By capturing the continuous becoming, co-constitution and co-evolution of human and nonhuman actors and the territories forming and focusing on what their assembling can do, it permits a kind of veritable and immanent analysis of tourism phenomena that is often difficult to do otherwise.

References

Anderson, J. (2012). Relational places: The surfed wave as assemblage and convergence. *Environment and Planning D, 30*, 570–587.

Anderson, B., & MacFarlane, C. (2011). Assemblage and geography. *Area, 43*(2), 124–127.

Anderson, B., Kearnes, M., MacFarlane, C., et al. (2012). On assemblages and geography. *Dialogues in Human Geography, 2*(2), 171–189.

Ateljevic, I., Pritchard, A., Morgan, N. (Eds.) (2007). *The Critical Turn in Tourism Studies; Innovative Research Methods*. Elsevier.

Barry, K. (2017). Diagramming: A creative methodology for tourist studies. *Tourist Studies, 17*(3), 328–346.

Bell, D., Holliday, R., Jones, M., et al. (2011). Bikinis and bandages: An itinerary for cosmetic surgery tourism. *Tourist Studies, 11*(2), 139–155.

Bell, D., Holliday, R., Ormond, M., et al. (2015). Transnational healthcare, cross-border perspectives. *Social Science & Medicine, 124*, 284–289.

Bianchi, R. V. (2009). The 'critical turn' in tourism studies: A radical critique. *Tourism Geographies, 11*, 484–504.

Bonta, M., & Protevi, J. (2004). *Deleuze and Geophilosophy: A Guide and Glossary*. Edinburgh University Press.

Briassoulis, H. (2017). Tourism destinations as multiplicities: The view from assemblage thinking. *International Journal of Tourism Research, 19*(3), 304–317.

Briassoulis, H. (2018). Tourism destinations as multiplicities: The governance challenges. Proceedings, International Conference on Tourism (ICOT2018) "Emerging Tourism Destinations: Working towards Balanced Tourism Development" (pp. 16–36).Kavala, Greece, 27–30 June 2018.

Bruttomesso, E. (2018). Making sense of the square: Facing the touristification of public space through playful protest in Barcelona. *Tourist Studies, 18*(4), 467–485.

Buchanan, I. (Ed.) (n.d.). *Deleuze Connections*. Edinburgh University Press. https://edinburghuniversitypress.com/series-deleuze-connections)

Burrai, E., Mostafanezhad, M., & Hannam, K. (2017). Moral assemblages of volunteer tourism development in Cusco, Peru. *Tourism Geographies, 19*(3), 362–377.

Butler, R. W. (1980). The concept of a tourist area cycle of evolution: Implications for management of resources. *Canadian Geographer, 24*, 5–12.

Chee, H. L., Whittaker, A., & Por, H. H. (2017). Medical travel facilitators, private hospitals and international medical travel in assemblage. *Asia Pacific Viewpoint, 58*(2), 242–254.

Coffin, J. (2021). Posthuman phenomenology: What are places like for nonhumans? In D. Medway, G. Warnaby, & J. Byrom (Eds.), *A Research Agenda for Place Branding* (pp. 183–200). Edward Elgar.

Cohen, E., & Cohen, S. A. (2012). Current sociological theories and issues in tourism. *Annals of Tourism Research, 39*(4), 2177–2202.

Cousquer, G. O., & Beames, S. (2013). Professionalism in mountain tourism and the claims to professional status of the International Mountain Leader. *Journal of Sport & Tourism, 18*(3), 185–215.

Crang, M. (2014). Cultural geographies of tourism. In A. A. Lew, C. M. Halll, & A. M. Williams (Eds.), *The Wiley-Blackwell Companion to Tourism* (pp. 66–77). Wiley.

Crofts Wiley, S. B., Sutko, D. M., & Becerra, T. M. (2010). Assembling social space. *The Communication Review, 13*(4), 340–372.

Davis, S., Munger, L. A., & Legacy, A. J. (2020). Someone else's chain, someone else's road: U.S. military strategy, China's Belt and Road Initiative, and island agency in the Pacific. *Island Studies Journal, 15*(2), 13–36.

Debbage, K. G., & Ioannides, D. (2012). The economy of tourism spaces: A multiplicity of 'critical turns'? In J. Wilson (Ed.), *Routledge Handbook of Tourism Geographies* (pp. 149–156). Routledge.

DeLanda, M. (2002). *Intensive Science and Virtual Philosophy*. Continuum.

DeLanda, M.. (2006). *A New Philosophy of Society: Assemblage Theory and Social Complexity*. Continuum.

DeLanda, M. (2011). *Philosophy and Simulation: The Emergence of Synthetic Reason*. Continuum.

Deleuze, G., & Guattari, F. (1984). *What is Philosophy?* Columbia University Press.

Deleuze, G., & Guattari, F. (1987). *A Thousand Plateaus: Capitalism and Schizophrenia*. University of Minnesota Press.

Duncan, T. (2012). The 'mobilities turn' and the geography of tourism. In J. Wilson (Ed.), *Routledge Handbook of Tourism Geographies* (pp. 113–119). Routledge.

Edensor, T. (2000). Staging tourism: Tourists as performers. *Annals of Tourism Research, 27*(2), 322–344.

Foley, R. (2014). The Roman-Irish bath: Medical/health history as therapeutic assemblage. *Social Science & Medicine, 106*, 10–19.

Franklin, A. (2004). Tourism as an ordering: Towards a new ontology of tourism. *Tourist Studies, 4*(3), 277–301.

Gale, T. (2012). Tourism geographies and post-structuralism. In J. Wilson (Ed.), *Routledge Handbook of Tourism Geographies* (pp. 37–45). Routledge.

Haanpää, M., Salmela, T., García-Rosell, J.-C., & Äijälä, M. (2021). The disruptive 'other'? Exploring human-animal relations in tourism through videography. *Tourism Geographies, 23*(1–2), 97–117.

Hall, C. M., & Page, S. J. (2009). Progress in tourism management: From the geography of tourism to geographies of tourism – A review. *Tourism Management, 30*, 3–16.

Hall, C. M., & Page, S. J. (2012). From the geography of tourism to geographies of tourism. In J. Wilson (Ed.), *Routledge Handbook of Tourism Geographies* (pp. 9–25). Routledge.

Hammelman, C., & Saenz-Montoya,A. (2020). Territorializing the urban-rural border in Medellín, Colombia: Socio-ecological assemblages and disruptions. *Journal of Latin American Geography, 19*(2), 36–59.

Heikkinen, H. I., Acosta Garcia, N., & Sarkki, S. (2016). Context-sensitive political ecology to consolidate local realities under global discourses: A view for tourism studies. In S. Nepal, & J. Saarinen (Eds.), *Political Ecology and Tourism* (pp. 211–224). Routledge.

Holliday, R., Bell, D., Cheung, O., et al. (2014). Brief encounters: Assembling cosmetic surgery tourism. *Social Science and Medicine, 124*, 298–304.

Holloway, J. (2010). Legend-tripping in spooky spaces: Ghost tourism and infrastructures of enchantment. *Environment and Planning D, 28*, 618–637

Ioannides, D., & Debbage, K. G. (2014). Economic geographies of tourism revisited: From theory to practice. In A. A. Lew, C. M. Hall, & A. M. Williams (Eds.), *The Blackwell-Wiley Companion to Tourism* (pp. 107–119). Wiley.

Ivanova, M., & Buda, D.-M. (2020). Thinking rhizomatically about communist heritage tourism. *Annals of Tourism Research, 84*, 103000.

Jóhannesson, G. T. (2005). Tourism translations: Actor–network theory and tourism research. *Tourist Studies, 5*(2), 133–150.

Kadfak, A., & Knutsson, P. (2017). Investigating the waterfront: The entangled sociomaterial transformations of coastal space in Karnataka, India. *Society & Natural Resources, 30*(6), 707–722.

King, R., & Dovey, K. (2013). Interstitial metamorphoses: Informal urbanism and the tourist gaze. *Environment and Planning D, 31*, 1022–1040.

Lapointe, D. (2024). *Alterity, mobility and territory: Conceptualising tourism space in a world in flux.* In J. Wilson & D. K. Müller (Eds.), *The Routledge Handbook of Tourism Geographies.* 2nd edition. Routledge.

Lew, A. A., Hall, C. M., & Williams, A. M. (Eds.) (2014). *The Wiley-Blackwell Companion to Tourism.* John Wiley and Sons.

Li, T. M. (2007). Practices of assemblage and community forest management. *Economy and Society, 36*(2), 263–293.

Li, T. E., & Chan, E. T. H. (2020). Diaspora tourism and well-being over life-courses. *Annals of Tourism Research, 82*, 102917.

McCann, E. (2011). Veritable inventions: Cities, policies and assemblage. *Area, 43*(2), 143–147.

Mostafanezhad, M. (2014). Volunteer tourism and the popular humanitarian gaze. *Geoforum, 54*, 111–118.

Mostafanezhad, M., & Promburom, T. (2018). 'Lost in Thailand': The popular geopolitics of film-induced tourism in northern Thailand. *Social & Cultural Geography, 19*(1), 81–101.

Mostafanezhad, M. (2020). Tourism frontiers: Primitive accumulation, and the "free gifts" of (human) nature in the South China Sea and Myanmar. *Transactions Institute British Geographers, 45*(2), 434-447.

Mostafanezhad, M., Azcárate, M. C., & Norum, R. (Eds.) (2021). *Tourism Geopolitics: Assemblages of Infrastructure, Affect, and Imagination.* The University of Arizona Press.

Müller, M., & Schurr, C. (2016). Assemblage thinking and actor-network theory: Conjunctions, disjunctions, cross-fertilisations. *Transactions of the Institute of British Geographers, 41*, 217–229.

Oakes, T. (2012). Touring modernities: Disordered tourism in China. In C. Minca, & T. Oakes (Eds.), *Real Tourism* (pp. 103–122). Routledge.

Obrador, P. (2003). Being-on-holiday: Tourist dwelling, bodies and place. *Tourist Studies, 3*(1), 47–66.

Ormond, M., & Kaspar, H. (2018). Medical travel/tourism and the city. In I. Vojnovic, A. L. Pearson, G. Asiki, et al. (Eds.), *Global Urban Health* (pp. 182–200). Routledge.

Podoshen, J. S. (2017). Trajectories in Holocaust tourism. *Journal of Heritage Tourism, 12*(4), 347–364.

Ren, C. (2010). Assembling the socio-material destination. In G. Richards, & W. Musters (Eds.), *Cultural Tourism Research Methods* (pp. 198–206). CABI.

Rhoden, S., & Kaaristo, M. (2020). Liquidness: Conceptualising water within boating tourism. *Annals of Tourism Research, 81*, 102854.

Robbins, P., & Marks, B. (2010). Assemblage geographies. In S. Smith, S. A. Marston, & R. Pain (Eds.), *The Sage Handbook of Social Geographies* (pp. 176–194). Sage.

Rosin, C., Dwiartama, A., Grant, D., & Hopkins, D. (2013). Using provenance to create stability: State-led territorialisation of Central Ottago as assemblage. *New Zealand Geographer, 69,* 235–248.

Saarinen, J. (2004). 'Destinations in change': The transformation process of tourism destinations. *Tourist Studies, 4*(2), 161–179.

Saldanha, A. (2006). Vision and viscosity in Goa's psychedelic trance scene. *ACME, 4*(2), 172–193.

Saxena, G. (2015). Imagined relational capital: An analytical tool in considering small tourism firms' sociality. *Tourism Management, 45,* 109–118.

Sheller, M., & Urry, J. (Eds.) (2004). *Tourism Mobilities: Places to Play, Places in Play*. Routledge.

Sommer, C. (2019). What begins at the end of urban tourism, as we know it? *The Urban Transcripts Journal, 2*(1).

Sommer, C., & Kip, M. (2019). Commoning in new tourism areas: Co-performing evening socials at the Admiralbrócke in Berlin-Kreuzberg. In T. Frisch, C. Sommer, L. Stoltenberg, & N. Stors (Eds.), *Tourism and Everyday Life in the Contemporary City* (pp. 211–231). Routledge.

Sutherland, L.-A., & Calo, A. (2020). Assemblage and the 'good farmer': New entrants to crofting in Scotland. *Journal of Rural Studies, 80,* 532–542.

Van der Duim, R. (2007). Tourismscapes: An actor-network perspective. *Annals of Tourism Research, 34*(4), 961–976.

Van der Duim, R., Ren, C., & Jóhannesson, G. T. (2013). Ordering, materiality, and multiplicity: Enacting actor–network theory in tourism. *Tourism Studies, 13*(1), 3–20.

Waitt, G., & Duffy, M. (2010). Listening and tourism studies. *Annals of Tourism Research, 37*(2), 257–277.

Wilson, J. (2012). Tourism geographies in a post-disciplinary age. In J. Wilson (Ed.), *Routledge Handbook of Tourism Geographies*. (pp. 251–254). Routledge.

Wilson, S., Chambers, D., & Johnson, J. (2019). VW campervan tourists' embodied sonic experiences. *Annals of Tourism Research, 76,* 14–23.

8

CRITICAL STUDIES OF GENDER EQUALITY AND SUSTAINABLE DEVELOPMENT IN TOURISM GEOGRAPHIES

Xu Honggang and Xiaojun Fan

Introduction

Tourism is usually regarded as a gendered landscape, and naturally, an understanding of tourism cannot be separated from the gender perspective. Tourism is a product of gendered societies, and its processes are gendered in their construction, presentation and consumption (Pritchard & Morgan, 2000). Awareness of the gender–tourism nexus has gradually grown in academic society, resulting in gender becoming one of the most important issues in tourism geography studies today.

On the practical side, gender, tourism and sustainable development have been linked to the UN Sustainable Development Goals (SDGs). In the UN's publication on the SDGs, gender equality is one of the 17 mentioned goals (UN, 2014). The 2030 Agenda on Sustainable Development approved by the United Nations (UN) has created a new path aimed at sustainable development, including eradicating poverty, protecting the planet and ensuring the prosperity of humanity. These 17 SDGs are aimed at creating a world not only for current but also for future generations and SDG 5 aims to achieve gender equality and empowerment of women and girls. SDG 5 lists six targets to achieve this:

(1) End all forms of discrimination against all women and girls everywhere;
(2) Eliminate all forms of violence against all women and girls in public and private spheres, including trafficking and sexual and other types of exploitation;
(3) Eliminate all harmful practices, such as child, early and forced marriage and female genital mutilation;
(4) Recognise and value unpaid care and domestic work through the provision of public services, infrastructure, and social protection policies and the promotion of shared responsibility within the household and the family as nationally appropriate;
(5) Ensure women's full and effective participation and equal opportunities for leadership at all levels of decision making in political, economic and public life;
(6) Ensure universal access to sexual and reproductive health and reproductive rights.

DOI: 10.4324/9781003286301-10

Gender equality is basically essential for achieving the other goals (Boluk et al., 2019), especially SDG 6: Clean Water and Sanitation, and SDG 8: Decent Work and Economic Growth (Alarcon & Cole, 2019).

Tourism is also regarded as an important path toward sustainable development. It is referenced in three of the Global Goals: SDG 8: Decent Work and Economic Growth, SDG 12: Responsible Consumption and Production, and SDG 14: Life Below Water. Tourism also has a direct impact on efforts to achieve other goals, such as eradicating poverty, advancing gender equality and protecting the environment. The global context of sustainability, reflected in the SDGs, makes it imperative to integrate gender equality within the SDGs and tourism development (Alarcon & Cole, 2019).

This chapter aims to examine the SDGs at the intersection of gender, tourism and sustainable development. It does not intend to provide a systematic review of all the papers related to gender, tourism and sustainable development. Instead, we attempt to discuss some key issues in this area and draw attention to extant selected research findings to advance knowledge in this field.

Current Under-researched Topics

Gender analysis of tourism emerges from the development of general discussions on gender. Margaret Swain(1995) edited a special issue of *Annals of Tourism Research* on *Gender in Tourism*, which included 17 papers and addressed two main parameters of gender relations in tourism: ideologies of masculinity and femininity in host and guest populations; and social divisions of labour, power and sexuality. The empirical focus was on countries such as the United States, Japan, Indonesia, Ecuador, Greece, Cyprus, Mexico, Spain and Jamaica. Subsequently, Sinclair (1997) edited a volume examining the interfaces between gender, work and tourism through comparisons across international contexts to consider how tourism, patriarchal systems and capitalism constrain and enable women's empowerment. In this case, geographical coverage extended to northern Cyprus, Bali, Mexico and the Philippines, South-East Asia and Japan.

Xu and Gu (2018) undertook a review of studies on gender and tourism development in China and found that the number of such studies is rising. Most agree that tourism contributes to gender equality within communities, but in diverse patterns. In 2018, Anatolia published a special issue comprising five papers on *Gender and the Tourism Academy*. The objective was to unpack the nature, roles and effects of gender (in)equality in our tourism academy and to call for greater self-reflexivity to achieve a more just and equitable tourism academy. In 2022, the *Journal of Sustainable Tourism* published a special issue on gender and tourism sustainability organised by Claudia Eger, Ana Maria Munar and Cathy Hsu, which contained 15 research papers. The geographic areas covered by the papers included Iran, Turkey, the UK, Nepal, Canada and the United States. The special issue took a critical approach to consider the gendered complexity of the world and engage in contextualising knowledge production. In particular, the traditional binary approach to gender, culture and nature, knowledge and emotion was critically examined.

All these special issues point out the commonality of the paucity of gender and tourism studies and a severe drought in studies on gender and sustainable tourism development. For this chapter, we conducted a search of the papers in six well-known journals on this subject: *Annals of Tourism Research (ATR), Tourism Management (TM), Journal of Travel Research (JTR), Journal of Sustainable Tourism (JOST), Current Issues in Tourism (CIT), and Tourism Geographies (TG)*. We conducted the subject search focusing on the themes of gender, tourism and SDGs in these six journals and found only 13 papers. Furthermore, 12 out of these 13 papers were published in JOST with the other paper from JTR.

Next, we broadened the research into the specific keywords listed within the SDGs, such as tourism and gender equality, gender, tourism and water, gender, tourism and poverty alleviation, and the 17 SDGs (see Table 8.1 for the results). All the journals have published papers on gender, tourism and sustainable development, with JOST being the leader in this field. TG does not perform particularly well in this area, as the journal only has 13 papers on gender and

Table 8.1 Summary of Gender and Tourism Topics in Six Major Tourism Journals

SDG goals	Theme	TM	JOST	ATR	CIT	JTR	TG
1 no poverty	gender, tourism, poverty alleviation	0	3	0	0	0	0
2 zero hunger	gender, tourism, hunger	0	1	0	0	0	0
2 zero hunger	gender, tourism, food	1	3	1	1	0	0
3 good health and well-being	gender, tourism, health	6	4	4	6	1	3
3 good health and well-being	gender, tourism, well-being	1	1	1	1	0	0
4 quality education	gender, tourism, education	9	10	5	6	2	1
5 gender equality	gender, tourism, equality	2	14	6	3	1	1
6 clean water and sanitation	gender, tourism, water	0	3	1	0	1	1
7 affordable and clean energy	gender, tourism, energy	0	0	0	0	0	0
8 decent work and economic growth	gender, tourism, work	13	20	16	10	0	4
8 decent work and economic growth	gender, tourism, economic growth	1	21	2	3	2	5
9 industry, innovation and infrastructure	gender, tourism, industry	19	18	3	8	2	3
10 reduced inequalities	gender, tourism, inequalities	3	5	4	0	1	5
11 sustainable cities and communities	gender, tourism, urban	2	1	2	2	0	1
11 sustainable cities and communities	gender, tourism, community	2	20	11	5	0	8
12 responsible consumption and production	gender, tourism, consumption	1	3	4	0	1	1
12 responsible consumption and production	gender, tourism, production	1	1	3	1	0	2
13 climate action	gender, tourism, climate	1	2	1	0	0	0
14 life below water	gender, tourism, marine life	0	0	0	0	0	0
15 life on land	gender, tourism, ecology	0	5	1	0	0	1
16 peace, justice and strong institutions	gender, tourism, institution	0	1	1	0	0	2
16 peace, justice and strong institutions	gender, tourism, justice	0	4	0	0	0	0
16 peace, justice and strong institutions	gender, tourism, power	2	17	4	2	0	6
17 partnerships for the goals	gender, tourism, partnership	0	0	0	0	0	0
	gender, tourism, sustainable development goals	0	12	0	0	1	0
Total (excluding redundancy)		39	60	43	32	9	13

Source: Compiled by Authors.

sustainable development in tourism. Despite Xu's (2018) call for more studies on gender and sustainable development from a tourism geographies perspective, little progress has been made in this regard.

The majority of these papers are qualitative studies. A few studies also adopted an economic modelling approach to test the relationship between tourism development and gender equality. For instance, in their analysis of available data from the United Nations Development Programme (UNDP), Mitra et al. (2022) found that tourism generally supports gender equality with an exception at the 10th quantile, where tourism has a negative and significant correlation to gender inequality. Low-income countries with low levels of tourism development are unable to derive the benefits of tourism's potential to reduce gender inequality. Zhang and Zhang (2020) also arrived at similar conclusions based on data from 36 Asian countries collected from 2006–2018, using a system-generalised method of moments estimation approach. Their statistical modelling provided very positive results. In both studies, aggregate indexes are used.

Gender inequality used in Zhang and Zhang's (2020) research is a composite of four fundamental categories: economic participation and opportunity, educational attainment, political empowerment, and health and survival. By contrast, Mitra et al. (2022) employ a composite index reflecting inequality of achievement between women and men in three dimensions: reproductive health, empowerment and the labour market. These aggregated data sets do not really reflect gender inequality in practice. Such research findings require further investigation and further qualitative research to obtain an in-depth understanding of gender inequality data that otherwise is not revealed through statistical sets.

In order to achieve this, we narrowed our focus to a few key themes considered relevant to gender and sustainable development from the perspective of tourism geographies. These include: (1) gender equality in the workplace, particularly in the hospitality sector; (2) gender equality in community tourism; (3) gender issues in consumption practices; (4) knowledge, technology and gender equality. These selected themes were chosen considering the SDGs that touch on gender equality and studies on tourism and sustainability.

Gender equality in the workplace

Home space and workspace are traditional gender-segregated domains, as women are required to work in the kitchen and at home. Therefore, working in formal businesses has the potential to free women from households and expose them to a wider social and public life. Similarly, working in formal tourism enterprises, especially in hospitality, can be considered the first step toward gender equality. Nevertheless, research findings reveal a complex picture of mixed results.

Women are prominent in the tourism workforce, but mainly in low and precarious positions. Women are severely underrepresented in management or on boards and face obstacles and challenges in pursuing higher career opportunities. Local women, in particular, have difficulty accessing the new development opportunities offered by tourism. Benefits tend to flow to immigrant groups and men. The prevailing conditions of culture, education, race, gender, politics, history, location, mobility, socio-economy, tourism skills and knowledge constitute major barriers (Schellhorn, 2010). A study of the number of women on the boards of large Spanish hotel chains (Carrasco-Santos et al., 2020) found that a large number of women work in the hotel, yet gender equality has not yet been achieved at higher echelons since senior management positions are dominated by males. This is due to the fact that gender inequalities are reflected in religious teachings, cultural habits and outdated ways of thinking, as well as in the workplace.

A year-long qualitative study of the Women's Mentoring Programme in the UK hospitality industry, based on data from 71 interviews with 13 mentors and 14 mentees, revealed that there has always been a gender barrier when women negotiate their career path with hotel chains. There are issues in the hospitality industry regarding employee motivation, commitment and retention, as well as an entrenched glass ceiling that limits career opportunities for many women. Mentoring programmes can help to support and develop staff, and it is imperative to provide individual support and challenge gendered discourse in the hospitality profession to help women overcome gender barriers in career development and in confronting gender inequality in the hospitality industry (Dashper, 2019). However, women's associations can provide great support to female professionals through training, courses and meetings, give visibility to gender issues and actions and expand professional networks and access to resources (Freund and Hernandez-Maskivker, 2021).

Women also face gender-related emotional and psychological problems in the workplace. Sexual harassment is widespread in the service industry, especially in tourism and hospitality (Morgan & Pritchard, 2019; Poulston, 2008). There are several distinct characteristics of jobs in the tourism and hospitality industry that can trigger sexual harassment. These include long working hours, fluctuating shifts, a highly gendered environment and a pronounced power imbalance (Gilbert et al., 1998; Ineson et al., 2013).

Sexual harassment is considered a barrier to gender equality. A survey of 221 Turkish tour guides reveals that female tour guides' experience of sexual harassment negatively impacts on their job satisfaction and psychological well-being (Alrawadieh et al., 2021). The possibility of customer sexual harassment has a negative impact on the social and economic sustainability of the industry, including employee well-being, and can lead to employee turnover. A qualitative phenomenological study of interviews at Australian spas suggests that some therapists feel obliged to hide their true feelings when experiencing sexual harassment from clients, which can lead to mood disorders and emotional harm (Frost et al., 2021).

Gender stereotypes are prevalent in tourism and play a role in preventing gender equality. A study based on a survey of 392 employees at 4- and 5-star hotels in Jordan (Koburtay & Syed, 2019) shows that employees stereotype successful leaders as masculine rather than feminine and believe that women, besides their feminine traits, should have masculine traits too, while gender equity practices and leadership development programmes (LDPs) can mitigate biased evaluations of women leaders. The same conclusion is drawn in a qualitative analysis exploring gender roles, stereotypes, tourism productivity and spatial relationships through interviews with 49 Mexican women from 2015 to 2018 (Díaz-Carrión & Vizcaino, 2022). By analysing how emotion mediates women's life experience in sexualised rural tourism work, increasing women's participation in rural tourism to cope with or resist traditional gender roles and stereotypes is suggested.

Overall, studies on the issue of gender equality in the workplace remain limited. A comparison of 102 academic papers and 122 industry documents reveals that both types of study lack cultural and individual dimensions that focus on the gendered process, instead focussing explicitly on organisational systems and developed economies (Je et al., 2020).

Gender Equality in Community Tourism

Community tourism plays a key role in sustainable tourism. The community enables women to integrate their domestic service and tourism. Tourism development has a significant impact on women, in providing employment opportunities, entrepreneurship opportunities and so on. Until recently, most studies on gender and community tourism were carried out through a development framework, in that they attempted to explore the relationships between women's participation in tourism and improvements in their status. Women participating in such community-based tourism projects experience significant benefits at all levels of control, awareness-raising, access and welfare (Reimer & Walter, 2013, Xu and Gu 2018). By participating in cultural tourism, Botswanan women significantly obtain a sense of empowerment expressed in terms of freedom from economic dependency on men and society (Moswete & Lacey, 2015). A study investigating women's participation in a community-based tourism project in northern Vietnam highlights a more equitable division of labour, increased income, self-confidence and community involvement, and new leadership roles for women. Social class inequality, childcare and violence against women still prominently persist (Tran & Walter, 2014). Furthermore, recent studies adopt a more critical view on the relationship between community tourism and gender equality, since gender equality has to be considered in relation to the woman herself, the household structure, community structure, the tourism industry and society as a whole. It is always the intersection of these factors which leads to the overall outcome between tourism and gender.

The intersection of women, family and community in developing countries also leads to different conclusions from community tourism development. Since women are associated with household well-being, improving household welfare and the economic situation does not automatically lead to gender equality (Xu & Gu, 2018). Development projects are often designed so that women are empowered to work for development rather than to improve their status and well-being. They become instruments in these development projects and policies (Tucker & Boonabaana, 2012). The business goals they pursue are largely limited to generating a subsistence income to enable women to fulfil their obligations of caring for immediate and extended family members and community associations (Kimbu & Ngoasong, 2016). Studies also show an apparent gendered division of labour in community tourism; in some situations, traditional gender roles can even be strengthened. For instance, cooks employed in tourism development projects are women, and women also clean, wash clothes, make beds and do much of the 'emotional labour' of entertaining guests at host homes and hotels (see also Roelofsen, Chapter 25 of this volume). Men perform most of the other paid jobs generated by ecotourism projects: they work as tour guides, accountants, administrators, patrol officers, boatmen, motorcycle drivers, ox-cart drivers, garbage collectors, construction workers, porters and furniture manufacturers. This discriminatory gender division of labour is actually a typical pattern in ecotourism community development projects, where cultural and gender norms in patriarchal societies, as well as the threat of violence from men, still exist(Reimer & Walter, 2013). Thus, women can be forced to bear the 'double burden' of cooking and cleaning again at home after work (Reimer & Walter, 2013).

Other studies have arrived at similar conclusions. In Kazakh nomad families, female participation in tourism has led to a double burden on women; they have to provide service to both guests and their families (Talinbayi et al., 2018). A study conducted on the coast of Galicia (Spain) discovered that this activity might provide additional income for the family unit, but the practice also reinforces traditional female roles, assigning them homemaker tasks (cleaning, cooking, customer service, etc.), burdening them with a more significant workload and extending their workday (LaPan et al., 2021). The time women could devote to leisure and maintaining social relationships is significantly reduced. However, women in developing countries do not necessarily see men or children as oppressive forces in their lives, and they tend to associate their well-being with that of the whole family. Their well-being can be improved with the social mobilities of their families. Thus, they do not necessarily attempt to break with the gender norms in their society.

The relationship between community tourism development and gender equality is dynamic and diverse. Tourism development has influenced not only women in the community but also men. In the case of Mosuo's matrilineal ethnic community, tourism development means that men obtain more chances to participate in public affairs than women and have begun to play more important roles than before (Liu, 2001). This is because the tourist industry outside the community is patriarchal and men are often expected to participate in meetings and activities with the other stakeholders.

Nevertheless, women participating in tourism clearly show their self-development in many aspects. Indeed, participation in the tourism business, with new job opportunities and higher incomes, has changed women's individual and gender awareness (Tang & Zhu 2007). Tourism development also provides the conditions for the formation of feminine subjectivity, where hidden identities, feelings and negotiations are slowly being revealed. Participation in community tourism provides conditions for the formation of feminine subjectivities. Furthermore, the feminine subjectivity is iterated from the individual to the collective level (Wang & Sun, 2022). In a Bai village, in China, it was found that Bai women have become the main business operators and repositories of Bai culture (Yin, 2015). With their greater involvement in business, their activities now extend beyond domestic affairs, with more engagement in the public affairs of their community.

Through managing the business and seeing their work valued in tourism, Kazakh women also tend to be aware of the value of their domestic work (Talinbayi et al., 2018). Both craftswomen and Kazakh women operating yurk tourism activate displays of pride and resentment to stress their position and status as working women with a certain level of interdependence, as opposed to the stereotype of the passive 'fine lady', economically dependent on her husband (Jimenez-Esquinas & Guadalupe, 2017). Their attitudes towards raising their children have also changed (Yin, 2015, Moswete & Lacey, 2015).

Leadership is one of the most critical indicators reflecting gender equality. Through continuous efforts and struggle, women can gain leadership positions. For instance, Nepali female entrepreneurs were found to be more sensitive and in being so, could successfully overcome the challenges of intersections between gender, entrepreneurship and emancipation. They negotiate obstacles and upend the dominant paradigm (Hillman & Radel, 2022). Leadership traits displayed by women are quite different from men. Women have evolved from passive to active participation in culture-related tourism ventures (Wang, 2008; Chen & Zhang, 2015; Moswete & Lacey, 2015).

Women are often social entrepreneurs, and they combine social transformation and business goals while meeting the needs of specific tourism communities. Women and the small businesses they manage are typically social cobblers and/or social constructors. By using locally available resources, they identify locally discovered opportunities and create small tourism businesses. These businesses provide valuable solutions to local social problems. In this process, they shift from the 'traditional' role of the housewife to the owner-manager in a male-dominated environment and employ orphans, widows and young girls.

Studies also point out potential policies that could empower them at all levels. In comparison with the actions listed in the SDGs, actions are often addressed to undertake reforms to give women equal rights to economic resources, as well as access to ownership and control over land and other forms of property, financial services, inheritance and natural resources, in accordance with national laws. Policies are suggested that provide education programmes to enhance women's personal capacity and financial programmes targeted at women; decentralise tourism attractions; and enable women to access attractions (Moswete and Lacey, 2015). Alternative and small-scale tourism programmes are favoured for women to start their own businesses. Men should be encouraged to become facilitators and partners in women's tourism (Moswete & Lacey, 2015).

Gender and Tourism Consumption

Tourism is different from other industries in that consumption and production are integrated. The understanding of gender equality cannot be separated from the perspective of consumption. First of all, gender is one input factor in the production process of tourism products and places and therefore becomes a target of consumption. This gender consumption is clearly seen in the tourism landscape as a destination image. Examining the nature of the western-dominated global advertising industry and evaluating its impact on destination self-presentation reveals the construction of the 'male' gaze, favouring the privileged gaze of white, heterosexual males. Post-structuralist analysis uncovers the patriarchal construction in tourism, such as marketing the destination for the benefit of the 'heterosexual male gaze' (Pritchard & Morgan, 2000).

Studies also show that men are constructed to attract women through economic power, while others indicate that gender is integrated into the production process by business organisations. In the aviation industry, which is interdependent with tourism, gender representation is shaped by organisational culture and specific business interests. Deconstructing the airline's complex gender representation on Instagram through a feminist post-structuralist approach, shows that airlines continue to objectify female employees and hyper-feminised cabin spaces (Smith et al., 2021).

Comparably, sex tourism is a well-researched area in gender and tourism geography. Various forms of sex tourism have been explored, including those with or without money exchange, with or without intercourse, male tourists or female tourists, tourists and hosts or between tourists. The initial study of sex tourism was conducted in contexts in which wealthy male tourists from developed countries meet poor hosts in developing countries for sexual interactions. In these cases, monetary exchanges and exploitation are often found. Studies on Western women's tourist–local relationships have diverse results; either they are similar to the overtly exploitative relationships of heterosexual male sex tourists, or they are differ as they involve a softer, caring element of romance (Bauer, 2014; Herold, Garcia & DeMoya, 2001; Jacobs, 2009; Ryan & Kinder, 1996; Weichselbaumer, 2012; Xu & Ye, 2016). In a recent study, Kock (2021) applied theories from evolutionary ecology and carried out an empirical study to test whether a local male-skewed sex ratio can affect men's attitude toward sex tourism. Sex tourism serves as compensatory behaviour for same-sex competition for mates. However, reducing attitudes to sex tourism to biological causes should be questioned and carefully examined.

Gender equality issues are also observed in tourism consumption. Women are involved in tourism consumption differently to men (Swain 1995; Xu & Wang, 2021). Gender impacts their subjective well-being (Pyke et al., 2019). An obvious example is girlfriend getaway tourism, which gives women opportunities to enjoy leisure time and mental benefits away from the household. It is the cultural expectation that women should take on more household and caregiving responsibilities that limits their participation in leisure activities. As a result, some women take vacations to shake off these responsibilities. Girlfriend getaways show how women adopt feminist practice and gain power in the tourism industry. It also allows women to free themselves from gender constraints and gain autonomy by creating a space where women can talk, listen and share emotions and experiences. Women experience freedom, relaxation and empowerment through intimacy. Beyond therapeutic relaxation devoid of gender oppression, women expand self-care practices to manage gender issues and foster group solidarity (Kong et al., 2022).

The intersection of age and gender can also lead to other conclusions. Xu and Wang's study (2021) shows that retired women can adjust nimbly to their situation as they enjoy travelling and leisure. They can also make friends easily at the destination. In comparison, most men cannot adjust easily to retired life and find a lack of meaning in their leisure time. In addition, their social capacity with strangers or other travellers is low and they feel lonely while travelling. Overall, retired women derive more well-being than men from seasonal mobilities.

The absence of men on these tours provides an opportunity for women to be more comfortable and less conscious as they can express their genuine selves, free from the male gaze. They can enjoy themselves instead of paying attention to standards of femininity such as make-up and clothing. This is supported by research on 11 middle-aged women who participated in a one-day all-women tour in Iran (Nikjoo et al., 2021). The findings reveal that the mood of women undergoing later life transitions may benefit from such commonplace tours with their peer group. The women's well-being can be improved by distance from family responsibilities, daily life and gender-related restrictions, as can their social and personal self-advancement. From this perspective, women's right to travel also falls within the context of travel sustainability. Attention should be paid to the travel rights of women in the light of United Nations World Tourism Organisation (UNWTO) perspectives on the SDGs (2015).

However, tourist behaviours cannot occur without public spaces, including travel spaces and destinations. Public spaces are not innocuous and objectively defined; rather, they are politicised, sexualised, subjective and gendered. In the case of certain destinations in Islamic countries, women may be socially prohibited from accessing the same leisure activities in tourist destinations as men (Zaytoun et al., 2010). Even when female tourists are allowed in tourist public spaces, their experiences are highly influenced and often involuntarily altered by unwanted male attention and sexual harassment (Brown & Osman 2017).

Knowledge Production and Gender Equality

Awareness and knowledge are the first steps to reducing gender inequality. Academic studies could play a critical role in identifying gender issues in tourism, unpacking mechanisms and proposing policies to address gender inequality in tourism. This importance and lack are very much emphasised by Pritchard (2018). However, gender awareness and gender inequality are also prevalent in the tourism academic society.

The marginalisation of gender studies in tourism results directly from the under-representation of female scholars in leadership and gatekeeping positions. According to the survey by Munar et al. (2015), which extensively maps gender gaps in tourism academia, half of tourism academics are women, but only 25 percent of travel journal editors, acting as knowledge gatekeepers in the field, are women. Further research (Walters, 2018) also reveals that women hold only 11 percent of professorial positions in the UK, 16 percent in Australia and 12 percent in New Zealand. The study highlights gender inequality in two roles, keynote speakers and honorary committee members, by analysing 53 global academic conferences on tourism, hotel and leisure in 2017.

Similar findings are also seen in Chinese tourism academia. With the rapid development of domestic tourism, female tourism scholars have gradually emerged, and female PhD graduates outnumber male graduates. Based on data published in *Tourism Tribune*, the top Chinese journal on tourism studies, Liu et al. (2019) found that research results for female tourism scholars have grown from rapid growth to stable development. However, gender inequality still exists in academia, in terms of the output per scholar and the first author ratio. Yet gender is not a well-recognised topic for both genders. Males continue to dominate tourism research in China.

Gender inequality impacts women scholars and knowledge creation in academic fields. At present, research on the representation of tourism knowledge and gender inequality in production shows that tourism scholars' awareness of gender issues is relatively low in tourism and tourism academic society in China (Xu et al., 2017, Liu et al., 2019). They accept the negative gender norms and unconsciously adopt biased social mores. The academic tourism community has also failed to lead in promoting the SDG on gender equality and, in particular, ensuring women's full and effective participation and equal leadership opportunities. Exploring the 121 events organised and/or held by the UNWTO in 2017 shows that non-Western women remained the least represented group in the UNWTO's knowledge production space, even though tourism is widely promoted as a tool for the empowerment and employment of women and girls, especially in non-Western developing countries (Khoo-Lattimore et al., 2019).

Tourism education institutions are also gendered places where gender is constructed, replicated and potentially challenged. Dashper et al. (2020) introduced a Ketso method to help faculty members break down barriers and discuss gender inequality issues. This exercise also reveals deep-seated gender power dynamics and structural and institutional barriers to reform (Dashper et al., 2020). Although such experiences cannot solve the problem, attempts to use the approach to begin such discussions should be encouraged.

Discussion and Conclusions

Gender equality is essential in sustainable development. The existing literature demonstrates that academically significant achievements have been obtained since papers on gender equality and sustainable development have increased, as an important topic in tourism studies. More and more papers appear in top tier journals. Existing studies clearly show there is no linear relationship between tourism development and gender equality. Gender equality also goes beyond what is listed in SDG 5. Gender and sustainable tourism are complicated and constitute a rich research field full of potential.

However, compared to the significance of the prevalence of the gender (in)equality issue in tourism and society, the studies are far from sufficient. Compared to other sectors, there is a general lack of debate, policy and activity on gender and sustainable development from the tourism scholars (Ferguson & Alarcon, 2015). This is especially astonishing among tourism geographers, who have a tradition of commitment to addressing key issues in sustainable development through their knowledge contribution.

Tourism geographers hold a strong and long-term interest in sustainability, including many of the core issues in the SDGs, such as poverty alleviation and inclusive development (SDGs 1 and 10), responsibility in tourism (SDG 12), tourism and climate change mitigation and adaptation (SDG 13) and sustainable use of natural resources in tourism (SDGs 14 and 15) (Saarinen, 2020). In addition, tourism geographers often adopt an intersectional approach to tourism, sustainable development, gender, race, and so forth. This intersectional approach to gender studies is vital in tourism and gender studies since it rejects reductionist views of women's experiences in tourism and the attendant power relationships that such an approach (re)produces (Chambers, 2022). Thus, tourism geographers should be able to contribute significantly to gender and sustainable tourism research. Here, we also propose a number of areas for future research.

In terms of research fields, several are particularly relevant. First, studies are needed to investigate the intersection of gender, tourism and the SDGs. Basically, both gender and tourism are relevant to all the SDGs and, therefore, the intersection of gender, tourism and the other 16 SDGs would provide a rich area of research. This has been shown in the studies of the gender tourism-water nexus. A review of the intersection of water, tourism and gender shows a dearth of research (Cole et al., 2020). The study explores the tourism–water nexus, comparing and contrasting literature published in English, Chinese and Spanish. Securing access to safe water for continued tourism development is a common theme and the vast majority of work has focused on hotels, including water pricing, water-saving practices and innovative management methods. On all the continents struggles are apparent and the unsustainability of tourism is having an impact on water quantity and quality. Studies also show that these impacts are disparately shared by gender.

Women tend to bear more of the costs of tourism than men (Cole, 2012; Cole, 2014; Cole, 2017; Cole & Browne, 2015; Cole & Ferguson, 2015). Future research and practice are encouraged to employ an intersectional lens for a more thorough and robust exploration of diversity issues. Tourism research on climate change (SDG 13) could also benefit from gender considerations (SDG 5), which influence people's experience of and resilience to climate change (Nunkoo et al., 2021). While society and the tourism world are particularly shaped by virtual technology, the intersection of tourism, gender equality and technology should also be investigated. Technology is not gender-free and can have different impacts on gender equality (Pritchard, 2018). This area has been largely ignored so far.

Second, the intersection of gender, race, age and gender minorities needs to be critically examined. It is important to include gender identity minorities to achieve sustainability. Although the SDGs do not explicitly discuss inclusivity in the context of LGBTIQ communities (Kaygalak-Celebi et al., 2022), they do identify the importance of inclusion in building peaceful and prosperous societies for all. SDGs 5 and 10 aim to reduce inequalities among minorities and within countries, while SDG 16 aims to promote peaceful societies where people feel safe despite their differences. However, a sense of belonging needs to be created to achieve the SDG of community inclusion for LGBTIQ minorities (Ong et al., 2021).

Third, we also agree with Pritchard (2018) that future research should pay particular attention to sexual harassment issues in tourism work and tourism consumption places. Tourism places have been reported to be a hotbed of sexual harassment but serious research is lacking.

Current studies also emphasise critical research into the context in which gender equality is explored. Currently, most research is concentrated in a few areas: Western countries, China, Indonesia, Thailand and Central America. A more widespread geographical area could be explored. Future studies should also borrow from studies in feminist theories. Over the last few years, geographers have begun to draw heavily on feminist theories with a focus on the body, embodiment and individual practices to show that individuals can challenge the power relationship through their embodied practices. These are also reflected in tourism studies. Tourism scholars need to adopt feminist scholarship more purposefully in order to address gender (in) equality in academia. Important initiatives in this regard include raising our consciousness about our implicit gender biases, self-reflexivity and the actual implementation of emancipatory policies (Munar, 2017).

In practice, Scheyvens (2018: 341) called for tourism geographers in particular and scholars in general "to consider how we might utilise the SDGs to analyse the linkages between tourism and sustainable development in a wide range of contexts and at different scales". Responding to the call, tourism geographers could also work with UN and national SDG systems to give our knowledge and research more progressive visions of development that pay more attention to gender issues (Liverman, 2018). However, there are many contradictions in SDGs, including the contradictions of gender and other goals (Tucker 2020). These contradictions need to be further clarified and debated. Alternative development visions could take gender into account and admit the inevitable trade-offs and contradictions between growth, gender equality and other goals. New sets of indicators, methods and strategies for assessing and supporting progress toward more gender-equitable SDGs could also be adopted. But of course, all these have to be built on the efforts of tourism geographers to carry out studies on sustainable tourism and gender from their own perspective.

References

Alarcon, D.M. & Cole, S. (2019). No sustainability for tourism without gender equality. *Journal of Sustainable Tourism* 27(7–9), 903–919.

Alrawadieh, Z., Alrawadieh D.D., Olya, H.G.T., Bayram G.E. & Kahraman, O.C. (2021). Sexual harassment, psychological well-being, and job satisfaction of female tour guides: the effects of social and organizational support, *Journal of Sustainable Tourism* 30(7), 1639–1657. https://doi.org/10.1080/09669 582.2021.1879819

Bauer, I.L. (2014). Romance tourism or female sex tourism? *Travel Medicine and Infectious Disease* 12(1), 20–28.

Boluk, K.A., Cavaliere, C.T. & Higgins-Desbiolles, F. (2019). A critical framework for interrogating the United nations sustainable development goals 2030 agenda in tourism. *Journal of Sustainable Tourism* 27(7–9), 847–864. https://doi.org/10.1080/09669582.2019.1619748

Brown, L. & Osman, H. (2017). The female tourist experience in Egypt as an Islamic destination. *Annals of Tourism Research* 63, 12–22. https://doi.org/10.1016/j.annals.2016.12.005

Carrasco-Santos, M.J., Rodríguez, C.C. & Rodríguez, E.R. (2020). Why is the spanish hotel trade lagging so far behind in gender equality? A sustainability question. *Sustainability* 12(11), 1–20. https://doi.org/ 10.3390/su12114423

Celebi, S.K., Ozeren, E. & Aydin, E. (2022). The missing link of the Sustainable Development Goals (SDGs) in tourism: A qualitative research on Amsterdam Pride. *Tourism Management Perspectives* 41. https://doi. org/10.1016/j.tmp.2022.100937

Chambers, D. (2022). Are we all in this together? Gender intersectionality and sustainable tourism. *Journal of Sustainable Tourism* 30(7), 1586–1601. https://doi.org/10.1080/09669582.2021.1903907

Cole, S. (2012). A political ecology of water equity and tourism. A case study from Bali. *Annals of Tourism Research* 39(2), 1221–1241. https://doi.org/10.1016/j.annals.2012.01.003

Cole, S. (2014). Tourism and water: From stakeholders to rights holders, and what tourism businesses need to do. *Journal of Sustainable Tourism* 22(1), 89–106. https://doi.org/10.1080/09669582.2013.776062

Cole, S. (2017). Water worries: An intersectional feminist political ecology of tourism and water in Labuan Bajo, Indonesia. *Annals of Tourism Research* 67, 14–24. https://doi.org/10.1016/j.annals.2017.07.018

Cole, S. & Browne, M. (2015). Tourism and water inequity in Bali: A social-ecological systems analysis. *Human Ecology* 43(3), 439–450. https://doi.org/10.1007/s10745-015-9739-z

Cole, S. & Ferguson, L. (2015). Towards a gendered political economy of water and tourism. *Tourism Geographies* 17(4), 511–528. https://doi.org/10.1080/14616688.2015.1065509

Cole, S.K.G., Mullor, E.C., Ma, Y. & Sandang, Y. (2020). Tourism, water, and gender — an international review of an unexplored nexus. *Wiley Interdisciplinary Reviews: Water* 7(4). https://doi.org/10.1002/ wat2.1442

Chen, L.Q. & Zhang, X.W. (2015). The local knowledge of Li nationality in ethnic tourism and the growth of female tourism elites. *Journal of Haihan Normal University* 28(2), 117–121.

Dashper, K. (2019). Mentoring for gender equality: Supporting female leaders in the hospitality industry. *International Journal of Hospitality Management* 88, 102397. https://doi.org/10.1016/ j.ijhm.2019.102397.

Dashper, K., Turner, J. & Wengel, Y. (2020). Gendering knowledge in tourism: Gender (in)equality initiatives in the tourism academy. *Journal of Sustainable Tourism* 30(7), 1621–1638. https://doi.org/10.1080/09669 582.2020.1834566

Díaz-Carrión, I.A. & Vizcaino, P. (2022). Mexican women's emotions to resist gender stereotypes in rural tourism work. *Tourism Geographies* 24(2–3), 244–262. https://doi.org/10.1080/14616688.2020.1867886

Ferguson, L. & Alarcon, D.M. (2015). Gender and sustainable tourism: Reflections on theory and practice. *Journal of Sustainable Tourism* 23(3), 401–416. https://doi.org/10.1080/09669582.2014.957208.

Freund, D. & Hernandez-Maskivker, G. (2021). Women managers in tourism: Associations for building a sustainable world. *Tourism Management Perspectives* 38. https://doi.org/10.1016/j.tmp.2021.100820

Frost, J.H., Ooi, N. & Van Dijk, P.A.(2021). 'Is he going to be sleazy?' Women's experiences of emotional labour connected to sexual harassment in the spa tourism industry. *Journal of Sustainable Tourism* 30(12), 2765–2784. https://doi.org/10.1080/09669582.2021.1942892

Gilbert, D., Guerrier, Y. & Guy, J. (1998). Sexual harassment issues in the hospitality industry. *International Journal of Contemporary Hospitality Management* 10(2), 48–53. https://doi.org/10.1108/0959611981 0207183

Herold, E., Garcia, R. & DeMoya, T. (2001). Female tourists and beach boys: Romance or sex tourism? *Annals of Tourism Research* 28(4), 978–997.

Hillman, W. & Radel, K. (2022). The social, cultural, economic and political strategies extending women's territory by encroaching on patriarchal embeddedness in tourism in Nepal. *Journal of Sustainable Tourism* 30(7), 1754–1775. https://doi.org/10.1080/09669582.2021.1894159

Ineson, E.M., Yap, M.H. & Whiting, G. (2013). Sexual discrimination and harassment in the hospitality industry. *International Journal of Hospitality Management* 35, 1–9. https://doi.org/10.1016/j.ijhm.2013.04.012

Jacobs, J. (2009). Have sex will travel: Romantic 'sex tourism' and women negotiating modernity in the Sinai. *Gender Place & Culture* 16(1), 43–61.

Je, J.S., Khoo, C. & Yang, E.C.L. (2020). Gender issues in tourism organisations: Insights from a two-phased pragmatic systematic literature review. *Journal of Sustainable Tourism* 30(7), 1658–1681. https://doi.org/ 10.1080/09669582.2020.1831000

Jimenez-Esquinas, G. (2017). "This is not only about culture": On tourism, gender stereotypes and other affective fluxes. *Journal of Sustainable Tourism* 25(3), 311–326. https://doi.org/10.1080/09669 582.2011.622769

Khoo-Lattimore C., Chiao Ling Yang, E. & Sanggyeong Je, J. (2019). Assessing gender representation in knowledge production: A critical analysis of UNWTO's planned events. *Journal of Sustainable Tourism* 27(7), 920–938. https://doi.org/10.1080/09669582.2019.1566347

Kimbu, A.N. & Ngoasong, M.Z. (2016). Women as vectors of social entrepreneurship. *Annals of Tourism Research* 60, 63–79. http://dx.doi.org/10.1016/j.annals.2016.06.002

Koburtay, T. & Syed, J. (2019). A contextual study of female-leader role stereotypes in the hotel sector. *Journal of Sustainable Tourism* 27(1–3), 52–73. https://doi.org/10.1080/09669582.2018.1560454

Kock, L. (2021). The behavioral ecology of sex tourism: The consequences of skewed sex ratios. *Journal of Travel Research* 60(6), 1252–1264.

Kong, S., Guo, J., Huang, D. et al. (2022). The girlfriend getaway as an intimacy. *Annals of Tourism Research* 92. https://doi.org/10.1016/j.annals.2021.103337

Lapan, C., Morais, D.B., Wallace, T., Barbieri, C. & Floyd, M.F. (2021). Gender, work, and tourism in the Guatemalan Highlands. *Journal of Sustainable Tourism* 30(12), 2839–2859. https://doi.org/10.1080/ 09669582.2021.1952418

Liu, Y.Q. (2001). *Family and public domain: The influence of tourism development on the social gender relations of the Mosuo – a case study of Mosuo matrilineal ethnic community in Ninglang county]* (Doctoral dissertation). Yunnan University, Kunming.

Liverman, D.M. (2018). Geographic perspectives on development goals: Constructive engagements and critical perspectives on the MDGs and the SDGs. *Dialogues in Human Geography* 8(2), 168–185. https://doi. org/10.1177/2043820618780787

Liu, F.F., Wang, H. & Xu, H.G. (2019). Gender differences in academic output of Chinese tourism scholars: Based on literature published in Tourism Tribune 34(12), 109–119.

Morgan, N. & Pritchard, A. (2019). Gender matters in hospitality ('luminaries' special issue, International Journal of Hospitality Management). *International Journal of Hospitality Management* 76, 38–44. https:// doi.org/10.1016/j.ijhm.2018.06.008

Moswete, N. & Lacey, G. (2015). "women cannot lead": Empowering women through cultural tourism in Botswana. *Journal of Sustainable Tourism* 23(4), 600–617. https://doi.org/10.1080/09669582.2014.986488

Mitra, S.K, Chattopadhyay, M. & Chatterjee, T.K. (2022). Can tourism development reduce gender inequality. *Journal of Travel Research* 62(3), 563–577. DOI: 10.1177/00472875211073975 journals.sagepub.com/ home/jtr

Munar, A.M. (2017). To be a feminist in (tourism) academia. *Anatolia* 28(4), 514–529. DOI: 10.1080/ 13032917.2017.1370777

Munar, A.M., Biran, A., Budeanu, A. et al. (2015). The gender gap in the tourism academy: Statistics and indicators of gender equality. While Waiting for the Dawn. Online source: https://research.cbs.dk/en/publi cations/2a3a86f6-838c-4e27-8cfa-ed4668b348b4 [accessed February 2024]

Nikjoo, A., Zaman, M., Salehi, S. & Hernandez-Lara, A.B. (2021). The contribution of all-women tours to well-being in middle-aged Muslim women. *Journal of Sustainable Tourism* 30(7), 1720–1735. https://doi. org/10.1080/09669582.2021.1879820

Nunkoo, R., Sharma, A., Rana, N.P., Dwivedi, Y.K. & Sunnassee, V.A. (2021). Advancing sustainable development goals through interdisciplinarity in sustainable tourism research. *Journal of Sustainable Tourism* 31(3), 735–759. https://doi.org/10.1080/09669582.2021.2004416.

Ong, F., Lewis, C. & Vorobjovas-Pinta, O..(2021). Questioning the inclusivity of events: The queer perspective. *Journal of Sustainable Tourism* 29(11–12), 2044–2061. https://doi.org/10.1080/09669582.2020.1860072

Poulston, J. (2008). Metamorphosis in hospitality: A tradition of sexual harassment. *International Journal of Hospitality Management* 27(2), 232–240. https://doi.org/10.1016/j.ijhm.2007.07.013

Pritchard, A. (2018). Predicting the next decade of tourism gender research. *Tourism Management Perspectives* 25, 144–146. https://doi.org/10.1016/j.tmp.2017.11.014

Pritchard, A. & Morgan, N.J. (2000). Constructing tourism landscapes – gender, sexuality and space. *Tourism Geographies* 2(2), 115–139. https://doi.org/10.1080/14616680050027851

Pyke, J., Pyke, S. & Watuwa, R. (2019). Social tourism and well-being in a first nation community. *Annals of Tourism Research* 77(Jul.), 38–48. https://doi.org/10.1016/j.annals.2019.04.013

Reimer, J.K. & Walter, P. (2013). How do you know it when you see it? Community-based ecotourism in the Cardamom Mountains of southwestern Cambodia. *Tourism Management* 34(Feb.), 122–132. https://doi.org/10.1016/j.tourman.2012.04.002

Ryan, C. & Kinder, R. (1996). Sex, tourism and sex tourism: Fulfilling similar needs? *Tourism Management* 17(7), 507–518.

Saarinen, J. (2020). Tourism and sustainable development goals: Research on sustainable tourism geographies. In J. Saarinen (Ed.) *Tourism and Sustainable Development Goals* (pp. 1–10). Routledge. https://doi.org/10.1201/9780429324253-1

Schellhorn, M. (2010). Development for whom? social justice and the business of ecotourism. *Journal of Sustainable Tourism* 18(1), 115–135. https://doi.org/10.1080/09669580903367229

Scheyvens, R. (2018). Linking tourism to the sustainable development goals: A geographical perspective. *Tourism Geographies* 20(2), 341–342. https://doi.org/10.1080/14616688.2018.1434818

Sinclair, M.T. (1997). *Gender, work and tourism.* Routledge.

Smith, W.E., Kimbu, A.N., de Jong, A. & Cohen, S. (2021). Gendered Instagram representations in the aviation industry. *Journal of Sustainable Tourism* 31(3), 639–663. https://doi.org/10.1080/09669582.2021.1932933

Swain, M.B. (1995). Gender in tourism. *Annals of Tourism Research* 22(2), 247–266. https://doi.org/10.1016/0160-7383(94)00095-6

Talinbayi, S., Xu, H. & Li, W. (2018). Impact of yurt tourism on labor division in nomadic kazakh families. *Journal of Tourism and Cultural Change* 17(3), 339–355. https://doi.org/10.1080/14766825.2018.1447949

Tang, X.Q. & Zhu, H. (2007). Gender issues in tourism research. *Tourism Tribune* 22(2), 43–48.

Tran, L. & Walter, P. (2014). Ecotourism, gender and development in northern Vietnam. *Annals of Tourism Research* 44(Jan.), 116–130. http://dx.doi.org/10.1016/j.annals.2013.09.005

Tucker, H. (2020). Gendering sustainability's contradictions: Between change and continuity, *Journal of Sustainable Tourism* 30(7), 1500–1517. https://doi.org/10.1080/09669582.2020.1839902

Tucker, H. & Boonabaana, B. (2012). A critical analysis of tourism, gender and poverty reduction. *Journal of Sustainable Tourism* 20(3), 437–455. https://doi.org/10.1080/09669582.2011.622769

Walters, T. (2018). Gender equality in academic tourism, hospitality, leisure and events conferences. *Journal of Policy Research in Tourism, Leisure and Events* 10(1), 17–32. https://doi.org/10.1080/19407963.2018.1403165

Wang, X.L. (2008). *The impact of ethnic tourism development on minority female – A case study of Huangdu Dongzu Stokabed-village in Hunan Province* (Doctoral dissertation). Guilin University of Technology, Guangxi.

Wang, X.Y. & Sun, J.X. (2022). Embodiment of feminine subjectivity by women of a tourism destination. *Journal of Sustainable Tourism* 31(6), 1447–1463. https://doi.org/10.1080/09669582.2022.2053858

Weichselbaumer, D. (2012). Sex, romance and the carnivalesque between female tourists and Caribbeanmen. *Tourism Management* 33(5), 1220–1229.

Xu, H. (2018). Moving toward gender and tourism geographies studies. *Tourism Geographies*, 20(4), 721–727. https://doi.org/10.1080/14616688.2018.1486878

Xu, H. & Gu, H.(2018). Gender and Tourism Development in China, *Journal of China Tourism Research* 14(4), 393–404. https://doi.org/10.1080/19388160.2018.1539426

Xu, H. & Wang, Y. (2021). The impacts of gender on seasonal retirement mobility and wellbeing. *Ageing and Society* 41(1), 187–207. https://doi.org/10.1017/S0144686X19001004

Xu, H., Wang, K. & Ye, T. (2017). Women's awareness of gender issues in Chinese tourism academia. *Anatolia* 28(4), 553–566. https://doi.org/10.1080/13032917.2017.1370780

Xu, H. & Ye, T. (2016). Tourist experience in Lijiang-The capital of Yanyu. *Journal of China Tourism Research* 12(1), 108–125.

Yin, Q. (2015). Analysis of the impact of tourism development on the changing role of women – Taking Dali Zhou Cheng Bai women as an example. *Journal of Kunming University of Science and Technology* 15(2), 102–108.

Zaytoun, M., Heiba, A. & Abdelhakim, M. (2010). *Implications of the global financial and economic crisis on the tourism sector in Egypt.* Cairo: International Labour Organisation. Online source [accessed February 2024]. www.ilo.org/africa/information-resources/publications/WCMS_243798/lang--en/index.htm

Zhang, J.K. & Zhang, Y. (2020). Tourism and gender equality: An Asian perspective. *Annals of Tourism Research* 85. https://doi.org/10.1016/j.annals.2020.103067

9

TOURISM, PLACE, SPACE AND QUEER SEXUALITY

Gustav Visser

Introduction

Ten years ago, Gordon Waitt (2012, 88) argued that queer perspectives on tourism geographies have been grappling with the gradual normalisation of particular forms of homosexuality within the neoliberal market place, and the implications for more radical imperatives of queer politics to blur and reveal the inconsistencies of sexual and gender categories. Since then however, the dynamics of these types of argument have changed. The core of the argument remains prescient, although the focus is perhaps now more concerned with making visible minorities within the non-heterosexual minority. Over the past three decades there has been considerable debate concerning the various relationships between tourism, place, space and sexuality (Doan, 2015). These debates have demonstrated a particular developmental trajectory in that they begin with the intersections between male gay sexual identities in relation to place and space and then slowly draw in other non-heteronormative identities over time (Gorman-Murray, 2007; Waitt & Markwell, 2006).

The LGBTQ or, perhaps more inclusively, queer community[1] is diverse, and it is thus problematic to refer to it as a unified grouping or collectively. Historically, it has been argued that popular gay travel destinations exist because they demonstrate certain characteristics and serve several key functions, for example, they typically have permissive or liberal attitudes towards queer identified travellers and they provide a friendly infrastructure. It is also highlighted that travel to gay destinations affords opportunities to socialise with similarly identified individuals who provide a sense of community. The majority of investigations explore the development of gay male tourism in various well-known cities with large gay resident communities. This chapter argues that the connections that currently exist between place, space, sexuality and tourism demonstrate an evolution in both form and function (Anderson & Knee, 2021; Brown, 2009; Llewellyn, 2022). These changes have mainly been viewed from the vantage point of leisure-space and various forms of spatially consolidated "white gay male community(ies)". On the whole, however, few of the queer/gay markers remain relevant as analytic categories. It is argued that this is owing to concepts such as queer simply becoming too diffuse and diverse in meaning. This investigation explores this claim relative to tourism linked to sexual identity. In so doing, this chapter first provides some brief insights which illustrate the rise of debates regarding the intersections of gay identification, place and space. The long-established nature of these debates is highlighted and some form of

DOI: 10.4324/9781003286301-11

movement in discourse to the idea of homonormalisation in a changing heterosexual society (in some places) is discussed. The most significant contribution of this chapter is located in the final section whereby consideration is given to new directions of enquiry that might be fruitful in the foreseeable future in the investigation of the intersection(s) between tourism, place, space and sexuality.

Sexuality, Place, Space and Tourism Clusters

Debates concerning queer sexualities, place, space and tourism began with reference to gay men in the early 1960s (while really very present-minded and certainly ignorant of other cultural contexts differently understood, see Luongo, 2007) that academically came to the fore in the 1990s (Waitt & Markwell, 2006). In this style of literature it was generally suggested that gay men use space to adopt separate identities – identities often assumed away from home and work places, encouraging the consumption of leisure space (for example, Hughes, 1997, 1998, 2006). These investigations were set against the backdrop of mainly large Western cities (see Brown and Browne (2016) for an expansive exploration of this literature). It was argued that urban areas facilitate the assumption of a gay self in many ways, not least through the number and concentration of gay leisure places available. Pioneers in this line of argument (Binnie, 1997; Knopp, 1995) noted that such leisure-related freedoms have been widely embraced as a validation of gay identity, if only as a reaction to a more general powerlessness. Nevertheless, it was argued that the use of space may have the effect of transforming those spaces so that it becomes coded as gay (Note that little was really said about lesbian spaces).

Drawing on these observations, and subsequently expanded on by Hughes (2006) as well as Waitt and Markwell (2006), Visser (2003a: 171) noted that "leisure tourism by gay men at the time may have been the ultimate manifestation of the use of space in order to separate identities." This argument suggested that studies of gay leisure spaces interpret the clustering of such activity as part of a homosexual challenge to the heterosexual coding of public space, empowering those who are currently excluded and disenfranchised (see also Binnie & Valentine, 2000). At the time, Pritchard et al. (1998: 274) suggested:

> These are places which enable not only the display of behaviour and affection but also access to a variety of gay services and facilities including shops, bars, housing and legal and medical services. … Gay spaces, in essence, provide community and territory as well as a sense of order and power. They are sites of cultural resistance with enormous symbolic meaning, providing cultural and emotional support for a political movement comprising an increasingly diverse and geographically scattered community."

Drawing on Jansen-Verbeke (1994), Visser (2003a, 2003b) noted that, historically, not only do gay leisure spaces tend to cluster, but so too does tourism activity. It was observed that tourism in a city is related to a combination of its primary and secondary tourism elements. It was suggested that the activity place is composed of attractions that characterise most cities, such as museums and historic sites, whereas the leisure setting is the physical and socio-cultural context within which the attractions are set: the overall spatial structure of the city and its ambience. The secondary element in Jansen-Verbeke's (1994) view includes shops, cafés, restaurants and bars, hotels and entertainment.

Primary elements that are proximate to each other, as clusters, may have a greater appeal to the tourist than non-proximate elements. Hughes (1998) argued that several tourist clusters, spatially separate and with distinct features, may exist in a city to give it polycentricity. The identification of clusters ultimately depends on tourist use. These elements are, however, usually consumed by both tourists and local residents. The overlapping use of gay leisure cluster(s) and primary, as well as secondary, tourism elements may also cause a city, in whole or in part, to be transformed both materially and symbolically (Visser, 2003a; 2003b). These observations were made more than two decades ago and much has subsequently changed, if not entirely globally, then at least significantly so in many developed Western countries and a range of countries in the Global South.

Current Debates on the Link between Queer Sexualities, Place, Space and Tourism

During the ensuing decades, many developments occurred, particularly for gay men. There are many reasons for this, and these form the focus of this part of the investigation. Arguably, the zenith of queer space development in the form of physical infrastructure was reached by the end of the first decade of the new millennium. Since then the amount and concentration of gay infrastructure that supported gay leisure spaces and tourism started to decline across the global North, as well as in those places in the global South where it did register as a physical phenomenon. Anderson and Knee (2021) argue that there are at least three intersecting factors which account for this trend.

The first issue relates to the positioning of a gay (though not all members of the queer cohort) identity in an increasingly changing heterosexual leisure world. Visser (2007; 2008a; 2008b; 2010; 2013; 2016) hinted some time ago that these changes were already visibly underway in South Africa for instance. In Western society, scholars such as Ghaziani (2011; 2015), have confirmed that the LGBTQ community (in its broadest sense) has become more accepted and that 'queer enclaves' are not as necessary as they were in the past – the so-called demise of the gayborhood (Bitterman, 2021; Coffin, 2021; Doan, 2021; Gorman-Murray, 2021; Hess, 2021). Collectively, these investigators note the declining (perhaps mixed) importance of the fixed spatialisation of sexual orientation in the lives of queer community(ies) and a desire to assimilate and participate in the general public social spaces that once denied them (Doan & Higgins, 2011). These claims, I would suggest (Visser, 2016), are true for only a handful of countries, regions and cities globally, although those locations have formed the backbone of gay tourism scholarship and theorisation. Within these claims are embedded suggestions of a 'post-gay' society – a construct that needs to be considered with caution. Some of these ideas are located in the role of technology in 'queer community' formation (Collins & Drinkwater, 2017; Llewellyn, 2022; Miles, 2021).

The second factor, as some have argued, is that the increased use of technology to form community(ies) provides a further explanation to the decline of queer spaces (see Anderson and Knee, 2021: 120). For some time there have been conversations about the role of online platforms as a replacement for dedicated physical queer spaces (Miles, 2021). This, in particular, rings true for younger generations (Collins & Drinkwater, 2017). Smartphones have fundamentally reimagined the physical and virtual space interface. Renninger (2019) explored these ideas in the provocatively entitled "Grindr killed the gay bar, and other attempts to blame social technologies for urban development". Anderson and Knee (2021) argue that the proliferation of location-based social applications in the past decade has allowed the queer population to become comfortable creating and maintaining social connection virtually. The importance of these applications has surged against the backdrop of the current COVID-19 pandemic (Miles et al., 2021).

Thirdly, there are debates that have essentially argued that gay-led gentrification, not technology, has undermined the maintenance of queer places and spaces. 'Gay villages/spaces' were historically located in less-desirable areas where the investment of various forms of capital often led to urban renewal (Ghaziani, 2021; Visser, 2016). This, overlaid with the greater acceptance of queer individuals in mainstream society, correlates to the commercialisation of a once queer space and a resulting attraction of new heterosexual residents seeking 'in vogue' physical locations centred on consumptive practices. This phenomenon in many popular gay tourist destinations has led to the in-migration of heterosexual cohorts and, in some places, tourists. This has additionally resulted in the degaying of formerly gay neighbourhoods, making them less friendly to queer individuals, as well as leading to businesses favouring a wider audience attraction (Orne, 2016; Visser, 2016). Anderson and Knee (2021: 120), nevertheless, observe that "the changing nature of queer leisure spaces and ... the decline of some, does not diminish the role that such spaces play in the lives of queer persons" (also see Mattson, 2019a; 2019b). These concerns have focused on gay male spaces, but certainly do not ring true for lesbian spaces. Although historically there has not been a particularly strong spatialisation of lesbian leisure (compared with gay men) to dedicated physical spaces, in recent years, looking towards the United States of America over the past decade, there has been an unmitigated decline in lesbian infrastructure in the form of bars and social venues (The Lesbian Bar Project, 2021).

Some New Exploratory Ideas

Returning to the opening idea of Gordon Waitt (2012: 88) that queer perspectives on tourism geographies have been grappling with the gradual normalisation of particular forms of homosexuality and/or non-heteronormative identities within neoliberal market places and the implications for more radical imperatives of queer politics to blur and reveal the inconsistencies of sexual and gender categories. This notion has changed in significant ways, and not necessarily positively, depending on the approach taken to address such questions. The key concern is the relationship between the increasingly/registered sexuality identity(ies) markers to specific places/spaces and their/its relationship to tourism. One thing that has been bubbling at the surface recently is the decrease in physical queer infrastructure and actual tourism clustering in tourism destinations. In contrast to claims made by Visser (2003a), among others, that historically, not only do gay leisure spaces tend to cluster, but so too do tourism activities, this tendency seems no longer to be the case.

Queer places/spaces have increasingly become momentary inscriptions of the queer body on sexually ambiguous public/heterosexual places, enabled through interactions on Internet platforms, which is now common practice. Dedicated queer (but mainly meaning gay) places and spaces are now fewer and not necessarily clustered. In this regard there is much to be considered. Motivators for queer leisure, what that might constitute, and travel echo much of what has already been highlighted. Many of the observations to follow are not confined to the queer community(ies), but relate to broader leisure activities and tourism accommodation that have developed over at least the past decade.

In terms of entertainment built around nightlife in 'party locations' such as London, Paris, Amsterdam, New York, and so forth, there has been a general decline in club and bar culture, which holds true for both the queer and mainstream heterosexual communities (Anderson, 2007; Anderson, 2009 International National Music Summit, 2021). The spatially static club/party/bar, or accommodations, restaurants, and so forth, are more fluid now and 'sexuality tribe identity' has for some time become increasingly rare to find in one consolidated physical space (The Lesbian

Bar Project, 2021). In addition, temporality in terms of physical engagement is not governed by the rhythm of the week/weekend of the past, nor climate seasons of summer, and so forth. Collectively, this is dissonant with former gay discourses concerning gay place, space and tourism.

Another issue that has entered the leisure space/sexual identity/tourism debate, though not remarked upon by Waitt (2012) a decade ago and which is not limited to non-heteronormative cohorts concerning those kinds of links, is that the imprint of neoliberal marketisation of leisure in what was once described as 'queer community space' has come to bear on queer space development and hence tourism potential. Property development with better returns has become a key factor in the lack of stationary queer spaces, which in turn halts the development of permanent clustering and potential tourism linked to other non-heteronormative identities. This is significantly tied to the role of digital mobile technologies.

The popular use of social media has changed the notion of 'clubbing or cruising bars'. In this statement is located a number of issues concerning the intersections of sexuality, place, space and tourism. From an academic point of view, the importance of alternative data points come to mind – shifting from more traditional qualitative and quantitative techniques in gaining information that concerns issues dealing with intersections between sexuality, place and space, and tourism and the digital realm. This new data source of online information now ranges considerably from relatively public to extremely private; hidden in far-flung corners of the Internet. This information is not universally accessible, queer or otherwise. Physical place and space, as a context in which 'identity learning' took place three or four decades ago, has now evolved into a very uneven platform of information. New knowledge about queer communities are emerging and alternative 'clubbing venues' have emerged, however these 'places and spaces' are not very visible. In addition, the ethics of these investigations need consideration as the manner in which information is obtained becomes inadvertently more covert.

Mobility of club and party places and spaces and the demise of megaclubs in favour of more bespoke, intermittent music and identity sharing experiences are becoming important analytic classes (Faber, 2019; Kimbrell, 2020; *The Economist*, 2020). They do not 'speak to' the former notions of consolidated place/space and the monetisation thereof for tourism – these spaces are now fluid and 'placeless'. The re-emergence of the 'private party' and transient places and spaces of an earlier age (or even still part of many places in the Muslim world, for example, Luongo [2007]), seems relevant to current realities. Club and party scenes in tourism hotspots such as Ibiza, Amsterdam, Berlin or San Francisco, for example, are being re-thought, along with the notion of gay mass tourism. There are many reasons for this, cost being one and changes in Millennials' and Generation Z's relationship to the notion of leisure being another (International National Music Summit, 2021). This appears to be particularly pertinent to gay social/sexual interaction (Miles, 2021). Again, in the diminished number of bars and clubs available, a greater social acceptance of queer persons feeds the further declines of dedicated queer spaces (Doan, 2021).

Related to 'historic' concerns surrounding dedicated gay accommodation or restaurants (gay spaces generally, see Miller, 2021), the physical and symbolic landscape has changed dramatically since, for example, Hughes (2006); Waitt (2012) and Waitt and Markwell (2006) wrote about queer (mainly gay) perspectives on tourism geographies. Currently, in mainstream queer tourist destinations, 'gay-friendly' designations do not exist. Broadly speaking, homonormalisation has made that type of vocabulary redundant and even offensive. In this regard, one has to return to the role of technology in the sexuality, tourism, space and place interface within the contemporary context.

Developments in mobile digital technologies are disrupting conventional understandings of space and place for smartphone users. One way in which location-based media are reconfiguring previously taken-for-granted spatial traditions is via GPS-enabled online dating and hook-up apps (Miles, 2021). For sexual minorities, these apps can reconfigure any street, park, bar or home into a queer space through a potential meeting between mutually attracted individuals, but what does this signify for already-existing queer spaces? Smartphone apps including Grindr, Tinder, and Blued synthesise online queer encounters with offline physical space in order to create a new hybrid terrain predicated on availability, connection and encounter (Miles, 2021). This terrain is able to sidestep established gay neighbourhoods entirely which, in turn, impacts older, physically rooted gay neighbourhoods and the role that these neighbourhoods have traditionally played in brokering social and sexual connection for sexual minorities. Few would deny that location-based apps have come to play a valuable role in multiplying opportunities for sexual minorities. However, the stratospheric rise of these technologies also provokes questions about their influence on embodied encounter, queer community, and a sense of place.

Finally, there is the issue of disentangling the queer debate. This chapter has mainly followed debates on gay male sexuality, place, space and tourism intersections, which historically has been most remarked upon in the academic press – it purposefully does not address subsets of 'LBTQ+'. What should be underscored is that non-heteronormative and/or non-binary identities have become obsolete as a collective noun for spatial analysis vis-à-vis tourism (Bitterman, 2021). We are talking about minorities within minorities and their relationship to place, space, leisure and tourism, whether or not these investigations take place in large cities or smaller urban places (Browne, 2009; Forstie, 2020), it cannot be generalised, methodologically or epistemologically and probably should not be. Leading on from this observation is a critical need for far greater sensitivity regarding inter-generational changes in broadly queer leisure and various cohorts' relationship to place, space and tourism. What was in the 1960s and 1970s, regarding the search for shared space of 'the queer', and the major consolidating effects of AIDS in the 1980s, has changed. The 1990s and early 2000s were different in terms of what binds the non-heterosexual leisure seekers and tourism, a focus within the gay and lesbian cohorts as some form of a community. From industry sources, rather than the academic press, it has become apparent that different age cohorts within the broad denotation of 'queer' are fundamentally relating to place, space, leisure and sexuality in very different ways and, hence, tourism as well. Binding links between these cohorts are fleeting at best, both in meaning and place.

The seemingly inevitable fractured nature of these intersections reflected through the prism of a broadly speaking empirical base of Northern and Southern understandings of the relationship between sexuality, place, space and tourism holds several major challenges (Visser, 2015). This harks back to the notion of a methodology and the manner in which scholars access information as aforementioned. Additionally, it relates to the epistemic issue of what would qualify as research questions and comparable data points across different locational realities and building theory(ies) (Coffin, 2021). How comparable are the realities of a seemingly 'online' dating space and perhaps resultant tourism between generally uniform Northern contexts and significantly more diverse technological possibilities in the full range of countries currently 'clumped' into the Global South categorisation? There is gay tourism, albeit not always visible, in the Global South, regardless of whether this categorisation is not particularly useful in queer thinking about sexuality and tourism. The realities of Southern contexts such as South Africa, Brazil and Morocco are, for example, vastly different empirical contexts and will therefore demonstrate a variation of possible relationships to sexuality, place, space and tourism. Finally, it is necessary to beg the question, what does this mean to the ontology, and what queer space and place are and the relationship of tourism to aspire to? These questions aside, it is undeniably clear that the nature of queer place/space has changed, and it is left to us to unravel the new articulations of queer sexuality, tourism and community in this new age.

Note

1   Following Anderson and Knee (2021, 119), I use 'queer' as an umbrella term inclusive of all minoritised sex, gender and sexual identities that draws distinction from hegemonic/heteronormative society. The acronym 'LGBTQ' (lesbian, gay bisexual, transgender, queer) is also used as a whole or in part throughout the chapter to specify segments of the community.

References

Anderson, A., & Knee, E. (2021). Queer isolation or queering isolation? Reflecting upon the ramifications of COVID-19 on the future of queer leisure spaces. *Leisure Sciences, 43*(2), 118–124.

Anderson, C. (2007). *The Long Tail: How Endless Choice is Creating Unlimited Demand.* Random House.

Anderson, T. L. (2009). Better to complicate, rather than homogenize, urban nightlife: A response to Grazian. *Sociological Forum, 24*(4), 918–925.

Binnie, J. (1997). Coming out of geography: Towards a queer epistemology? *Environment and Planning D: Space and Society, 15*(2), 223–237.

Binnie, J., & Valentine, G. (2000). Geographies of sexuality – a review of progress. *Progress in Human Geography, 23*(2), 175–187.

Bitterman, A. (2021). Understanding generational gaps in LGBTQ+ communities: Perspectives about gay neighbourhoods among heteronormative and homonormative generational cohorts. In A. Bitterman, & D. Hess (Eds.), *The Life and Afterlife of Gay Neighbourhoods: Renaissance and Resurgence* (pp. 307–338). Springer.

Brown, G. (2009). Thinking beyond homonormativity: Performative explorations of diverse gay economies. *Environment and Planning A: Economy and Space, 41*(6), 1496–1510.

Brown, G., & Browne, K. (Eds.) (2016). *The Routledge Research Companion to Geographies of Sex and Sexualities.* Routledge.

Browne, K. (2009). Imagining cities, living the other: Between the gay urban idyll and rural lesbian lives. *The Open Geography Journal, 1*(1), 25–32.

Coffin, J. (2021). Plateaus and afterglows: Theorizing the afterlives of gayborhood as post-places. In A. Bitterman, & D. Hess (Eds.), *The Life and Afterlife of Gay Neighbourhoods: Renaissance and Resurgence* (pp. 371–389). Springer.

Collins, A., & Drinkwater, S. (2017). Fifty shades of gay: Social and technological change, urban deconcentration and niche enterprise. *Urban Studies, 54*(3), 765–785.

Doan, P., & Higgins, H. (2011). The demise of queer space? Resurgent gentrification and the assimilation of LGBT neighbourhoods. *Journal of Planning Education and Research, 31*(1), 6–25.

Doan, P. (2015). *Planning and LGBTQ Communities: The Need for Inclusive Queer Spaces.* Routledge.

Doan, P. (2021). After the life of LGBTQ urbanism and generation from Atlanta and Istanbul. In A. Bitterman, & D. Hess (Eds.), *The Life and Afterlife of Gay Neighbourhoods: Renaissance and Resurgence* (pp. 261–285). Springer.

Faber, T. (2019). Is London's club scene really under threat? Available at: www.ft.com/content/9e45e870-f668-11e9-bbe1-4db3476c5ff0 [Accessed 11 October, 2021]

Forstie, C. (2020). Theory making from the middle: Researching LGBTQ communities in small cities. *City and Community, 19*(1), 153–168.

Ghaziani, A. (2011). Post-gay collective identity construction. *Social Problems, 58*(1), 99–125.

Ghaziani, A. (2015). *There goes the Gaybourhood?* Princeton University Press.

Ghaziani, A. (2021). Why gayborhoods matter: The street empirics of urban sexualities. In A. Bitterman, & D. Hess (Eds.), *The Life and Afterlife of Gay Neighbourhoods: Renaissance and Resurgence* (pp. 87–113). Springer.

Gorman–Murray, A. (2007). Rethinking queer migration through the body. *Social & Cultural Geography, 8*(1), 105–21.

Gorman-Murray, A. (2021). Recovering the gay village: A comparative historical geography of urban change and planning in Toronto and Sydney. In A. Bitterman, & D. Hess (Eds.), *The Life and Afterlife of Gay Neighbourhoods, Renaissance and Resurgence* (pp. 239–260). Springer.

Hess, D. (2021). Who are the people in your gayborhood? Understanding population change and cultural shifts in LGBTQ+ neighborhoods. In A. Bitterman, & D. Hess (Eds.), *The Life and Afterlife of Gay Neighbourhoods: Renaissance and Resurgence* (pp. 3–39). Springer.

Hughes, H. (1997). Holidays and homosexual identity. *Tourism Management, 18*, 3–7.

Hughes, H. (1998). Sexuality, tourism and space: The case of gay visitors to Amsterdam: In D. Tyler, M. Roberson, & Y. Guerrier (Eds.), *Managing Tourism in Cities: Policy, Process and Mobility* (pp. 163–178). Wiley.

Hughes, H. (2006). *Pink Tourism: Holiday of Gay Men and Lesbians.* CABI.

International National Music Summit. (2021). Where is club culture headed? Available at: www.youtube.com/watch?v=sk5Yosev4vk [Accessed February 2024].

Jansen-Verbeke, M. (1994). The synergy between shopping and tourism. In W. Theobald (Ed.), *Global Tourism: The Next Decade* (pp. 347–362). Butterman-Heidemann.

Kimbrell, G. (2020). 4 Predictions for nightlife after Covid-19. Available at: www.rollingstone.com/culture-council/articles/predictions-nightlife-after-covid-1091856/ [Accessed February 2024].

Knopp, L. (1995). Sexuality and urban space: A framework for analysis. In D. Bell, & G. Valentine (Eds.), *Mapping Desire* (pp. 149–161). Routledge.

Llewellyn, A. (2022). A space where queer is normalized: The online world and fanfictions as heterotopias for WLW. *Journal of Homosexuality, 69*(13), 2348–2369.

Luongo, M. (2007). *Gay Travels in the Muslim World.* Harrington Park Press.

Mattson, G. (2019a). Small city gay bars, big city urbanism. *City and Community, 19*(1), 76–97.

Mattson, G. (2019b). Are gay bars closing? Using business listings to infer rates of gay bar closure in the United States, 1977–2019. *Socius, 5*, 204–223.

Miles, S. (2021). Let's (not) go outside: Grindr, hybrid space and digital queer neighourhoods. In A. Bitterman, & D. Hess (Eds.), *The Life and Afterlife of Gay Neighbourhoods, Renaissance and Resurgence* (pp. 203–220). Springer.

Miles, S., Coffin, J., Ghaziani, A., Hess, D. B., & Bitterman, A. (2021). After/lives: Insights from the COVID-19 pandemic for gay neighbourhoods. In A. Bitterman, & D. Hess (Eds.), *The Life and Afterlife of Gay Neighbourhoods, Renaissance and Resurgence* (pp. 393–418). Springer.

Miller, S. (2021). Commemorating historically significant gay places across the United States. In A. Bitterman, & D. Hess (Eds.), *The Life and Afterlife of Gay Neighbourhoods, Renaissance and Resurgence* (pp. 339–370). Springer.

Orne, J. (2016). *Boystown: Sex and Community in Chicago.* University of Chicago Press.

Pritchard, A., Morgan, N., Sedgely, D., & Jenkins, A. (1998). Researching out to the gay tourist: Opportunities and threat in an emerging market segment. *Tourism Management, 19*(3), 273–282.

Renninger, B. (2019). Grindr killed the gay bar, and other attempts to blame social technologies for urban development: A democratic approach to popular technologies and queer sociality. *Journal of Homosexuality, 55*(2), 1736–1755.

The Economist (2020). Berliners fear "club death". Available at: www.economist.com/europe/2020/01/30/berliners-fear-club-death [Accessed 10 October, 2021]

The Lesbian Bar Project (2021). The lesbian bar project. Available at: www.lesbianbarproject.com/ [Accessed 21 August, 2021]

Visser, G. (2003a). Gay men, leisure space and South African cities: The case of Cape Town. *Geoforum, 34*(1), 123–137.

Visser, G. (2003b). Gay men, tourism and urban space: Reflections on Africa's 'gay capital'. *Tourism Geographies, 5*(2), 168–189.

Visser, G. (2007). Homonormalising (white) heterosexual middle-class leisure spaces: The case of white gay men in Bloemfontein, South Africa. *International Journal of Diversity in Organisations, Communities and Nations, 7*(1), 217–228.

Visser, G. (2008a). The homonormalisation of white heterosexual leisure spaces in Bloemfontein, South Africa. *Geoforum, 39*(3), 1347–1361.

Visser, G. (2008b). Exploratory notes on the geography of black gay leisure spaces in Bloemfontein, South Africa. *Urban Forum, 19*(4), 413–423.

Visser, G. (2010). Leisurely lesbians in a small city in South Africa. *Urban Forum, 21*(2), 171–185.

Visser, G. (2013). Challenging the gay ghetto in South Africa: Time to move on? *Geoforum, 49*, 268–274.

Visser, G. (2015). Thinking beyond exclusionary gay male spatial frames in the developing world. In P. Doan (Ed.), *Planning and the LGBTQ Community: The Need for Inclusive Queer Spaces* (pp. 81–93). Routledge.

Visser, G. (2016). Urban leisure and tourism-led redevelopment frontiers in central Cape Town since the 1990s. *Tourism: An International Interdisciplinary Journal, 64*(4), 397–408.

Waitt, G. (2012). Queer perspectives on tourism geographies. In J. Wilson (Ed.), *The Routledge Handbook of Tourism Geographies* (pp. 82–89). Routledge.

Waitt, G., & Markwell, K. (2006). *Gay Tourism: Culture and Context.* Hartworth.

10

PERFORMATIVITY, SPACE AND TOURISM

Gunnar Thór Jóhannesson

Introduction

Tourist behaviour and activities and how tourism affects and shapes places and environments remain leading topics in tourism research. During the last two decades, increased attention has been paid to the concepts of performance and space in describing and understanding tourism dynamics and their repercussions. This chapter examines performativity and space in tourism research. It begins with the performance turn that surfaced at the beginning of the century, promoting embodied experiences and the view that reality is 'done and enacted'. This ontological stance, elaborated on in different fields (e.g., actor-network theory, practice theory and material phenomenology), has opened new vistas for describing and comprehending spatial relations and the implications of tourism. This chapter will discuss three examples of scholarship that highlight how the interplay of performativity and space matters in understanding and describing tourism dynamics, namely by approaching the destination as a topological space, to consider affect, atmosphere and placemaking and by attending to the more-than-human and earthly relations of tourism. Each case includes relevant literature and brings forth the practical implications of these conceptual approaches.

Performance, Performativity and Tourism Research

The concepts of performance and performativity, which became increasingly prominent in tourism studies in the late 1990s and early 2000s, are currently well-established conceptual tools for tourism research (see Jonas Larsen's chapter in the first edition of this handbook). Already in the 1970s, Dean MacCannell (1976) drew on Goffman's dramaturgical conception of everyday social interaction (Goffman, 1971) in his writings on staged authenticity and frontstage and backstage performances by tourism workers and operators. Even though the metaphor of performance seems suitable for describing tourism, there were few other examples of the metaphor's use until the beginning of the century, when a reinvigorated interest in performance culminated in the performance turn (Coleman & Crang, 2002; Larsen & Urry, 2011). Apart from Goffman's dramaturgical approach, tourism scholars have mainly drawn inspiration from two other sources: Judith Butler's linguistic concept of performativity (Butler, 1990; 1993) and non-representational theory,

DOI: 10.4324/9781003286301-12

which stresses the embodied practice of everyday life (Thrift, 1996; 1999). Although they are interconnected, there are key differences between these approaches.

Goffman described social interactions as performances in which individual subjects perform particular roles for an audience, either imaginary or real. Thus, the self is a performed character – put on stage and carefully managed. It follows that in daily life, people move between a frontstage, where they play out their character, and a backstage, where it is possible to remove the mask used for frontstage appearances. For Goffman, the self is intentional, calculating and conscious prior to any performance. The idea of a backstage (authentic) self and a frontstage (performed) self were crucial to MacCannell's notion of staged authenticity in tourism (1976).

Butler's concept of performativity disrupts this idea of stable identity existing prior to social interaction and is "an attempt to find a more embodied way of rethinking the relationships between determining social structures and personal agency" (Nash, 2000: 654). For Butler, the individual subject does not place itself on a stage. Rather, it is produced through routinised performance and ritualised practice (Larsen, 2005). Within this framework, people's everyday performances, that is, what people do, say or act, are the "doing" of a discourse that "cites already established formations of knowledge and it is this citation which produces social subjects" (Gregson & Rose, 2000: 436). Performativity refers to these "citational practices which reproduce and/or subvert discourse and which enable and discipline subjects and their performances" (Gregson & Rose, 2000: 434). Importantly, this understanding does not mean that subjects are performed or that people follow a social script without any deviation; instead, it means that performances and performativity are intrinsically connected and saturated with power. Even though discourses discipline subjects and shape and direct people's actions and behaviours, since any discourse is citational, individual subjects have the power to resist or disrupt a discourse. Its repetition can go wrong, and it is possible to play with it and destabilise it.

At the risk of oversimplification, non-representational theory emphasises creativity, play and the embodiment of everyday practices more than the other two approaches. According to non-representational theory, the social is produced and reproduced through performance. It stresses embodied practices, not only language and the world-making powers of discourse but also the senses and bodily enactments (Larsen, 2005). Thrift noted, for instance, that much of everyday performance is neither coded with prior meaning nor choreographed. It is habitual, non-cognitive and practical. Non-representational theory, inspired by actor-network theory, also acknowledged that actors could include animals, material objects and natural elements in addition to humans. This approach thus opened up and problematised the common understanding of the concepts of agency, actors and the social and highlighted the entanglement of the social and the material, which became an important dimension in the performance turn (Larsen & Urry, 2011).

Scholars following the performance turn have applied these three conceptual pillars in various ways and have related them to other currents of thought. Larsen and Urry (2011) outlined the main characteristics of research associated with the performance turn, including a focus on bodily doings and practices rather than representations and language, the use of performative metaphors, attention to choreography, the scripted nature of tourism performances and the creativity and agency of tourists. Places are often framed as unstable and open-ended and accomplished through relational or networked practices and power is seen as relational and distributed. This approach implies that tourism is not an "isolated and exotic island" (p. 1113) but as tightly connected to people's everyday social practices.

An early example of research representative of the performance turn is Tim Edensor's study of tourism at the Taj Mahal (Edensor, 1998; 2000; 2001). In his article 'Staging Tourism: Tourists as Performers' (2000), Edensor argued that "tourism is a process which involves the ongoing (re)construction of praxis" (p. 322–323). He describes different practices and performances that tourists engage in at the Taj Mahal, as well as different stages and choreographies that, to a greater or lesser extent, shape tourists' performances. This latter point is particularly important. Drawing on Butler's notion of performativity, Edensor claimed that social performances are citational. They must relate to a performative discourse about what is, for example, appropriate behaviour and the strategic management of the various stages on which tourism occurs to become meaningful in the situation.

Edensor also convincingly demonstrated that tourists do not necessarily follow the intended tourism script anticipated by other tourists or tourism providers. Rather, "performances vary enormously and depend upon the regulation of the stage and the players, and the relationship between the players" (Edensor, 2000: 324). Edensor indicated how tourism performance involves complex, ambivalent relations between following a script and creativity and between the intentional and the unintentional. Hence, Larsen (2005: 421; original italics) highlighted that "tourist bodies are simultaneously preformed *and performing*". This leads us to the potential of performance and performativity for understanding and describing the spatial relations of tourism.

Spatial Relations of Tourism

The concepts of performativity and performance have contributed to a new understanding of space in tourism geographies, framing it as performative and relational (Larsen, 2012). Gregson and Rose (2000) built on Butler's understanding of performance when they contended that not just social actors, but "the spaces in which they perform", were produced by power (p. 441):

> [W]e maintain that performances do not take place in already existing locations: the City, the bank, the franchise restaurant, the straight street. These 'stages' do not preexist their performances, waiting in some sense to be mapped out by performances; rather, specific performances bring these spaces into being.
>
> (Gregson & Rose, 2000: 441)

Consequently, spaces are performative, shaped by and shaping social relations through power. Space is produced through actors' performances, but it is not a passive playground for acting. This stance has important implications. If space is performed, it can be done in multiple ways, meaning that multiplicity and space are co-constitutive (Massey, 2005). Its performative character also implies that space is becoming. As Massey wrote, "Perhaps we could imagine space as a simultaneity of stories-so-far" (Massey, 2005: 9).

For tourism research, this alternative concept of space has opened fruitful avenues for research on tourist destinations and how tourism contributes to placemaking. In tourism research, the concept of place often appears in the conceptual framework of the destination, an organising principle of much discourse on tourism development, policy and planning. The common understanding of destinations is that they are territorially bounded or demarcated entities. Bærenholdt et al. (2004) noted that discussions on destinations often focus on organisational and marketing strategies and links to regional development strategies, such as cluster building. As a result, the destination is viewed as a container of attractions, activities and facilities necessary for producing a tourist experience. This optic, though, underlines a split between the passive stage of the destination and active engagement of tourists.

Studies emphasising tourist performance have disturbed this spatial image of the destination or the tourist place as a clearly bounded entity connected to tourism mobilities. They have also emphasised its relationality and illustrated how a tourism place is enacted and accomplished through performances, always under construction, potentially unstable and 'in play' (Sheller & Urry, 2004), proposing a move towards a relational ontology of tourist places. Tourist places are cast as relational rather than discrete or bounded. As the Taj Mahal example illustrates, their performance can be contested at any time. Additionally, tourist places are not described as social constructs. As Bærenholdt et al. asserted, "Tourist places are simultaneously places of the *physical environment, embodiment, sociality, memory,* and *image*" (Bærenholdt et al., 2004: 32, original emphasis). Multiple mobilities of people, technologies, things, memories and narratives contribute to the performance of place, further blurring the traditional distinction in social sciences between the social and the material. The destination is "an event of 'thrown-togetherness' (Massey, 2005) of heterogeneous parts where humans as well as more-than-humans are assigned creative capacities" (Huijbens & Jóhannesson, 2019: 285).

I will return to the spatial connotations of emphasising performance and performativity shortly, but first, a few words on methodology.

Methodology

The focus on performativity has unsurprisingly resulted in much emphasis on following tourist performances and practices. Arguably, the performance turn has promoted qualitative methodology in tourism research and creativity in methodological approaches. Scholars have employed various methods of tracing what people do, with whom, and how and what implications their performances have. These include various ethnographic methods, such as participant observation and interviews, and methods intended to better grasp the embodied character of performances, such as visual and walking methods.

In considering methodology's relation to performance, it is noteworthy that the understanding that tourism is performative blurs the usual separation between ontology and epistemology. As Franklin observed, tourism has come "to be a heterogeneous assemblage 'at large' in the world, remaking the world anew as a touristic world; a world to be seen, felt interpellated and travelled" (Franklin, 2004: 277). As previously discussed, various actors contribute to this "remaking", and the performance turn has widened the scope of whom and what is considered a relevant tourism actor. If this relational ontology is accepted, the implication is that researchers are not just describing the realities they encounter but playing an active role in composing them. It follows that researchers are confronted with the question of which realities they might enact (Law & Urry, 2004). Engaging in research is, therefore, inherently a question of ontological politics. Method, then, is performative but not merely descriptive. Studies emphasising performance are often "intent on seeking *relational* rather than representational understandings" (Thrift, 1999: 304, original italics). The focus is on connecting, composing and relating through research practices.

Four crucial implications of this understanding should be emphasised. First, the positionality of the researcher is altered. Any research is seen as a relational accomplishment in which the researcher is 'in the midst of things', as opposed to holding an external position from which she or he can describe a field of study from afar (Jóhannesson et al., 2018). Second, the research field is not a predefined entity or a clearly defined location. According to de Laet, it "should rather be considered as the object's *range*: A space that is performed by the travel of objects and an observer on-the-move" (de Laet, 2000: 167–168, original italics). Third, the knowledge produced is phronetic, not epistemic or universal. It underscores that knowledge is situated, value-based, practical, and created through performative relations involving the researcher engaging with the world (Flyvbjerg, 2001). Fourth, the usual understanding of whom to do research with is altered. More-than-human actors, such as animals, plants and rocks, appear as potentially valuable research collaborators (see e.g., Valtonen, Rantala, & Salmela, 2020; Valtonen, Salmela, & Rantala, 2020). The next section will further illustrate these points.

Performing Tourism Space

The interdependence of performativity and space continues to be a fruitful ground for tourism scholars to understand and describe tourism dynamics. Here, I briefly present three examples of scholarship that apply and develop further ideas and insights related to the performance turn. The first concerns the destination as a topological space, while the second is an example of affect, atmosphere and placemaking. The third centres on the more-than-human and 'earthly' relations of tourism.

The destination as a topological space

As previously mentioned, the performance turn has, along with other approaches, such as the new mobilities paradigm and actor-network theory, forwarded a relational view of space through which the destination can be described as a 'throwntogetherness' (Massey, 2005) of heterogeneous material and social entities and performances. Considering place as a relational effect makes approaching the destination as a topological space possible. Topology offers ways of grasping spatialities of constant movement, fluctuations and instabilities and a "plastic sense of space and time" (Harvey, 2012: 415). Topology highlights the existence of alternative spaces to Euclidian space and refers to how continuity, for instance, an object's shape, is retained through processes of change or movement through time and space (Law & Mol, 2001; Lury et al., 2012). Topological approaches focus on how power relations "compose the spaces of which they are a part" (Allen, 2011: 284) rather than view them as positioned in space. Ties, connections and links of transport, communication or resources are not understood "as lines on a map that cut across territories but rather as intensive relationships which create the distances between powerful and not so powerful actors" (Allen, 2011: 284). Instead of framing destinations as stable nodes or locations in a geometric gridwork that are separated in space but linked through transportation infrastructure to source markets, they can be conceptualised as ordering effects of particular power topologies that shape and move places. Places, including tourist destinations, "do not stay in one location but move about within networks of agents, human and non-human" (Hetherington, 1997: 185) and can take diverse and intersecting topological forms.

Using an example from the Strandir region in rural Iceland, Lund and Jóhannesson (2014) described how roads, buildings, humans, animals and weather contribute to a tourist place. In this example, a dirt road that twists and turns in tandem with an area's landscape plays a crucial role as a technical line of connection enacting the place in Euclidian space, allowing for drawing durable lines of metric distances. It also "draws together diverse temporalities of the Strandir region and [...] works as an interface of continuity and change" (Lund & Jóhannesson, 2014: 448). It provides alternative possibilities for performing place and tourism, which bring past times close to present experiences and highlights the multiplicity of destinations. For example, for many tourists driving through the area, the road invokes a feeling of driving into the past. One must ensure not to hit large stones protruding from its surface. For many of the region's inhabitants, the road's condition confirms that central authorities have left them behind in a position on margins and of little interest to the rest of the country. Constructed in the 1960s, the road was once a sign of progress and modernity. While its status has changed, it continues to affect and shape the performance of the place in multiple ways that sometimes cut across proximity and distance in time and space. As a destination, the Strandir region is a topological space "of ordering and continuity of transformation in which past and present coexist" (Lury, 2013: 129).

Affect, atmospheres and placemaking

The focus on tourist performances indicates how tourism is enacted through embodied encounters with things, objects and actors who are human and more-than-human. One of the directions research on tourist performance has taken is towards what may seem more ephemeral and immaterial phenomena, notably affect and atmospheres. In recent years, affect has gained much attention in geography and the social sciences in general (Bille & Simonsen, 2019). In its broadest sense, affect "consist[s] of bodily capacities to affect *and* to be affected" (Anderson, 2014: 9 original italics). It can be described as a relational force flowing through human and more-than-human bodies and the physical world (d'Hauteserre, 2015), creating a field of intensities and potential excess that may disturb any present order. Edensor (2010: 236) explicated,

> [i]n decentring the individual from analysis, it prompts us to think about how different configurations of objects, technologies, energies, non-human life forms, spaces, forms of knowledge and information combine to form 'affective fields' that are distributed across particular geographical settings.

Edensor further describes affective fields as relational configurations of energy and feeling that generate affective atmospheres. Affect, then, is generated through bodily immersion in an atmospheric environment that folds subject and space together (Edensor, 2010). An example is the effect of weather, light and darkness on how people experience and understand places when they "pervade space and the bodies that perceive them" (Edensor, 2010: 236).

The design and production of affective atmospheres is a notable area of study related to research in tourism and other fields (Edensor, 2012; Edensor & Sumartojo, 2015). Affect theory has also been used to explain why tourist places exhibit changing appeal, why some attractions fail to attract returning visitors (d'Hauteserre, 2015) and how human and more-than-human actors choreograph the tourist's experience (Jóhannesson & Lund, 2017). In the present context, though, it is important to stress the performative character of affect and how it emerges through (tourist) performances as part of becoming spatiality. The general notion of affect as a flow or a "sense of push in the world" (Thrift, 2004: 60) can undermine practice and embodied performances. Bille and Simonsen observed that "[a]ffect in this sense is an autonomous and self-generating movement, primarily thought of as processes of circulation flow, transmission or contagion" (Bille & Simonsen, 2019: 295). The image of autonomous flow or a contagious transmission of affect reproduces a spatial distinction between individual subjects and their embodied encounters with their (atmospheric) environments. Bille and Simonsen held that affect should be conceptualised as situated in embodied practices and "that these affective practices are spatially embedded and felt phenomena'' (ibid.: 296). Rather than seeing atmosphere and affect "as something *in* the relation between people, place and things, they unfold *as* the relation in what we call atmospheric practices" (ibid.: 296, original italics). Attention is once again given to the inherent relationality of performances that contribute to placemaking and the production of space.

Earthly Performances

Performativity and space are integral in writings about tourism, the Anthropocene and earthly tourism (Gren & Huijbens, 2016; Huijbens, 2021; Huijbens, Chapter 19 of this volume). The focus is on the performativity of tourism and how, as an ordering (Franklin, 2004), it affects the Earth and is partly responsible for the present-day climate emergency. The focus on tourist performance and the performativity of space has contributed to blurring the boundaries between the spheres of nature and the traditional definition of society. The notion of the Anthropocene further affirms how geologies are part of people's everyday lives. According to Huijbens (2021: 2),

> [W]ith the looming catastrophe of climate emergency the Earth has trans-muted from being perceived as a background surface for human actions and inscriptions to being a dynamically foregrounded matter of concern, arising both from the effects of the geoforce of humanity itself and our science-mediated attempts to map and exploit earthly functions.

The Anthropocene underlines the coexistence of humans and the Earth and the cocreation of geologies and biographies (Pálsson & Swanson, 2016). Human performances have ontological repercussions and entangle with what has usually been regarded as non-human laws of nature. The notion of the Anthropocene has been criticised for masking or reducing local and contextual differences and implying universalism. The multiple relations and performances through which the 'geoforce of humanity' is enacted therefore require scrutiny.

Pálsson and Swansson recommended using the concept of geosociality to explore multiple layers of relations through which Earth systems and human and more-than-human lives perform together. Geosocial performances bypass common distinctions between macro and micro levels or the global and local to provide openings to address and trace situated performances (Valtonen & Rantala, 2020). Studies that have attempted to do so have appeared in recent years. One example is Huijben's account (2020), which deals with the seeming contradiction evident in tourist performances in Iceland in that tourists perceive the landscape they experience as an untouched wilderness, even though it is the planned, modified landscape of geothermal energy production (see, e.g., Huijbens, 2021). Another example is Valtonen, Rantala and Salmela's (2020) study, which included rocks and minerals in their account of walking through a Finnish national park, exploring the entanglement of geological performances and human mobilities in the context of place becoming. In sum, the idea of the performativity of earthly relations highlights the politics of performance and the performativity of power relations through which tourism places are created.

Future Agenda

The focus on tourism's performativity and spatial relations has affected tourism geography research in many ways. The performance turn grew in prominence parallel to the increased attention given to mobility as an organising principle of social life at the turn of the century, drawing on different conceptual currents. A discernible move has occurred, from the focus on human performances in relation to landscapes, objects and things, towards more earthly or material heterogeneous relations through which tourism spaces are seen to emerge. It is difficult to separate the focus on performativity from a more general move towards post-humanist approaches in tourism research, including feminism, new materialism and material phenomenology. The future agenda for research is likely to relate to the three tropes of scholarship briefly discussed above. Tourism research on the climate emergency eliciting possible change processes and responsible tourism development must be prioritised. This type of research involves a continuous effort to grasp multiple topologies of the geosocial relations that tourism contributes to and is shaped by. Accordingly, the focus on lived practices and embodied encounters is also crucial for rendering the geoforce of humanity meaningful in the everyday lives of people on the move.

References

Allen, J. (2011). Topological twists: Power's shifting geographies. *Dialogues in Human Geography, 1*(3), 283–298.

Anderson, B. (2014). *Encountering Affect: Capacities, Apparatuses, Conditions.* Ashgate.

Bille, M., & Simonsen, K. (2019). Atmospheric practices: On affecting and being affected. *Space and Culture, 24*(2), 295–309.

Butler, J. (1990). *Gender Trouble.* Routledge.

Butler, J. (1993). *Bodies that Matter: On the Discursive Limits of Sex.* Routledge.

Bærenholdt, J. O., Haldrup, M., Larsen, J., & Urry, J. (2004). *Performing Tourist Places.* Ashgate.

Coleman, S., & Crang, M. (Eds.) (2002). *Tourism: Between Place and Performance.* Berghahn Books.

d'Hauteserre, A.-M. (2015). Affect theory and the attractivity of destinations. *Annals of Tourism Research, 55,* 77–89.

de Laet, M. (2000). Patents, travel, space: Ethnographic encounters with objects in transit. *Environment and Planning D: Society and Space, 18,* 149–168.

Edensor, T. (1998). *Tourists at the Taj: Performances and Meaning at a Symbolic Site.* Routledge.

Edensor, T. (2000). Staging tourism: Tourists as performers. *Annals of Tourism Research, 27*(2), 322–344.

Edensor, T. (2001). Performing tourism, staging tourism: (Re)producing tourist space and practice. *Tourist Studies, 1*(1), 59–81.

Edensor, T. (2010). Aurora landscapes: Affective atmospheres of light and dark. In K. Benediktsson, & K. A. Lund (Eds.), *Conversations with Landscape* (pp. 227–240). Ashgate.

Edensor, T. (2012). Illuminated atmospheres: Anticipating and reproducing the flow of affective experience in Blackpool. *Environment and Planning D: Society and Space, 30,* 1103–1122.

Edensor, T., & Sumartojo, S. (2015). Designing atmospheres: Introduction to special issue. *Visual Communication, 14*(3), 251–265.

Flyvbjerg, B. (2001). *Making Social Science Matter: Why Social Inquiry Fails and how it can Succeed again.* Cambridge University Press.

Franklin, A. (2004). Tourism as an ordering: Towards a new ontology of tourism. *Tourist Studies, 4*(3), 277–301.

Goffman, E. (1971). *The Presentation of Self in Everyday Life.* Penguin.

Gregson, N., & Rose, G. (2000). Taking Butler elsewhere: Performativities, spatialities and subjectivities. *Environment and Planning D: Society and Space, 18,* 433–452.

Gren, M., & Huijbens, E. H. (Eds.). (2016). *Tourism and the Anthropocene.* Routledge.

Harvey, P. (2012). The topological quality of infrastructural relation: An ethnographic approach. *Theory, Culture & Society, 29*(4/5), 76–92.

Hetherington, K. (1997). In place of geometry: The materiality of place. In K. Hetherington, & R. Munro (Eds.), *Ideas of Difference: Social Spaces and the Labour of Division* (pp. 183–199). Blackwell.

Huijbens, E., & Jóhannesson, G. T. (2019). Tending to destinations: Conceptualising tourism's transformative capacities. *Tourist Studies, 19*(3), 279–294.

Huijbens, E. H. (2021). Earthly tourism and travel's contribution to a planetary genre de vie. *Tourist Studies, 21*(1), 108–118.

Jóhannesson, G. T., & Lund, K. A. (2017). Aurora Borealis: Choreographies of darkness and light. *Annals of Tourism Research, 63,* 183–190.

Jóhannesson, G. T., Lund, K. A., & Ren, C. (2018). Making matter in the midst of things: Engaging with tourism imponderables through research. In C. Ren, G. T. Jóhannesson, & R. Van der Duim (Eds.), *Co-Creating Tourism Research: Towards Collaborative Ways of Knowing* (pp. 39–54). Routledge.

Larsen, J. (2005). Families seen sightseeing: Performativity of tourist photography. *Space and Culture, 8*(4), 416–434.

Larsen, J. (2012). Performance, space and tourism. In J. Wilson (Ed.), *The Routledge Handbook of Tourism Geographies* (pp. 67–73). Routledge.

Larsen, J., & Urry, J. (2011). Gazing and performing. *Environment and Planning D: Society and Space, 29*(6), 1110–1125.

Law, J., & Mol, A. (2001). Situating technoscience: An inquiry into spatialities. *Environment and Planning D: Society and Space, 19*(5), 609–621.

Law, J., & Urry, J. (2004). Enacting the social. *Economy and Society, 33*(3), 390–410.

Lund, K. A., & Jóhannesson, G. T. (2014). Moving places: Multiple temporalities of a peripheral tourism destination. *Scandinavian Journal of Hospitality and Tourism, 14*(4), 441–459.

Lury, C. (2013). Topological sense-making: Walking the Mobius Strip from cultural topology to topological culture. *Space and Culture, 16*(2), 128–132.

Lury, C., Parisi, L., & Terranova, T. (2012). Introduction: The becoming topological of culture. *Theory, Culture & Society, 29*(4/5), 3–35.

MacCannell, D. (1976). *The Tourist: A New Theory of the Leisure Class*. Schocken Books.

Massey, D. (2005). *For Space*. Sage.

Nash, C. (2000). Performativity in practice: Some recent work in cultural geography. *Progress in Human Geography, 24*(4), 653–664.

Pálsson, G., & Swanson, H. A. (2016). Down to Earth: Geosocialities and geopolitics. *Environmental Humanities, 8*(2), 149–171.

Sheller, M., & Urry, J. (Eds.). (2004). *Tourism Mobilities: Places to play, places in play*. Routledge.

Thrift, N. (1996). *Spatial Formations*. Sage.

Thrift, N. (1999). Steps to an ecology of place. In D. Massey, J. Allen, & P. Sarre (Eds.), *Human Geography Today* (pp. 295–322). Cambridge: Polity Press.

Thrift, N. (2004). Intensities of feeling: Towards a spatial politics of affect. *Geografiska Annaler: Series B, Human Geography, 86*(1), 57–78.

Valtonen, A., & Rantala, O. (2020). Introduction: Reimagining ways of talking about the Anthropocene. In A. Valtonen, O. Rantala, & P. D. Farah (Eds.), *Ethic and Politics of Space for the Anthropocene* (pp. 1–15). Edward Elgar.

Valtonen, A., Rantala, O., & Salmela, T. (2020). Walking with rocks – with care. In A. Valtonen, O. Rantala, & P. D. Farah (Eds.), *Ethic and Politics of Space for the Anthropocene* (pp. 35–50). Edward Elgar.

Valtonen, A., Salmela, T., & Rantala, O. (2020). Living with mosquitoes. *Annals of Tourism Research, 83*, 102945.

11

EMBODIED ENCOUNTERS IN TOURISM GEOGRAPHIES RESEARCH

Phoebe Everingham, Pau Obrador-Pons and Hazel Tucker

Introduction

In the first edition of this handbook, Chris Gibson highlighted that "[a]nalysis of tourism's encounters is now more attentive to how bodies and materials interact in fluid, complicated ways – and the spaces in which these encounters take place" (2012: 56). Indeed, tourism research has come a long way since the turn of the century when Franklin and Crang, in the opening editorial of the journal *Tourist Studies,* viewed it as "stale, tired, repetitive and lifeless" (2001: 5). Instead, they called for "the development of critical perspectives on the nature of tourism as a social phenomenon" (Franklin & Crang, 2001: 6). There is now an exciting body of work that speaks to the multi-layered phenomena of contemporary tourism spaces, mobilities and encounters from within and beyond the tourism geographies field. Much of this research is explicitly concerned with multisensorial notions of embodiment. As Gibson highlights, "beyond sight, researchers are now analysing the other senses and how encounters are experienced in an affective, embodied fashion, through touch, sound and taste" (2012: 57). Crucially, a clear link can be established between these conceptual departures and the changing nature of tourism at the turn of the twenty-first century, for a wider range of concepts and theories to reflect on an activity that was no longer the minor ritual of modern life that tourism theory founders MacCannell and Urry encountered, but "a significant modality through which transnational modern life is organised" (2001: 6–7).

These conceptual departures away from static notions of tourism encounters are embedded within the turn in social sciences towards ideas of embodiment, sensuality and performativity. In line with other fields, tourist scholars have paid increasing attention to "the agentive, embodied role of the tourist" (Crouch et al., 2001: 253) and considered how touristic encounters turn our attention to what tourists *do* – how the body is proactively engaged in and with space. Attention to what tourists do centres agency and process rather than prefigured and static notions of tourism spaces. Crouch, Aronsson and Walhströn (2001) examined the geographical implications of this turn, by going beyond theorising tourism spaces as just a destination or context in which tourism merely 'happens'. Tourism spaces are at least in part "constructed and signified by the tourist", a medium where "tourists negotiate [their] world, tourism signs and contexts, and may construct [their] own distinctive meanings" (Crouch et al., 2001: 254). Theorised in this way, tourism practices are contextualised through individuals, within encounters and 'body-practices' (Ibid., 2001).

DOI: 10.4324/9781003286301-13

The embodied turn has also led to the consideration of a wider range of bodies in tourism research beyond its traditional focus on male, white and middle-class figurations. Theorising the body in tourism geographies, we find gendered female bodies (Aitchinson, 2005; Brown et al., 2020), racialised bodies (Saldanha, 2002; Putcha, 2020), drunk euphoric bodies (Jayne, 2012; Tutenges, 2015), young bodies (Kimber et al., 2019; Garcia, 2015; Tucker, 2005), old bodies (O'Reilly, 2003; Holloway et al., 2011; Tucker, 2005), bodies at work (Harris, 2009; Veijola, 2009), LGBTQI+ bodies (Vorobjovas-Pinta & Robards, 2017; Binnie & Klesse, 2011), medicalised bodies (Cook, 2010; Bell et al., 2011), bodies having sex (Collins, 2007; Frohlick, 2008) and all sorts of active and thrilled bodies. Such a diverse range of bodies within the research field that analyses tourism, show the extent to which tourism provides "a unique space to explore the role of such expressive, sensual and illusory faculties of the body" (Obrador-Pons, 2003: 56).

In this chapter, we map the theoretical and ontological trajectories of *embodiment* within theorisations of tourism spaces and encounters and revisit some of the important conceptions of the body, beginning with the tourist gaze in considering the role of the senses in mediating tourism experiences and encounters. Going beyond the gaze and considering the importance of the 'multisensual', we also consider how more recent theorisations of embodiment, including non-representational analyses, relational materiality, affect, more-than-representational and more-than-human, can further this important trajectory in studies of tourism geographies now and in the future.

From the Tourist Gaze to the Multisensorial Body

John Urry's (1990) influential concept of the 'tourist gaze' set the stage for discussions on the body in theorising tourism spaces. The advent of the tourist gaze has substantially enriched our understanding of the constitution of the tourist subject, advancing tourist debates beyond a traditional focus on visual representations and the framing of destinations in the creation of tourist desire. However, the sort of research that Urry inspired often reproduced a mind/body dualism which portrayed tourists as detached, disembodied and passive observers, effectively reducing them to a disembodied pair of eyes. As Veijola and Jokinen (1994) pointed out, the focus on the gaze ironically tended to negate the holistic ways of understanding the tourist body and the multisensorial dimensions of the embodied tourist.

Scholars have continued to highlight the limitations of the tourist gaze: for how it epitomised a masculine subject position (Pritchard & Morgan, 2000; Veijola & Jokinen, 1994); for being Western-based (Agapito et al., 2013; Chaney, 2002), for not paying sufficient attention to the practices of seeing (Crang, 1997), for disregarding the multiple sensual configurations of tourism (Edensor, 2001; 2006), for narrowly defining the subject/object of the gaze, and for ignoring the complex social relations involved in viewing and gazing (Gillespie, 2006; Maoz, 2006). There is also a significant body of work that has furthered this trajectory of complicating and situating the tourist gaze; repositioning the dematerialised visualities of tourism within an embodied and sensual space. One of the most prominent contributions is Haldrup and Larsen's (2003) work on the family gaze, which highlights the extent to which tourist photography engages significant others and is part of what we might call the family theatre that enables people to enact and produce a sense of intimacy and togetherness. Relationality, performativity and power are further explored through work on tourists gazing at each other (Holloway et al., 2011), locals gazing at tourists (Chan, 2006), migrant workers gazing at migrant tourists (Moufakkir, 2019), and the enactment of the gaze in non-Western contexts (Zara, 2015). This trajectory of work not only engages with the embodied and multi-sensuous nature of gazing, but also the complex social relations and fluid power geometries constituting these numerous performances of gazing.

Inspired by the embodied and practice turn in social sciences, tourism research has increasingly challenged the occulocentrism of tourism research and its tendency to collude "in writing the body out of tourism" (Franklin & Crang, 2001; 14). The influence of the embodied turn in tourism research is felt the most in the sensory and performative explorations of the tourist experience (Cohen & Cohen, 2019), opening up novel trajectories in tourism research away from its traditional focus on vision. There is now a plethora of sensory studies in tourism research, many of which highlight a specific touristscape based on a sense other than sight (Agapito, 2020). For example, soundscapes have been articulated through the work of Garcia (2016) on techno tourism in Berlin, where electronic music is a thing to be toured itself. Saldanha's (2002) work on the psychedelic rave scene in Goa, which explores the ability of music to organise 'factions' of bodies. Researchers have also been sensitive to other sounds beyond music, including the embodied sonic experiences of camperavan travel (Wilson et al., 2019), where the relation between tourist and the machine is equivalent to the relation of the musician with his/her instrument. Taste is a sensescape that has received significant attention in tourism research, with many articles focussing on the role food and drink play in articulating the tourist experiences. For example, Bezzola and Lugosi (2018) look at how tourists negotiate a sense of home and away through the consumption of food and drink, while Everett (2009) uses food to draw a more heterogeneous sensory landscape where tourists are immersed in waves of taste, smell, sound and touch. Likewise, the alterations of tourist bodies with psychoactive substances are manifested in the work of Jayne et al. (2012) on alcohol, drinking and drunkenness within the context of backpacking. There are a plethora of sensory studies of wine tourism, the latest example of which is Brochado, Stoleriu and Lupu´s article (2021). While these kinds of sensory experiences in tourism may seem 'trivial', Tarulevicz and Ooi (2021) have pointed to the ways that the embodied sense of taste is in fact raced and hierarchical. Drawing on the role of food in Singapore, they describe how 'food safety' regulations shed light on the tensions between high and low culture. The connection between the senses and broader constructions of class has also been explored (De Jong & Varley, 2017). We hope to see more analyses of the connections between embodiment and the senses in relation to class culture and ethnicity in coming years.

Despite such advances in the critical reflection and understanding of tourism as underpinned by multisensory active doings and performances, the sense of touch remains relatively elusive. One of the most interesting contributions is Merchant's (2011) work on scuba diving which examines the rearrangement of the sensorium underwater. The importance of haptic sensualities in tourism has been successfully addressed in discussions relating to visually impaired people. The works of Hetherington (2003), Larsen and Savavo (2014) and Macpherson (2009) on the tactile experiences of visually impaired people in museums and the countryside are particularly interesting. In addition to the sensations of pressure and contact, the haptic senses include kinaesthesia (the sense of movement), proprioception (sense of bodily position), and the sense of balance. These other haptic sensations feature in tourism related discussions on speed, slow and thrill. For example, haptic sensations of kinaesthesia underpin the popularity of German Autobahns as a tourist destination in what Gross (2020) calls 'Speed Tourism'. A more thorough engagement with the complexities of touch would substantially enrich tourism research.

Traditionally, and perhaps influenced by anthropological approaches to examining culture, tourist research has focused on cultural representation of tourist bodies, giving priority to the image over the actions of the body. Indeed, tourism journals are packed with contributions exploring the extent to which elaborations and representations of the body in tourism are rooted in colonial (Kothari, 2015) and heteropatriarchal discourses (Pritchard & Morgan, 2005), reflecting on the close intimacy of tourism with colonialism (Hall & Tucker, 2004; Putcha, 2020). However, the body does not merely feature as a thing or a concept, a surface of inscription, a container of meaning or a plastic matter for sculpting. Rather, in tourism research the body is also an active, expressive and sensual force, a body-subject with the ability to configure the tourist experience; a body that is simultaneously situated and situational, object and subject. In this broader frame-work, there are many forms of tourism that do not prioritise the sense of sight. One clear example of the significance of the body in tourist experiences is that of the beach, with its capacity to reconfigure tourist bodies through a series of embodied pleasures, like swimming, sunbathing and building sandcastles (Diken & Laustsen, 2004; Obrador-Pons, 2007, 2009, 2012; Franklin, 2014), This focus on pleasure has been systematically ignored in tourism research. As Arun Saldnaha notes: "Don't tourists swim, climb, stroll, ski, relax, become bored perhaps, or ill; don't they go to other places to taste, smell, listen, dance, get drunk, have sex?" (2002: 43). In this quote, Saldanha reminds us that tourism is not so much a cognitive intellectual experience, but a matter of doing and seeking out fun and pleasurable sensations. Tourism experiences involve a practical sensual embodied engagement with the world. In the next section we consider how materialities are also embodied within tourism spaces.

From the Bounded Body to Relational Ontology/Materiality

An exploration of the geographies of the beach also opens up an exploration of the importance of materiality to touristic experiences as well as relationality and the capacity of the tourist body to be affected and disrupted by other agencies, human and non-human. A prominent example is Franklin's (2014) article on the bucket and spade as a foundational element in the relational materialism of the beach. Drawing on the work of Haraway, Ingold and Latour, Franklin's dis-cussion is pivotal in its post-humanist understanding of tourism as "a gathering together of a heterogeneous community of agencies and objects with whom humans must, as they do every-where, negotiate relationships" (2014: 263). Franklin thus points out that while the embodi-ment turn has been crucial in tourism/tourist studies, it runs "the risk of over-emphasising the human body at the expense of other non-human agencies in the environment" (2014: 267); Human action "should never of course be privileged over the action of the world it connects to" (2014: 268).

Expanding the scope of the senses and embodiment, a relational materiality perspective was brought to the fore in Walsh and Tucker's (2009) consideration as to how the material object of the backpack is implicated in the social world of backpackers. Walsh and Tucker argued there is a need to recognise that material objects do important things: "Beyond the realm of representation, the backpack is critical in backpackers' lived experience" (2009: 235). By highlighting "the affective relation between self and 'thing'" (2009: 225), Walsh and Tucker contend that their analysis "thus collapses any real distinction between sociality and materiality" (2009: 235). Hence, the adoption of the embodied turn has not only led to the expansion of our understandings of the role of the body in tourism, but to an ontological recognition of the essential materiality of the world and, in this, the important role of non-human actors in tourism.

The ontologies and methodologies of relational materialism have thus become prominent in tourism research, not only in terms of the extension and elaboration they offer to ideas of embodiment, but also in their decentring of the human in tourist studies. This is highlighted in Benali and Ren's (2019) discussion of head-lice in a volunteer tourism context in Nepal. Like Franklin, Benali and Ren (2019, 239) highlight the need to move beyond binary subject–object positions and to recognise the "ongoing ontological choreography also involving human and non-human actors, bodies and affects". Showing lice to be useful "analytical entry points to describe the disruptive power of bodies and affect", Benali and Ren (2019: 253) developed a clear understanding of how lice were an active element in shaping the emotions and experiences of volunteer tourists.

Indeed, the ability for the tourist body both to affect *and to be affected* was recognised in Fullagar's (2001) path-breaking analysis of Alphonso Lingis' travel writing. Fullagar (2001: 178) read Lingis as being troubled by the mediating effects of Western discourses and images that work to codify his experience, instead wanting "his flesh to be porous, open to the movements of the other's body". In this sense, travel, conceptualised as "an ethical encounter requires an openness to the affect of the other, at the risk of decentring the self" (Fullagar, 2001: 174). This idea was picked up in Waitt, Figueroa and McGee's (2007: 252) vivid account of the potential for tourist encounters "to be disrupted by the fleshiness of the body" and likewise in Tucker's (2009) discussion of the postcolonial potentiality of 'moral discharge' emanating from the body in tourism encounters. The importance of paying attention to affect in tourism research is highlighted by Duff's (2010: 881) argument, pertaining to the 'affective turn' in geography, that "to experience place is to be *affected by place*". Such a relational materialist view which "blurs and recasts conventional understandings of subjectivities and power relations" (Benali & Ren, 2019: 240), as well as understanding the tourist body as a relational and situated site of affects, is an ontological position which we hope will feature more prominently in studies of tourism. Likewise, the acceptance of the porosity of the body and its vulnerability to the disruptive power of affect is itself an ethical position which signals the need for future work that is more clearly engaged with tourism and ethics. This point links next to acknowledging the non-representational thinking that has helped further affective trajectories in tourism research.

From Non-representational Thinking to Affective Emergences

In 2003, Obrador-Pons introduced non-representational theories into tourism research to further centre the importance of the dynamic processes of meaning making that occur within encounters in tourism. This approach, Obrador-Pons argued, is key when seeking to move beyond the "essentialised and decontextualised meta-narratives of being in modernist and postmodernist theories that leave lived and situated subjects underscrutinised", thus overcoming "Cartesian divisions between subjects and objects, material and spiritual, facts and fetish" (2003: 48). Drawing on Thrift, Obrador-Pons (2003: 55) situates embodiment not just as 'a body', but as a 'body-subject' engaging in "joint body practices of becoming". This framing opens up new opportunities for conceptualising the tourist subject as embedded in relationalities and practices. Non-representational style of thinking, later renamed as more-than-representational by Lorimer (2005), allows an analysis of the body-subjects experiences which are expressive, sensual and perhaps most importantly, elusory. Crucially, a non-representational analysis pays attention to the tourism experiences that cannot be codified into the discursive realms, unsettling the presupposition of 'a stable and ordered subject' (Obrador-Pons, 2003: 56). It is within these more-than-representational realms of experiences that allow for analyses "to make new connections and assemblages", to understand the capacities of embodiment to create "meaning in such a way that refuses the objectifying gaze and the control of the discourse" (2003: 56).

These expansive notions of embodiment have provided tourism research with valuable opportunities to draw in more-than-rational notions of what constitutes tourism experiences. The notion of pleasure, for example, is something that cannot be made sense of merely through discursive analysis, instead it requires an attunement to embodied practices, feelings and dispositions. Indeed, the pleasures of tourism are rarely an intellectual or semiotic game but a creative exercise of engaging with a material and changing environment. There are many echoes of more-than-representational thinking in tourism research, a good example of which is the work of Prince (2018) on bodily movements and mundane practices of the craft-artists in Bornholm, Denmark. Drawing on Lorimer′s (2005) 'more-than-representational' analysis of the everyday 'taskscape'. Prince's work shows the extent the tourist landscape is the product of embodied skills and techniques rather than simply arising out of the spectacular eye. However, and following Lorimer (2005: 89), Prince looks at the 'fleshiness and pliability of bodies' as a way to unsettle and extend the scope of the tourist subject rather than as something completely opposed to representation. There is an emerging presence of more-than-representational theories in tourism research that (re)configures life as "a series of infinite 'ands' which add to the world rather than extract stable representations from it" (Cadman, 2009: 456). Nevertheless, there remains the challenge as to how the unconscious and non-discursive can contribute to "a dominant disciplinary paradigm of power exercised for political resistance and progressive emancipation" (Lorimer, 2005: 90). While 'more-than-representational' analysis should continue to unsettle discursive realms that fix notions of power, these analyses also need to recognise the discursive domains that shape affective experiences.

There is therefore a danger that exists when developing insights into sensual and affective bodies of losing a political dimension of tourism. As Cohen and Cohen (2019) explain, the study of sensory experiences of tourists is dominated by psychological approaches and rarely examines the link between embodiment and capitalism within the context of tourism. Exceptions to this include Minca's (2009) reflection on the relationship between work, bodies and space in bio-political terms and the life, emotional and performative encounters between host and guest. Modlin et al. (2011) use affect to politicise the storytelling narratives at plantation house museums in the United States, where slavery is misrepresented or even ignored while planter-class families are foregrounded and represented with emotionally evocative accounts by tour guides. This is a process they refer to as *affective inequality*. More broadly within tourism research, the work of Mostafanezhad (2013; 2018) and Crossley (2012) have paved the way for developing connections between affect and broader structures of power in volunteer tourism in theorising the links between the moral economies of neoliberalism and geographies of compassion and empathy. This work reminds us that tourism mobilities are deeply unequal, with tourism embedded within broader practices of colonial development and neoliberalising forces. Debates around the affective realms of tourism spaces must acknowledge the broader politics these encounters are embedded within.

A 2015 special issue on volunteer tourism brings some of these issues and tensions between broader structural power dynamics and the micro world of embodied encounters to the fore. In the introduction to the special issue, Sin et al. (2015: 124) ask "[h]ow do we critique development or neoliberalism while not being dismissive of meaningful or affective experience? How do we link the personal and the social?". Griffiths (2015: 206) brings some important insights to these issues by arguing that affective experiences do not always "defer to forms of power but may actually emerge from the 'dynamism' of the body and its capacity to affect and be affected". Even when volunteer tourists lack critical perspectives on development, neo-colonialism and neoliberalism, the intensities that are experienced on the ground, within encounters cannot "so readily ascribe to constellations of power" (Griffiths, 2015: 216). Indeed, the capacity to affect and be affected can work to reinforce as well as disrupt some formations of power in volunteer tourism. However, attention to affect can give researchers insights into possibilities for decolonising these encounters (Everingham & Motta, 2022), broadening the scope of what is deemed 'political'. In exploring the connections between the structural and discursive relationships to affect, theorists should take seriously the ambivalences that arise from bodily disruptions and consider both the affective closures (that are shaped by broader structures of power) and affective openings (which contain possibilities to exceed and /or be autonomous from them (Everingham & Motta, 2022).

The more-than-representational realms of affect, then, can draw attention to transformative possibilities for subverting broader structural power dynamics within tourism encounters. An article by Bawaka Country et al. (2017) brings the affective dimensions of human and more-than-human agencies to the fore, describing the inter-relationalities between Bawaka Country and humans in an indigenous-led tourism venture in north east Arnhem Land in Australia. Here, opportunities exist for tourists to experience a different way of being in the context of a colonial settler state; to "co-become with country" becoming "part of its ongoing co-constitution". Emotions and sensory experiences therefore "actively shape and enable transformative learning for tourists" (Ibid.: 446). Here, the more-than-representational includes agencies of more-than-human and indigenous ontologies. In terms of using this lens to understand how tourism can enable transformative experiences, this analysis captures how more-than-human agencies can "unsettle pre-conceived understandings of other people and other places, producing potentially transformative feelings of uneasiness, awe and wonder, and/or opening moral and ethical dilemmas" (2017: 445).

Also taking a non-Western lens to embodiment, Zhang, Cai and Qiu (2019: 63) explore how collective memory and emotion "play an important role in the maintenance of modern life", particularly when negotiating 'strangeness and otherness'. Their study focuses on Chinese tourists in Japan to "investigate how the collective memory of Japan has shaped the Chinese Cohort 60's emotions and views of Japan, which subsequently influenced their travel decision-making in regard to Japan" (2019: 81). Focusing on the personal and the embodied, important insights are gained between how theorists understand 'memory, emotion, and time' and the ways in which the Chinese negotiate with the 'outside word' (including ideological friends and opponents, trade partners and competitors, and former war enemies). Exploring the subjectivities of Chinese outbound tourists offers Tourism Studies, a non-Eurocentric lens to explore "everyday geopolitical encounters between hosts and guests" (Mostefanezhad, 2018: 343), considerations of national and ethnic identities (Ooi, 2005) and contested urban landscapes related to insider/outsider identities (Chang, 2000). Attention to the mediated historical geographies of race and geoeconomics relations brings much needed insights into the political geographies of tourism (Mostefanezhad et al., 2020). We hope to see more of these insights over the coming years. Attention to embodiment in tourism in a time of broader geo-political shifting will be crucial, particularly in relation to ensuring equity and inclusion of various bodies.

Moving Forward

In this chapter, we have reminded readers of what theorising embodiment can bring to understanding encounters in tourism spaces, and specifically how attention to embodiment moves analysis away from fixed and static notions of culture and power, towards dynamic interplays between bodies and more-than-human modalities. We have paid special attention to expanding notions of embodiment, thus incorporating the non-human into affective entanglements. Tourism research has not only incorporated the relationalities of how sights, smells and sensations are central to the 'tourism experience', it has also shown how non-representational theories and the realms of the more-than-representational can be utilised to address gaps in human experiences that cannot solely be explained through discursive representation. In other words, being attentive to these affective realms creates the possibilities for transformative tourism encounters and experiences that disrupt discursive understandings of power and agency.

The situation of tourism is now changing rapidly, with growing restrictions to international mobility and leisure, from Brexit to climate change. Indeed, the disruption to the normal functioning of tourism brought by the COVID-19 pandemic has been unprecedented. There is a danger that *bodies* are marginalised as increasing attention is paid to the problems and limitations of tourism in the context of global challenges such as COVID-19, climate change and shifting power dynamics between nation states. Future research in tourism should address the changing nature of tourism without losing sight of an expansive, affectual, multi-sensual body. However, the future conceptual trajectories of embodiment will have to draw on different, probably more biopolitical, articulations of embodiment. Thus recognising the materiality of the world as tourism encounters will take place in a fast changing and increasingly precarious environment, in a context of shifting geo-politics and global hegemonies. The relational and multi-sensorial analyses of embodiment that we have reviewed in this chapter provide a platform from which to explore the interplay of tourist bodies in and with this fast-changing world.

References

Agapito, D. (2020). The senses in tourism design: A bibliometric review. *Annals of Tourism Research, 83*, 1–15.

Agapito, D., Mendes, J., & Valle, P. (2013). Exploring the conceptualization of the sensory dimension of tourist experiences. *Journal of Destination Marketing and Management, 2*(2), 62–73.

Aitchison, C. (2005). Feminist and gender perspectives in tourism studies: The social-cultural nexus of critical and cultural theories. *Tourist Studies, 5*(3), 207–24.

Country, B., Wright, S., Lloyd, K., Suchet-Pearson, S., Burarrwanga, L., Ganambarr, R., Maynuru, D., & Tofa, M. (2017). Meaningful tourist transformations with Country at Bawaka, North East Arnhem Land, northern Australia. *Tourist Studies, 17*(4), 443–467.

Bell, D., Holliday, R., Jones, M., Probyn, E., & Taylor, J. S. (2011). Bikinis and bandages: An itinerary for cosmetic surgery tourism. *Tourist Studies, 11*(2), 139–155.

Benali, A., & Ren, C. (2019). Lice work: Non-human trajectories in volunteer tourism. *Tourist Studies, 19*(2), 238–257.

Bezzola, T., & Lugosi, P. (2018). Negotiating place through food and drink: Experiencing home and away. *Tourist Studies, 18*(4), 486–506.

Binnie, J., & Klesse, C. (2011). 'Because it was a bit like going to an adventure park': The politics of hospitality in transnational lesbian, gay, bisexual, transgender and queer activist networks. *Tourist Studies, 11*(2), 157–174.

Brochado, A., Stoleriu, O., & Lupu, C. (2021). Wine tourism: A multisensory experience. *Current Issues in Tourism, 24*(5), 597–615.

Brown, L., de Coteau, D., & Lavrushkina, N. (2020). Taking a walk: The female tourist experience. *Tourist Studies, 20*(3), 354–370.

Chan, Y. W. (2006). Coming of age of the Chinese tourists: The emergence of non-Western tourism and host—guest interactions in Vietnam's border tourism. *Tourist Studies, 6*(3), 187–213.

Cadman, L. (2009). Non-representational theory/non-representational geographies. In R. Kitchin, & N. Thrift (Eds.), *International Encyclopaedia of Human Geography* (pp. 456–63). Elsevier.

Chaney, D (2002). The power of metaphors in tourism theory. In M. Crang, & S. Coleman (Eds.), *Tourism: Between Place and Performance* (pp. 193–206). Berghahn.

Chang, T. C. (2000). Singapore's Little India: A tourist attraction as a contested landscape. *Urban Studies, 37*(2), 343–366.

Cohen, S. A., & Cohen, E. (2019). New directions in the sociology of tourism. *Current Issues in Tourism, 22*(2), 153–172.

Collins, D. (2007). When sex work isn't 'work': Hospitality, gay life, and the production of desiring labor. *Tourist Studies*, *7*(2), 115–139.

Cook, P. S. (2010). Constructions and experiences of authenticity in medical tourism: The performances of places, spaces, practices, objects and bodies. *Tourist Studies*, *10*(2), 135–153.

Crang, M. (1997). Picturing practices: research through the tourist gaze. *Progress in human geography*, *21*(3), 359–373.

Crossley, É. (2012). Poor but happy: Volunteer tourists' encounters with poverty. *Tourism Geographies*, *14*(2), 235–253.

Crouch, D., Aronsson, L., & Wahlström, L. (2001). Tourist encounters. *Tourist Studies*, *1*(3), 253–270.

De Jong, A., & Varley, P. (2017). Food tourism policy: Deconstructing boundaries of taste and class. *Tourism Management*, *60*, 212–222.

Diken, B., & Laustsen, C. B. (2004). Sea, sun, sex and the discontents of pleasure. *Tourist Studies*, *4*(2), 99–114.

Duff, C. (2010). On the role of affect and practice in the production of place. *Environment and Planning D: Society and Space*, *28*(5), 881–895.

Edensor, T. (2001). Performing tourism, staging tourism: (Re)producing tourist space and practice. *Tourist Studies*, *1*(1), 59–81.

Edensor, T. (2006). Sensing tourist space. In C. Minca, & T. Oakes (Eds.), *Travels in Paradox: Remapping Tourism* (pp. 23–46). Rowman & Littlefield.

Everett, S. (2008). Beyond the visual gaze? The pursuit of an embodied experience through food tourism. *Tourist Studies*, *8*(3), 337–358.

Everingham, P., & Motta, S. C. (2022). Decolonising the 'autonomy of affect' in volunteer tourism encounters. *Tourism Geographies*, *24*(2–3), 223–243.

Franklin, A. (2014). On why we dig the beach: Tracing the subjects and objects of the bucket and spade for a relational materialist theory of the beach. *Tourist Studies*, *14*(3), 261–285.

Franklin, A., & Crang, M. (2001). The trouble with tourism and travel theory? *Tourist Studies*, *1*(1), 5–22.

Frohlick, S. (2008). Negotiating the public secrecy of sex in a transnational tourist town in Caribbean Costa Rica. *Tourist Studies*, *8*(1), 19–39.

Fullagar, S. (2001). Encountering otherness: Embodied affect in Alphonso Lingis' travel writing. *Tourist Studies*, *1*(2), 171–183.

Garcia, L. M. (2016). Techno-tourism and post-industrial neo-romanticism in Berlin's electronic dance music scenes. *Tourist Studies*, *16*(3), 276–295.

Gillespie, A. (2006). Tourist photography and the reverse gaze. *Ethos*, *34*(3), 343–366.

Gibson, C. (2012). Geographies of tourism: Space, ethics and encounter. In J. WIlson (Ed.), *The Routledge Handbook of Tourism Geographies* (pp. 72–77). Routledge.

Griffiths, M. (2015). I've got goose bumps just talking about it!: Affective life on neoliberalized volunteering programmes. *Tourist Studies*, *15*(2), 205–221.

Gross, M. (2020). Speed tourism: The German Autobahn as a tourist destination and location of "unruly rules". *Tourist Studies*, *20*(3), 298–313.

Haldrup, M., & Larsen, J. (2003). The family gaze. *Tourist Studies*, *3*(1), 23–46.

Hall, C. M., & Tucker, H. (Eds.) (2004). *Tourism and Postcolonialism: Contested Discourses, Identities and Representations*. Routledge.

Harris, C. (2009). Building self and community: The career experiences of a hotel executive housekeeper. *Tourist Studies*, *9*(2), 144–163.

Hetherington, K. (2003). Spatial textures: Place, touch, and praesentia. *Environment and Planning A*, *35*(11), 1933–1944.

Holloway, D., Green, L., & Holloway, D. (2011). The intratourist gaze: Grey nomads and 'other tourists'. *Tourist Studies*, *11*(3), 235–252.

Jayne, M., Gibson, C., Waitt, G., & Valentine, G. (2012). Drunken mobilities: Backpackers, alcohol, 'doing place'. *Tourist Studies*, *12*(3), 211–231.

Kimber, S., Yang, J., & Cohen, S. (2019). Performing love, prosperity and Chinese hipsterism: Young independent travellers in Pai, Thailand. *Tourist Studies*, *19*(2), 164–191.

Kothari, U. (2015). Reworking colonial imaginaries in post-colonial tourist enclaves. *Tourist Studies*, *15*(3), 248–266.

Larsen, J., & Svabo, C. (2014). The tourist gaze and "Family Treasure Trails" in museums. *Tourist Studies*, *14*(2), 105–125.

Lorimer, H. (2005). Cultural geography: The busyness of being 'more-than-representational'. *Progress in Human Geography, 29*(1), 83–94.

Macpherson, H. (2009). The intercorporeal emergence of landscape: Negotiating sight, blindness, and ideas of landscape in the British countryside. *Environment and Planning A, 41*(5), 1042–1054.

Maoz, D. (2006). The mutual gaze. *Annals of Tourism Research, 33*(1), 221–239.

Merchant, S. (2011). Negotiating underwater space: The sensorium, the body and the practice of scuba-diving. *Tourist Studies, 11*(3), 215–234.

Minca, C. (2009). The island: Work, tourism and the biopolitical. *Tourist Studies, 9*(2), 88–108.

Modlin Jr., E. A., Alderman, D. H., & Gentry, G. W. (2011). Tour guides as creators of empathy: The role of affective inequality in marginalizing the enslaved at plantation house museums. *Tourist Studies, 11*(1), 3–19.

Mostafanezhad, M. (2013). The geography of compassion in volunteer tourism. *Tourism Geographies, 15*(2), 318–337.

Mostafanezhad, M. (2018). The geopolitical turn in tourism geographies. *Tourism Geographies, 20*(2), 343–346.

Mostafanezhad, M., Cheer, J. M., & Sin, H. L. (2020). Geopolitical anxieties of tourism: (Im)mobilities of the COVID-19 pandemic. *Dialogues in Human Geography, 10*(2), 182–186.

Moufakkir, O. (2019). The liminal gaze: Chinese restaurant workers gazing upon Chinese tourists dining in London's Chinatown. *Tourist Studies, 19*(1), 89–109.

Obrador-Pons, P. (2003). Being-on-holiday: Tourist dwelling, bodies and place. *Tourist Studies, 3*(1), 47–66.

Obrador-Pons, P. (2007). A haptic geography of the beach: Naked bodies, vision and touch. *Social & Cultural Geography, 8*(1), 123–141.

Obrador-Pons, P. (2009). Building castles in the sand: Repositioning touch on the beach. *The Senses and Society, 4*(2), 195–210.

Obrador-Pons, P. (2012). Touching the beach. In M. Paterson, & M. Dodge (Eds.), *Touching Space, Placing Touch* (pp. 47–70). Routledge.

O'Reilly, K. (2003). When is a tourist? The articulation of tourism and migration in Spain's Costa del Sol. *Tourist Studies, 3*(3), 301–317.

Ooi, C. S. (2005). State-civil society relations and tourism: Singaporeanizing tourists, touristifying Singapore. *Sojourn: Journal of Social Issues in Southeast Asia, 20*(2), 249–272.

Prince, S. (2018). Dwelling in the tourist landscape: Embodiment and everyday life among the craft-artists of Bornholm. *Tourist Studies, 18*(1), 63–82.

Pritchard, A., & Morgan, N. J. (2000). Constructing tourism landscapes: Gender, sexuality and space. *Tourism Geographies, 2*(2), 115–139.

Pritchard, A., & Morgan, N. (2005). 'On location': Re(viewing) bodies of fashion and places of desire. *Tourist Studies, 5*(3), 283–302.

Putcha, R. S. (2020). After eat, pray, love: Tourism, orientalism, and cartographies of salvation. *Tourist Studies, 20*(4), 450–466.

Saldanha, A. (2002). Music tourism and factions of bodies in Goa. *Tourist Studies, 2*(1), 43–62.

Sin, H., Oakes, T., & Mostafanezhad, M. (2015). Traveling for a cause: Critical examinations of volunteer tourism and social justice. *Tourist Studies, 15*(2), 119–131.

Tarulevicz, N., & Ooi, C. S. (2021). Food safety and tourism in Singapore: between microbial Russian roulette and Michelin stars. *Tourism Geographies, 23*(4), 810–832.

Tucker, H. (2005). Narratives of place and self: Differing experiences of package coach tours in New Zealand. *Tourist Studies, 5*(3), 267–282.

Tucker, H. (2009). Recognizing emotion and its postcolonial potentialities: Discomfort and shame in a tourism encounter in Turkey. *Tourism Geographies, 11*(4), 444–461.

Tutenges, S. (2015). Pub crawls at a Bulgarian nightlife resort: A case study using crowd theory. *Tourist Studies, 15*(3), 283–299.

Urry, J. (1990). *The Tourist Gaze: Leisure and Travel in Contemporary Societes*. Sage.

Veijola, S. (2009). Introduction: Tourism as work. *Tourist Studies, 9*(2), 83–87.

Veijola, S., & Jokinen, E. (1994). The body in tourism. *Theory, Culture & Society, 11*(3), 125–151.

Vorobjovas-Pinta, O., & Robards, B. (2017). The shared oasis: An insider ethnographic account of a gay resort. *Tourist Studies, 17*(4), 369–387.

Waitt, G., Figueroa, R., & McGee, L. (2007). Fissures in the rock: Rethinking pride and shame in the moral terrains of Uluru. *Transactions of the Institute of British Geographers, 32*(2), 248–263.

Walsh, N., & Tucker, H. (2009). Tourism 'things': The travelling performance of the backpack. *Tourist Studies, 9*(3), 223–239.

Wilson, S., Chambers, D., & Johnson, J. (2019). VW campervan tourists' embodied sonic experiences. *Annals of Tourism Research, 76*, 14–23.

Zara, C. (2015). Rethinking the tourist gaze through the Hindu eyes: The Ganga Aarti celebration in Varanasi, India. *Tourist Studies, 15*(1), 27–45.

Zhang, Y., Cai, L., & Qiu, S. (2019). The Chinese Cohort 60s and Japan: A journey of emotions. *Tourist Studies, 19*(1), 62–88.

PART III

Place Perspectives on Tourism Geographies

12

LANDSCAPE PERSPECTIVES ON TOURISM GEOGRAPHIES

Theano S. Terkenli

Introduction

The fact there can be no tourism or study of tourism without geography has been well established and is considered broadly indisputable (Hall & Page 2009, 2014; Pearce 1995; Squire 1994; Terkenli et al. 2021; Wilson, 2012). Geography provides not only the background and container of the tourism experience, but also the pleasures and attractions themselves, while tourism feeds on the novelty and value of nature and culture, invariably celebrating humanity. In this sense, it constitutes the most global and massive way of face-to-face interpersonal means of human contact with the world (Terkenli, 2018). Furthermore, the most immediate and direct forum/platform of this interchange between tourism and geography is landscape; one of the central and most basic conceptual units and pillars of geography. Tourism capitalises on and appropriates landscape – whether 'real', virtual/iconic or imaginary (Adler 1989; Urry 1995) – and, in doing so, through its face-to-face encounter with the world, becomes "the pleasure of practicing geography" (Terkenli 2018: 1), in the context of landscape.

Notwithstanding the centrality of landscape to tourism and vice versa, much remains to be explored and understood regarding their interrelatedness (Cartier & Lew 2005; Jiménez-García et al., 2021; Terkenli, 2014); as the authors of the prequel to this chapter claim in the previous edition of the Handbook (Knudsen et al., 2012), "landscape has only recently become an heuristic tool for understanding tourism" (p. 202). In this context, this chapter addresses, recounts and assesses efforts towards a better scientific understanding of the significance of landscape to tourism and of tourism-induced – actual and potential – changes to the landscape. Thus, it aims to explore recent and state-of-the-art inroads and advances in the interdisciplinary scientific area of the tourism-landscape interrelationship, while highlighting their range and significance for both tourism and the landscape, as well as for landscapes of tourism, in terms of theory, empirical practice, approach, policy, ethics and future prospects.

DOI: 10.4324/9781003286301-15

Theoretical/Conceptual Underpinnings of the Landscape–Tourism Relationship

Beginning once again with the definition of landscape according to the European Landscape Convention (Council of Europe 2000), it is significant that this definition encapsulates the whole trajectory of approaches and resulting body of knowledge and understanding of landscape, historically driven by geography and the social sciences: Landscape is "an area, as perceived by people, whose character is the result of the action and interaction of natural and/or human factors" (Council of Europe 2000, Chapter 1, Article 1: 9). Nonetheless, it is helpful in any analysis of the tourism–landscape relationship to trace its theoretical underpinnings, through its recent development in geographic thought. Furthermore, as regards the distinctive properties of landscape versus other geographical units of analysis (i.e., place, space, etc.), those primarily lie in landscape's property of 'relationality' (the ways in which humans interrelate with space, in body-mind-spirit) with their surrounding world. Landscape is distinct from place, as an ever-evolving, mutually informing medium of human association with the world, because it connotes a stage set and view of the world (notions of opposite/facing/across and against), with which the observer/user/spectator/tourist engages in variable ways.

During the past three to four decades, the socio-cultural constitution of landscapes has been steadily gaining ground in all scientific fields pertaining to the landscape (Cosgrove, 1998; Nassauer, 1988: 973–977; Tress & Tress, 2001; Zube & Pitt, 1981: 69–87), thus opening the way for the study of the implications of landscape for tourism and vice versa. In this broader and growing theoretical context, landscape has been conceptualised as a literal and/or metaphorical image of, and stage set for, the humanised environment (Cosgrove, 1998; Daniels & Cosgrove, 1988:1–10; Olwig, 2019; Sauer, 1925). Moreover, landscape reflects the societies that created it, like a mirror, in which an observer may read human–land interrelationships and engage in experiences of self-determination and cultural identity construction. "As a portion of land which the eye can comprehend at a glance" (Jackson, 1984: 3), landscape has been widely viewed as representing both a medium and an outcome of human perception, experience, and action, where vision predominates (Lothian, 1999: 177–198), and where people's landscape notions and preferences for certain types and characteristics of landscape are largely culturally driven, socially contested and politically/institutionally articulated (Burley and Machemer 2016; Burton 2012:51–71; Cosgrove 1998; Jackson 1984; Knudsen et al. 2012; Meinig 1979; Olwig 2019; Roe 2014).

During the 1980s and 1990s, critical approaches to landscape conceptualisation and analysis were firmly established in geography, with an emphasis on addressing the representational qualities of landscape, the contestation within landscape, and landscape as a mirror or a text to be read (Knudsen et al., 2012; Terkenli, cited in Mathewson et al., 2019: 295). More-than-representational perspectives into landscape geographies have recently revisited and further instigated interactive, performative and experiential ways of relating to landscape (Dewsbury et al., 2002: 437–440; Lorimer, 2005: 83–94; Wylie, 2007), cognitively, emotionally, and behaviourally. These ontological and epistemological advances in geography have laid the groundwork for the growth and development of an interdisciplinary scientific area of study of the interrelationship between tourism and landscape.

Knudsen et al. (2012) stress the co-constructional attribute of tourism that a "landscape approach" to tourism captures.

Landscapes, besides being material objects, are the intentional or unintentional symbols of power relations and ideology from which locals and tourists seek to make meaning wherein in making meaning those very same locals and tourists alter the underlying power relations and ideologies.

(p. 206)

Thus, landscape is key to the development, marketing/promotion, and consumption of tourism destinations, to triggering and sustaining tourism markets, and to enticing tourist dreams, fantasies, and behaviours (Terkenli, 2020). In other words, landscape is at the basis of all tourism activities; it mirrors and figures prominently in a broad spectrum of tourism practices ranging from 'sightseeing' to the overall spatial planning and management of a destination for tourism development. Indeed, all types of landscapes and places (whether spectacular or ordinary) may potentially hold interest for some type of visitor, for purposes of consumption of goods, services, activities, experiences, and so forth (Terkenli, 2020). Moreover, landscapes provide a long series of time-place-culture contingent tourist/visitor services and experiences, and so forth, pleasure, change, relaxation, excitement, education, inspiration, well-being and so forth (Crouch, 1999; Terkenli et al., 2001).

Consequently, landscapes of tourism (or 'tourism landscapes') reflect and stage recreational trends, multifunctional livelihood systems, conflicts and opportunities for employment and income generation, as well as human, cultural and natural resource management and use (Terkenli, 2020). Furthermore, landscapes of tourism represent the stage and medium through which the tourism side and the local side come together. The interface of this encounter engenders the construction, reproduction, consumption and promotion of both place/destination/landscape identity and human/ cultural identity for the different sides involved (tourists and locals) (Figure 12.1) (Terkenli, 2014).

This leads us to the confusion that exists in scholarly tourism literature around the terms 'tourism' versus 'tourist' in conjunction with the concept of landscape (Skowronek et al., 2018; Terkenli et al., 2021). The term 'tourism landscape' is intended to refer to the processes through which a landscape, activity, development and so forth is shaped to serve purposes of tourism (also,

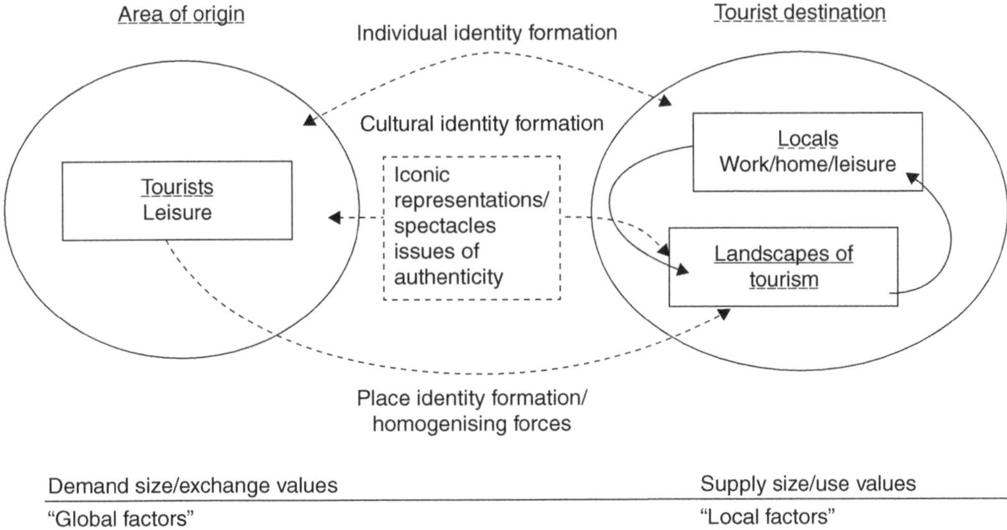

Figure 12.1   The landscapes of tourism model (after Terkenli, 2014).

i.e., 'landscapes of tourism'). The term 'tourist landscape' (or 'touristed' landscapes, according to Knudsen et al., 2012: 2) indicates the ways/reasons/processes in/through which such landscapes are substantiated and/or appropriated via the phenomenon of tourism. In other words, the term 'tourism landscapes' implies the ways in which these landscapes are produced, whereas 'tourist landscapes' rather implies the ways in which these landscapes are consumed (Terkenli et al., 2021). In the previous *Routledge Handbook of Tourism Geographies*, first edition, chapter on landscape (Knudsen et al., 2012), the term 'touristed landscapes' carries the same latter connotation and is employed in similar ways as the term 'tourist landscapes', although the authors place heavier emphasis here on the resultant state of such landscapes, stemming from tourism development (Knudsen et al., 2012).

The interrelatedness between tourism and landscape may be illustrated with the aid of a series of concepts and tools, such as ideology, power relations, performance, authenticity, semiotics, place mythology/symbolism, identity, embodied experience, and so forth (Knudsen et al., 2012). In this regard, attention may be drawn to two metaphors which have long served geographers when decoding and elucidating the multiple and fluid interpretations of this complex interrelationship, namely the metaphors of *theatre* and *mirror*. These metaphors have played a historically significant role in Western landscape representations and experiences. They may also be aptly applied to the realm of recreation/tourism practices and norms vis-à-vis landscape (Mathewson et al., 2019; Olwig, 2019), in terms of both theoretical schemata and empirical life experiences.

In the context of the tourism industry, place, landscape, culture are not only consumed; they are also (re)produced poetically, physically, emotionally, performatively, bodily (Squire, 1994), so that the world is offered and "perceived as a stage commanded by the eye of the spectator" (Olwig, 2019: 180). Accordingly, "the scenic landscape is something that is performed upon, like a stage, within a scaled, hierarchical, spatial structure of authority" (Olwig, 2019: 16), as in the context of a guided tour, in the context of a holiday package offered by a tour operator. Indeed, tourism/recreation landscapes may be considered as "most significant contemporary cultural grounds on which much of today's socio-cultural difference is increasingly created, contested and claimed, inviting the viewer/visitor to 'see' or walk or 'do' the landscape in various ways" (Terkenli, cited in Mathewson et al., 2019: 295). In the course of 'seeing', 'doing', 'experiencing' and so forth in the landscape, the tourist reflexively ascertains one's own identity vis-à-vis that of the destination, as if through a mirror that serves to reposition one experientially and existentially in the world (Adler, 1989).

Notwithstanding the myriad types of tourism and ways of performing tourism (e.g., more or less conscious or involved kinds of tourism), tourism may generally be seen as a practice of seeing and doing the landscape ('sightseeing'). Therefore, an analysis of the metaphors of landscape as theatre and landscape as 'mirror' may reveal the relevance, interrelatedness and historical continuity of the themes of vision/image, sight/spectacle, performance and speculation in decoding the tourist-landscape relationship. Terkenli (Mathewson et al., 2019: 295) thereby argues that:

The construction and consumption of the tourist experience greatly stems from and rests on the visual and performative character of the tourist landscape as enriched by its spectacular (theatrical) and specular (reflective) properties and enhanced by the participation of all human faculties (cognitive, emotional, experiential). In the performative – albeit ambivalent and contested – (re)construction and consumption of place and landscape through tourism, the dramaturgical metaphor of landscape as theater elicits multiple ways of individually and collectively produced 'stages' for tourism performances and tourist role-playing, just as the mirror metaphor continues to capture the infinite possibilities of reflection and representation of the human being and the tourism landscape in each other.

(Olwig, 2019: 44; 112; 116)

Indeed, the significance of reflective knowledge of the self vis-a-vis the other and the idea of identity, whether personal, social/cultural or place/landscape, has already been well developed and argued in the social sciences and tourism studies, either through issues of identity development or in reference to discussions on 'authenticity'.

Methodological and Epistemological Approaches

As already noted elsewhere, landscape ought to be approached more in its multidisciplinarity and multiplicity –"multiformity, multitextuality, multivocality, multisemity, and so on" (Terkenli, cited in Knudsen et al. 2012: 5). Accordingly, contemporary scientific approaches to the tourism-landscape relationship consciously signal multiple and shifting points of view, in the context of leisure economy production and consumption (Cartier & Lew, 2005; Terkenli & d'Hauteserre, 2006). The interface between these two broad and complex areas of scientific study – tourism and landscape – has elicited a variegated body of research, in terms of its nature, focus and approach, for example, analysis of aspects of tourism destinations as cultural landscapes, or in the context of sustainable development.

Nonetheless, there is, as of yet, no comprehensive and cohesive conceptual/theoretical framework to support this increasingly interdisciplinary body of work. However, interest in the study of landscapes of tourism and tourism has been growing, especially in the last decade, as reflected in the increase of publications and research questions addressed to this area of study (Terkenli, 2021). Finally, there are few (teams of) researchers specialising and conducting research consistently in this field, although such publications tend to be widely dispersed among relevant publication outlets (Jiménez-García et al. 2021).

Two substantial efforts to organise theory and approaches in this area of study have been undertaken in recent years, aiming at highlighting key issues and empirical advances, as regards the state of the art and future horizons. The first such undertaking was the chapter 'Landscape perspectives on tourism geographies' by Daniel C. Knudsen, Jillian M. Rickly-Boyd and Michelle M. Metro-Roland in the first edition of the *Routledge Handbook of Tourism Geographies*, edited by Julie Wilson (2012), to which the present chapter serves as the sequel, building on, updating and complimenting its review of the subject matter. Specifically, the authors establish the significance of the study of landscape as an approach to tourism, point to the multi-methodological context of such study, and describe the main research trends in this field in (a) the materiality of landscapes as cultural artefacts; (b) the representation of these landscapes through various media; and (c) the interpretation of landscapes as meaningful places (2012).

The second undertaking was the chapter 'Landscapes of Tourism', by Theano S. Terkenli in *The Wiley-Blackwell Companion to Tourism*, edited by Alan A. Lew, C. Michael Hall and Allan M. Williams (2014), which also serves as a point of departure for the present chapter. Here, the author posits that, in the previous decade, not only did landscape assume a more central role in social perspectives on tourism; it also increasingly crept into various types of tourism analysis. This earlier chapter also summarised predominant themes in this area of study, such as corporeality/materiality, place identity, consumption, and heritage. It also provided an overview of recent trends and advances in the field, using a framework of two chronological periods in the development of ontological and epistemological research perspectives to landscapes of tourism (Terkenli, 2014).

Overall, few attempts have been made towards tourism (or visited) landscape typologies, aiming to combine, on the one hand, the different landscape features and dimensions and on the other hand, different types of tourism and destination development (Gkoltsiou et al., 2013; Skowronek et al., 2018). Many challenges are involved in this task. From the landscape perspective, visited landscapes not only represent different cultures, but also comprise different natural, semi-natural or rural ecosystems and settings, offering a wide range of recreational possibilities, with positive, negative or other implications for the destination, for visitors and for the tourism industry. From the recreation/tourism side, more or less generic or unique attractions and respective recreational infrastructures/amenities run the range of highly to minimally developed and/or sought-after destinations.

These facts point to the significance of place-based tourism experiences, which inadvertently unfold in/through the landscape terrain/medium. Relevant literature increasingly describes and cautions against the despoliation of landscape by tourism – often destroying the very basis of its development – as well as both positive and negative challenges posed by trends of globalisation to landscapes of tourism, such as practices of branding destination landscapes or risks induced by overtourism (see also de Cássia Ariza da Cruz, Chapter 3).

Significant scientific advances at the interface of landscapes and/of tourism have come about in the past couple of years. In this regard, attention should be drawn to a special journal issue which introduces research efforts pushing the frontier of state-of-the-art inroads and advances in this study area. The special issue on *Landscape and Tourism, Landscapes of Tourism*, recently published in *Land* (Terkenli, 2021), aims "to enhance the interdisciplinary scientific dialogue on these issues and challenges, while highlighting their range and significance for tourism and landscape, in terms of theory, empirical practice, approach, policy, ethics, and future prospects" (Terkenli et al. 2021: online). It contains 16 research articles which contribute to the latter objective by covering a broad spectrum of relevant scholarly advances. The contributions come from and/ or represent applications in a series of countries, such as Spain, the Netherlands, China, Slovakia, Poland, Japan and Italy. Three contributions especially stand out on the basis of their achievement, in terms of the objective of this collective issue, and of the prospects of the interdisciplinary field of tourism-landscape study, more generally.

According to the editorial (Terkenli, 2021), the first one delves into the theorisation of the tourism–landscape interrelationship by performing a bibliographic analysis of scholarly contributions to the nexus and conceptualisation of 'tourist landscapes' and other relevant terms, in order to map different ways of defining and understanding this complex emergent interrelationship (Meneghello 2021). The valuable findings of this study enrich scientific reflections on this relationship, providing new definitional contributions and conceptual frameworks able to influence coherently both theory and practice. The bibliometric analysis undertaken brings up three main research topics predominating the field: (a) planning and governance, (b) situated spatial-social-symbolic interrelations, and (c) impact evaluation.

The study unveils a recent increase in the awareness of terminological issues, despite a more general lack of attention to the actual use of specific terms. It also illustrates reflections on the relational dimension of landscape, as the latter matured in tourism studies around 2010, but have only very recently begun to consolidate and increasingly guide pertinent theoretical research. The second contribution by Terkenli, Skowronek & Georgoula (2021) represents an in-depth investigation of the relationship between landscape and tourism in a comprehensive and integrated way, on the basis of a broad questionnaire-based survey of European landscape and tourism experts. The latter's notions and perceptions of the reciprocal landscape–tourism interrelationship are analysed in terms of their (a) understanding of and visions for future optimisation of the tourism–landscape interrelationship, (b) conceptualisations of 'landscapes of tourism', and (c) assessment of the prospects/opportunities and challenges/threats of the tourism–landscape relationship, both for the tourism industry and for local societies. The study points to the emergence of a definition for 'landscapes of tourism'. Moreover, the study's findings point to the reportedly high significance of the tourism–landscape relationship, vis-à-vis both its positive and negative aspects, but reveal a leaning towards its negative aspects. However, also revealed are experts' expressions of crucial socio-environmental concerns regarding the tourism–landscape interrelationship, alongside their support for the principles of sustainability, locality and participatory governance, while calling for appropriate future governmental planning.

The third significant contribution to this line of research by Mercedes Jiménez-García, José Ruiz-Chico and Antonio Rafael Peña-Sánchez (2021) was also motivated by the lack of a theoretical framework, as well as of deficient knowledge of the structure and organisation of the existing body of relevant knowledge. The article constitutes a first attempt at mapping this research topic through the application of bibliometric techniques, using VOSviewer and Science Mapping Analysis Software Tool (SciMAT) for a total of 3,340 articles from journals indexed in Web of Science. The results obtained confirm that interest in the study of these concepts has been growing, especially in the past decade. The main contribution of this work lies in the identification of research themes which have been basic to the construction of the field but are currently in decline, such as cultural heritage and other promising themes, that is, national parks and geotourism. Most significantly, the article points to sustainability – as a term and as related issues – which features as a transversal presence and value throughout this scientific research area. This research also showed that the concept and issues revolving around and/or relating to sustainability appeared in cross-reference to other currently emerging themes, such as planning and environment; a finding that highlights and reinforces an already acknowledged trend (Terkenli et al., 2021).

Challenges and Prospects for the Post-COVID-19 Pandemic Era

The intertwined relationship between tourism and landscape comes with a series of costs and benefits, as well as opportunities and threats, in the context of tourism landscapes. Admittedly, landscapes of tourism reflect and stage recreational trends, multifunctional livelihood systems, conflicts and opportunities for employment and income generation, as well as human, cultural, and natural resource management and use (Coccossis, 2004; Terkenli, 2018). As such, they are increasingly coming into the foreground of current debates about the future of the planet, in conjunction with various human and environmental crises (e.g., economic depressions, climate change, and the COVID-19 pandemic). These crises offer significant opportunities but also carry a serious bearing on the realms of both tourism and landscape.

One positive trend in this direction is the enormous current proliferation of a broad range of alternative, place-based and special-interest forms of tourism/leisure, variably (and often, intricately) connected to visited landscapes. The main goals of such endeavours tend to be increasingly compatible with sustainable/'green' development for the landscape, for the local societies and for tourism, while catering to a variety of broadly accessible tourism/leisure pursuits and activities (Coccossis et al., 2020). Meanwhile, rising rates and globalising patterns of mobility and consumption have continued to reinforce detrimental consequences of climatic change, already bearing a heavy social, economic, ethical and ecological imprint on the tourism industry.

As part of and driver of 'globalisation' in the culture of consumption, tourism plays a crucial role in landscape transformation. Globalisation produces and circulates social knowledge about places, as well as symbols of consumption, often referred to as processes of 'Disneyfication' (Gottdiener, 2001) and 'McDonaldisation' (Knudsen et al., 2012: 2044; Ritzer, 1993). Leisure and tourism experiences increasingly inform and substantiate new types of landscapes of tourism and leisure, in terms of their nature, scale and geographical characteristics, cutting across much of the more traditional typologies of tourist environments. The segregation, for instance, of leisure from home life that modernisation instilled becomes more and more tentative and irrelevant in the post-modern Western world, rendering tourism a ubiquitous practice in the landscape.

Practices that we understand as leisure/tourism today merge with other areas of life (Crouch, 1999) in our societies of spectacle, where specific pleasures are not place-bound and objects of delight proliferate. Increasingly, individualised tourism motives and demands are rapidly changing contemporary mobility patterns and practices, into special-purpose forms of tourism. This results in a growing spatial de-differentiation between leisure/tourism, shopping, work, culture, satisfaction of basic needs, comfort, play, familiarity and so forth. Thus, in the context of this new cultural economy of space (Terkenli & d'Hauteserre, 2006), all landscapes increasingly tend to assume characteristics of tourism/leisure landscapes, while distinctions between geographically/historically/culturally diverse landscapes are also increasingly blurred.

All of these trends and developments are currently becoming more and more challenged and complicated through the multifold, interactive and ever-multiplying repercussions of the climate crisis, which affect landscape, tourism, and human lives in multiple, vital and mutually reinforcing ways, with catastrophic prospects for humanity. Longstanding exigencies are in the process of becoming ever more pressing, in terms of human and environmental losses, grave risks to human culture and ways of life, the onslaught of pathogenic organisms (i.e., the COVID-19 pandemic), economic disasters, and so forth.

The COVID-19 pandemic affected patterns of leisure and tourism mobility at all geographical scales, signalling an uncertain course for tourism in the foreseeable future. In the meantime, precious landscape wealth, diversity and value of both ordinary and extraordinary landscapes (i.e., historical sites, coastal settlements, wetlands and beaches) are under grave risk of despoliation and loss. On the other hand, contemporary societies hit by the pandemic and concomitant governmental measures with an impact on human mobilities have been re-discovering 'their' landscapes, from which they had been cut off, during lockdown and quarantine conditions. Neighbourhood landscapes, landscapes of everyday life, 'green'/outdoor urban spaces, countryside and leisure/tourism/recreation landscapes are, thus, presently reclaimed and re-valorised for their enormous significance in human life, well-being, cultural and environmental qualities and multiple therapeutic properties. Landscape, in general, seems to be regaining its place, role and purpose in human lives – not the least in the realms of tourism and recreation.

In the tentative, fluid and ambivalent context of leisure/tourism mobilities in the post-COVID-19 pandemic era, therefore, it becomes crucial to develop and renew our theoretical and empirical understanding of distinctive tourism trends and patterns, and their impact generally on the landscape – as well as on tourism landscapes, in particular. Some of these emerging trends and patterns point to a favouring of physical/natural and 'protected/safe' aspects of destinations, but also to newer trends in tourism mobilities, such as domestic (proximity), solo, reclusive, slow, silent, 'green', and so forth. It may be too early to postulate as to future trends in the tourism industry, as all of the above crises continue to evolve. Nevertheless, the combination of the physical environment's lure, continued interest in cultural attractions and the pursuit of less crowded/reclusion/place-based motives seem to point to tourists' growing preferences for smaller cities/

towns and/or rural/semi-rural destinations, where landscape (combining all of these aspects) seems poised to figure very prominently.

Concluding Thoughts

It is important to conclude by conceding that "landscape as an approach to tourism is not a mutually exclusive perspective but is quite compatible with numerous other theories, frameworks, and perspectives used to understand tourism, thus making a landscape perspective a robust tool of inquiry" (Knudsen et al., 2012: 202).

In this same piece of research, the authors had concluded by sketching three most fruitful future trajectories for landscape-based research in tourism geography, (a) the need to link interpretation and embodied experience, (b) the need to bring to bear the theoretical insights afforded by landscape to the processes by which tourism sites are chosen, cultivated and marketed, and (c) the need to understand better the ways in which landscape and material culture intersect in tourism (Knudsen et al. 2012: 206). All of these lines of thought and inquiry have since been the object of scientific research in this area of study, and are poised to continue being so, in the future. This chapter, loosely following in the steps of the corresponding chapter in the first edition of the Handbook (Knudsen et al., 2012: 202), purported to explore further key concepts of landscape significant in the study of tourism, to update definitions of tourist/ tourism landscapes, (tourism representations, tourist landscape experience) and to elaborate on developing trends in tourism-landscape research. Rather than emphasising the production, reproduction and consumption of tourism landscapes (often referred to also as 'touristed' landscapes) (Knudsen et al. 2012), this chapter aimed to develop a better scientific understanding of the significance of landscape to tourism and of tourism-induced changes to the landscape, with a view also towards an unstable and uncertain future, in light of contemporary global crises (climate change, pandemics, etc.).

In conclusion, the context within which we are presently called to analyse and evaluate the tourism-landscape interrelationship is one of new prospects and emerging challenges. By providing the principles/values, theory/tools and character/objectives, now more than ever, landscape perspectives may hold significance and relevance for destinations, local societies and the tourism industry in tourism and (tourism/tourist) landscape planning and stewardship for a better future. However, in order for landscape to become such a catalyst for an emerging new model for human life, further, renewed and more in-depth scientific investigation into the landscape (attractions) sought by visitors and to the role of landscape in visitor/tourist experiences is certainly necessary and called for.

References

Adler, J. (1989). Origins of sightseeing. *Annals of Tourism Research 16*(1), 7–29.

Burley, J.B., & Machemer, T. (2016). *From Eye to Heart: Exterior Spaces Explored and Explained,* 1st ed. Cognella Academic Publishing.

Burton, R.J. (2012). Understanding farmers' aesthetic preference for tidy agricultural landscapes: A Bourdieusian perspective. *Landscape Research 37*(1), 51–71.

Cartier, C., & Lew, A.A. (Eds.) (2005). *Seductions of Place: Geographical Perspectives on Globalization and Touristed Landscapes*. Routledge.

Coccossis, H. (2004). Sustainable development, landscape conservation and tourism in the small islands of Greece: The Nature Conservation-Society Interface. In D. Martin, & J. van der Straaten (Eds.), *Cultural Landscapes and Land Use* (pp. 111–124). Kluwer Academic Publishers.

Coccossis, H., Tsartas, P., & Griba, E. (2020). *Special and Alternative Forms of Tourism. Demand and Supply of New Tourist Products*. 2nd ed. Kritiki.

Cosgrove, D. (1998). *Social Formation and Symbolic Landscape*. University of Wisconsin Press.

Council of Europe (2000). *European Landscape Convention*. www.coe.int/en/web/landscape [accessed February 2024]

Crouch, D. (1999, Ed.). *Leisure/Tourism Geographies: Practices and Geographical Knowledge*. Routledge.

Daniels, S., & Cosgrove, D. (1988). Introduction: Iconography and landscape. In D. Cosgrove, & S. Daniels (Eds.), *The Iconography of Landscape: Essays on the Symbolic Representation, Design and Use of Past Environments* (pp. 1–10). Cambridge University Press.

Dewsbury, J.D., Harrison, P., Rose, M., & Wylie, J (2002). Introduction: Enacting Geographies. *Geoforum 33*, 437–440.

Gottdiener, M. (2001). *The Theming of America: American Dreams, Visions, and Commercial Spaces*. Westview Press.

Gkoltsiou, A., Terkenli, T.S., & Koukoulas, S. (2013). Landscape indicators for the evaluation of tourist landscape structure. *International Journal of Sustainable Development and World Ecology 20*(5), 461–475.

Hall, C.M., & Page, S.J. (2009). Progress in tourism management: From the geography of tourism to geographies of tourism – a review. *Tourism Management 30*(1), 3–16.

Hall, C.M. & Page, S.J. (2014). *The Geography of Tourism and Recreation*. 4th ed. Routledge.

Jackson, J.B. (1984). *Discovering the Vernacular Landscape*. Yale University Press.

Jiménez-García, M., Ruiz-Chico, J., & Peña-Sánchez, A.R. (2021). Landscape and Tourism: Evolution of Research Topics. *Land 9*, 488–505.

Knudsen, D.C., Rickly-Boyd, J.M., & Metro-Roland, M.M. (2012). Landscape perspectives on tourism geographies. In J. Wilson (Ed.), *The Routledge Handbook of Tourism Geographies* (pp. 201–206). Routledge.

Lorimer, H. (2005). Cultural geography: The busyness of being 'more-than-representational'. *Progress in Human Geography 29*(1), 83–94.

Lothian, A. (1999). Landscape and the philosophy of aesthetics: Is landscape quality inherent in the landscape or in the eye of the beholder? *Landscape and Urban Planning 44*(4), 177–198.

Mathewson, K., Mels, T., Terkenli, T.S., Waterman, T., Minca, C., Jones, M., & Olwig, K.R. (2019). The meanings of landscape: Essays on place, space, environment and justice. *The AAG Review of Books 7*(4), 291–304.

Meinig, D.W. (1979). The beholding eye: Ten versions of the same scene. In D.W. Meinig (Ed.), *The Interpretation of Ordinary Landscapes: Geographical Essays* (pp. 33–48). Oxford University Press.

Meneghello, S. (2021). The tourism–landscape nexus: Assessment and insights from a bibliographic analysis. *Land 10*, 237–262.

Nassauer, J.I. (1988). The aesthetics of horticulture. *HortScience 23*, 973–977.

Olwig, K.R. (2019). *The Meanings of Landscape: Essays on Place, Space, Environment and Justice*. Routledge.

Pearce, D. (1995). *Tourism Today: A Geographical Analysis*. 2nd ed. Longman.

Ritzer, G. (1993). *The McDonaldization of America*. Pine Forge.

Roe, M. (2014). Exploring future cultural landscapes. In M. Roe, & K. Taylor (Eds.), *New Cultural Landscapes* (pp. 241–269). Routledge.

Sauer, C.O. (1925). The morphology of landscape. *University of California Publications in Geography 193*(2), 19–54.

Skowronek, E., Tucki, A., Huijbens, E., & Jóźwik, M. (2018). What is the tourist landscape? Aspects and features of the concept. *Acta Geographica Slovenica 58*(2), 73–85.

Squire, S.J. (1994). Accounting for cultural meanings: The interface between geography and tourism studies re-examined. *Progress in Human Geography 18*(1), 1–16.

Terkenli, T.S., Tsalikidis, I. & Tsigakou, F.M. (2001). The physical landscape of Greece in nineteenth-century painting: An exploration of cultural images. In G.L. Anagnostopoulos (Ed.), *Art and Landscape* (pp. 618–632). Panagiotis & Effie Michelis Foundation.

Terkenli, T.S. (2014). Landscapes of tourism. In A.A. Lew, C.M. Hall, & A.M. Williams (Eds.), *The Wiley-Blackwell Companion to Tourism* (pp. 282–293). John Wiley & Sons.

Terkenli, T.S. (2018). Tourism: Beyond the pleasure of practising geography. *Tourism Geographies 20*(1), 170–171.

Terkenli, T.S. (2020). Landscape. In D. Buhalis (Ed.), *Encyclopedia of Tourism Management and Marketing*. Edward Elgar Publishing.

Terkenli, T.S. (Guest ed.) (2021). Special issue 'Landscape and Tourism, Landscapes of Tourism'. *Land 10*(9), 944–950.

Terkenli, T.S., & d'Hauteserre, A.-M. (2006, Eds.). *Landscapes of a New Cultural Economy of Space.* Springer.

Terkenli, T.S., Skowronek, E., & Georgoula, V. (2021). Landscape and tourism: European expert views on an intricate relationship. *Land 10*, 221–236.

Tress, B. & Tress, G. (2001). Capitalising on multiplicity: A transdisciplinary systems approach to landscape research. *Landscape and Urban Planning 57*(3–4), 143–157.

Urry, J. (1995). The 'consumption' of tourism. *Sociology 24*, 23–35.

Wilson, J. (2012, Ed.). *The Routledge Handbook of Tourism Geographies.* Routledge.

Wylie, J. (2007). *Landscape.* Routledge.

Zube, E.H., & Pitt, D.G. (1981). Cross-cultural perceptions of scenic and heritage landscapes. *Landscape and Urban Planning 8*, 69–87.

13

PLACE MAKING AND UNMAKING IN TOURISM

T.C. Chang

Introduction

An intersecting concern in tourism geography and tourism studies is the making, management and marketing of *places*. The concept of 'place making' first emerged as an urban development strategy in the 2000s aimed at improving the design of cities and the quality of life of their inhabitants (Friedmann, 2010). In tourism, the concept of place making has been incorporated into destination planning and management, often with varying outcomes across space and time (Dupre, 2019).

This chapter critically unpacks the 'tourism place making' concept by first exploring its origins within the urban development field, and more specifically its intersections with tourism. Many types of place making exist upon which tourism exerts an influence in different ways. One area in which tourism scholars have contributed significantly is 'creative place making'. Debates on the role of arts, culture, heritage and creative experiences in tourism planning and development will thus be reviewed specifically. However, just as places are 'made', they may also be contested, resisted and 'unmade'. We must therefore critically engage with the politics of tourism place making and its implications for social and spatial divisions between people. The chapter ends by outlining areas that remain to be studied in place making, particularly in the realm of critical tourism scholarship.

Tourism and Place Making

As a planning concept, the term place making refers to participatory and collaborative planning in which inhabitants are involved in the design of public urban spaces. The inspiration may be traced back to Jane Jacobs's (1961) initial call for a more people-centred approach in development, encapsulated in the Project for Public Spaces, a non-profit organisation established in 1975 to expand on the work of William Whyte, an American urbanist and Jacobs's mentor (Project for Public Spaces, 2010). The goal was to ensure a greater sense of attachment to and accountability for the urban spaces within which people live, work and play.

In the context of tourism, Alan Lew (2017) identifies a number of sub-concepts in place making. He differentiates between 'placemaking' and 'place-making' by explaining that while the former refers to large-scale, often costly, intentional planning of tourist destinations by

DOI: 10.4324/9781003286301-16

government and large private enterprises (e.g., development of festival marketplaces and water-front zones), the latter includes more organic, bottom-up involvement as lay people go about their everyday lives in "making" places (e.g., neighbourhood centres and community spaces). While the former offers "safe, known, predictable, familiar" spaces catering to mass tourists, the latter promises a sense of "risk, uncertainty, surprise, escape" appealing to alternative travellers (Lew, 2017: 451). More often than not, however, most tourist destinations manifest 'mixed place makings', combining elements from both ends of the spectrum, catering to a broad array of visitors (Lew, 2017).

Tourism destinations also manifest standard and strategic place making (Lew, 2017; Mansilla & Milano, 2022). While the former refers to routine place maintenance such as street repairs, up-keep of amenities like bus stops and public seating, the latter includes development efforts stra-tegically aimed at bringing increased tourism attention to the site. Curated programmes like food and music events and art, as well as features such as light and water installations in public spaces help to brand places strategically as safe, trendy and appealing to businesses, tourists and leisure consumers.

Both tangible and intangible elements are also incorporated into tourism place making. The former refers to the material components of a place, such as buildings, street furniture and amen-ities, while the latter includes non-material dimensions such as one-off events, street life and place histories. While developing infrastructural hardware will not inject life and soul into a place, nor can introducing software alone overcome tangible shortcomings. Both the tangibles and intangibles of a place must be harnessed for a site to appeal to tourists and tourism investments, as well as local residents. Place *management* must thus be balanced with place *activation* for a tourism/rec-reational project to be successful and sustainable.

In his review of academic works on place making between 1991 and 2016, Dupre (2019) notes that research focuses mainly on the Global North, particularly North America. The literature spans four key issues, all of which prominently feature urban tourism. They include: (a) the nexus of placemaking and globalisation (e.g., effects of global-city planning on tourism); (b) the role of peoples' participation in place making (e.g., how tourists might be involved in providing feedback on urban development projects); (c) challenges and conflicts in place making (especially between stakeholders with differing claims to space); and (d) strategies in place making (from high-profile events to sustainable projects). Dupre (2019: 115) argues for greater attention to the "emerging interdisciplinary research area at the intersection of place-making and urban development and tourism", with special emphasis on how tourism might implicate critical issues relating to sustain-ability, quality of life and policy formulation. The need to widen place making scholarship beyond the West should also be noted.

Writing in 2010, Friedmann (2010: 149) observed that the "art of place-making" had not informed planners in newly industrialising regions in Asia where the "principal preoccupation has been with the branding of cities and the advanced infrastructure required by global capital". Not surprisingly, therefore, extant literature on tourism place making in non-Western countries is sparse compared to the West. Staiff and Bushell (2017) explore the role of film festivals as a place-making strategy in Luang Prabang, Laos, while Richards (2020) highlights Thailand's emerging creative tourism status through its community-based rural creative programmes. More recently, Chang and Oh (2022) note that tourism place making has emerged in Singapore with the advent of suburban walking tours aimed at intrepid travellers and local heritage buffs. The trend of researching tourism place making beyond the West should be encouraged, allowing us to better appreciate different place making efforts across the globe, as well as their variable outcomes.

Creative Place Making and Cultural Tourism

A particular area to which tourism scholars have contributed significantly is "creative place making". Indeed, the roles of arts, culture, heritage and creative experiences in tourism place making are well established (Nicodemus, 2013; Richards, 2020; Nieuwland & Lavanga, 2021). The term creative place making was coined by the National Endowment for the Arts (or NEA, an independent federal agency that funds arts and cultural activities in the United States in a 2010 report (Markusen & Gadwa, 2010a) encouraging support for arts and culture through multi-agency cooperation (Webb, 2014). It is defined as a planning practice uniting "partners from public, private, non-profit, and community sectors [to] strategically shape the physical and social character of a neighbourhood, town, city, or region around arts and cultural activities" (Markusen & Gadwa, 2010a, cited in Markusen & Gadwa, 2010b: 3). As a philanthropic arm of the federal government, NEA's 'Our Town' grant programme, for example, had supported over 460 projects in all 50 US states to the tune of more than US$36 million by late 2017.

The mutual relationship between tourism and creative place making is clear. Not only are local arts and culture appealing to leisure visitors, but many tourists and consumers are also enthusiastic about co-creating their own cultural experiences for personal enrichment and enjoyment. Indeed, a key component in tourism place making is to take charge of one's activities and experiences (Chang & Oh, 2021); such a form of co-created experiences is the cornerstone of creative tourism. The nexus between creative tourism, place making and co-creation is explained by Richards (2020: 4) as follows: "A placemaking perspective on creative development has important implications for tourism. Tourists become essential actors in the co-creation of place, re-negotiating meanings of place that attract them." Today, many tourists "increasingly seek more active participation in their experiences and more meaningful contact with local people" (Richards, 2020: 3).

The intersection of creative tourism and place making yields distinctive urban geographies, which include the emergence of creative clusters and zones like Beijing's 798 zone and Shanghai's M50, which are popular with both art-seeking tourists and residents (Ning & Chang, 2022); also, development of spectacles such as iconic buildings, spectacular architecture and hallmark events and festivals (Bilbao's Guggenheim museum and the Edinburgh Arts Festival come to mind); and creative tourism workshops and courses that showcase the creativity of a destination (Richards, 2020). In cities as varied as Beijing, Liverpool, Luang Prabang, Rotterdam and 's-Hertogenbosch, a range of projects such as film festivals, pop-culture events and arts projects have been inaugurated to establish these cities' credentials as creative tourism sites (e.g., Richards, 2017; Staiff & Bushell, 2017; Zhang, 2018; Wise et al., 2019; Nieuwland & Lavanga, 2021).

Although predominantly urban-centric, creative place making has also made inroads into rural locales, with more travellers seeking alternative sites and novel experiences, the countryside has opened up possibilities for new forms of tourism. Duxbury's (2017) report on the 'CREATOUR project – Creative Tourism Destination Development in Small Cities and Rural Areas in Portugal' – and Richards' case study (2020) of Thailand's "Designated Areas for Sustainable Tourism Administration" (DASTA) reveal not only new opportunities for place development but also a focus on empowerment of and life enhancement for rural communities (see also Hultman & Hall, 2012; Aquilino et al., 2021).

The emphasis on individual travellers and their experiences casts the spotlight on the agentic powers of tourists. It should be noted that place making is not just the responsibility of tourism planners, enterprises and local service providers, but also within the control of individuals. Seen in this light, place making must be understood as a highly personalised process in which tourists negotiate the physical environments wherever they might be, to create meaningful place experiences for and by themselves. As Thurlow and Jaworski (2014) explain, "people never simply visit places but are always actively shaping and making these places" (p. 459); tourists are therefore more than just 'viewing subjects' but also 'doing subjects', and it is through their actions and bodily movements that spaces are invested with meaning and memories. Embodied activities such as participating in food preparation courses, volunteering while backpacking, and going on walking/hiking trails are among the many different place making strategies undertaken by tourists either travelling on their own or with a group (e.g., Everett, 2012; Iaquinto, 2020; Wang et al,. 2020; Witte, 2021).

Tourism and the Unmaking of Places

Just as places may be 'made' for and by tourism, they can also be contested, resisted and 'unmade'. The scholarship on tourism place making must therefore be attentive to the negative repercussions it might engender. The notion that tourism place making is welcome by all is therefore to be challenged. This challenge has come about in at least two ways: (a) a realisation that creative place making entails place unmaking for certain segments of the community; and (b) a critique of the social and spatial displacements arising from tourism place making. As Lew (2017: 457) notes, different stakeholders – state planners, business owners, tourists and local residents – all "have their own worldviews through which they define their taken-for-granted lifeworld". This gives rise to different conceptions of 'place' and what constitutes good place making. While some might regard tourism as introducing positive changes to a place, others see it as "violating, expropriating, and commodifying the lifeworld of local residents" (Lew, 2017: 458).

As noted earlier, creative place making entails locals and tourists deploying the arts to create a sense of place for themselves. When arts and culture are instrumentalised for tourism, the original purposes for which they are intended might be diluted. The question thus arises as to whether creative place making should be undertaken for artistic and cultural purposes or for the economic value they bring to a place, and how best to balance both ostensibly laudatory goals. In funding creative place projects, a criterion used by ArtPlace (a ten-year collaboration among federal agencies, foundations and financial institutions in the United States) (ArtPlace, 2020) is how the arts can contribute to 'vibrancy indicators' by way of attracting tourists, cultural consumers and businesses, and raising property values in a location (Nicodemus 2013). Similarly, the aforementioned NEA made use of 'livability indicators' as a way to monitor the effects of creative place making. Such indicators include resident attachment to community, quality of life, level of arts activity and economic conditions (Frenette, 2017).

According to Nicodemus (2013: 218), practitioners are often confused by the "mixed messaging" in place making programmes. As the author asks "should initiatives try to raise property values and attract tourists? Can the field reaffirm a commitment to the intrinsic value of arts amidst these instrumental outcomes?" (Nicodemus, 2013: 219), many other questions persist. Who are (or should be) the ultimate beneficiaries of creative place making – tourists, local residents, artists, property developers or planning authorities? Does place making for one group lead to place unmaking for another? Indeed a key challenge in place making is how best to balance the economic, social and cultural goals of place making, especially in the midst of the competing agendas of different practitioners.

The case of Beijing's 798 Art District is both instructive and a foreshadowing of the dangers of competing interests between place stakeholders. In what she terms "discursive constructions of places – place-frames", Zhang (2018: 91) explains that development projects are often mobilised according to the dominant place-frames ascribed to them. How a site is 'framed' is a political exercise determining the fate of its place making (or unmaking). Should the former 798 factory space be framed as an architectural zone of industrial heritage, a site for artistic experimentation, or a tourist attraction offering cultural and retail delights? In 2006, the municipal and district government officially designated (framed) the 798 district as a 'cultural and creative industry cluster' and recognised its industrial heritage as an important attraction to tourists (Zhang 2018). Thus framed, the 798 Art District underwent an intensive place-making programme that saw its eventual enshrinement as an arts and cultural site offering architectural and industrial aesthetics to a trendy middle-class leisure market (see also Ning & Chang, 2022, for a similar case with Shanghai's M50 Art Zone).

Missing from the above place framing are questions on how the 798 district should/could be developed for the many artists who called the place home before redevelopment began. Zhang (2018: 96) argues that this omission was deliberate because of the Chinese state's limited "acceptance and tolerance of contemporary art". Arising from concerns that artists are prone to social and political critique, and therefore a potentially destabilising force for the city/country, the emphasis on tourism, entertainment and lifestyle is a political sleight of hand aimed at highlighting "the benefits of the arts district to the government (historical value and city image)" while downplaying 'troubling aspects' associated with contemporary artists and artistic critique (Zhang, 2018: 96).

The case of Beijing's 798 district ominously foreshadows other examples of marginalisation and displacement that frequently attend place making projects. Urban tourism gentrification, for example, may be similarly interpreted as a case of place re-making in which tourism-oriented enterprises displace original residents and activities, be it in the Vieux Carre (French Quarter) of New Orleans (Gotham, 2005) or the redeveloped waterfront of the Singapore River (Chang & Huang, 2009; 2011). As new life is introduced to "old" buildings and environments, the transplanted vibrancies of tourism activities and cultural events strike some critics as evidence of "place faking" rather than place making (Courage, 2017: 56). With more urban environments sporting chain stores, public art and stylised ('Instagram-friendly') architectural features all in the name of place making, tourism geographies have become increasingly and ironically plagued by "homogenization or sameness between places", the very antithesis of place making (Lew, 2017: 458).

In their diatribe against urban gentrification, Burns and Berbary (2021) argue that urban changes often conceal acts of social and spatial displacement, and "every placemaking is always already an unmaking of something, and that these unmakings perpetuate racialized and socioeconomic injustices under the guise of a collaborative, participatory process of 'revitalization' and 'progress'". They urge caution in viewing place making as 'revitalisation' or as creating 'new opportunities'. Indeed the very terminology of place making presupposes places as "empty and void of meaning, ready to be made more meaningful through placemaking" (Burns & Berbary, 2021: 646).

The complicit role of tourism in place making must thus be questioned, particularly in the way(s) tourism development might privilege the needs of a leisure class and/or particular service sector, at the expense of the local community and non-tourism businesses. Burns and Berbary's example of Goudies Lane (Kitchener, Canada), and Speake and Kennedy's work (2019) on Valleta (Malta) document how urban rejuvenation projects certainly advantage some segments of society while disadvantaging others. It is therefore the role of critical scholars – in urban, development and tourism studies – to question the "aesthetic common sense" (Speake & Kennedy, 2019: 2) that belies contemporary place making projects as an unqualified good.

Conclusion

Place making and Critical Tourism Studies

The discussion on tourism place making touches many issues that go beyond both 'tourism' and 'places'. The interrogation of various place-making strategies and their outcomes across different time-spaces suggests a research field rich with possibilities. As we conclude this chapter, the links between tourism place making and Critical Tourism Studies (CTS) are outlined. CTS is principally concerned with the application of critical social-cultural theory in tourism scholarship (Chambers, 2007). It seeks to uncover socio-economic inequalities in destination areas and advises on how tourism can help to empower host communities, among other matters (Gale, 2012). In epistemological terms, CTS is also concerned with "just relations between researchers and tourism stakeholders" and focuses on participatory research methodologies as a way for researchers to be more reflexive and culturally sensitive (Higgins-Desbiolles & Whyte, 2014: 91). An expressed goal in CTS is to "challenge the field's dominant discourse", traditionally founded on the "masculine tradition of [W]estern thought", and to accept different perspectives and approaches as a better way to comprehend a complex world (Ateljevic et al., 2007: 4).

Scholarship on tourism place making can contribute to the CTS cause in various ways. As noted earlier, the literature on place making has been dominated by works on and from Western countries (Courage et al., 2021). What works in one part of the world might not have the same results elsewhere, as evinced by tourism place making in gentrified neighbourhoods in New Orleans, Shanghai and Singapore (e.g., Gotham, 2005; Chang & Huang, 2009; 2011; Chang, 2016; Ning & Chang, 2022). By highlighting cases from both the Global North and South from the perspectives of different authors and stakeholders, a more holistic understanding of place making processes, outcomes and challenges may be obtained. One such example is *The Routledge Handbook of Placemaking* (Courage et al., 2021), which comprises a mix of empirical cases from around the world, pushing place making scholarship "from a centralized and nearing-homogeneous practice" to embracing "contributions from indigenous researchers, practitioners, and communities" (2021: vi). A varied mix such as this allows for place making to be truly recognised as a "global, intersectional, and multidisciplinary field" (2021: lvi).

Urban place making may also have an unequal effect on different segments within a society. According to Burns and Berbary (2021: 3), critical scholars must be encouraged to contribute "disruptive narratives about placemaking to leisure studies" in order to "continue exposing placemaking's complicated complicity in problematic processes of erasure, displacement, destruction, and targeting of Black, Brown, Indigenous, and under-waged peoples". Critical discourse dismantles the orthodox views of positive place making, uncovering issues of dispossession, displacement and even racism that might arise through tourism development. How tourism can also contribute to possible courses of remediation via CTS should also be acknowledged.

The questioning spirit of CTS is well aligned with parallel research enquiries relating to place making. As exemplified above, place making is a contentious exercise because making for one person might result in unmaking for another. To interrogate the politics of place making, Lew (2017: 461) asks five questions:

(a) "[W]ho defines success?" (the government, business, residents, or tourists?); (b) "[W]hose story is being told through placemaking?" (i.e., [W]hose heritage and social history is being recognised); (c) "[W]ho owns the new touristed placemaking landscape?" (with the emergence of new tourism businesses, [W]hat happens to former land owners?); (d) "[H]ow do tourists contribute to place making?" (quite apart from planners and business owners, [H]ow do end-users make places for themselves?); and (e) "[H]ow do people know, experience, and engage with place makings?" ([W]hat are the effects of place making on anybody?).

We need to keep these questions in mind if we are to align place making research and practice with the tenets of critical scholarship.

Much research remains to be done on place making, be it in the field of tourism, urbanisation or cultural studies. Certainly the centrality of 'place' as a fundamental idea in both geography and tourism ensures that, regardless of the vagaries of conceptual turns, empirical trends and even place making terminologies, continued ruminations on the manifold geographies of tourism will persist into the future.

References

Aquilino, L., Harris, J. & Wisec, N. (2021). A Sense of Rurality: Events, Placemaking and Community Participation in a Small Welsh Town. *Journal of Rural Studies* 83, 138–145.

ArtPlace (2020). *Introduction*. Online source (consulted June 2024): https://www.artplaceamerica.org/about/introduction

Ateljevic, I., Morgan, N. & Pritchard, A. (2007). Editor's Introduction: Promoting an Academy of Hope in Tourism Enquiry. In I. Ateljevic, I., A. Pritchard & N. Morgan (Eds.) *The Critical Turn in Tourism Studies. Innovative Research Methodologies* (pp. 1–8). Elsevier.

Burns, R. & Berbary, L.A. (2021). Placemaking as Unmaking: Settler Colonialism, Gentrification, and the Myth of 'Revitalized' Urban Spaces. *Leisure Sciences* 43(6), 644–660.

Chambers, D. (2007). Interrogating the 'Critical' in Critical Approaches to Tourism Research. In I. Ateljevic, I., A. Pritchard & N. Morgan (Eds.) *The Critical Turn in Tourism Studies. Innovative Research Methodologies* (pp. 105–119). Elsevier.

Chang, T.C. (2016). 'New Uses Need Old Buildings': Gentrification Aesthetics and the Arts in Singapore. *Urban Studies* 53(3), 524–539.

Chang, T.C. & Huang, S. (2009). Geographies of Everywhere and Nowhere. Place-(un)making in a World City. *International Development Planning Review* 30(3), 227–247.

Chang, T.C. & Huang, S. (2011). Reclaiming the City: Waterfront Development in Singapore. *Urban Studies* 48(10), 2085–2100.

Chang, T.C. & Oh, F. (2022). Discovering the 'Heart' in Heartland Tourism. *Geografiska Annaler: Series B, Human Geography* 104(2), 75–92.

Courage, C. (2017). *Arts In Place. The Arts, the Urban and Social Practice*. Routledge.

Courage, C., Borrup, T. & Rosario Jackson, M. (2021, Eds.). *The Routledge Handbook of Placemaking*. Routledge.

Dupre, K. (2019). Trends and Gaps in Place-making in the Context of Urban Development and Tourism. 25 Years of Literature Review. *Journal of Place Management and Development* 12(1), 102–120.

Duxbury, N. (2017). CREATOUR: Creative Tourism Destination Development in Small Cities and Rural Areas. *Revista Turismo & Desenvolvimento* 27–28(2), 47–48.

Everett, S. (2012). Production Places or Consumption Spaces? The Placemaking Agency of Food Tourism in Ireland and Scotland. *Tourism Geographies* 14(4), 535–554.

Frenette, A. (2017). The Rise of Creative Placemaking: Cross-sector Collaboration as Cultural Policy in the United States. *The Journal of Arts Management, Law, and Society* 47(5), 333–45.

Friedmann, J. (2010). Place and Place-making in Cities: A Global Perspective. *Planning Theory and Practice* 11(2), 149–165.

Gale, T. (2012). Tourism Geographies and Post-Structuralism. In J. Wilson (Ed.) *The Routledge Handbook of Tourism Geographies* (pp. 37–45). Routledge.

Gotham, K.F. (2005). Tourism Gentrification: The Case of New Orleans' Vieux Carre (French Quarter). *Urban Studies* 42(7), 1099–1121.

Higgins-Desbiolles, F. & Whyte, K.P. (2014). Critical Perspectives on Tourism. In A. Lew, C.M. Hall & A. Williams (Eds.) *A Companion to Tourism* (pp. 88–97). Blackwell Publishing.

Hultman, J. & Hall, C.M. (2012). Tourism and Placemaking. Governance and Locality in Sweden. *Annals of Tourism Research* 39, 2547–2570.

Iaquinto, B.L. (2020). Understanding the Place-making Practices of Backpackers. *Tourist Studies* 20(3), 336–353.

Lew, A. (2017). Tourism Planning and Place making: Place-making or Placemaking. *Tourism Geographies* 19(3), 448–466.

Mansila, J.A. & Milano, C. (2022). Becoming Centre: Tourism Placemaking and Space Production in Two Neighbourhoods in Barcelona. *Tourism Geographies* 24(4–5), 599–620.

Markusen, A. and Gadwa, A. (2010a). *Creative placemaking.* Washington DC: National Endowment for the Arts NEA.

Markusen, A. & Gadwa, A. (2010b). Arts and Culture in Urban or Regional Planning: A Review and Research Agenda. *Journal of Planning Education and Research* 29, 379–391.

Nicodemus, A.G. (2013). Fuzzy Vibrancy: Creative Placemaking as Ascendant US Cultural Policy. *Cultural Trends* 22(3–4), 213–222.

Nieuwland, S. & Lavanga, M. (2021). The Consequences of Being 'the Capital of Cool'. Creative Entrepreneurs and the Sustainable Development of Creative Tourism in the Urban Context of Rotterdam. *Journal of Sustainable Tourism* 29(6), 926–943.

Ning, Y. & Chang, T.C. (2022). Production and Consumption of Gentrification Aesthetics in Shanghai's M50. *Transactions of the Institute of British Geographers* 47(1), 184–199. https://doi.org/10.1111/tran.12482

Project for Public Spaces (2010) *William H. Whyte* https://www.pps.org/article/wwhyte (accessed June 2024).

Richards, G. (2017). From Place Branding to Placemaking: The Role of Events. *International Journal of Event and Festival Management* 8(1), 8–23.

Richards, G. (2020). Designing Creative Places: The Role of Creative Tourism. *Annals of Tourism Research* 39(2), 102922. https://doi.org/10.1016/j.annals.2020.102922

Staiff, R. & Bushell, R. (2017). The "Old" and the "New": Events and Placemaking in Luang Prabang, Laos. *International Journal of Event and Festival Management* 8(1), 55–65.

Speake, J. & Kennedy, V. (2019). Changing Aesthetics and the Affluent Elite in Urban Tourism Place Making. *Tourism Geographies* 24(6–7), 1197–1218. https://doi.org/10.1080/14616688.2019.1674368

Thurlow, C. & Jaworski, A. (2014). 'Two Hundred Ninety-four': Remediation and Multimodal Performance in Tourist Placemaking. *Journal of Sociolinguistics* 18(4), 459–494.

Wang, X., Xie, J. & Sun, J. (2020). Travellers' Meaning-Making of the Sichuan-Tibet Highway: From Space of Flows to Place. *Tourism Geographies* 25(1), 23–43. http://DOI:10.1080/14616688.2020.1819402

Webb, D. (2014). Placemaking and Social Equity: Expanding the Framework of Creative Placemaking. *Artivate* 3(1), 35–48 https://doi.org/10.1353/artv.2014.0000

Wise, N., Melis, C. & Jimura, T. (2019). Liverpool's Urban Imaginary: The Beatles and Tourism Fanscapes. *The Journal of Popular Culture* 52(6), 1433–1450.

Witte, A. (2021). Revisiting Walking as Mobile Place-making Practice: A Discursive Perspective. *Tourism Geographies* 25(1), 334–356. https://doi.org/10.1080/14616688.2021.1878269

Zhang, A.Y. (2018). Thinking Temporally When Thinking Relationally: Temporality in Relational Place-making. *Geoforum* 90, 91–99.

14

RURAL CREATIVE TOURISM GEOGRAPHY FOR COMMUNITY REVITALISATION

Meng Qu

Introduction

A Relational, Geographical and Creative Integration of Art, Tourism and Rurality

In rural social-ecological systems, the creative transformation and new developments of art tourism are making the study of related patterns more complicated and boundaryless. However, the focus on creativity in the field of geography has emphasised urban contexts and focused more on economic urban strategies (Wilson, 2012b; Gibson, 2010a). Rural society, often seen by mainstream society as uncreative soil, is receiving more attention from the creativity literature (Gibson, 2010b). Exploring creative industries in remote and socioeconomically disadvantaged regions means that researchers cannot take context for granted, which is a necessary difference from urban-based theories (ibid).

The implementation, transformation and re-articulation of creativity from urban creative clusters into relatively isolated rural destinations (Woods, 2012) introduced the concept of 'creative rural' and focused attention on relational flows connecting the regional and the global, the rural and the urban (Prince, 2018). Unlike Gibson's 'creative industries research' (2010a), Woods emphasises the 'indirect' measurement of the creative rural economy compared to urban areas (Woods, 2012). This further proves the need for "new ways in which creative industries research can be enlivened and made social, by not assuming a capitalist-oriented language of firms, growth, employment and export and instead valuing the communitarian purposes to which creativity can be put" (Gibson, 2010a, 7–8). In creative tourism, this approach is needed to examine creative small rural communities in transition from a resource-based economy to one that uses its cultural, place-specific and endogenous knowledge to provide visitors with place-based experiences (Scherf, 2021).

Rural decline in the Global North reflects another side of radical urbanisation and globalisation, and exploring the uneven distribution and connections among these phenomena through art and creativity can help to better understand and address the conditions that caused rural decline in the first place (Duxbury & Campbell, 2011; Qu, 2019). Recently, there has been an increase in popularity of contemporary art intervention in rural areas, such as rural art festivals in regional contexts of the Global North (Dunphy, 2009; Duxbury & Campbell, 2011; Gibson & Connell, 2011), aimed

DOI: 10.4324/9781003286301-17

at place-making and developing art tourism. Representative cases span from creative urban spaces, such as the Bilbao effect of Guggenheim Museums, to the urban-rural connection of the Museum of Old and New Art (MONA) effect in Hobart, Tasmania (Booth et al., 2021, Franklin, 2018; Ryan, 2016). Simply put, both the Bilbao and MONA effects are about creating cultural Meccas and visual wonder through contemporary art (Franklin, 2018; Ryan, 2016). Subsequently, the trend has spread to the rural context, such as Echigo Tsumari in Japan (Borggreen & Anemone, 2020). It is worth noting that peripheral and remote islands such as Fogo in Canada, Setouchi in Japan, Shetland in Scotland and Bornholm in Denmark are also becoming a focus of creative and tourism research and attracting worldwide attention (Prince et al., 2021b).

Rural Japan, in particular, has developed a significant branch of socially engaged art interventions closely associated with regional tourism development in connection with community revitalisation (Prince et al., 2021b; Qu, 2022; Qu et al., 2022; Qu & Cheer, 2021; Tu, 2022). Through social engagement, creativity and a site-specific approach, these interventions attempt to re-establish rural communities' cultural roots while also re-shaping each community for creative tourism development. This type of large-scale, socially engaged rural art festival has been considered a successful model for tourism-based rural revitalisation by the Japanese government (Qu, 2019; Qu & Funck, 2021; Tu, 2022). As of 2017, art festivals had been organised in more than 80 locations in rural Japan (Qu, 2019). Among these festivals, a 2014 Organisation for Economic Co-operation and Development (OECD) report described the Setouchi Triennale as a representative case study of creative tourism connecting art projects' social engagement, public-private partnerships, civic participation and art tourism development (OECD, 2014). In 2017, China also started inviting rural art place-makers from the Setouchi Triennale to collaborate on art rural revitalisation projects and share best practices of rapid commercial implementation.

These large, trans-regional art projects link creative tourism and relational geography (Qu, 2022). The integration of art and creative tourism in rural settings establishes cross-national/ regional exchanges of population, culture and economic resources between urban and rural areas, but also creates new social issues such as unequal urban-rural exchanges, exogenous impacts, and development ethics (Qu, 2019; Qu & Funck, 2021). For example, the issue of 'borrowed art exhibited on borrowed land' (Qu, 2019) inevitably raises critical voices about the relationship between globalisation and regional culture. The biggest conceptual gap can be identified between art and social science scholars, with endless debates about the difference between art 'in' society and art 'for' society (Qu, 2020).

In art tourism, the desire to travel is motivated by art appreciation rather than by the tourism industry offer (Franklin, 2018). The emerging types of socially engaged art interventions, however, show their power as a driving force for rural tourism development (Prince et al., 2021a) by enhancing global-urban-rural linkages through cultural and resource exchange (Qu & Funck, 2021). This facilitates the creation of a diversified rural tourism and creatively enhanced community places (Mitchell, 2013; Qu, 2022) as well as triggering rural in-migration, including attracting newcomers establishing creative businesses (Qu et al., 2022). In some cases, these businesses also benefit their destination's communities and play an additional role as social assets (Qu et al., 2022).

Consequently, a new interdisciplinary research framework suggested by the current development in tourism geography (Wilson, 2012a) is urgently needed to avoid partial interpretations of the role of rural art interventions and the unresolved debates between art in society, society in art, art in tourism, tourism in art, tourism in rural areas and rurality in tourism. Moreover, the growing interconnections between art tourism and rural dynamics make it increasingly difficult to differentiate the disciplinary boundaries between visual arts and aesthetics, cultural and tourism geographies, planning and development, sociology and anthropology of art, and rural studies. The intersection of creative/relational tourism (Duxbury, 2021; Duxbury & Richards, 2019), relational geography (Massey, 2005), and creative geography (Gibson, 2010b; Hawkins, 2015) does not merely offer one field as a model or source of critique for the others, but rather demands that new research moves beyond the current horizons of those disciplines.

There is a need to adapt geography's spatial thinking to demonstrate the relationality and creativity of rural art place-making, art tourism and rural revitalisation. This chapter attempts to establish a new framework connecting art, tourism and rurality in their geographical sub-disciplines and their recent development from creative (contemporary art) geographies (Hawkins, 2015; Price & Hawkins, 2018), creative/relational tourism (Duxbury, 2021; Duxbury & Richards, 2019; Richards & Wilson, 2006), tourism geography (Williams, 2009; Williams & Lew, 2014; Wilson, 2012a) and creative rural spaces (Mitchell, 2013; Woods, 2012). This discussion does not only benefit tourism geographers but it also connects socially engaged art historians and other rural development and planning scholars more closely, integrating them into a new creative social site.

Contemporary Art's Social Transformation, Creative (Art) Geography and Rural Art Tourism

Although tourism triggered by art is not planned by professional tourism agencies but created within the art world (Franklin, 2018), newly emerging urban art lovers going to see art in rural areas are not so much art connoisseurs as diversified urban tourists attracted by art (Funck & Chang, 2018). Both art, tourism and rurality are highly relational in different ways. A recent comparative study shows that traditional craft-art tourism in Bornholm and contemporary art in Naoshima both owe their distinctive reproduction of global processes to their spatial identities as locations for the pleasure and consumption of artistic and rural experiences (Prince et al., 2021b).

Traditional craft-art tourism connected to its rural context, such as the craft-artists of Bornholm, also shows that community-scapes are formed by social practices in art (Prince, 2018). Furthermore, when urban and global-based contemporary art moves out from urban art museums and galleries into the rural territory, it often carries a more strongly socially engaged mission rather than a purely aesthetic value (Qu, 2019, 2020). This provides opportunities for organisers and artists to turn local communities into hubs for this type of socially engaged and relational art site by including community as part of the source of artistic expression or the stage for artistic creation.

In socially engaged art place-making, culture and art can play a key role in the participatory process of community adaptation, development and revitalisation (Duxbury, 2021; Duxbury & Campbell, 2011; Qu & Cheer, 2021; see also Chang, Chapter 13). Creative geography studies (Hawkins, 2015) have confirmed that research on 'art in society' should not be limited to the artistic side but also focus on art's impact and interaction with society. The political 'doing' of contemporary art itself is highly geographically engaged (ibid). Creative geography shows the role of art as agency, medium and trigger to bring changes that combine social and aesthetic impacts in one (ibid). This is the reason why the focus is not on art geography but more on creative geography. This entails considering a wider geographical perspective on socially engaged art, from the artwork to the relational art site (Qu, 2020; 2022). Instead of discussing art and society as separate entities, it is more fruitful to mix art, community and subsequent tourism phenomena within a diverse social space for discussion. Thus, the geographical intervention of contemporary art, art tourism, festivals and the additional social, cultural and environmental impacts of art tourism on a locality are closely related to tourism and rural geographies.

Although one cannot rely on art alone to provide solutions to all rural community issues, art can trigger opportunities for solving them (Duxbury & Campbell, 2011). Art itself does not have a social enterprise function. However, the resulting art tourism can facilitate rural structural change. Art tourism is also shifting towards more site-specific and socially engaged approaches in regional contexts (Franklin, 2018; Qu & Cheer, 2021). Art tourism and micro-entrepreneurs trigger relational social interactions; for example, artists interact with tourists and residents, shaping the locally embedded cultural landscape (Prince, 2018). In some positive cases, art tourism shows its power to strengthen cultural capital such as sense of place and social belonging (Duxbury & Campbell, 2009; Gibson & Connell, 2011), establish local and extra-local social networks (Balfour et al., 2018; Duxbury & Campbell, 2009) and enhance community resilience (Mackay et al., 2018; Qu & Cheer, 2021).

However, although art place-makers usually make commitments to the local community, more often than not they aim to attract as many tourists as possible (Franklin, 2018). The rapid development of art tourism has had a number of sudden impacts on local communities, such as 'art washing' and tourism gentrification that causes artists to move away from their art district due to over-commercialisation (Franklin, 2018). Ironically, socially engaged art tourism (SEAT) in rural art interventions advances a philosophy of anti-urbanisation, but in practice it exhibits artworks collected from the world's top-ranking artists, through postmodernist contemporary art experiences.

In Japan, socially engaged art strongly focuses on rural spaces and issues such as rural decline and counter-urbanisation (Prince et al., 2021b) by "enhancing local culture and society and bolstering economic retention on island level" (Qu, 2020, p. 263). In the case of the Setouchi Triennale, all co-organisers emphasise the importance of social responsibility, but at the same time fail to recognise the impact brought by large numbers of tourists on small islands (Prince et al., 2021b; Qu & Funck, 2021). This type of place-making process also carries the risk of producing art tourism development outcomes that have little to do with community revitalisation, such as "theme park" type tourism development and cultural colonisation by private art foundations (Qu, 2019). Development strategies for art tourism should consider the difference not only in the relationship between art and tourism (Franklin, 2018) but also between art tourism development and regional revitalisation (Qu & Cheer, 2021). In the rural context, art tourism policy should focus more on cultural development and revitalisation rather than on tourism-oriented development through simply building more luxury art hotels and contemporary art museums. Geography as a discipline provides a framework to examine differences between 'development' and 'revitalisation' among different communities under the same art tourism brand.

Large-scale Creative and Art Tourism Place-making in Small Peripheral Areas

As an expanding interdisciplinary subject, tourism geography (Williams, 2009; Williams & Lew, 2014) provides a baseline framework to explore the relationship between creative art geography and creative tourism geography. Art tourism itself sits at the intersection between cultural tourism and creative tourism. Creative tourism represents a more relational (Richards, 2013) and potentially more sustainable form of cultural tourism (Richards & Wilson, 2007). Rural art tourism is both creative and relational, facilitating co-creation and consumption processes that trigger urban-rural linkages, diversified social interactions and networks that can connect residents with regional, national and global visitors (Duxbury & Richards, 2019). As a development strategy, creative place-making has shown its potential to build creative districts and cultural quarters in urban centres (Richards, 2020; see also Chang, Chapter 13).

Art place-making can be performed in rural destinations as well, where it offers visitors a means to connect with 'the local' through social interaction as a new type of creative and relational community-based tourism in rural areas. The discussion of rural creativity, however, is often framed in opposition to perspectives examining urban-based creativity with its larger scale industries and models, which are very different from regional strategies aiming at reactivating inherent local creativity and creating new economic opportunities (Woods, 2012). Creativity and innovation are not limited to urban places, but perform differently in small-scale peripheral destinations (Richards & Duxbury, 2021; Scherf, 2021) and island communities (Baixinho et al., 2020), which can be seen as 'creative outposts' (Brouder, 2019). Creative place-making strategies such as networking, co-creation, producer collaboration, digital technologies and indigenous cultures are considered extremely valuable in smaller places (Richards, 2021). By establishing an endogenous model of creativity (Wilson, 2012b) and long-term sustainability through creative outposts, micro-businesses appear to be more resilient (Brouder, 2019; Qu & Cheer, 2021).

The concepts of creative tourism and art tourism are similar in that they are an active type of cultural experience and consumption pattern (Richards, 2011). Like creative tourism, SEAT is based on co-creation between visitors, in-migrants and long-term residents. The creative outcome of SEAT is not limited to tourism products or experiences but centres on its social promise of community revitalisation. In this sense, art becomes a social practice that connects with tourism (Qu, 2019), tourism becomes a tool to facilitate urban-rural cultural and socio-economic capital exchange and peripheral communities become reactors for tourist-artist-resident interaction and resource exchange. In the case of rural Scandinavia, the artistic network shows the ability to revitalise a declining area and enhance both art making and the art business (Prince et al., 2021a). Some of the Setouchi Triennale's small islands are similarly experiencing a sudden transition from peripheral declining regions to ranking among the top four travel destinations in Japan for international and domestic tourists (Milner et al., 2020). This trend also reflects a transition in rural areas, from the home village from which young people escape to new 'art islands' (Prince et al., 2021b; Qu, 2020) attracting young visitors and new residents. Artwork that pays homage to specific islandscapes and uses local history and cultural elements as a fundamental tactic for community co-creation can be found at smaller-scale community art festivals like Shiosai (Qu & Cheer, 2021).

According to Richards' shift from culture to creativity (Richards, 2011), as shown in Figure 14.1, SEAT establishes a linkage between art tourism, creative tourism and a community's cultural heritage and culture. In the Setouchi Triennale, for example, in addition to the elite art that attracts urban tourists, some artworks and projects have been found to have the power to revitalise the community's cultural heritage and encourage the development of local folk art and crafts (Qu, 2020; Qu & Funck, 2021). Often, these socially engaged artworks can bring social interaction and local-to-extra-local knowledge and capital exchange (Qu, 2020; Qu & Cheer, 2021). On the art islands where revitalisation has been most effective, art businesses also play a non-tourism role as social assets for ageing residents (Qu, 2020; Qu et al., 2022). This makes them highly relational and matches the creative tourism theory of co-creation (Duxbury & Richards, 2019).

However, creative tourism and rural art tourism differ in certain attributes. Unlike endogenous creative outposts, socially engaged art and its related tourism usually develop through neo-endogenous processes – understood as action for change 'from within' rural places, led by rural people themselves but enabled and steered by policy actors from the top down (Ray, 2006) – rather than from purely bottom-up efforts (Prince et al., 2021a,b; Qu, 2020). The dualistic top-down versus bottom-up perspective that is frequently associated with rural development activities is something that the neo-endogenous paradigm aims to avoid. Neo-endogenous processes result from the constant dialectic interaction of endogenous and exogenous factors, blending external

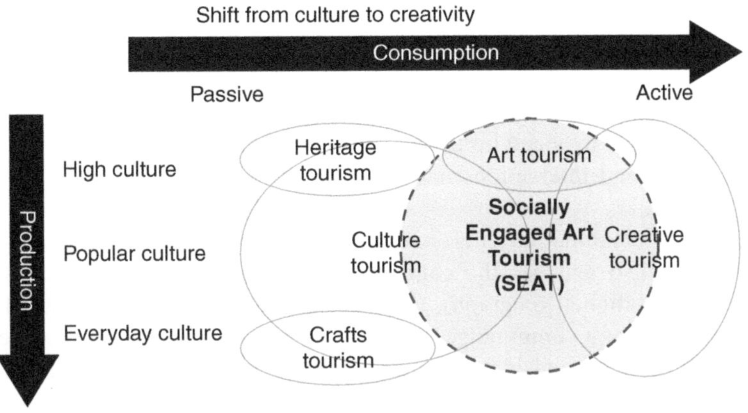

Figure 14.1   Socially engaged art tourism (SEAT) in a creative tourism framework.
Source: Author's Illustration (Adapted from Richards, 2011).

factors with intrinsic regional potential. The evaluation of neo-endogenous development outcomes depends heavily on the degree of "local control of the development process" (Bosworth et al., 2016, p. 427) from the community perspective. Without art place-makers from outside, the likelihood of these initiatives arising through purely locally initiated efforts is low, especially in places characterised by socio-economic and demographic decline (Muramatsu & Akiyama, 2011). In such cases, co-creation, learning and relational aspects depend on how extra-local creativities engage, transform and re-integrate with the community. Thus, SEAT is a new type of art tourism integration in rural contexts through social engagement and practice, with the following characteristics:

(1) Neo-endogenous festival-community sustainable co-development/revitalisation (Qu & Cheer, 2021; Qu & Funck, 2021) includes both exogenous contemporary art and rural cultural tourism development, as well as endogenous community active engagement and rural-to-extra-rural co-creation, and is characterised by a clear mission of rural revitalisation (Prince et al., 2021b; Qu, 2020).
(2) The bigger its scale, the harder it is to organise under a single management agency; therefore it is often managed through diverse public-private partnerships (Qu et al., 2022).
(3) As shown in Figure 14.1, it condenses elite art, popular culture and the everyday community way of life into one tourism experience, which attracts young urban visitors who are not necessarily a professional art audience (Funck & Chang, 2018). At the same time, its origins in urban-based art tourism influence rural contexts (Franklin, 2018; Qu, 2019) and link cultural tourism elements and creative tourism experiences (Duxbury, 2021; Qu & Funck, 2021).
(4) It has the power to trigger a community response and attract newcomers creating new businesses (Duxbury & Campbell, 2011; Qu et al., 2022). In the ideal scenario, these businesses also play a non-tourism role as social assets for the community (Qu et al., 2022).

Revitalisation: Art Islandness, Creative/Relational Ruralities and Creative Enhancement

In cultural terms, the separation between rural and urban is one of the oldest geographical concepts (Woods, 2011). However, the definition of 'rural' has "become increasingly difficult and contested", testing the boundaries of disciplines such as rural geography (Woods, 2009, p. 429). Rural spaces are becoming more and more diversified in terms of stakeholder networks and social capital exchange flows with extra-local entities (Heley & Jones, 2012). Many rural areas are highly connected with nearby urban centres (Bosworth et al., 2016). In this context, SEAT becomes a trigger to speed up this co-creation process between places, landscapes and people. Connections and interactions have blurred the boundaries between urban and rural, local and extra-local.

The power of the combination of art and tourism to create a dynamic population exchange flow in rural areas also matches the concepts of constant re-shaping of rural-urban spatial linkages found in relational geography (Massey, 2005). Through tourism, art is constantly shaping the landscape of a community. In the case of creative art in urban spaces, the MONA effect has recently attracted considerable attention. It not only brought significant social change for Tasmania, but also generated a two-sided debate as to whether it was a new version of 'the Bilbao of the South' or a 'subversive Disneyland' (Ryan, 2016; Booth et al., 2021). In other words, although art has the potential to transform a backward area into a cultural district, it also carries the risk of revealing the nature of over-commercialisation. One important finding is that the community-level impacts of art are limited to those with cultural capital (Booth et al., 2021). In Japan's rural contexts, the Setouchi Triennale gave rise to twelve distinctive 'art islands' (Prince et al., 2021b; Qu, 2020). The risk of becoming an art island and art theme park is also similar (Qu, 2022).

However, the difference between the MONA effect in urban Tasmania and the Naoshima Effect in the Setouchi rural islands is the community revitalisation aspect. The newfound 'art islandness' refined the community's original sense of peripherality and established a new creative identity, one that also attracts urban-rural in-migrants. These new residents are not just moving to a new physical rural location, but also connecting their networks to – and unleashing their capacity for innovation in – the local community. The interaction and knowledge exchange and integration among artists, tourists, and new and old residents also accelerate the establishment of art islandness. Researchers such as Duxbury and Campbell (2011) have suggested the success of revitalisation initiatives be evaluated on the basis of whether the community can retain and engage youth, attract new residents and businesses and maintain community vitality and identity. In the best case, the cultural and social exchange attached to these initiatives brings a new way of life through newcomers while at the same time preserving traditional values and refashioning them for a new era.

SEAT has the power to build new art islands and their 'art islandness', but this does not ensure the success of social commitments to the community. As such, the results of art interventions can be divided into community-friendly creative enhancement and tourist-centred creative destruction (Mitchell, 2013). The success of creative rural revitalisation relies heavily on whether the involvement of community stakeholders in art projects has a deeper significance. In the case of Setouchi Triennale's art islands, four distinct outcomes can be described: (a) art tourism co-development/revitalisation, with the establishment of newcomers with businesses that benefit the local population; (b) art tourism development with less local involvement and more outsiders (e.g. seasonal workers and business owners) who do not interact with the community; (c) community development with less consideration for art tourism; (d) lack of community-level power to bring about change (Qu et al., 2022).

Outcome (a) can be seen as the ideal revitalisation outcome, characterised by a mechanism of joint operation between endogenous (top-down) and exogenous (bottom-up) efforts, which should further be integrated into a neo-endogenous way of conceptualising art revitalisation (Qu, 2020).

As a new type of integrated rural development that differs from urban creativity and relies heavily on the integration of regional knowledge, resources and collaboration at the community level, "local control of the development process" (Bosworth et al., 2016, p. 427) is the key to evaluating neo-endogenous development outcomes. Other outcomes, such as result (b), do not support the community in overcoming its original issues. Type (a) outcomes are also closer to the concept of creative enhancement, characterised by the co-creation of tourism destinations where the community can maintain multiple functions to ensure local livelihoods in parallel with art tourism development. Distinguishing between different outcomes can be best performed through a comparison of different communities under the same art revitalisation effort, keeping in mind their geographical and socio-cultural characteristics as well.

Future Trajectory and Methodological Framework for Rural Creative Tourism Geography

'Rural creative tourism geography' (RCTG) or 'rural art tourism geography' (RATG) are even more complex social-ecological systems that not only integrate socially engaged art, creative tourism, rural place-making and relational global-regional exchange flows, but also re-shape the field of art tourism geography. Due to their neo-endogenous attributes, the activities of in-migrant and social entrepreneurs in creative tourism also represent a local push factor for revitalisation. Therefore, RCTG is a more accurate expression than RATG.

As discussed above, socially engaged art tourism has a clear social mission: community revitalisation, a goal that is difficult to measure though standards and indicators. More research taking an interdisciplinary approach should be conducted to understand SEAT's mode of operation and effectiveness (Wilson, 2012a). New developments from relational art sites (Qu, 2020; 2022), relational tourism (Richards, 2013) and relational geography (Massey, 2005) are indivisible for discussing relational and creative rural places. To reach an effective outcome of sustainable community revitalisation (Qu & Cheer, 2021), a balanced approach in RCTG between art and tourism development, as well as the community revitalisation agenda, should be carefully considered.

By adapting the research framework for Setouchi Triennale, which covers the three keywords of art, tourism and community revitalisation (Qu & Funck, 2021), a large number of empirical findings through mixed methods have been refined for the RCTG framework (Figure 14.2) (Qu, 2019, 2020, 2022; Qu & Cheer, 2021; Qu et al., 2022).

The present chapter has discussed socially engaged art place-making, creative tourism in smaller peripheral and regional contexts, how it shapes creative rurality, art islandness, and its outcomes for art tourism revitalisation. The interlinking fields in creative geography, relational art/creative tourism and relational (local to extra-local) geography highlight the creativity, relational and geographical characteristics. Previous research indicates that excessive art and tourism development may cause a shift away from social promises to the community (Franklin, 2018; Qu, 2019). A neo-endogenous type of creative tourism should maintain its sustainable development goals without forgetting that the locality's peripherality is the prerequisite for the survival of its tourism products and the source of inspiration for artistic creation.

This chapter focuses on RCTG in smaller peripheral communities, which differ from creative urban clusters and stakeholder networks (Woods, 2012), as well as from urban-based art interventions such as MONA effects (Franklin, 2018; Ryan, 2016; Booth et al., 2021). A comparative study of multiple destinations under one SEAT product is needed to evaluate the advantages and disadvantages and demonstrate current issues. The trajectory of changes in community

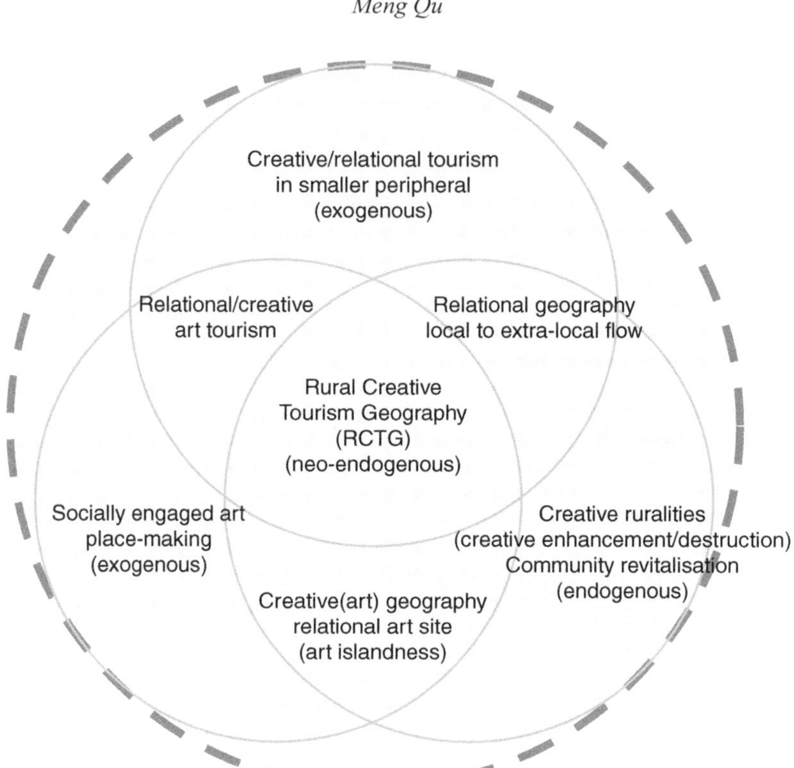

Figure 14.2   A research framework for rural creative tourism geography for community revitalisation.
Source: (Author's own illustration).

art ruralscape and art islandscape can be shown through Mitchell's creative enhancement and destruction concepts (2013). Due to the uniqueness and limitations of contemporary art itself, it is impossible to demand that art tourism agencies provide the same artwork and resource-based opportunities to all communities. However, the dialogue between sustainable tourism management and community revitalisation needs to be constantly improved. The diversification and common landscape transition of communities is one of the benchmarks for evaluating the degree of art washing and tourism gentrification (Qu, 2019).

In conclusion, socially engaged art/festival tourism is a concept of increasing interest for rural areas in the Global North and can lead to the development of locally based versions of creative rural revitalisation. Therefore, the possible outcomes of these projects and their related indicators need to be better understood and defined. Since each intervention should adapt to local conditions in order to support social engagement, understanding the difference between creative destruction through excessive art tourism development and neo-endogenous creative enhancement becomes fundamental to community revitalisation through socially engaged art tourism. To this end, ongoing inter- and trans-disciplinary exchanges should be established between researchers, art place-makers and community stakeholders.

The rise of physical mobility and digitalisation, which helps to connect urban and rural areas, will have a significant impact on rural society's creative turn in the upcoming decades. Urban creative clusters will also discover a more welcoming climate for their work in the countryside.

The diversity of rural communities offers the ideal setting for creative entrepreneurs and artists to create communities, creative social enterprises, art tourism and projects that are socially involved in many ways. As a result, academics are more likely to be interested in rural creativity via more transdisciplinary, spatial, relational and non-urban-based typologies. The countryside offers a creative platform for artists with greater potential on a smaller scale, particularly following the COVID-19 pandemic. Thus, "small is the new big", in the sense that it signifies a more diverse environment and resilient relational social spaces for social studies and the arts.

Acknowledgements

This research was funded in part by the Japanese Society for the Promotion of Science (JSPS) Grant No. 22K13251.

References

Baixinho, A., Santos, C., Couto, G., Albergaria, I. S. De, Silva, L. S. da, Medeiros, P. D., & Simas, R. M. N. (2020). Creative tourism on islands: A review of the literature. *Sustainability, 12*(24), 1–25.

Balfour, B., W-P Fortunato, M., & Alter, T. R. (2018). The creative fire: An interactional framework for rural arts-based development. *Journal of Rural Studies, 63*, 229–239.

Booth, K., Ragaini, B. S., & Hardy, A. (2021). A Mona effect: How place discourse constitutes culture-led change. *Geographical Research, 59*(1), 16–28.

Borggreen, G., & Anemone, P. (2020). Autonomy and collectivity at the Echigo-Tsumari art Triennale in Japan. In B. Eriksson, B. Valtysson, & C. Stage (Eds.), *Participatory Cultures – Arts, Media and Politics* (pp. 30–50). Routledge.

Bosworth, G., Annibal, I., Carroll, T., Price, L., Sellick, J., & Shepherd, J. (2016). Empowering local action through neo-endogenous development: The case of leader in England. *Sociologia Ruralis, 56*(3), 427–449.

Brouder, P. (2019). Creative tourism in creative outposts. In N. Duxbury, & G. Richards (Eds.), *A Research Agenda for Creative Tourism* (pp. 57–68). Edward Elgar.

Dunphy, K. (2009). *Developing and Revitalizing Rural Communities Through Arts and Creativity: Australia.* Creative City Network of Canada.

Duxbury, N. (2021). Cultural sustainability, tourism, and development: Articulating connections. In N. Duxbury (Ed.), *Cultural Sustainability, Tourism and Development* (pp. 1–18). Routledge.

Duxbury, N., & Campbell, H. (2009). *Developing and Revitalizing Rural Communities through Arts and Creativity: A Literature Review.* Creative City Network of Canada.

Duxbury, N., & Campbell, H. (2011). Developing and revitalizing rural communities through arts and culture. *Small Cities Imprint, 3*(1), 111–122.

Duxbury, N., & Richards, G. (Eds.) (2019). *A Research Agenda for Creative Tourism.* Edward Elgar.

Franklin, A. (2018). Art tourism: A new field for tourist studies. *Tourist Studies, 18*(4), 399–416.

Funck, C., & Chang, N. (2018). Island in transition: Tourists, volunteers and migrants attracted by an art-based revitalization project in the Seto Inland Sea. In D. K. Müller, & M. Więckowski (Eds.), *Tourism in Transitions: Recovering Decline, Managing Change* (pp. 81–96). Springer.

Gibson, C. (2010a). Creativity in 'peripheral' places: Redefining the creative industries. *Australian Geographer, 41*(1), 1–158.

Gibson, C. (2010b). Guest editorial—creative geographies: Tales from the 'margins'. *Australian Geographer, 41*(1), 1–10.

Gibson, C., & Connell, J. (2011). *Festival Places: Revitalising Rural Australia.* Channel View Publications.

Hawkins, H. (2015). *For Creative Geographies.* Routledge.

Heley, J., & Jones, L. (2012). Relational rurals: Some thoughts on relating things and theory in rural studies. *Journal of Rural Studies, 28*(3), 208–217.

Mackay, M., Fountain, J., & Cradock-Henry, N. (2018). Festivals as devices for enhancing social connectivity and the resilience of rural communities. In J. Mair (Ed.), *The Routledge Handbook of Festivals* (pp. 214–222). London.

Massey, D. (2005). *For Space*. Sage.

Milner, R., Bartlett, R., Bender, A., Forge, S., McLachlan, C., Morgan, K., O'Malley, T., Richmond, S., Tang, P., Walker, B., & d'Arc, T. S. (2020). *Lonely Planet Japan*. 16th ed. Lonely Planet.

Mitchell, C. J. A. (2013). Creative destruction or creative enhancement? Understanding the transformation of rural spaces. *Journal of Rural Studies, 32*, 375–387.

Muramatsu, N., & Akiyama, H. (2011). Japan: Super-aging society preparing for the future. *The Gerontologist, 51*(4), 425–432.

OECD. (2014). *OECD Studies on Tourism: Tourism and the Creative Economy*. OECD.

Price, L., & Hawkins, H. (Eds.) (2018). *Geographies of Making, Craft and Creativity*. Routledge.

Prince, S. (2018). Dwelling in the tourist landscape: Embodiment and everyday life among the craft-artists of Bornholm. *Tourist Studies, 18*(1), 63–82.

Prince, S., Petridou, E., & Ioannides, D. (2021a). Art worlds in the periphery: Creativity and networking in rural Scandinavia. In K. Scherf (Ed.), *Creative Tourism in Smaller Communities: Place, Culture, and Local Representation* (pp. 259–282). University of Calgary Press.

Prince, S., Qu, M., & Zollet, S. (2021b). The making of art islands: A comparative analysis of translocal assemblages of contemporary art and tourism. *Island Studies Journal, 16*(1), 235–254.

Qu, M. (2019). Art Interventions on Japanese Islands: The promise and pitfalls of artistic interpretations of community. *The International Journal of Social, Political and Community Agendas in the Arts, 14*(3), 19–38.

Qu, M. (2020). Teshima – from Island art to the art Island. *Shima: The International Journal of Research into Island Cultures, 14*(2), 250–265.

Qu, M., & Cheer, J. M. (2021). Community art festivals and sustainable rural revitalisation. *Journal of Sustainable Tourism, 29*(11–12), 1756–1775.

Qu, M., & Funck, C. (2021). Rural art festival revitalizing a Japanese declining tourism island. In N. Duxbury (Ed.), *Cultural Sustainability, Tourism and Development (Re)articulations in Tourism Contexts* (pp. 51–68). Routledge.

Qu, M., McCormick, A. D., & Funck, C. (2022). Community resourcefulness and partnerships in rural tourism. *Journal of Sustainable Tourism, 30*(10), 2371–2390.

Qu, M. (2022). Socially engaged art tourism, in-migrants micro-entrepreneurship, and peripheral island revitalization. *Wakayama Tourism Review, 3*, 17–19.

Ray, C. (2006). Neo-endogenous rural development in the EU. In P. Cloke, T. Marsden, & P. Mooney (Eds.), *The Handbook of Rural Studies* (pp. 278–291). SAGE.

Richards, G. (2011). *Creative Tourism: Development: Trends and opportunities* [PowerPoint slides].

Richards, G. (2020). Designing creative places: The role of creative tourism. *Annals of Tourism Research, 85*, 102922.

Richards, G. (2021). Creative tourism in smaller communities. In K. Scherf (Ed.), *Creative Tourism in Smaller Communities: Place, Culture, and Local Representation* (pp. 283–297). University of Calgary Press.

Richards, G. (2013). *Creating Relational Tourism through Exchange*. Paper presented at the ATLAS Annual Conference, Malta, November 2013. November.

Richard, G., & Duxbury, N. (2021). Trajectories and trends in creative tourism: Where are we headed?. In S. Albino, & C. P. de Carvalho (Eds.), *Creative Tourism: Activating Cultural Resources and Engaging Creative Travellers* (pp. 53–58). CABI.

Richards, G., & Wilson, J. (2006). Developing creativity in tourist experiences: A solution to the serial reproduction of culture? *Tourism Management, 27*(6), 1209–1223.

Richards, G., & Wilson, J. (Eds.) (2007). *Tourism, Creativity and Development*. Routledge.

Ryan, L. (2016). Re-branding Tasmania: MONA and the altering of local reputation and identity. *Tourist Studies, 16*(4), 422–445.

Scherf, K. (Ed.) (2021). *Creative Tourism in Smaller Communities: Place, Culture, and Local Representation*. University of Calgary Press.

Tu, S. H. S. (2022). Island revitalization and the Setouchi Triennale: Ethnographic reflection on three local events. *Okinawan Journal of Island Studies, 3*, 21–40.

Williams, S. (2009). *Tourism Geography: A New Synthesis*. Routledge.

Williams, S., & Lew, A. A. (2014). *Tourism Geography: Critical Understandings of Place, Space and Experience*. Routledge.

Wilson, J. (Ed.) (2012a). *The Routledge Handbook of Tourism Geographies*. Routledge.

Wilson, J. (2012b). Tourism, creativity and space. In J. Wilson (Ed.), *The Routledge Handbook of Tourism Geographies* (pp. 143–149). Routledge.

Woods, M. (2009). Rural geography: Blurring boundaries and making connections. *Progress in Human Geography, 33*(6), 849–858.

Woods, M. (2011). *Rural*. Routledge.

Woods, M. (2012). *Creative Ruralities*. Paper at the 'Creativity on the Edge' Symposium, Moore Institute, National University of Ireland Galway, June.

15

GEOGRAPHIES OF FESTIVAL AND EVENT SPACES AND PLACES

Bernadette Quinn

Introduction

Festivals and events did not merit a dedicated chapter in the 2012 edition of this Handbook. Indeed, as recently as 2014, Hall and Page (2014) described event studies as a young study area. However, the geographic literature on festivals and events has expanded rapidly since then. What now seems like an enormous expansion in research output on the topic reflects the fact that as Frost (2016: 569) notes, events like "festivals and carnivals are significant contemporary phenomena in societies across the world". In very recent pandemic times, one could argue that the significance of festivals and events has become even more apparent. Somewhat ironically, this is because their relative absence or at least their otherwise constrained or sharply altered materialisation since early 2020, has heightened awareness of their multiple values. The cancellation, postponement or curtailment of events, particularly large-scale events like the Olympic Games, translated into huge economic losses. Restrictions on social gatherings made people more aware of the importance of festivals and events in creating opportunities for social encounters, making space meaningful and memorable, and facilitating cultural expression. Never before has it been clearer that festivals and events create public space; and access to public space, as experience of the pandemic has taught us, matters greatly for human health and well-being and for the liveability of towns and cities. This is unequivocally acknowledged in sustainability debates, in the UN's sustainable development goals, especially goal 11, and its 2016 New Urban Agenda.

Festivals and events are also obvious examples of the kinds of activities that endow space with the significance that turns it into place, the kind of places where we actively want to spend time, often with significant others, engaging in activities that promote well-being, facilitate self-expression and enable the formation of identities. Implicit in these ideas is the fact that festivals and events inform the dynamism of place, causing it to be continuously in motion and constantly re-made (Sheller and Urry 2004). In transforming space from the everyday into the extraordinary, they represent transformative experiences and facilitate passage into the kind of liminal spaces (Crampton et al., 2020) or 'other spaces' (Foucault, 2008) that allow people to refresh and re-energise so as to be better able to engage with the everyday. However, events are inherently political and potentially subversive (Waterman, 1998). They usually involve "both tensions and contradictions" (Falconi, 2011: 12), creating all sorts of disruptions but also potentials (Pløger,

DOI: 10.4324/9781003286301-18

2010) not only at the level of the host places but at wider geographical scales and in broader political and cultural contexts.

So, for geographers, festivals and events have an obvious relevance that has been long recognised. For several decades now, geographical research has focused on trying to understand the complex relationships between festivals and events, and the core geographic concepts of place, space and to a lesser extent, environment. At this point, there have been several 'state of the art' reviews of festival and event research (Duffy & Mair, 2021) and some of these adopt geographical perspectives (e.g., Cudny, 2014; Finkel & Platt, 2020). This chapter adds to this literature by reflecting on how recent geographic work has contributed to theorising and furthering understandings of the phenomena, and by considering emerging developments in the literature. The discussion is structured under the headings of place, space and environment, notwithstanding the blurring of distinctions between these concepts.

Place

Festival and event activity is extremely varied in nature, type, size and scale, and geographic work is correspondingly diverse, asking an array of questions about: festivals of all kinds, set in urban and rural contexts; communal social gatherings like parades, protests and carnivals set in urban contexts, and large-scale sports events associated with metropolitan cities. The long-established understanding that festivals and events contribute to making place (Ryan & Wollan, 2013; Wynn, 2015; McClinchey, 2020) fundamentally draws upon seminal contributions from geographers like Relph (1976) who explained that place is space that people have imbued with meaning through lived experience. As numerous geographers have explained, festivals and events represent activities through which "people interpret themselves and are also interpreted by others according to the place where they live" (DeBres & Davis, 2001: 327). They inform the everyday practices of people's lives (Ryan & Wollan, 2013), creating opportunities for people to come together, interact, and create (Bakas et al., 2019). Through involvement in festivals and events, social networks and connections are built (Duffy & Waitt, 2011) and attachments and belonging to others and to place are forged (Quinn, 2003). Identities of varying kinds are expressed, negotiated and contested in the process of producing and consuming festivals and events (Finkel & Platt 2020). All of this work illustrates what McClinchey (2020: 2028) means when she says that 'place-making is how community groups imprint their values, perceptions, memories, and traditions on a landscape and give meaning to geographical space'.

The term 'placemaking' as used in urban planning, however, has a number of meanings which Wychoff (2014) describes as 'sometimes confusing and contradictory'. These all deliberately and strategically involve strategists, policy-makers and other actors fashioning place in particular ways. They have long been of critical interest to geographers. An enormous literature in geography and elsewhere (e.g., urban studies, sociology) critically investigates how festivals, but particularly events, are strategically instrumentalised in the interests of placemaking (Quinn et al., 2020). Wynn (2015) comprehensively discusses how and why cities and communities use them as mechanisms for investing in, using and replenishing, economic, spatial, socio-cultural and symbolic resources in the interest of making and remaking place. For many, the outcomes of this are negative (Duignan and McGillivray, 2019), and result in mega-events subjecting places to the "strictures and structures of an increasingly global neo-liberal political economy" (Gaffney, 2010: 27). Indeed, over time, the literature on mega-events has come to underscore a mountain of contentious outcomes linked to the reconfiguration, privatisation and commoditisation of space (Smith, 2014; 2016), social exclusion and the displacement and marginalisation of local communities (Wise, 2017; Casaglia, 2018), both while the event is ongoing, and afterwards (Ferreri, 2020). Müller (2015) has argued that planning for mega-events appropriates resources, monopolises public attention, can suspend the normal rule of law, and often rewrites urban and regional development plans. Nevertheless, until fairly recently, mega events continued to be unproblematically promoted by urban coalitions as tools of urban development because they serve so many powerful interests. Tourism is one of those interests. This new *Routledge Handbook of Tourism Geographies* accommodates a chapter on festivals and events in no small part because the latter have proliferated as cities have recognised and actively promoted their tourism potential at every opportunity. Large-scale events, in particular, are used to attract tourists, those "semiotic armies who move around the world to gather signs of local identity" (Ma & Lew, 2012: 18). In similar vein, festivals and events are instrumentalised in the interest of global positioning, branding cities, and creating urban imaginaries. This has been equally widely and critically written about (Smith, 2005; Broudehoux, 2017), with McGillivray and Frew (2015: 2650) viewing the city as "a canvas willingly given over to the territorialising tendencies of brandscaping where parks, squares and buildings are opened up to commodification processes".

However, very recently there has been a shift in urban policy thinking about mega events, as large-scale opposition critical of spiralling staging costs and associated risks, sceptical about the long-term benefits actually achieved through event legacies, and concerned about environmental and social impacts has materialised (Lauermann, 2019). As Lauermann (2019) goes on to explain, this opposition has manifested itself in the relatively new phenomenon of widespread protests against mega-event planning, even if political protests during mega events, as at COP26 in Glasgow in 2021, are not new (Cottrell & Nelson, 2011). It seems likely that this development will stimulate new kinds of geographical enquiries into mega events in the future. It is also likely that the near future will see geographers ask more questions about space, identity and power in the context of events as protest. Civic mobilisation is on the rise (Brechenmacher et al., 2020), as evidenced in recent years by civil society taking to the streets to lay claim to space and contest hegemonies linked to a range of globally-reaching cultural and political issues including gender, race, freedom of individual and political expression and climate change.

Space

The above discussion started with place but ended in the global, clearly illustrating Massey's (1994: 4) conceptualisation of place as "neither a bounded enclosure (nor) a privileged site of meaning-making, but rather a subset of the interactions which constitute space, a local articulation within a wider whole". The proliferation of festivals and events in recent decades is of course, a strong manifestation of globalisation in action (Hannam et al., 2016). The spatial transformations that they effect derive from activities that are influenced both by place and by a set of dynamic performances, activities and forces that are fluid and highly mobile (Falconi, 2014). These transformations are enormously complex and a preoccupation in the literature has been to investigate how small and large events alike can change how space is used, looks, sounds and feels; how routine mobilities are altered; and how districts and even entire cities can be taken over as festivals and events occupy space.

While there is a huge literature on mega and large events and transformations in large city contexts, (e.g., Johansson & Kociatkiewicz, 2011; Müller, 2015, Duignan & McGillivray, 2019), recent years have seen an increase in the attention paid to how smaller events transform space in smaller places (e.g., Quinn & Wilks, 2017; Mahon & Hyyryläinen, 2019; Fisker, Kwiatkowski & Hjalager, 2019). The latter authors critique extant festival literature for paying scant attention to social and spatial theory, arguing that little is known about the role festivals play in on-going processes of socio-spatial change. In calling for a translocal perspective to understand how festivals function as social practices, they advocate conceiving of events as part of the network society (Castells, 2010). In particular, they highlight the usefulness of Richards' (2015) ideas about how different types of events, termed iterative and pulsar, function differently as forces of fixity and continuity, and change and fluidity, in the network society. The onset of the mobility turn in the social sciences (Cresswell, 2006) is another development that has begun to influence understandings of how festivals and events bring about spatial transformations that are "multi scalar and varied, involving the human and the non-human, the social and the material" (Sweeney et al., 2018: 573). Hannam, Mostafanezhad and Rickly (2016), for example, offer conceptual avenues for thinking translocally by applying mobilities thinking. They argue that festivals and events can be conceived of in terms of interruptions, transitions, framings and materialities. Recurring regularly, as they often do, and attracting loyalty among audiences, spectators and participants who repeatedly return either physically or virtually, they can also be thought of in terms of cyclical routines and circularities. In a wider and longer-term spatial-temporal context, Hjalager (2009: 267) has discussed how festivals are intimately bound up with a much wider musical economy, explaining that "festivals are not isolated events, rather they contribute to, and establish links in, rather complex value chains and innovation systems". Chaney (2019) makes a similar point, commenting on the role that rock festivals play relative to the recorded music industry. Implicit in his analysis is a multitude of linkages, connections and circular movements of tangible and intangible entities between festivals and other parts of the rock music industry. Similarly, in an empirical study of traditional music festivals, Quinn and Wilks (2013: 26) offer insights into "the highly mobile, fluid and internationally connected circuit of [material and immaterial connections and] flows that make up the world of festivals". All of this literature prompts many more research questions that could be furthered by thinking translocally and in terms of mobilities. Future work will require that careful attention be paid to the inevitable immobilities, displacements, absences and exclusions that will characterise festivals and events into the future, especially in an era when heightened uncertainties, risks and fears linked to climate change and the spread of viruses, and humanitarian crises are likely to exacerbate global inequalities. Being alert to these problems will be difficult if the work being produced on the geographies of festivals and events comes from just a handful of places in the world, as it predominantly does at the moment.

The idea that the particular space-time of festivals and events has a transformative, liminal quality, offers a temporary release from the strictures of the everyday and creates opportunities to temporarily unsettle or subvert the established social order is long established in the social sciences (Duvignaud, 1976). Turner (1982: 50) associates liminality with the development of communitas, which he describes as an 'alternative and more liberated way to be socially human'. In festival space-time, people have an opportunity to cultivate a sense of belonging with others; to experiment with identities and with normal social ordering by immersing themselves temporarily in what Foucault (2008) describes as a heterotopia. As counter-sites or 'other spaces', festivals are a clear example of heterotopia. They exist in real, physical space-time and only make sense in relation to the ordinary spaces and times alongside which they exist. While the subversive qualities of festivals seem to be time bound, they actually offer much potential for revitalising, re-energising and re-imagining not only the self, but also cultural practices, social relations and aspects of social and spatial ordering beyond the moment of the festival (Quinn & Wilks, 2017).

Elsewhere, taking references from the new cultural geography, a growing body of work thinks about transformations in terms of the materialities and corporealities through which festivals are produced and consumed (e.g., Cidell, 2016; Duffy et al., 2011). This work recognises that festivals spatially concentrate and vigorously mobilise a wide range of artistic practices, sensory experiences, and embodied practices (Kingsbury, 2016). They are performative sites where social engagement is performed. For Dundjeravic and Martínez Sánchez 2018: 462), performativity makes it possible for "spatial boundaries to be more flexible and for unpredictable behaviours to ensue". In studying festivals and events in this way, geographers have been noting how sensuous bodies (Duffy & Waitt, 2011), bodily rhythms and movements between different actors create "dynamic, … varying and emerging relations to each other" (Giovanardi et al., 2014: 113). Again, a key theme in this literature is that space, thus transformed, offers opportunities for individuals and groups to negotiate and contest everyday cultural politics played out on grounds like race, sexuality, gender and class, as well as normative ideas about who/what belongs in particular spaces.

More recently, an interest in emotion and the affective qualities of festivals as sites of performativity has increased, rounding out still further understandings of how social worlds are constituted and drawing links between the complex politics played out at societal level and those experienced and embodied by individuals in event settings (Johnston & Waitt, 2021). With protests and gatherings on the rise with increasing civic mobilisation, the scope for extending this line of enquiry seems very evident. The substantial body of work on events like Pride parades epitomises the kind of research currently drawing on performativity ideas and contributing to geographies of affect (e.g., Johnston & Waitt, 2015) but there is a need for further intersectional work encompassing many different kinds of cultural identities. In this regard, Platt and Finkel's (2020) edited collection highlighting the gendered nature of interactions and practices in festival and event settings represents important work. It queries whether the subversive potential of these phenomena could possibly be gender balanced in light of prevailing patriarchal structures and controls. Thus, there is now a small but growing, and very welcome body of work bringing feminist and critical perspectives into festival and events studies and geographies. More is needed.

Environment

More attention also needs to be paid to understanding the nexus between festivals and events, and the environment. There is a growing body of literature on this area, but somewhat inexplicably, it remains very small, leaving a great deal of scope existing for further research. As Duffy and Mair (2021) note, festivals and events have clear potential to promote environmental awareness. For example, Marks et al. (2014) found that in encouraging the 're-imagination of one's surrounds' festivals can prompt changes in environmental behaviour, enhance a sense of place and of environmental ownership. If further developed, this is one clear example of an area where festival and event geographers could influence policy and practice in the field of environmental awareness. The idea of festivals and events as places of learning, especially in respect of encouraging more sustainable lifestyles has also recently emerged, with Rowan (2020) raising the possibility that overtly labelled transformative festivals like the 'Burning Man' could potentially inspire more responsible tourism behaviours.

Perhaps more obviously, festivals and events, especially large-scale events, represent a clear environmental liability. This is becoming ever more obvious in the growing literature being produced by researchers across a wide variety of disciplines. However, with some notable exceptions, relatively little work is being produced by geographers on the topic. Notable exceptions include Andrea Collins (e.g., Collins & Cooper, 2017; Collins & Potoglou, 2019) who has published widely on the environmental problems associated with festivals and events, including on the problems associated with travelling to events. However, this is a vital area of work where more research is very obviously needed.

The Future

Back in 2012, Getz (2012: 382) envisaged seven different propositions for the future of the event industry. One of these was that "virtual events will gain in frequency and importance ..., but they will be in addition to, and not a substitute for, live event experiences". The advent of digitally enhanced events has very definitely arrived and the incorporation of digital technologies will have very significant implications for how festivals and events intersect with place, space and environment into the future. As yet, there is no clear research agenda signalling how future work on digital developments might usefully unfold. Very obviously there is the fact that production and consumption will now increasingly occur, and be variously mediated, in and through virtual space. However, what this means is not at all fully understood. Geographers will reach back into their disciplinary tool boxes to find approaches that help advance understanding. Duignan et al. (2017) find liminoid a useful concept for investigating how social media plays a role in constituting new and reconstituted spaces of consumption. Panopticism is a basic concept in Crampton et al.'s (2020) work on the rapid introduction of smart surveillance technology into festival settings which finds that smart surveillance changes people's experience of place (see also Dudley, Chapter 27 of this volume). Understanding people's connections, experiences and behaviours in festival and event places will involve further emphasis on embodiment, immersion and the sensuous aspects of events and festival activities. However, it will also probably bring about a return to engagements with the visual, as in areas like the geohumanities. All of this will require an expansion of methodological approaches to analysing data gathered in digital space. Understanding how physical and virtual spaces interweave to create new and re-invented kinds of social and spatial orderings will be an important avenue of enquiry. In line with the spatial reconfigurations spurred on by the COVID-19 pandemic, cities are already intensifying the presence of technological infrastructure in public spaces, thereby altering the function, appearance and sound of the city, as well as effecting change in the urban 'affective atmosphere' (Gandy, 2017). Future work in this vein can draw on the extensive research carried out by Edensor and various colleagues on light festivals (e.g., Edensor & Sumartojo, 2018).

While the disruptive effect of digital technologies will likely excite research interest and open up new lines of enquiry, it is very important not to overlook fundamental questions about the role that festivals and events play in creating sustainable places for people to live in. For a start, the uneven production of knowledge in this field is itself an issue. Most of the literature reviewed in this chapter comes from geographers working in advanced capitalist societies. Yet festival and event activity has been on the rise across the world for some time now. Ma and Lew (2012), for example, note a great increase in the number of cultural festivals in China linked to the aim of place branding and regional economic development. Van Zyl (2011: 181) notes a phenomenal increase in the number of festivals in South Africa post-apartheid, justified on "artistic, social, and cultural grounds as well as economic ones". Issues related to the unevenness of festival and event consumption and production, questions about social justice and social equity are already underexplored in geography. These will be even more important in the future as the challenges and risks posed by harmful events like the recent pandemic and recurring climate change related natural disasters will affect regions of the world unevenly.

Conclusion

As already noted, the last edition of this Handbook did not dedicate a chapter to this field. Yet in 2024, producing a single chapter that does justice to the geographies of festivals as well as events feels an impossible task. Researchers working on these phenomena have developed a very sizeable body of knowledge and understanding. Any review of geographical literature on festivals and events must firstly point to the enormous expansion of output on this topic in the last decade. There is now an expansive breadth to the work that could be interpreted as fragmentation (Finkel & Platt, 2020), but equally it could be seen as a maturation of this area of geographical enquiry into more easily identifiable areas of specialist enquiry. Previous review style articles noted concerns about under-theorisation (Hall & Page, 2014), but the literature discussed in this chapter points to work that is strongly characterised by critical thinking. It is engaging with contemporary geographical debates and progressing in line with several major developments within the discipline. Of course, there are many gaps in knowledge and definite scope for closer alignment with theories and approaches more associated with 'mainstream' geographies, but this is still a field of geography in the making.

Very generally, one dimension of the field that seems problematic is the question of scale. From the smallest of neighbourhood gatherings to the largest of mega events staged in the world's largest cities, geographers gain insight into how festival and event practices and interactions make places, and about the potential of festivals and events to unsettle normative ideas about space. While all of this work is productive, deeper understanding of the different spatial transformations associated with these wildly varying scales of operation would add more coherency to this body of knowledge. Waitt (2008) underlines the importance of festivals and events for geographic study when writing about how established geographies of all kinds can be challenged as these activities temporarily suspend social relations and sustain playful practice. However, Allen and McCreary (2021: 51) note that geographers tend to focus "on enduring material landscapes rather than temporary enactments such as festivals and parades". Certainly, within the wider field of geography, festivals and events are not exactly mainstream, however, this may change. As the concept of public social gathering becomes more problematic because of public health and safety concerns, there is a sense that we will have to find new ways of socialising without fear, and of reconfiguring social spaces so as to alleviate health risks and fears about being in proximity with others. This will create the need for further research. As part of this, geographers might be prompted to think about advocating landscapes of care (Maddrell, 2020) through the staging of festivals and events in public places, in parks and squares as well as in institutions like libraries, museums and galleries. At the height of the pandemic, when normal social gatherings were curtailed/not allowed, people created public space and performed community in spontaneous, grassroots fashion, by making the simplest of gestures (e.g., lighting candles, displaying thank you messages for front line workers) at the smallest of domestic and local scales (in front windows/on garden railings). These socially sustaining events embodying solidarity and compassion could inspire ways forwards for festivals and events as well as for future research into the intersections between geographies of festivals and events, and geographies of care.

All indications are that festival and events geographies will continue to grow strongly. A boost may come from the fact that many human geographers are becoming more interested in temporary urbanism and interim space use (Harris, 2015), developments very closely associated with festival and event activity. Furthermore, there are signs that the ephemerality long accepted as a defining characteristic of events of all kinds is becoming less pronounced and that the material and symbolic landscapes produced through these activities may become more pronounced. Cities now justify the hosting of mega events more in terms of their legacy programmes than in terms of their actual spectacle (Pereira, 2018), arts festivals are becoming increasingly institutionalised and starting to function as cultural producers year-round, and the acceleration of digitisation will significantly alter the production and consumption of events, including the temporality of the spectacles produced. For all of these reasons it becomes hard to imagine that the next edition of this Handbook will be justified in confining the geography of festivals and events to just one chapter.

References

Allen, D. L. & McCreary, T. (2021). Performing black life: The FAMU Marching 100 and the Black aesthetic politics of disruption, presence and affirmation. *Cultural Geographies* 28(1), 41–55.

Bakas, E, Duxbury, N., Remoaldo, P. C. & Matos, O. (2019). The social utility of small-scale art festivals with creative tourism in Portugal. *International Journal of Event and Festival Management* 10(3), 248–266.

Brechenmacher, S., Carothers, T. & Youngs, R. (2020). Civil society and the coronavirus: Dynamism despite disruption, carnegie endowment for international Peace. Available at: https://carnegieendowment.org/files/Brechenmacher_Carothers_Youngs_Civil_Society.pdf [accessed February 2024).

Broudehoux, A. M. (2017). *Mega-Events and Urban Image Construction: Beijing and Rio De Janeiro.* Routledge.

Casaglia, A. (2018). Territories of struggle: Social centres in Northern Italy opposing mega-events. *Antipode* 50(2), 478–497.

Castells, M. (2010). *The Information Age: Economy, Society, and Culture.* Wiley-Blackwell.

Chaney, D. (2019). Rock festivals as marketplace icons. *Consumption Markets & Culture*, 23(3), 215–222. https://doi.org/10.1080/10253866.2019.1571490

Cidell, J. (2016). Time and space to run. In K. Hannam, M. Mostafanezhad & J. Rickly (Eds.) *Event Mobilities: Politics, Place and Performance* (pp. 82–94). Routledge.

Collins, A. & Cooper, C. (2017). Measuring and managing the environmental impact of festivals: The contribution of the ecological footprint. *Journal of Sustainable Tourism* 25(1), 148–162.

Collins, A. & Potoglou, D. (2019). Factors influencing visitor travel to festivals: Challenges in encouraging sustainable travel. *Journal of Sustainable Tourism* 27(5), 668–88.

Cottrell, M. P. & Nelson, T. (2011). Not just the games? Power, protest and politics at the olympics. *European Journal of International Relations* 17(4), 729–753.

Crampton, J. W., Hoover, K. C., Smith, H., Graham, S. & Berbesque, J. C. (2020). Smart festivals? Security and freedom for well-being in urban smart spaces. *Annals of the American Association of Geographers* 110(2), 360–370.

Cresswell, T. (2006). *On the Move: Mobility in the Modern Western World.* Routledge

Cudny, W. (2014). Festivals as a subject for geographical research. *Geografisk Tidsskrift-Danish Journal of Geography* 114(2), 132–42.

De Bres, K., & Davis, J. (2001). Celebrating group and place identity: A case study of a new regional festival. *Tourism Geographies*, 3(3), 326–337. https://doi.org/10.1080/14616680110055439

Duffy, M., & Mair, J. (2021). Future trajectories of festival research. *Tourist Studies*, 21(1), 9–23. https://doi.org/10.1177/1468797621992933

Duffy, M. & Waitt, G. (2011). Rural festivals and processes of belonging. In C. Gibson & J. Connell (Eds.) *Festivals Places: Revitalising Rural Australia.* Channel View Publications.

Duffy, M., Waitt, G., Gorman-Murray, A. & Gibson, C. (2011). Bodily rhythms: Corporeal capacities to engage with festival spaces. *Emotion, Space and Society*, 4, 17–24.

Duignan, M. & McGillivray (2019). Disorganised host community touristic-event spaces: Revealing Rios fault lines at the 2016 Olympic Games. *Leisure Studies* 38(5), 692–711.

Duignan, M., Everett, S., Walsh, L. & Cade, N. (2017). Leveraging physical and digital liminoidal spaces: The case of the #EATCambridge festival. *Tourism Geographies* 20(5), 858–879.

Dundjerovic, A. S. & Martínez Sánchez, M. J. (2018). Festival and the city: Performativity of sexual acts in public spheres. *Architecture and Culture* 6(3), 457–471.

Duvignaud, J. (1976). Festivals: A sociological approach. *Cultures* 3, 13–25.

Edensor, T. & Sumartojo, S. (2018). Reconfiguring familiar worlds with light projection: The Gertrude street projection festival, 2017. *GeoHumanities*. DOI: 10.1080/2373566X.2018.1446760

Falconi, J. P. (2011). Spaces and festivalscapes. *Platform. Communities Perform* 5, 5–18.

Falconi, J. P. (2014). The Festivals Internacional de Teatro de la Habena (FITH) and the Festival of Mexico (FMX): between place and placelessness. *Latin American Theatre Review* 48, 181–193.

Ferreri, M. (2020). Learning from temporary use and the making of on demand communities in Londons Olympic "fringes". *Urban Geography* 41(3), 409–427.

Finkel, R. & Platt, L (2020). Cultural festivals and the city. *Geography Compass* 14, e12498.

Fisker, J. K., Kwiatkowski, G. & Hjalager, A. M. (2019). The translocal fluidity of rural grassroots festivals in the network society. *Social & Cultural Geography*. DOI: 10.1080/14649365.2019.1573437

Foucault, M. (2008). Of other spaces (1967) (L. De Cauter & M. Dehaene, Trans.). In M. Dehaene & L. De Cauter (Eds.) *Heterotopia and the City: Public Space in a Post-Civil Society* (pp. 13–29). Routledge.

Frost, N. (2016). Anthropology and festivals: Festival ecologies. *Ethnos* 81(4), 569–583.

Gaffney, C. (2010). Mega-events and socio-spatial dynamics in Rio de Janeiro, 1919–2016. *Journal of Latin American Geography* 9(1), 7–29.

Gandy, M. (2017). Urban atmospheres. *Cultural Geographies* 24(3), 353–374.

Getz, D. (2012). *Event Studies. Theory, Research and Policy for Planned Events.* Routledge.

Giovanardi, M., Lucarelli, A. & Decosta, P. L. E. (2014). Co-performing tourism places: The "Pink Night" festival, *Annals of Tourism Research* 44, 102–115.

Hannam, K., Mostafanezhad, M. & Rickly, J. (2016). Introduction: towards an agenda for event mobilities research. In K. Hannam, M. Mostafanezhad & J. Rickly (Eds.) *Event Mobilities* (pp. 1–14). Routledge.

Hall, C. M. & Page, S. J. (2014). Geography and the study of events. In S. J. Page & J. Connell (Eds.) *The Routledge Handbook of Event Studies* (pp. 208–228). Routledge.

Harris, E. (2015). Navigating pop-up geographies: Urban space–times of flexibility, interstitiality and immersion. *Geography Compass* 9/11, 592–603.

Hjalager, A.M. (2009). Cultural Tourism Innovation Systems – The Roskilde Festival. *Scandinavian Journal of Hospitality and Tourism*, 9(2-3), 266–287. DOI: 10.1080/15022250903034406

Johansson, M. & Kociatkiewicz, J. (2011). City festivals: Creativity and control in staged urban experiences. *European Urban and Regional Studies* 18, 392–405.

Johnston, L & Waitt, G. (2015). The spatial politics of gay pride parades and festivals: Emotional activism. In D. Paternotte & M. Tremblay (Eds.) *Ashgate Research Companion on Lesbian and Gay Activism* (pp. 105–119). Ashgate.

Johnston, L., & Waitt, G. (2021). Play, protest and pride: Un/happy queers of Proud to Play in Auckland, Aotearoa New Zealand. *Urban Studies*, 58(7), 1431–1447. https://doi.org/10.1177/0042098020905513

Kingsbury, P. (2016). Rethinking the aesthetic geographies of multicultural festivals: A Nietzschean perspective. *Annals of the American Association of Geographers* 106(1), 222–241.

Lauermann, J. (2019). The politics of mega-events: Grand promises meet local resistance. *Environment and Society: Advances in Research* 10, 48–62.

Ma, L. & Lew, A. (2012). Historical and geographical context in festival tourism development. *Journal of Heritage Tourism* 7(1), 13–31.

Maddrell, A. (2020). Bereavement, grief, and consolation: Emotional-affective geographies of loss during COVID-19. *Dialogues in Human Geography* 10(2), 107–111.

Mahon, M. & Hyyryläinen, T. (2019). Rural arts festivals as contributors to rural development and resilience. *Sociologia Ruralis* 59(4), 612–635.

Marks, M., Chandler, L., & Baldwin, C. (2014). Re-imagining the environment: Using an environmental art festival to encourage pro-environmental behaviour and a sense of place. *Local Environment*, 21(3), 310–329. https://doi.org/10.1080/13549839.2014.958984

Massey, D.B. (1994). *Space, Place, and Gender.* University of Minnesota Press.

McClinchey, K. A. (2020). Contributions to social sustainability through the sensuous multiculturalism and everyday place-making of multi-ethnic festivals. *Journal of Sustainable Tourism* 29(11–12), 2025–2043.

McGillivray, D. & Frew, M. (2015). From fan parks to live sites: Mega events and the territorialisation of urban space. *Urban Studies* 52(14), 2649–2663.

Müller, M. (2015). The mega-event syndrome: Why so much goes wrong in mega-event planning and what to do about it. *Journal of the American Planning Association* 81(1), 6–17.

Pereira, R.H.M. (2018). Transport legacy of mega-events and the redistribution of accessibility to urban destinations. *Cities* 18, 45–60.

Platt, L. & Finkel, R. (2020). *Gendered Violence at International Festivals: An Interdisciplinary Perspective.* Routledge.

Pløger, J. (2010). Presence-experiences: The eventalisation of city space. *Environment and Planning: Society and Space* 28, 848–866.

Quinn, B. (2003). Symbols, practices and myth-making: Cultural perspectives on the Wexford festival opera. *Tourism Geographies* 5(3), 329–349.

Quinn, B. & Wilks, L. (2013). Festival connections: People, place and social capital. In G. Richards, M. de Brito & L. Wilks (Eds.) *Exploring the Social Impacts of Events* (pp. 15–30). Routledge.

Quinn, B. & Wilks, L. (2017). Festival heterotopias: Spatial and temporal transformations in two small-scale settlements. *Journal of Rural Studies* 53, 35–44.

Quinn, B., Columbo, A., Lindström, K., McGillivray, D. & Smith, A. (2020). Festivals, public space & cultural inclusion: Public policy insights. *Journal of Sustainable Tourism* 29(12), 1875–1893.

Relph, E. (1976). *Place and Placelessness.* Pion.

Richards, G. (2015). Festivals in the network society. In C. Newbold, C. Maughan, J. Jordan & F. Bianchini (Eds.) *Focus On Festivals: Contemporary European case studies and perspectives* (pp. 245–254). Goodfellow Publishers. http://dx.doi.org/10.23912/978-1-910158-15-9-2645

Rowen, I. (2020). The *transformational festival* as a subversive toolbox for a transformed tourism: Lessons from burning man for a COVID-19 world. *Tourism Geographies* 22(3), 695–702.

Ryan, W. A. & Wollan, G. (2013). Festivals, landscapes, and aesthetic engagement: A phenomenological approach to four Norwegian festivals. *Norsk Geografisk Tidsskrift – Norwegian Journal of Geography* 67(2), 99–112.

Sheller, M. & Urry, J. (2004, Eds.). *Tourism Mobilities: Places to Play, Places in Play.* Routledge.

Smith, A. (2005). Conceptualizing city image change: The re-imaging of Barcelona. *Tourism Geographies* 7(4), 398–423.

Smith, A. (2014). Leveraging sport mega-events: New model or convenient justification? *Journal of Policy Research in Tourism, Leisure and Events* 6(1), 15–30.

Smith, A. (2016). *Events in the City: Using Public Spaces as Event Venues.* Routledge.

Sweeney, J., Mee, K. & McGuirk, P. (2018). Assembling placemaking: Making and remaking place in a regenerating city. *Cultural Geographies* 25(4), 571–587.

Turner, V. (1982). Introduction. In V. Turner (Ed.) *Celebration: Studies in Festivity and Ritual* (pp 11–32). Smithsonian Institution Press.

Van Zyl, C. (2011). A model for positioning arts festivals in South Africa. *South African Theatre Journal* 25(30), 181–96.

Waitt, G. (2008). Urban festivals: Geographies of hype, helplessness and hope. *Geography Compass* 2(2), 513–537.

Waterman, S. (1998). Carnival for elites: The cultural politics of arts festivals. *Progress in Human Geography* 22, 54–74.

Wise, N. (2017). In the shadow of mega-events: The value of ethnography in sports geography. In N. Koch (Ed.) *Critical Geographies of Sport: Space, Power and Sport in Global Perspective* (pp. 220–234). Routledge.

Wyckoff, M. A. (2014). Definition of placemaking: Four different types. *Planning & Zoning News*, January, available at: www.canr.msu.edu/uploads/375/65814/4typesplacemaking_pzn_wyckoff_january2014.pdf (accessed February 2024)

Wynn, J. R. (2015). *Music/City: American Festivals and Placemaking in Austin, Nashville, and Newport.* University of Chicago Press.

16

TOURISM AND URBANISATION PROCESSES

Towards an Encounter between Tourism Geographies and Urban Theory?

Mathis Stock and Valérian Geffroy

Introduction

Tourism is known for creating specific qualities of inhabited space by associating mobile and temporary inhabitants to specific places. In this process, urban locales are produced by the concentration of inhabitants, the establishment of urban amenities and an economy of services oriented toward leisure. This is congruent with Lefebvre's thesis of the association of industrialisation and urbanisation: "[T]he leisure industry combines with the construction industry to extend cities and urbanisation along coasts and in mountainous areas" (Lefebvre, 2001: 152). Tourism therefore plays a key role in these urbanisation processes, due to its ever-expanding importance: The double process of emergence of seaside resorts and mountain resorts over the last 200 years and re-urbanisation of cities through tourism has been interpreted as a 'double urban revolution' powered by tourism (Stock & Lucas, 2012). Some authors even go as far as considering tourism an inherently urban phenomenon, arguing that it was historically led by urban citizens and that tourism practices and infrastructure turn all kinds of places into urban space (Anton Clavé & Wilson, 2017; Coëffé & Stock, 2021).

Yet, tourism studies and urban studies have evolved along parallel trajectories (Coëffé & Stock, 2021; Stock, 2019), with the only intersection being so-called 'urban tourism', that is, tourism within the context of a city (Ashworth & Page, 2011). This leads to a series of questions: Is there an anti-urban bias in tourism geographies? Does theory-laden observation on tourism phenomena – be it from political economy, practice theory or other post-structuralist frameworks – sufficiently integrate urban perspectives? Is there anything to gain should tourism phenomena be interpreted as urban processes? What would the consequences be for tourism geographies if a more thorough conversation were engaged in with urban studies literature?

Tackling these questions requires unpacking what 'urban' and 'urbanisation' are meant to designate. This is an important issue to determine the analytical power and limitations of these concepts. Brenner & Schmid (2015) call for a 'new epistemology of the urban' which would involve 'decentering [the] perspective' (Schmid, 2018), that is to say recognising urban forms and urbanisation processes beyond the sole territorial form of the city. As Gandy (2013: 86) stresses,

DOI: 10.4324/9781003286301-19

"cities are just a particular form of urbanisation". Consequently, the corollary notion of 'urbanity' rather than the notion of 'city' is key. Urbanity comes in different intensities or degrees (from hypo-urban to hyper-urban), takes different territorial forms and comprises symbolic as well as material components (Lévy, 2014). The 'urban' is here defined as a concept which connotes diverse degrees of monumentality, centrality, density, diversity and, most importantly, encounter and suspension of distance between human and non-human actants (Lefebvre, 1991; Lévy, 2014; Schmid, 2013), stemming from "transfers of urbanity" (Lefebvre, 1968: 120). These elements are relational, or as Merrifield (2013: 42) describes centrality: "[A]lways relative, never fixed, always in a state of constant mobilisation and negotiation". As inhabited space, urban space is produced by everyday practices as well as institutional processes. What would we gain if tourism places were to be interpreted as a result of transfers of urbanity or as emerging touristic centralities, where encounters with the Other take place; where the urban form of monumentality is key, and thus a specific 'touristic urbanity' can emerge?

Consequently, the concept of 'urbanisation' does not connote (only) the growth of cities (be it through concentration or sprawl), but the emergence and complexification of urban nodes and urban ways of life (Brenner, 2019; Brenner & Schmid, 2015). "Urbanisation must then be understood [...] as the production of specific and quite heterogeneous spatio-temporal forms embedded within different kinds of social action" (Harvey, 2013: 62)[1]. As a process, it denotes change towards a more complex assemblage of materialities and practices, but it is also reversible as de-urbanisation or re-urbanisation. What would be gained if tourism places were framed as provisory and temporary states of becoming within processes towards more or less complex forms of urbanity? (See also Lapointe, Chapter 6; Briassoulis, Chapter 7; Everingham et al., Chapter 11).

The contemporary period is characterised by a specific context, which renders the conversation between urban and tourism studies ever more urgent. As understood clearly through the COVID-19 pandemic, tourism places depend on a specific 'global mobility regime' (Glick Schiller & Salazar, 2013; Sheller, 2018). There is therefore a need to link urban theory and tourism theory, following the so-called 'mobilities turn' (Sheller & Urry, 2006) in social sciences, as temporary inhabitants play an increasingly important role in the shaping of urban places.

Destinations[2] as processes evolving across various social fields

Abundant literature has been dedicated, on the one hand, to framing destination development with models, be it through life-cycle (Butler, 2006), restructuring (Agarwal, 2012a) or changing place qualities (Saarinen, 2004; Équipe MIT, 2011), and to criticising these models on the other (Anton Clavé & Wilson, 2017; Gale, 2012; Stock et al., 2021). This literature has focused particularly on resorts because of their ideal-typical quality as places wholly dedicated to, and dependent on, tourism.

Resorts therefore develop a specific kind of urbanity, characterised by a smaller resident population, yet a certain monumentality, leisure-oriented public space, and a specific social and cultural diversity among tourists and residents which goes hand in hand with seasonal patterns of presence and absence of tourists and workers; in brief, a 'touristic urbanity' (Équipe MIT, 2002; Coëffé & Stock, 2021). However, the underlying issue is the reconstruction of development processes over a long period of time, which leads to highly differential and changing qualities, forms and functions of places; and all destinations, not only resorts, experience such processes. Whereas some globally renowned destinations show continued success in attracting visitors over a long period of time, others decline rapidly, and yet others undergo a complete transformation in which tourism no longer plays a key role (Clivaz et al., 2011; Stock et al., 2021): How can we understand the whole spectrum of development paths – from sustained tourism-oriented development to the complete metamorphosis of resorts into non or less tourism-dependent places?

A first key to this issue might lie in a 'touristic capital' mobilised by destination actants and the inherent urban qualities of destinations. This concept helps link tourism trajectories to urban characteristics of places evolving in an ever-changing touristic field. Stock et al. (2021: 211) define the concept as follows: "Touristic capital can be defined as an ensemble of physical, economic, social, political, urban and symbolic characteristics that permits a place – and the actors living in it, developing it, or exploiting it – to position itself in a tourism field and, in comparison to other tourist places, to gain advantages. This concept aims at understanding the differential capacity of tourist places to maintain (or to lose) touristic quality – relative to the modification of the tourism field – over a certain period of time".

Such a conceptualisation of a localised capital draws from both Marxian political economy (implying both the ownership of means of production and the process of accumulation), and Bourdieusian sociology. The latter extended the notion to the sum of the various resources agents engage with in social relations; it is the relational expression of capabilities and power relations within a specific field (Bourdieu, 1980, 1984). Hence, touristic capital makes sense *only* in a touristic field; not necessarily in other fields. Conceived as such, capital is assigned to places rather than actors, following the propositions of a 'territorial capital' put forward by Camagni and Capello (2009). Harvey's notion also seems useful, concerning a

> collective *symbolic capital* which attaches to names and places like Paris, Athens, New York, Rio de Janeiro, Berlin and Rome is of great import and gives such places great economic *advantages* relative to, say, Baltimore, Liverpool, Essen, Lille, Glasgow.
>
> (2001b: 405, *our emphasis*)

It adds a symbolic dimension to built environment infrastructure and real estate, seen as one of the main fixtures and solutions of capitalist accumulation ('spatial fix'). From an urban–theoretical point of view, the question of the centrality of these places within specific social fields is raised: Does long-term tourism development imply keeping up with a touristic centrality within the touristic field?

A second key lies in urbanisation and urbanity, key features of development processes. For instance, the Équipe MIT (2011) explicitly conceives the link between long-term success of a destination and invested urban capital: "[T]he spatial fix of the invested economic capital and the urban capital are a token of durability, especially because of the evolution potential towards other forms of activities" (Équipe MIT, 2011: 235). Specific kinds of resorts – that is, fully tourism-dependent settlements – undergo urbanisation processes through diversification, concentration and sprawl, but also the development of new centralities that turn them into cities or even metropolises, such as Brighton & Hove, Nice and other 'resort cities' (Judd & Fainstein, 1999), where tourism ends up being one among many other activities. Such resorts gain more complex functions – and thus increase their urbanity – for instance when they start hosting mobile, creative resident populations (Romero-Padilla et al., 2016). Especially within complex urban structures, destinations can be understood through their 'path plasticity' (Anton Clavé & Wilson, 2017), that is, their potential to engage in multiple development paths, incrementally, with an emphasis on flexibility and adaptability. Some remain relatively small, albeit with a global symbolic capital, such as Zermatt (Sauthier, 2016), while others evolve into small cities, such as Montreux (Guex, 2016). Urbanisation – understood as the emerging centrality and diversity of urban functions – seems to be essential for understanding the inherent scale, scope and variety of development processes.

If urbanisation processes can be recognised, de-urbanisation processes linked to de-touristification can also be observed when touristic capital is eroded without being replaced, thus diminishing urban diversity or even causing urban shrinkage (see also Lapointe, Chapter 6). As Bærenholdt et al. (2004: 46) note, "the complexities of images, materialities, businesses, tourists and locals may become destabilised, so that networks fade away and tourist places change or become 'dead' places of the past". Seaside resorts lose their popularity and become places of social exclusion (Agarwal & Brunt, 2006), poorly integrated into local economies (Agarwal, 2012b). Winter sports resorts struggle not only with less snow cover, but also with out-of-fashion architecture and an ageing infrastructure (Fablet, 2013). Specific islands see a sharp decline in tourism attractiveness and lose a share of their social and professional relationships and entertainment opportunities (Canavan, 2021). Thus, a number of studies on the 'decline' of tourism resorts raise the question of renewal (Agarwal, 2012a) or revitalisation (Cooper, 2018). Through the lens of urban theory, the problem could be framed as involving issues of centrality, and de- or re-urbanisation processes, as well as a specific urban capital that redevelopment strategies could leverage.

Destinations as places and (urban) place-making

Place serves as a concept to designate material and meaningful locations (Cresswell, 2013) to which multiple and sometimes controversial meanings, values, memories and imaginaries are assigned. It also serves to indicate how societies or collectives suspend distance (Lévy, 2003) by assembling things together in one location and creating co-presence of human and non-human actants (see also Briassoulis, Chapter 7). It is therefore a relational concept, which allows for a deeper understanding of an incessant state of 'becoming' (Massey, 2005; Pred, 1984), defined by an ever-renewing set of activities, mobilities and symbols – 'spatial semantics' are especially important for the socio-political construction of destinations (Pott, 2007; Saretzki, 2019). It conceptualises the connection of a specific location to networks "by which 'hosts, guests, buildings, objects and machines' are continually brought together to perform certain performances" (Hannam et al., 2006: 13). This has consequences for the analysis of destination development as place-making, since the "clear distinction [that] is often drawn between places and those travelling to such places" (ibid.) has been superseded: "Thus activities are not separate from the places that happen contingently to be visited. Indeed, the place travelled to depends in part upon what is practised within them" (Sheller & Urry, 2006). Place-making has indeed emerged as a major issue (see also Chang, Chapter 13), where this constantly renewing, re-commenced co-production of place is analysed as practice in progress (Dupre, 2019; Lew, 2017; Wang, 2021), including the contestation and negotiation of uses as well as meanings of place. As such, "tourism depends on these diverse but also overlapping *notions* of place" (Baerenholdt et al., 2004: 32, *our emphasis*).

Thus the notion of co-production is key, especially for cities. Dupre (2019) argues for the relevance of place-making perspectives to analyse the different dimensions of place development, in particular to uncover potential competition or synergy between tourism and urban development. The place-making approach emphasises bottom-up, vernacular and small-scale practices that contribute to shaping space into place. Dupre (2019) notes that such practices are still too often overlooked in tourism planning and development, with the risk of failing to ground it within the urban fabric. A new form of urban development and tourism development as relatively more intertwined has been unpacked under the heading of 'new urban tourism' (Füller & Michel, 2014; Novy, 2010; Novy, 2017). New, previously predominantly residential neighbourhoods become touristified in addition to the traditional tourist centres within the city, hence the everyday life of permanent and temporary inhabitants can become interdependent (Frisch et al., 2019). This contrasts with earlier findings on 'tourist bubbles' (Judd & Fainstein, 1999), that is, districts primarily devoted to tourism or leisure consumption and carefully demarcated from the rest of the city, which have typically – though not exclusively – been developed via massive investment in leisure infrastructure and activities, focused on poorer, formerly industrial central districts of North American cities.

Cities, as the foremost complex setting of urbanity, develop specific urban intensities, which are in themselves an attractor: The quality of 'urbanity' in itself, as a combination of density and diversity, that is, the multiplicity of opportunities, is a central feature of attraction for many tourists, and increasingly so as the focus of mainstream urban tourism has progressively shifted from cultural attractions to everyday urban life (Ashworth & Page, 2011; Maitland, 2010). Far from the conception of tourism as flight from the city, the significance of cities as destinations is evident today (Duhamel & Knafou, 2007); but that is at least in part the result of an economic movement of widespread de-industrialisation, and of the related 'recreational turn' of cities towards tourism (Stock, 2007). The touristification of cities has become a widespread phenomenon and a new urban quality of the city has emerged: "*[B]elle, propre, festive et sécurisante*[3]" (Gravari-Barbas, 1998: 175), but also heritage-accentuated, atmosphere-imbued and pedestrian-friendly; 'dressed up' and rebranded for visitors (Gotham, 2007). In this sense, a new relationship of society and individuals to the city has clearly emerged.

Moreover, tourism and leisure have been part of the contemporary evolution of leading global cities: alongside the ever-increasing concentration of financial and decisional power, which results in business and conference trips and the MICE industry, culture and leisure also contribute to the development of global centralities (Ashworth & Page, 2011; Maitland & Newman, 2009). Institutional and corporate actors remain a major force in the production of tourism places. Destination marketing organisations, in particular, play a central role in the diffusion of place image, as Rabbiosi (2015) reminds us in her study of leisure shopping in Paris as a central part of urban branding campaigns. Rabbiosi also shows that the image of Paris as a capital of style and fashion has long preceded urban marketing efforts. Thus, the production of this powerful symbol is in large part due to retailers and urban shoppers and symbolic capital is co-produced by multiple actors far beyond the tourism industry. However, the recent re-investment of such a 'brand' for tourism promotion purposes bears witness to a shifting balance of power towards a more intensive exploitation of urban resources through visitor economies and consumer capitalism.

Other examples show how citizens can affect the balance between tourism development and urban public goods. Cultural heritage is an obvious example of a public good – bearing strong identity values – that is also a key component of tourism attractiveness. The case of Vienna's *Museumsquartier* (De Frantz, 2018) shows that tourism marketing and planning, too often reduce and simplify the meanings of this heritage, and that a pluralist political debate is necessary to achieve an interpretation of a city's heritage that is both locally anchored and communicative to international visitors. Another example of an original form of community-based tourism development can be found in Japanese cities (Horita, 2018). The development of urban tourism has raised residents' awareness of their living areas and incited them to manage their neighbourhoods in a way that caters to visitors as well. *Kankō machizukuri*, a term combining tourism (*kankō*) and community development (*machizukuri*), refers to the set of activities offered by local people utilising local resources. This is an example of a collective investment in both tourism development and everyday urban life.

Yet this touristification of the city, despite its cultural and economic significance and its urbanisation effects, remains contested. A recent trend towards socio-spatial, cultural and economic contestation has targeted the negative effects of tourism in some cities; indeed, the debate has achieved public status under the banner of *overtourism*, and has been framed as 'overdose' or resistance (Novy & Colomb, 2020). In the most extreme cases, tourism comes to be judged, at least by a significant part of the resident population, as excessive, as a disruption of 'normal' urban life (Blanco-Romero et al., 2019 and Blàzquez et al., Chapter 5 of this volume). Controversies typically develop in cities where tourism, once only one among the multiple elements of urban life, starts threatening residents' practices and/or livelihoods. Colomb and Novy (2016) provide a rich synthesis on the matter and a detailed analysis of the main causes of protest and resistance. They identify commercial and residential gentrification, overcrowding and the appropriation of public space and community resources, as well as issues relating to loss of cultural identity and deterioration of sense of belonging/community. In particular, gentrification (see Buhr & Cocola-Gant, Chapter 31 of this volume; Cocola-Gant, 2018; Gravari-Barbas & Guinand, 2017) seems to be at the centre of the contestation. From an urban-theoretical point of view, overtourism could be seen as a loss of diversity and thus of urbanity, as tourism specialisation homogenises and simplifies both the practices and populations of a given place, while reducing the potential for encounter and otherness. That having been said, such controversies should not mask the diversity of practices among mobile inhabitants and the diversity of stances towards tourism that can exist (Eggli, 2021).

Issues

What if tourists moved throughout a highly differentiated urbanised space and, by their mere presence, produced urban space? Not only is the visitor economy a major element of tourism in cities: Coastal, mountainous regions and countryside beyond the city also develop tourism-based urban economies. Despite the myths of 'eco-tourism' and 'national parks' as wilderness and pristine nature, we argue that such spaces, when inhabited by tourists, can to some extent be understood as 'urbanised': because of the urban origin of the contemporary gaze upon nature, and because urban amenities and urban sprawl are produced through resort planning and second homes. The question of a (planetary) urbanisation through tourism is therefore raised (Cometta & Stock, 2021). 'Planetary urbanisation' is a highly contested expression and framing of a phenomenon first coined by Henri Lefebvre in the 1970s (Lefebvre, 1991). It posits that urban characteristics expand beyond the city and reorganise the urban at different scales; from the local to the global (Brenner, 2019; Brenner & Schmid, 2015). Tourism is one of the vectors through which this production of space is carried out in a capitalist society and can be seen as an example of capitalism's 'spatial fix' (Fletcher, 2011; Harvey, 2001a), that is, a way to solve or prevent crises by expanding investment in new geographical locations, thus triggering urbanisation. The massive-scale urbanisation of coastlines, in Spain for example (Blázquez-Salom et al., 2019), offers evidence that tourism became a privileged outlet for capital over-accumulation. On a material level, urbanised surfaces in the Balearic Islands have increased fivefold since the 1950s (Pons & Rullan, 2014). In many cases, tourism urbanisation in rural areas serves solely the tourism industry's needs and may "undermine other economic activities and forms of occupying space" (Brooks, 2018: 30).

This theory of ubiquitous urbanisation is also linked to the contemporary theoretical context of the 'mobilities turn' (Sheller & Urry, 2006), which heightens attention on mobilities and mobility regimes (Sheller, 2018). Mobility in all its forms contributes to the process of urbanisation through the growing importance of 'presential economies' (Talandier, 2013; Terrier et al., 2005) or 'residential economies' (Ruault, 2017; Segessemann & Crevoisier, 2016) in many places and regions. Such economies centred on services related to the mere presence of all kinds of mobile inhabitants are indeed greatly expanded by tourists, excursionists, long-distance commuters, digital nomads (see also Hannonen, Chapter 30 of this volume) and other highly mobile subjects – that is, more or less temporary inhabitants – especially in high-amenity environments (Mollard et al., 2007).

Second homes in coastal areas, countryside and mountain regions are also part of this phenomenon (Hall & Müller, 2018), yet the link to urban sprawl still needs to be explicitly acknowledged. Despite numerous analyses of capitalist development and sustainability (Büscher & Fletcher, 2017; Fletcher et al., 2019), no links to urbanisation processes are clearly drawn. There is an urgent need to engage in a conversation with the literature on planetary urbanisation and ask to what extent tourism can contribute to it.

Secondly, in related terms, tourism geographies have long considered the urbanite's 'flight from the city' as the core of tourism practices (Krippendorf, 1975). As Ashworth & Page (2011) note, an "anti-urban bias" historically led tourism research to see everyday urbanity as contrary to the need for rest and recreation, and thus to consider tourism predominantly as a 'land-use' of high-amenity rural areas. Today, beaches, mountains, countryside, resorts, rivers, oceans, forests and national parks are still largely seen as non-urban setting, places where the 'over-civilised urbanite' comes to rest[4]. This romanticised vision of 'wilderness' (Fletcher, 2014) supports practices of 'ecotourism' seen as both an antithesis and solution to the overcrowding and increasing artificiality associated with mass tourism (Fennell, 2020). However, it can also be interpreted as contributing to planetary urbanisation as it can lead to even more intensive mobilities and even wider diffusion of tourists and tourist infrastructure. This specific geographical imagination fostered by destination marketing organisations and the media is problematic if mirrored by scientific investigations in tourism geography.

Therefore, we ask if there is indeed an anti-urban bias in tourism geographies, which means we fail to identify those urban processes triggered and shaped by tourism? If this is the case, does it reflect a specific moral point of view on the 'urban', seen as inherently problematic? Or is it simply a reflection of the (out of date) traditional scientific discourse on a conceptual distinction between the city as the sole urban place and the rest of the ecumene?

The need to rid tourism research of this anti-urban bias, and to rethink tourism from the perspective of urban theory, is made particularly urgent by the current ecological crises. These crises are indeed closely intertwined with tourism urbanisation in at least two major aspects. First of all, second homes and resort developments play a significant role in the continued urban sprawl. Second, the boom in aero-mobilities between major cities and destinations across the world is an important driver of climate change. Both contribute to planetary urbanisation. Only by acknowledging this, and adopting an urban theory perspective, can future tourism research unpack the actual ecological implications of tourism, among the many other challenges it needs to tackle. This also includes the solutions urbanity can provide for a less resource-intensive tourism, such as centrality, density, commons, and so forth. In sum, urban theory greatly expands the scope of tourism geographies, by uncovering the multiple ties of the tourism phenomenon to the major evolutions of contemporary space.

Notes

1 Even in its traditional form as concentration of density and diversity, urban development has been observed as corollary to the development of tourism. Dubbed 'tourism urbanisation' (Mullins, 1991), it leads to touristic conurbations or 'resort regions' (Soane, 1993), relying on a diverse urban economy with

its urban problems such as congestion and service management and where gentrification issues comparable to those of cities (Herrera et al., 2007) can occur.

2 The notion of 'destination' is widely used in tourism research, especially in economy-oriented analyses, where it refers to a delimited area to develop, manage and market. However, the risk with such a notion is that it may reduce places to their tourist function; thus, essentialising geographical places as destinations and underestimating their complexity (see Équipe MIT [2002] on tourist places as common places). There is a reflection to develop on the comparative advantages of the notion of 'tourist place' and 'place-making' respective to 'destination'.

3 "beautiful, clean, festive and safe"

4 "Thousands of tired, nerve-shaken, over-civilized people are beginning to find out that going to the mountains is going home, that wildness is a necessity, and that mountain parks and reservations are useful not only as fountains of timber and irrigating rivers, but as fountains of life" (John Muir, "The Wild Parks and Forest Reservations of the West", *Atlantic Monthly* 81 (January 1898) in Runte (1979))

References

Agarwal, S. (2012a). Relational spatiality and resort restructuring. *Annals of Tourism Research*, *39*(1), 134–154.

Agarwal, S. (2012b). Resort economy and direct economic linkages. *Annals of Tourism Research*, *39*(3), 1470–1494.

Agarwal, S., & Brunt, P. (2006). Social exclusion and English seaside resorts. *Tourism Management*, *27*(4), 654–670.

Anton Clavé, S., & Wilson, J. (2017). The evolution of coastal tourism destinations: A path plasticity perspective on tourism urbanisation. *Journal of Sustainable Tourism, 25*(1), 96–112.

Ashworth, G., & Page, S. J. (2011). Urban tourism research: Recent progress and current paradoxes. *Tourism Management, 32*(1), 1–15.

Bærenholdt, J. O., Haldrup, M., Larsen, J., & Urry, J. (2004). *Performing Tourist Places*. Ashgate.

Blanco-Romero, A., Blàzquez-Salom, M., Morell, M., & Fletcher, R. (2019). Not tourism-phobia but urbanphilia: Understanding stakeholders' perceptions of urban touristification. *Boletín de la Asociación de Geógrafos Españoles*, 83. https://doi.org/10.21138/bage.2834

Blázquez-Salom, M., Blanco-Romero, A., Vera-Rebollo, F., & Ivars-Baidal, J. (2019). Territorial tourism planning in Spain: From boosterism to tourism degrowth? *Journal of Sustainable Tourism, 27*(12), 1764–1785.

Bourdieu, P. (1980). Le capital social. *Actes de la recherche en sciences sociales, 31*(1), 2–3.

Bourdieu, P. (1984). *Questions de sociologie*. Ed. de Minuit.

Brenner, N. (2019). *New Urban Spaces: Urban Theory and the Scale Question*. Oxford University Press.

Brenner, N., & Schmid, C. (2015). Towards a new epistemology of the urban? *City, 19*(2–3), 151–182.

Brooks, S. (2018). Growth of tourism urbanisation and implications for the transformation of Jamaica's rural hinterlands. In P. Horn, P. Alfaro d'Alencon, & A. C. Duarte Cardoso (Éds.), *Emerging Urban Spaces: A Planetary Perspective* (pp. 129–148). Springer.

Büscher, B., & Fletcher, R. (2017). Destructive creation: Capital accumulation and the structural violence of tourism. *Journal of Sustainable Tourism, 25*(5), 651–667.

Butler, R. (2006). *The Tourism Area Life Cycle, Vol. 1: Applications and Modifications*. Channel View Publications.

Camagni, R., & Capello, R. (2009). Territorial capital and regional competitiveness: Theory and evidence. *Studies in Regional Science, 39*(1), 19–39.

Canavan, B. (2021). Pushed over the periphery: Downsides of degrowth on a small island – Experiences of tourism degrowth on the Isle of Man. In K. Andriotis (Éd.), *Issues and Cases of Degrowth in Tourism* (pp. 145–159). CABI.

Clivaz, C., Nahrath, S. & Stock, M. (2011). Le développement des stations touristiques dans le champ touristique mondial. *Espaces - Mondes du tourisme, n° hors série "Tourisme et mondialisation"*, 2011, 276–286.

Cocola-Gant, A. (2018). Tourism gentrification. In L. Lees & M. Phillips (Eds.), *Handbook of Gentrification Studies* (pp. 281–293). Edward Elgar.

Coëffé, V., & Stock, M. (2021). Tourism as urban phenomenon and the crux of "urban tourism". In M. Stock (Éd.), *Progress in French Tourism Geographies: Inhabiting Touristic Worlds* (pp. 185–202). Springer.

Colom, C. & Novy, J. (2016). *Protest and Resistance in the Tourist City*. Routledge.

Cometta, M., & Stock, M. (2021). Discursive construction of a destination: Urban transition through tourism in Ticino between 1980s and 2010s. *Mondes Du Tourisme, 19*. https://doi.org/10.4000/tourisme.3620

Cooper, C. (2018). Challenging tourism contexts for planning and policy: Revitalising failing destinations. In K. Andriotis, D. Stylidis & A. Weidenfeld (Eds.), *Tourism Policy and Planning Implementation* (pp. 78–93). Routledge.

Cresswell, T. (2013). *Place: A Short Introduction*. Blackwell.

De Frantz, M. (2018). Tourism marketing and urban politics: Cultural planning in a European capital. *Tourism Geographies, 20*(3), 481–503.

Duhamel, P., & Knafou, R. (Éds.). (2007). *Mondes urbains du tourisme*. Belin.

Dupre, K. (2019). Trends and gaps in place-making in the context of urban development and tourism: 25 years of literature review. *Journal of Place Management and Development, 12*(1), 102–120.

Eggli, F. (2021). *Living with Tourism in Lucerne: How People Inhabit a Tourist Place* [PhD thesis]. University of Lausanne.

Équipe MIT. (2002). *Tourismes 1*: Lieux communs. Belin.

Équipe MIT. (2011). *Tourismes 3: La révolution durable*. Belin.

Fablet, G. (2013). La croissance immobilière des stations de sports d'hiver en Tarentaise. *Journal of Alpine Research | Revue de géographie alpine*, 101–3.

Fennell, D. A. (2020). *Ecotourism*. Routledge.

Fletcher, R. (2011). Sustaining tourism, sustaining capitalism? The tourism industry's role in global capitalist expansion. *Tourism Geographies, 13*(3), 443–461.

Fletcher, R. (2014). *Romancing the Wild: Cultural Dimensions of Ecotourism*. Duke University Press.

Fletcher, R., Mas, I. M., Blanco-Romero, A., & Blázquez-Salom, M. (2019). Tourism and degrowth: An emerging agenda for research and praxis. *Journal of Sustainable Tourism, 27*(12), 1745–1763.

Frisch, T., Sommer, C., Stoltenberg, L., & Stors, N. (Éds.). (2019). *Tourism and Everyday Life in the Contemporary City*. Routledge.

Füller, H., & Michel, B. (2014). 'Stop being a tourist!' New dynamics of urban tourism in Berlin-Kreuzberg. *International Journal of Urban and Regional Research, 38*(4), 1304–1318.

Gale, T. (2012). Tourism geographies and post-structuralism. In J. Wilson (Éd.), *The Routledge Handbook of Tourism Geographies* (pp. 37–45). Routledge.

Gandy, M. (2013). Where does the city end? In N. Brenner (Éd.), *Implosions—Explosions: Towards a Study of Planetary Urbanization* (pp. 86–89). Jovis.

Glick Schiller, N., & Salazar, N. B. (2013). Regimes of mobility across the globe. *Journal of Ethnic and Migration Studies, 39*(2), 183–200.

Gotham, K. F. (2007). (Re)Branding the big easy: Tourism rebuilding in post-Katrina New Orleans. *Urban Affairs Review, 42*(6), 823–850.

Gravari-Barbas, M. (1998). Belle, propre, festive et sécurisante: L'esthétique de la ville touristique. *Norois, 178*(1), 175–193.

Gravari-Barbas, M., & Guinand, S. (2017). *Tourism and Gentrification in Contemporary Metropolises: International Perspectives*. Taylor & Francis.

Guex, D. (2016). *Tourisme, mobilités et développement régional dans les Alpes Suisses: Mise en scène et valeur territoriale. Montreux, Finhaut et Zermatt du XIXe siècle à nos jours*. Editions Alphil Presses universitaires suisses.

Hall, C. M., & Müller, D. K. (Éds.). (2018). *The Routledge Handbook of Second Home Tourism and Mobilities*. Routledge.

Hannam, K., Sheller, M., & Urry, J. (2006). Editorial: Mobilities, immobilities and moorings. *Mobilities, 1*(1), 1–22.

Harvey, D. (2001a). Globalization and the spatial fix. *Geographische Revue, 2*(3), 23–31.

Harvey, D. (2001b). *Spaces of Capital: Towards a Critical Geography*. Routledge.

Harvey, D. (2013). Cities or urbanization? In N. Brenner (Éd.), *Implosions—Explosions: Towards a Study of Planetary Urbanization* (pp. 52–66). Jovis.

Herrera, L. M. G., Smith, N., & Vera, M. Á. M. (2007). Gentrification, displacement, and tourism in Santa Cruz de Tenerife. *Urban Geography, 28*(3), 276–298.

Horita, Y. (2018). Urban development and tourism in Japanese cities. *Tourism Planning & Development, 15*(1), 26–39.

Judd, D. R. (1999). Constructing the tourist bubble. In D. R. Judd, & S. S. Fainstein (Éds.), *The Tourist City* (pp. 35–53). Yale University Press.

Judd, D. R., & Fainstein, S. S. (Éds.). (1999). *The Tourist City*. Yale Univ. Press.

Krippendorf, J. (1975). *Die Landschaftsfresser: Tourismus und Erholungslandschaft, Verderben oder Segen?* Hallwag.

Lefebvre, H. (1968). *Le droit à la ville*. Éditions Anthropos.

Lefebvre, H. (1991). *The Production of Space*. Blackwell.

Lefebvre, H. (2001). *Du rural à l'urbain*. La Découverte.

Lévy, J. (2003). Lieu. In J. Lévy, & M. Lussault (Éds.), *Dictionnaire de la géographie et de l'espace des sociétés* (pp. 612–613). Belin.

Lévy, J. (2014). Science + space + society: Urbanity and the risk of methodological communalism in social sciences of space. *Geographica Helvetica*, *69*(2), 99–114.

Lew, A. A. (2017). Tourism planning and place making: Place-making or placemaking? *Tourism Geographies*, *19*(3), 448–466.

Maitland, R. (2010). Everyday life as a creative experience in cities. *International Journal of Culture, Tourism and Hospitality Research*, *4*(3), 176–185.

Maitland, R. & Newman, P. (2009). *World Tourism Cities: Developing Tourism Off the Beaten Track*. Routledge.

Massey, D. B. (2005). *For Space*. Sage.

Merrifield, A. (2013). *The Politics of the Encounter: Urban Theory and Protest under Planetary Urbanization*. University of Georgia Press.

Mollard, A., Rambonilaza, T., & Vollet, D. (2007). Environmental amenities and territorial anchorage in the recreational-housing rental market: A hedonic approach with French data. *Land Use Policy*, *24*(2), 484–493.

Mullins, P. (1991). Tourism urbanization. *International Journal of Urban and Regional Research*, *15*(3), 326–342.

Novy, J. (2010). What's new about new urban tourism? And what do recent changes in travel imply for the 'tourist city' Berlin? In J. Richter (Ed.), *The Tourist City Berlin. Tourism and Architecture* (pp. 190–199). Braun.

Novy, J. (2017). 'Destination' Berlin revisited: From (new) tourism towards a pentagon of mobility and place consumption. *Tourism Geographies*, *20*(3), 418–442.

Novy, J., & Colomb, C. (2020). Overdosed, underplanned or what? Making sense of urban tourism's 'politicisation from below'. In J. A. Oskam (Éd.), *The Overtourism Debate* (pp. 75–94). Emerald Publishing.

Pons, A., & Rullan, O. (2014). The expansion of urbanisation in the Balearic Islands (1956–2006). *Journal of Marine and Island Cultures*, *3*(2), 78–88.

Pott, A. (2007). *Orte des Tourismus: Eine raum- und gesellschaftstheoretische Untersuchung*. transcript Verlag.

Pred, A. (1984). Place as historically contingent process: Structuration and the time-geography of becoming places. *Annals of the Association of American Geographers*, *74*(2), 279–297.

Rabbiosi, C. (2015). Renewing a historical legacy: Tourism, leisure shopping and urban branding in Paris. *Cities*, *42*, 195–203.

Romero-Padilla, Y., Navarro-Jurado, E., & Malvárez-García, G. (2016). The potential of international coastal mass tourism destinations to generate creative capital. *Journal of Sustainable Tourism*, *24*(4), 574–593.

Ruault, J.-F. (2017). Beyond tourism-based economic development: City-regions and transient custom. *Regional Studies*, *52*(8), 1122–1133.

Runte, A. (1979). *National Parks: The American Experience*. University of Nebraska Press.

Saarinen, J. (2004). 'Destinations in change': The transformation process of tourist destinations. *Tourist Studies*, *4*(2), 161–179. https://doi.org/10.1177/1468797604054381

Saretzki, A. (2019). Produktion lokaler Erinnerungsräume durch immaterielles Welterbe. *Geographische Zeitschrift*, *107*(1), 37–60.

Sauthier, G. (2016). Pouvoir local et tourisme Jeux politiques à Finhaut, Montreux et Zermatt de 1850 à nos jours. Éditions Alphil-Presses universitaires suisses. https://doi.org/10.33055/ALPHIL.03066

Schmid, C. (2013). Networks, borders, differences: Towards a theory of the urban. In N. Brenner (Éd.), *Implosions—Explosions: Towards a Study of Planetary Urbanization* (pp. 67–81). Jovis.

Schmid, C. (2018). Journeys through planetary urbanization: Decentering perspectives on the urban. *Environment and Planning D: Society and Space, 36*(3), 591–610.

Segessemann, A., & Crevoisier, O. (2016). Beyond economic base theory: The role of the residential economy in attracting income to Swiss regions. *Regional Studies, 50*(8), 1388–1403.

Sheller, M. (2018). *Mobility Justice: The Politics of Movement in an Age of Extremes*. Verso.

Sheller, M., & Urry, J. (2006). The new mobilities paradigm. *Environment and Planning A, 38*(2), 207–226.

Soane, J. V. N. (1993). *Fashionable Resort Regions: Their Evolution and Transformation with Particular Reference to Bournemouth,* Nice, Los Angeles and Wiesbaden. CAB International.

Stock, M. (2007). European cities: Towards a recreational turn? *Hagar: Studies in Culture, Polity and Identities, 7*(1), 115–134.

Stock, M. (2019). Inhabiting the city as tourists: Issues for urban and tourism theory. In T. Frisch, C. Sommer, L. Stoltenberg, & N. Stors (Éds.), *Tourism and Everyday Life in the Contemporary City* (pp. 42–66). Routledge.

Stock, M., Clivaz, C., Crevoisier, O., & Kebir, L. (2021). Rethinking resort development through the concept of "touristic capital" of Place. In M. Stock (Éd.), *Progress in French Tourism Geographies: Inhabiting Touristic Worlds* (pp. 203–222). Springer.

Stock, M., & Lucas, L. (2012). La double révolution urbaine du tourisme. *Espaces et sociétés, 151*(3), 15–30.

Talandier, M. (2013). Redéfinir l'enjeu de l'économie présentielle et le rôle des femmes dans les économies locales. *Journal of Alpine Research | Revue de géographie alpine, 101–1.*

Terrier, C., Sylvander, M., & Khiati, A. (2005, novembre). En haute saison touristique, la population présente double dans certains départements. *Insee Première, 1050.*

Wang, Y. (2021). *Tourism, Heritage, and the Transformation of the World Heritage Site of Honghe Hani Rice Terraces.* Ph.D. dissertation, Lausanne University.

17

HISTORICAL GEOGRAPHIES OF TOURISM AND HERITAGE

Dallen J. Timothy

Introduction

Since the earliest days of the Anthropocene, humans have travelled for many different purposes. Some of the world's earliest sacred sites provide archaeological evidence that people travelled away from their tribal lands or settlements to sacrosanct places to worship deity or to supplicate gods for material blessings. Early mobility patterns are also evident with hunter-gatherer groups and pastoralists who practised transhumance as a significant part of their livelihoods. Later, the domestication of plants and animals, and the resultant agricultural patterns and human settlements, emerged circa 10,000 years ago. The use of stone tools and later bronze and iron implements were hallmarks of human advancements. These and many other occurrences moulded the cultural landscape, formed the foundations of many cultures, saw the beginnings of long-distance travel, and were the forerunners of modern societies.

This history tells a great deal about the mobilisation of humans, humankind's imprints on the natural world to form unique cultural landscapes, and the motivations for travel – one of the oldest being religious or spiritual purposes. Much cultural heritage has roots in the events and occurrences noted here. Human heritage has a very long history and manifests in many ways in many places. Geography as a discipline has been at the forefront of understanding the past and its modern-day uses, particularly in the realm of heritage tourism (Graham et al., 2000; Johnson, 1996; Timothy, 2018). To illustrate these points, this chapter examines three geo-heritage settings of interest to historical geographies: tourism landscapes, human mobilities, and religious tourism.

Geographies of Heritage Tourism

Owing to the inseparability of time and space, historical geography is central to the discipline of geography. Many other subdisciplines, such as physical, cultural, social, political, regional, urban, and rural geographies, come together in historical geography to examine various phenomena on the Earth at different periods of time. Cultural and historical geographers are interested in the evolution of natural and human landscapes, humankind's uses of natural resources, human mobility, human settlement patterns and land use, and how power has been exercised over people and territory, and various manifestations of cultural heritage (Butlin, 1993). Historical

DOI: 10.4324/9781003286301-20

geography approaches history and geography from several perspectives. Among other concerns, it involves understanding how geography influenced historical events, how geography has changed through time, and how humankind has altered physical and cultural landscapes through various processes and in different temporal contexts. Historical geography, thus, plays an important role in understanding historically-rooted phenomena that manifest in sundry ways in many places. One display of historical geographies is tourism, particularly in relation to the nexus between tourism and migration and global diasporas, power relations, transportation systems, urbanisation and rural development, and cultural heritage (Timothy, 2012).

Although all types of tourism are of interest to geographers and are rooted in the historical geographies of places, perhaps the clearest manifestations of the crossovers between tourism and historical geography can be found within the realm of heritage (Graham et al., 2000). Heritage is what humankind inherits from the past, utilises and values in the present, and hopes to pass on to future generations (Timothy, 2021a). Based on the definition above, heritage is extremely broad in scope, and almost every type of tourism (e.g., including sport, shopping, sun-sea-and-sand, agri-tourism, culinary, dark, religious tourism/pilgrimage, and volunteer) manifests certain elements of heritage. Each one of these tourisms has unique histories and geographies, based upon destination characteristics and tourist demand. Heritage may be ancient or relatively modern. It may be tangible or intangible, and it may be extraordinary with global appeal, or ordinary and of local or regional importance (Timothy, 1997). Although the contribution of geography, historical geography in particular, to tourism studies is enormous, the remainder of this chapter describes only a few themes related to heritage tourism that are, or could be better, informed through a deeper historical geographical analysis.

Tourism–Landscape Relationships

One of the oldest traditions in historical and cultural geography is understanding the evolution and meanings of cultural landscapes, or the human imprint on the natural environment (Head, 2017; Lewis, 1983; Meinig, 1979). Although less has been written about tourism landscapes compared to other types of cultural landscapes, the topic has emerged as an important exercise in understanding how tourists and tourism modify and determine the forms, functions, and landscapes of places, as well as how landscapes are consumed and commoditised as tourism assets (Aitchison et al., 2014; Butler, 1992; Skowronek et al., 2018; Terkenli, 2021). Terkenli (2004) analysed the extant research on tourism landscapes, suggesting that most geographical research had theretofore focused on resort/destination morphology; modelling the outcomes of tourism activities on the landscape with its planning and land-use implications and landscapes as resources for development through tourism with implications for carrying capacities, impacts, and people's travel motives; and landscape changes through tourism.

Against the work of Terkenli (2004; 2021; see also Chapter 12 of this volume) and others, this section highlights three primary relationships between cultural landscapes and tourism: landscapes as tourist commodities; tourism as a modifier of landscapes; and tourism as landscape. It should be noted, however, that these three relationships are not exhaustive, nor are they mutually exclusive.

Landscape as tourist commodity

Perhaps the most pervasive relationship between tourism and landscapes is that of landscapes as tourist attractions. The importance of natural and cultural landscapes in tourism cannot be overstated, with some of the most impressive cultural and natural landscapes being among the most visited places on Earth (e.g., the Maasai cultural area in Tanzania and Kenya; the industrial landscapes of Europe; and the rice agricultural areas of the Philippines and China). Many have been inscribed on UNESCO's World Heritage List since 1992 because of their demonstrations of human-nature interaction and human ingenuity, which UNESCO categorises as landscapes created intentionally by human activity, organically evolved landscapes, and associative cultural landscapes (Aplin, 2007). Landscape values such as these underscore the policies of UNESCO and individual national conservation authorities and have spurred many tourism development efforts. Landscape values are ensconced within a combination of natural features and human imprints on nature, creating very attractive settings for tourism. Undisturbed natural settings are venues for much nature-based tourism and outdoor recreation. Modified landscapes are equally impressive and have become the centre of tourism development in many places, particularly those that demonstrate rural livelihoods, human settlements, heavy industry, and agricultural patterns. Increased awareness of the value of place uniqueness has elevated the importance of cultural landscapes as tourism assets during the past century (O'Hare, 1997), and the widespread growth of the correlative phenomena of heritage tourism, immersive tourism, and slow tourism has enabled more diverse landscapes to be valued as cultural landscapes and tourism assets.

As the values indicated earlier denote (Aplin, 2007), most natural and cultural attractions do not exist in isolation. Instead, they are an integral part of broader cultural landscapes and larger ecosystems. In this sense, the geographical principle of clustering or spatial concentration is evident in the idea of landscapes, where individual structures or cultural features on their own might not wield much touristic appeal, but when part of a larger integrated landscape, these features become something of far greater value than the sum of its parts. Themed trails and routes, agricultural and industrial landscapes, and vernacular architecture clusters are examples of this geographical principle that increases the heritage value and valorisation potential of cultural landscapes.

Tourism as a modifier of landscapes

Cultural landscapes become commoditised and altered as tourists seek novel places and as landscapes are interpreted for tourist consumption (O'Hare, 1997; Terkenli, 2004; Urry, 1995). The primary way in which tourism functions as a modifier of landscape is through direct and indirect physical impacts such as wear and tear, vandalism, and pollution (Mathieson & Wall, 1982). For instance, much has been written recently about trekking tourism in the Himalaya, suggesting that adventure tourism in Nepal (now resembling mass tourism), especially, has modified cultural and natural landscapes in irreparable ways. For example, the mass collection of firewood for heating and cooking has resulted in severe deforestation and the depletion of many floral species in the Himalaya range (Apollo & Andreychouk, 2022; Byers & Shrestha, 2022; Nepal, 2022). Trekking tourism has also caused many small mountain villages to evolve into towns whose sole purpose now is to provide services for tourists. Many traditional homes have been razed in favour of modern lodges, restaurants, and bars (Byers & Shrestha, 2022). This trend is not unique to the Himalaya but can be seen throughout the world where traditional settlements, homes, and rural livelihoods have been replaced by more profitable tourism services.

The notion of physical change has led much geographical thought in the area of tourism development. Some of the earliest tourism investigations by geographers focused on the spatial morphology of tourism destinations, suggesting that tourism is the primary determining variable in the geographical growth of resort communities and the forms and functions of those cities (Butler, 1980; Getz, 1993; Wall, 1977). Most research attention has focused on coastal and waterfront communities where tourism is the primary force in how destinations were planned or developed organically (Andriotis, 2003; Liu & Wall, 2009; Xie et al., 2013).

In addition to resort morphology, urban gentrification and waterfront development are examples of tourism-induced landscape modifications. Owing largely to tourism demand in urban areas, historic cities have become a focus of gentrification efforts, entailing investments in the renovation and redevelopment of heritage buildings and urban neighbourhoods. Waterfront development is a form of gentrification that changes the face of old factories, harbourfront, maritime landscapes, and industrial brownfields into usable leisure spaces, such as parks, museums, restaurants, theatres, and tourist accommodation. While lucrative for city governments and outside investors, the process of gentrification typically disenfranchises local residents because increasing property values mean they are priced out of the marketplace; they cannot afford to pay taxes or maintain their properties, and younger consumers and retirees can no longer afford to live in the city. In many places, urban inhabitants have fought adamantly against gentrification as they become disillusioned by tourism and the unaffordability of urban living and crowded conditions it brings in its wake. Gentrification in tourist cities such as Barcelona, Prague, and Budapest has permanently altered those cities' heritage landscapes and has become a significant point of contention among locals, tourists, tourism promoters, and real estate developers (Jover & Díaz-Parra, 2022; Smith et al., 2019).

Areas adjacent to international borders (borderlands) also have unique characteristics and spatial morphologies that parallel the demand for border tourism services (Gelbman & Timothy, 2019; Więckowski & Timothy, 2021). Except between Schengen countries in Europe, there is an ever-present landscape of security, passport and customs controls, and law enforcement at most international boundaries. These border landscapes of security and officialdom are typically enhanced by services established specifically to cater to the needs of border crossers: insurance brokers, currency exchange offices, information centres, SIM card vendors, and taxi stations, among others. Depending on the type of tourism associated with a specific border area, certain servicescapes evolve and reflect the destination's development focus.

In the case of the US–Mexico border, Arreola and Curtis (1993; Curtis & Arreola, 1991) examined how tourism(s) of vice (e.g., alcohol consumption, gambling, prostitution, and drugs) created unique spatial morphologies in Mexican border towns, with red light districts, or '*zonas de tolerancia*', developing organically to satisfy the touristic depravities of the past. Developing from the early 1900s until the 1990s, these zones were generally clustered adjacent to border crossings and along major pedestrian thoroughfares in almost all northern Mexican border cities. Today, the extent of these districts has diminished substantially, having been replaced overwhelmingly by medical services and pharmacies and repurposed into other commercial functions. The medical tourism phenomenon now dominates most of the northern Mexican border zone (Adams et al., 2019; Martínez Almanza et al., 2019) and is located in spatial clusters that have organically infiltrated and replaced the traditional tourist zones, or have been purposely built as new '*zonas de turismo médico*' (Boda & Harris, 2013; Cuevas Contreras, 2015; Cuevas Contreras & Zizaldra Hernández, 2023).

Tourism as heritage landscape

Intensive tourism development leads to the formation of unique servicescapes that cater to the needs of tourists above everyone else. In intensively commercialised destinations, urban (and sometimes rural) spaces emerge that are entirely tourism-oriented, similar to the urban morphology examples described earlier. However, these particular servicescapes are characterised by an abundance of accommodation facilities, eateries, tour agencies, banks, shops, and other enterprises that are often imprinted permanently in the urban structure of places. In intensively developed destinations, urban spaces emerge that are entirely meant to service tourists' needs. These situations often mean that tourism feeds off tourism; tourism itself and its leisurescapes and pleasurescapes attract even more tourists in a domino effect.

Thus, a rather unique way of viewing tourism as landscape is literally *tourism as the heritage* of a place. Weaver (2011) rightfully argues that tourism *is* the heritage of Las Vegas, United States, and Gold Coast, Australia. In century-old Las Vegas, tourism quickly became the city's leading industry after its founding in the early twentieth century and has developed since that time almost entirely around tourism. There, tourism itself and the physical landscape it created have in fact become the heritage of Las Vegas, feeding the further growth of tourism. Similar situations can be found in many tourist destinations throughout the world, such as famous coastal resorts in the United States (e.g., Atlantic City) and the UK (e.g., Brighton), which grew organically into tourist destinations because of their locations and recreational amenities and whose leisurescapes and servicescapes provide more appeal than their coastal locales.

Other localities were purposely planned to become tourist resort destinations, such as Cancun, Mexico, and some are a mix of both (e.g., Dubai). In all of these cases, tourism and its associated landscapes have become the local heritage that now dominates place-based cultural landscapes and creates an appealing environment for tourists.

Human Mobilities

Something of considerable interest from an historical geographical perspective is human mobility and its changes through time. The mobilities paradigm, which started to appear in the early 2000s as scholars began to realise that exclusive and singular divisions of travel and human mobility were too myopic, owing largely to their overlapping conceptual foundations and blurred boundaries in practice. Mobilities as seen today take a broader and more holistic view of human movement than any one manifestation of mobility alone (Hannam et al., 2006; Sheller & Urry, 2006, 2016). Multiple disciplines study mobility, and the phenomenon involves many mobility variables, not least of which are globalisation, corporeal versus virtual movement, technological change, transportation innovations, tourism, and the inseparability of time and space (Sheller & Urry, 2006). Recognising that tourism is only one manifestation of many mobilities, Hannam, Sheller and Urry (2006, p. 1) note that mobilities encompass "both large-scale movements of people, objects, capital and information across the world, as well as the more local processes of daily transportation, movement through public space and the travel of material things within everyday life". Thus, the mobilities paradigm recognises the commonalities between, and the inseparability of, tourism, migration, and other forms of travel in the movement of people across the Earth (Hall, 2015; Hannam et al., 2006; Timothy & Michalkó, 2016; Timothy et al., 2022).

Arbitrary boundaries have long been drawn between tourism and other mobilities. Hall (2008) argues that day trips, shopping, visits to local recreational spots, weekends away, study abroad experiences, and others are also manifestations of 'tourism' mobilities. He also suggests that long-term travel, such as migration, extended working holidays, missions, and military service are clear examples of human mobilities that have many characteristics in common with normative tourism, despite their not meeting the official UNWTO definition of temporary travel. Although refugees and permanent migrants are not tourists by definition, because they lack the temporariness of tourism, they have many things in common with tourists and use many of the same services, assets, and spaces. Thus, mobility is a broader concept that includes tourism, as well as other movements that may or may not resemble traditional tourism, even though they do utilise many of the same services and infrastructures, and adhere to legal and policy requirements (e.g., passports and visas).

Relatedly, there are plethoric misconceptions that tourism is a leisure-oriented or pleasure-motivated activity undertaken during leisure time, when in fact, tourism is much farther reaching than only leisure and pleasure. It also entails people seeking medical treatments, undertaking religious pilgrimages, shopping for utilitarian items, and volunteering abroad. This realisation has led many scholars to understand that traditional views of tourists and tourism (e.g., leisure and pleasure) are inadequate for understanding the breadth and depth of human mobility, much of which can be considered tourism.

From strictly a heritage perspective, migrants and refugees play an important part in tourism. That is, their mobility history often forms a foundation of tourism later in the form of diasporic tourism, return travel, personal heritage and roots tourism, cultural heritage tourism, and other forms of heritage tourism. Urban and rural immigrant neighbourhoods or communities (i.e., ethnic islands or ethnic enclaves) are extremely popular attractions throughout the world, such as Chinatowns, Little Italys, and Koreatowns. Additionally, immigrants and their progeny travelling back to their original homelands is an important form of heritage tourism, known variously as diaspora, roots, genealogy, or personal heritage tourism and is especially important for residents of settler societies that often travel to their ancestral homeland to experience a sense of belonging, to seek answers to questions about their personal identities, and to connect with their genetic forebears. These and other forms of tourism that derive from migration came into existence through the historical geographies of time, place, and mobility.

Pilgrimage Travel and Religious Tourism

As noted at the outset of this chapter, humans have travelled for spiritual or religious purposes since the beginning of time. Prehistoric humans journeyed for spiritual reasons, as archaeological evidence illustrates at places such as the Musical Stone of Gobustan in Azerbaijan (18,000–3,000 BC), Göbekli Tepe temple in Turkey (9,500–8,000 BC), and the Hypogeum of Ħal-Saflieni in Malta (4,000–2,500 BC). Later pilgrimages developed in South Asia during the Vedic Era, from about 1,500 to 500 BC, with early Hindu and Buddhist pilgrimages beginning around that time. Later, Christianity was formalised and spread through the Mediterranean region and throughout Europe and the Caucasus in the first few centuries after the time of Jesus (Metreveli & Timothy, 2010). Likewise, the present Islamic *Hajj* pilgrimage dates to around 632. Thus, throughout the Middle Ages, pilgrimages became one of the most pervasive forms of international travel and continues today, with pilgrimages constituting the largest tourist gatherings in the world (e.g., the *Kumbh Mela* and the *Hajj*) (Bielo & Ron, 2023; Olsen & Timothy, 2022; Ron & Timothy, 2019).

The movement of people through time and space for religious or spiritual reasons, makes pilgrimages an important topic of interest to geographers, and its evolution through time is of considerable interest to historical geographers and other heritage scholars. Initially, ancient and Mediaeval pilgrimages entailed people travelling in humble circumstances, as an act of worship and self-effacement, to show devotion to deity and to supplicate God for blessings and redemption from sin (Badone & Roseman, 2004; Stausberg, 2012; Vukonić, 1996). Formerly, religious travel, including pilgrimage, entailed very little by way of joviality, pleasure, or fun, as these emotions and experiences were seen as antithetical to the austere endurance of pilgrimage. In Catholicism and some other branches of Christianity, the pilgrimage journey was traditionally more important than arriving at the sacred destination, for it was the hardships of the journey that humbled the pilgrims and prepared them to commune with deity and to be worthy of godly blessings. The harshness of the journey in Islam and Hinduism also made the pilgrimage more effectual in the lives of the travellers (Nordin, 2011; Taylor, 2011).

Today, although some pilgrims choose to undertake very traditional journeys that include walking (sometimes crawling on their knees in extreme circumstances), horseback riding, camping along the way or staying in pilgrim rest houses, eating simple foods, and living humbly while in the final destination, increasing numbers of modern-day pilgrims choose a more luxurious 'vacation' experience. Technology, transportation innovations, globalisation processes, mass marketing, and modernisation tendencies in general, enable pilgrims, like other tourists, to travel in greater comfort, with many pilgrimage experiences now resembling mass tourism and luxury travel, with sightseeing and recreational activities often supplementing the rites and rituals of the pilgrimage (Aukland, 2018; Dimitrovski et al., 2021).

Through the process of pilgrimage 'touristification', landscapes have changed and people's religious practices have expanded to include non-spiritual or non-religious activities and practices. Thus, religious theme parks have been developed in India, other parts of Asia, and in North America (Rose, 2021; Ron & Timothy, 2019). Old neighbourhoods in Mecca have been demolished in recent years to make way for new luxury hotels and resorts in the vicinity of the Great Mosque (*Masjid al-Haram*), the holiest shrine in Islam, where only wealthy pilgrims are able to stay nearby and undertake the rituals and ceremonies of the *Hajj* in conjunction with luxury accommodation, paid guiding services, and elegant meals (Qurashi, 2017; Shirazi, 2016). These activities have wrought permanent changes to the sacred urban landscape of Mecca and transformed much of the city from a sacredscape into a leisurescape, which many pilgrims find distasteful and irreverent (Shirazi, 2016). Over-commercialised and extremely commoditised Christianity dominates the sacred landscapes of Jerusalem, Lourdes, Knock, Medjugorje, the Shrine of Our Lady of Guadalupe, and Santiago de Compostela (Ron & Timothy, 2019). From a destination marketing perspective, it matters little to the destination managers or entrepreneurs whether visitors classify themselves as pilgrims or tourists; both are big spenders and require the same hospitality services (Pourtaheri et al., 2012; Saayman et al., 2014), and many spiritual destinations are changing to capitalise on this lucrative form of tourism.

Changes in pilgrimage from pure devotion to religious tourism is of considerable interest to historical geographers and heritage specialists. Every element of pilgrimage/religious tourism is a manifestation of cultural heritage. Rituals, spiritual mobilities, pilgrim trails and routes, beliefs, dogmas, celebrations, sacraments, holy writ, music, sacred sites, poetry and recitations, sacrosanct attire and symbols, buildings and archaeological sites, eschatological rites and practices, funerary traditions, filial piety, and religious holidays are some of the commonest manifestations of religious and spiritual heritage involved in religious tourism, contributing to socio-spatial changes in time and place (Timothy, 2021b).

Conclusion

Heritage and its tourisms can be interpreted and understood through many geographical lenses, including those of historical geographies (Timothy, 2012; 2018). Geographers have a keen interest in the evolution of places through time, and the inseparability of time and space. Thus, heritage and its many forms and manifestations are of keen interest to historical and other geographers. Nearly all types of tourism have manifestations of heritage and historicity and provide substantial fodder for geographical analysis. The definition of heritage is broad, and its manifestations in tourism are of considerable interest to historical geographers. Although the relationships between historical geography, heritage, and tourism are manifold, only three were examined here: landscapes, mobilities, and religious tourism.

Tourism is closely related to various notions of landscape – landscape as attraction, tourism as a modifier of landscapes, and tourism as heritage landscape, to name only a few. Landscape has a long history in geographical studies, yet relatively little has been done to further our understanding of tourism landscapes as cultural heritage landscapes. Although cultural landscapes have long been studied for their touristic value, additional work on the crossovers between landscape and tourism would fruitfully bring to bear new insight about tourism as a medium of physical change and how different tourisms result in different imprints on the Earth through time.

Human mobility is a vast field of study. Tourism is but one part of mobilities studies, but it is incumbent upon scholars to situate tourism within broader conceptualisations of mobilities. Understanding the more ostensibly 'touristic' elements of non-tourism mobilities would provide a deeper understanding of the mobility experiences of different groups, including migrants, refugees, and long-term travellers. These less-temporary mobilities clearly overlap with tourism in many ways, and understanding their motives and experiences might help cater better to their needs. Work on the historical geographies of mobility would also be key in understanding different heritages created from different mobilities, for they are not all the same. For example, forced migration (e.g., slavery, refugees) nearly always results in dark heritage, whereas economic or social migration (e.g., seeking work, joining family members) often results in positive manifestations of heritage tourism.

Pilgrimage, a form of religious tourism, is of extreme interest for geographers because of its evolutionary history from a puritanical spiritual mobility to an increasingly leisure- and holiday-oriented experience, even among the most devout pilgrims, and the changes it effects on its destinations. Like other forms of tourism, pilgrimage has made unique imprints on the cultural landscapes of places through worship rites and the sanctification of ritual space and commercialisation, thus feeding an ever-growing religious mobility and increasing attendance at sacred sites by non-religious curiosity-seekers. Some of the most commercialised and overtouristed places today are simultaneously some of the most sacred sites on Earth. Every religion and faith tradition involves sacred mobilities, either through formalised distant pilgrimages or localised ceremonial mobility in situ.

These three phenomena – landscapes, mobilities, and pilgrimage – are all salient manifestations of heritage and possess clear connections with tourism. As a result, historical geographers are keenly aware of the need to study tourism more deeply from geographical perspectives. They are equipped with the necessary tools to analyse and understand how and why landscapes change through time, how tourism best fits within the broader notion of mobility, and the important role of pilgrimage as a pure form of heritage tourism and something that is growing in its popularity among the faithful and non-faithful alike.

References

Adams, K., Snyder, J., & Crooks, V. A. (2019). Narratives of a "dental oasis": Examining media portrayals of dental tourism in the border town of Los Algodones, Mexico. *Journal of Borderlands Studies, 34*(3), 325–341.

Aitchison, C., MacLeod, N. E., & Shaw, S. J. (2014). *Leisure and Tourism Landscapes: Social and Cultural Geographies*. Routledge.

Andriotis, K. (2003). Coastal resorts morphology: The Cretan experience. *Tourism Recreation Research, 28*(1), 67–75.

Aplin, G. (2007). World heritage cultural landscapes. *International Journal of Heritage Studies, 13*(6), 427–446.

Apollo, M., & Andreychouk, V. (2022). *Mountaineering Adventure Tourism and Local Communities: Social, Environmental and Economics Interaction.* Edward Elgar.

Arreola, D. D., & Curtis, J. R. (1993). *The Mexican Border Cities: Landscape Anatomy and Place Personality.* University of Arizona Press.

Aukland, K. (2018). At the confluence of leisure and devotion: Hindu pilgrimage and domestic tourism in India. *International Journal of Religious Tourism and Pilgrimage, 6*(1), 18–33.

Badone, E., & Roseman, S. R. (2004). *Intersecting Journeys: The Anthropology of Pilgrimage and Tourism.* University of Illinois Press.

Bielo, J. S., & Ron, A. S. (2023). *Landscapes of Christianity: Destination, Temporality, Transformation.* Bloomsbury.

Boda, P. J., & Harris, J. (2013). Root canals and crowns: An analysis of the spatial distribution of dental offices in Ciudad Juárez, Chihuahua, Mexico, 1996–2011. *International Journal of Geosciences, 4*(6), 38–43.

Butler, R. W. (1980) The concept of a tourist cycle of evolution: Implications for the management of resources. *Canadian Geographer, 24*, 5–12.

Butler, R. W. (1992). Tourism landscapes: For the tourist or of the tourist? *Tourism Recreation Research, 17*(1), 3–9.

Butlin, R. A. (1993). *Historical Geography: Through the Gates of Space and Time.* Edward Arnold.

Byers, A. C., & Shrestha, M. (2022). Conservation and restoration of alpine ecosystems in the Himalaya: New challenges for adventure tourism in the 21st century. In G. P. Nyaupane, & D. J. Timothy (Eds.), *Tourism and Development in the Himalaya: Social, Environmental, and Economic Forces* (pp. 55–73). Routledge.

Cuevas Contreras, T. (2015). An approach to medical tourism on Mexico's northern border. *Eurasia Border Review, 6*(1), 45–62.

Cuevas Contreras, T., & Zizaldra Hernández, I. (2023). Borders and healthcare: Medical mobility, globalization and borderland tourism. In D. J. Timothy, & A. Gelbman (Eds.), *Routledge Handbook of Borders and Tourism* (pp. 281–2959. Routledge.

Curtis, J. R., & Arreola, D. D. (1991). Zonas de tolerancia on the northern Mexican border. *Geographical Review, 81*(3), 333–346.

Dimitrovski, D., Ioannides, D., & Nikolaou, K. (2021). The tourist-pilgrim continuum in consumer behaviour: The case of international visitors to Kykkos Monastery, Cyprus. *Journal of Tourism and Cultural Change, 19*(5), 568–586.

Gelbman, A., & Timothy, D. J. (2019) Differential tourism zones on the western Canada-US border. *Current Issues in Tourism, 22*(6), 982–704.

Getz, D. (1993). Planning for tourism business districts. *Annals of Tourism Research, 20*(3), 583–600.

Graham, B. J., Ashworth, G. J., & Tunbridge, J. E. (2000). *A Geography of Heritage: Power, Culture and Economy.* Arnold.

Hall, C. M. (2008). Of time and space and other things: Laws of tourism and the geographies of contemporary mobilities. In P. M. Burns, & M. Novelli (Eds.), *Tourism and Mobilities: Local-Global Connections* (pp. 15–329). CABI.

Hall, C. M. (2015). On the mobility of tourism mobilities. *Current Issues in Tourism, 18*(1), 7–10.

Hannam, K., Sheller, M., & Urry, J. (2006). Mobilities, immobilities and moorings. *Mobilities, 1*(1), 1–22.

Head, L. (2017). *Cultural Landscapes and Environmental Change.* Routledge.

Johnson, N. C. (1996). Where geography and history meet: Heritage tourism and the big house in Ireland. *Annals of the Association of American Geographers, 86*(3), 551–566.

Jover, J., & Díaz-Parra, I. (2022). Who is the city for? Overtourism, lifestyle migration and social sustainability. *Tourism Geographies, 24*(1), 9–32.

Lewis, P. (1983). Learning from looking: Geographic and other writing about the American cultural landscape. *American Quarterly, 35*(3), 242–261.

Liu, J., & Wall, G. (2009). Resort morphology research: History and future perspectives. *Asia Pacific Journal of Tourism Research, 14*(4), 339–350.

Martínez Almanza, M. T., Guía Julve, J., Morales Muñoz, S. A., & Esparza Santillana, M. A. (2019). Border medical tourism: The Ciudad Juárez medical product. *Anatolia, 30*(2), 258–266.

Mathieson, A., & Wall, G. (1982). *Tourism: Economic, Physical and Social Impacts.* Longman.

Meinig, D. W. (Ed.) (1979). *The Interpretation of Ordinary Landscapes*. Oxford University Press.

Metreveli, M., & Timothy, D. J. (2010). Religious heritage and emerging tourism in the Republic of Georgia. *Journal of Heritage Tourism, 5*(3), 237–244.

Nepal, S. K. (2022). Moral geographies, porters, and social exclusion in trekking tourism in the Himalayan region. In G. P. Nyaupane, & D. J. Timothy (Eds.), *Tourism and Development in the Himalaya: Social, Environmental, and Economic Forces* (pp. 111–126). Routledge.

Nordin, A. (2011). The cognition of hardship experience in Himalayan pilgrimage. *Numen, 58*(5–6), 632–673.

O'Hare, D. (1997). Interpreting the cultural landscape for tourism development. *Urban Design International, 2*(1), 33–54.

Olsen, D. H., & Timothy, D. J. (Eds) (2022). *The Routledge Handbook of Religious and Spiritual Tourism*. Routledge.

Pourtaheri, M., Rahmani, K., & Ahmadi, H. (2012). Impacts of religious and pilgrimage tourism in rural areas: The case of Iran. *Journal of Geography and Geology, 4*(3), 122–129.

Qurashi, J. (2017). Commodification of Islamic religious tourism: From spiritual to touristic experience. *International Journal of Religious Tourism and Pilgrimage, 5*(1), 89–104.

Ron, A. S., & Timothy, D. J. (2019). *Contemporary Christian Travel: Pilgrimage, Practice, and Place*. Channel View Publications.

Rose, L. (2021). Religious theme parks. In D. H. Olsen, & D. J. Timothy (Eds.), *The Routledge Handbook of Religious and Spiritual Tourism* (pp. 179–190). Routledge.

Saayman, A., Saayman, M., & Gyekye, A. (2014). Perspectives on the regional economic value of a pilgrimage. *International Journal of Tourism Research, 16*(4), 407–414.

Sheller, M., & Urry, J. (2006). The new mobilities paradigm. *Environment and Planning A, 38*(2), 207–226.

Sheller, M., & Urry, J. (2016). Mobilizing the new mobilities paradigm. *Applied Mobilities, 1*(1), 10–25.

Shirazi, F. (2016). *Brand Islam: The Marketing and Commodification of Piety*. University of Texas Press.

Skowronek, E., Tucki, A., Huijbens, E., & Jóźwik, M. (2018). What is the tourist landscape? Aspects and features of the concept. *Acta Geographica Slovenica, 58*(2), 73–85.

Smith, M. K., Sziva, I. P., & Olt, G. (2019). Overtourism and resident resistance in Budapest. *Tourism Planning & Development, 16*(4), 376–392.

Stausberg, M. (2012). *Religion and Tourism: Crossroads, Destinations and Encounters*. Routledge.

Taylor, R. M. (2011). Holy movement and holy place: Christian pilgrimage and the Hajj. *Dialog, 50*(3), 262–270.

Terkenli, T. S. (2004). Tourism and landscape. In A. A. Lew, C. M. Hall, & A. M. Williams (Eds.), *A Companion to Tourism* (pp. 339–348). Blackwell.

Terkenli, T. S. (2021). Research advances in tourism-landscape interrelations: An editorial. *Land, 10*, 944.

Timothy, D. J. (1997). Tourism and the personal heritage experience. *Annals of Tourism Research, 34*(3), 751–754.

Timothy, D. J. (2012). Historical geographies of tourism. In J. Wilson (Ed.), *The Routledge Handbook of Tourism Geographies* (pp. 157–162). Routledge.

Timothy, D. J. (2018). Geography: The substance of tourism. *Tourism Geographies, 20*(1), 166–169.

Timothy, D. J. (2021a). *Cultural Heritage and Tourism: An Introduction*, 2nd ed. Channel View Publications.

Timothy, D. J. (2021b) Heritage and tourism: Alternative perspectives from South Asia. *South Asian Journal of Tourism and Hospitality, 1*(1), 35–57.

Timothy, D. J., & Michalkó, G. (2016). European trends in spatial mobility. *Hungarian Geographical Bulletin, 65*(4), 317–320.

Timothy, D. J., Michalkó, G., & Irimiás, A. (2022). Unconventional tourist mobility: A geography-oriented theoretical framework. *Sustainability, 14*(11), 6494.

Urry, J. (1995). *Consuming Places*. Routledge.

Vukonić, B. (1996). *Tourism and Religion*. Pergamon.

Wall, G. (1977). Recreational land use in Muskoka. *Ontario Geography, 11*, 11–28.

Weaver, D. B. (2011). Contemporary tourism heritage as heritage tourism: Evidence from Las Vegas and Gold Coast. *Annals of Tourism Research, 38*(1), 249–267.

Więckowski, M., & Timothy, D. J. (2021). Tourism and an evolving international boundary: Bordering, debordering and rebordering on Usedom Island, Poland-Germany. *Journal of Destination Marketing & Management, 22*, 100647.

Xie, P. F., Chandra, V., & Gu, K. (2013). Morphological changes of coastal tourism: A case study of Denarau Island, Fiji. *Tourism Management Perspectives, 5*, 75–83.

PART IV

Sustainability Transitions in Tourism Geographies

18

SUSTAINABILITY AND GEOGRAPHIES OF TOURISM

Jarkko Saarinen

Introduction

Over two decades ago, Swarbrooke (1999: 41) hesitantly wrote that "sustainable tourism is, perhaps, an impossible dream." A decade later, Sharpley (2009: xviii) furthered this critical view by stating "sustainable tourism is an idea whose time has now passed" and that we "need to move beyond sustainability" in tourism. Still today, however, despite these hesitations and increased criticisms, the connection between tourism and sustainability is perhaps stronger than ever. The institutional and terminological connection is rhetorically based on the widely referenced book *Our Common Future* outlined by the World Commission on Environment and Development (WCED, 1987), also known as the Brundtland report. The report followed the United Nations Conference on Environment and Development (UNCED) that was held in Rio de Janeiro in 1992. Based on the agreed need for greening the global economy identified during the Rio Earth Summit, the World Tourism Organization (UNWTO) outlined their definition for sustainable tourism the following year. According to the UNWTO (1993: 7), sustainable tourism represents tourism development that aims to meet "the needs of current tourists and host regions while protecting and enhancing opportunities for the future".

Since the Brundtland report and the Earth Summit, the idea of sustainable tourism has been in a central position in tourism development, planning, policy-making and research. In tourism geographies, the evolution of sustainable tourism has represented a paradigm shift, and as noted by Hall (2011: 649) this fundamental change in thinking on tourism and its developmental aspects has made sustainability "one of the great success stories of tourism research". This shift has also made tourism and tourism research an issue of high policy relevance in local and global development discussions and prospects. For example, tourism was integrated into achieving the United Nations Millennium Development Goals (UN MDGs) by 2015 (Saarinen & Rogerson, 2014; UNWTO, 2006; 2012). Currently, the global tourism industry is connected with the United Nations Agenda 2030: the Sustainable Development Goals (SDGs) (Scheyvens, 2018) aiming to end poverty, protect the planet, and ensure that by 2030 all people enjoy peace and prosperity (United Nations, 2015a). Integration of the tourism industry into such high goals is a demonstration of the policy success that is originally based on the created connection between tourism and sustainable development. However, it is also a great challenge for the industry, characterised mainly by small and medium size private businesses, to address such global development tasks. Furthermore, it creates an interesting and often highly contested research frontier for tourism geographies.

This chapter focuses on sustainable development studies in tourism geographies. First the background of sustainability thinking is briefly discussed with references to wise use, carrying capacity, crowding and limits to growth perspectives. After that, the emergence of sustainable tourism and its different perspectives are outlined, followed by the current connections of the tourism industry and SDGs. Finally, future prospects of sustainable tourism geographies are discussed.

Origins of Sustainable Tourism Geographies: 'New Labels on Old Bottles, Old Wine in New Bottles?'

In 1998, Hall and Lew published an edited volume titled *Sustainable Tourism: A Geographical Perspective*. The collection represents one of the early comprehensive and critical depictions of academic sustainable tourism geographies and studies, in general. The collection was a result of newly emerged interests in sustainable tourism among geographers associated with the International Geographical Union (IGU) and the Association of American Geographers (AAG). However, as the editors of the book state the following: "[G]eographers, and tourism geographers in particular, have long had interest in sustainable development and natural resource management" (Hall & Lew, 1998: xii). This long-term interest was highlighted in the titles of the two opening chapters that followed the introduction: Hall (1998) focused on the antecedents of sustainable development in tourism based on the 'new labels on old bottles' metaphor, while Butler (1998) discussed the same based on the expression 'old wine in new bottles'.

Indeed, there is a long history of sustainability thinking in geographical research on tourism and resource management. More generally, the basic principles of sustainable use of resources have origins in the nineteenth-century idea of 'sustained yield' in forestry in Germany and especially in the North American nineteenth-century conservation movement (see Enders & Remig, 2015) that resulted in the establishment of the world's first National Parks, in the United States and Canada (see Hall, 1998). In addition to the sustained yield approach, Gifford Pinchot, who was the first Chief of the United States Forest Service in the 1910s, introduced the so-called 'wise use' principle in natural resource management (Butler, 1998). These historical antecedents played important roles in the evolution of sustainable development thinking, but in the context of the current sustainable development framework they have their limitations (Saarinen & Gill, 2019). The sustained yield and wise use traditions focused on sustaining the resources for economic uses (i.e., economic sustainability and profitability) (McCarthy & Hague, 2004). In contrast to that, the conservation movement was largely interested in ecological aspects, but there were also economic elements (tourism and recreation) involved (Nash, 1967: 108). However, the social aspects of sustainable use of resources were not yet understood, as the conservation of the world's first national park was based on 'fencing animals in, natives out'. This resulted in a so-called fortress model in nature protection, which has been widely utilised by the global conservation movement in the nineteenth and twentieth centuries, creating conflicts between protected area management, tourism uses and local communities in various parts of the world. For tourism geographers this has offered numerous opportunities and cases for research (see Butler, 1991; Hall & Boyd, 2005; Fennel, 2022; Nepal, 2000).

In addition, early conservation and management processes created a new research avenue for tourism geographers: Carrying capacity studies that related to the land-use conflicts in tourism and recreation contexts. Carrying capacity research focused on an attempt to define the maximum number of visitors in a given space and time, without causing unacceptable changes in the physical, economic and socio-cultural environments and on visitors' satisfaction (Butler, 1999; 1996; Getz, 1983). In geography, recreational and tourism carrying capacity issues were initially raised by McMurray (1930) who published an article in the *Annals of the Association of American Geographers* on land use, the recreation industry and possible conflicts between them. By the 1960s and 1970s, carrying capacity research transformed to include conservation area visitors' satisfaction, social carrying capacity and crowding research (Butler, 2010; Lucas, 1964).

A move away from these site-specific issues and scales was largely based on the highly influential report, *The Limits of Growth*, by the Club of Rome in 1972, in which the idea of sustainability was first used in the current 'sustain' or 'maintain' meaning (Crober, 2015). The report stated this: "We are searching for a model output that represents a world system that is: 1. sustainable without sudden and uncontrolled collapse; and 2. capable of satisfying the basic material requirements of all of its people" (Meadows et al., 1972: 152). Later, the term was explicitly used in the World Conservation Strategy in 1980 that aimed "to ensure the sustainable utilisation of species and ecosystems [...], which support millions of rural communities as well as major industries (IUCN, 1980: 7). These and many other international agreements paved the way for establishing the World Commission on Environment and Development (WCED) in 1982, which outlined the idea of sustainable development and integrated it into the policy and academic uses as we now know it: "Sustainable development is development that meets the needs of the present generations without compromising the ability of future generations to meet their own needs" (WCED, 1987: 43).

Sustainable Tourism Geographies: Phases and Perspectives

Nowadays, a widely used (policy-level) definition for sustainable tourism is "tourism that takes full account of its current and future economic, social and environmental impacts, addressing the needs of visitors, the industry, the environment and host communities" (UNEP, 2005: 12). In academia, however, there are many different frameworks and conceptualisations for sustainable tourism (Butler, 1991; 1999) that can be complementary or conflicting (Saarinen, 2006; 2021a). First, an obvious reason for a diverse base of academic conceptualisation is the elusiveness of the term sustainable development itself. The idea of sustainable development is an ideological one and a result of a political compromise in the 1980s. Thus, there were numerous definitions for the term already in the early 1990s in academic research (see Redclift, 1992) and the number of 'variants' has increased over time (Redclift, 2005), indicating "the concept of sustainability is a still-developing framework for scientific research and environmental management" (Ruggerio, 2021: 147481).

Secondly, sustainability has been approached from different stakeholder perspectives, which opens up different kinds of emphases and definitions. Thus, answers to the following questions arise: Whose sustainability should we focus on? For whom are we responsible for making tourism sustainable? What should tourism specifically sustain (see McCool et al., 2001)? These questions potentially produce very different kinds of ideas and definitions of sustainable tourism, which depend on specific stakeholder perspectives and case contexts. Thirdly, the idea of sustainability in tourism has evolved over time, but early understandings and prejudices concerning what forms and activities of tourism have or do not have the capacity to be transformed towards sustainable tourism have not fully disappeared. According to Clarke (1997), sustainability was initially perceived as a potential target for alternative and small-scale tourism operations, while mass-scale tourism was regarded as incompatible with sustainable development goals. Although this kind of distinction is not as clear cut in current tourism research, there are still either explicit or

Table 18.1 Perspectives for Setting the Limits to Growth in Tourism Development (Modified from Saarinen and Gill, 2019a)

Backgrounds	Resource-Based	Industry-Based	Community-Based	System-Based
Orientation	Environment (physical)	Industry	Community	System
Origin/Manifestation	Carrying Capacity Model	Wise use, Product Cycle	Participatory Planning	Limits to growth
Limits to growth	Objective/ Measurable	Relative	Constructed/ Negotiated	Objective/ Measurable
Resource view	Static	Dynamic	Dynamic/Static	Dynamic/Static
Time scale	Long	Short	Short-medium	Long
Governance	Institutional	Market-driven	Collaborative planning	Regulative

implicit assumptions that small-scale tourism activities are automatically more sustainable for socio-ecological systems than larger operations (see Weaver, 2011). However, research has often demonstrated that larger business units, for example, have more resources, capacity and knowledge (and external pressures) to design their operations towards sustainable development needs (Kasim et al., 2014).

Finally, there are different scalar issues related to the constitution of sustainability in tourism (Saarinen, 2021a). Although a clear majority of tourism research has focused sustainability on local or destination scales (see Butler, 1999), there have been increasing calls for system-level approaches in sustainable tourism development (Gössling et al., 2020; Ólafsdóttir & Sæþórsdóttir, Chapter 20 of this volume). These calls are linked with previously mentioned limits of growth thinking by the Club of Rome and the fact that the Earth's resources are not infinite. Furthermore, the increased awareness of and research activity on climate change (Bauer Knowles & Scott, Chapter 21 of this volume), and the role of tourism in global warming, have intensified the need for system-level approaches beyond a contained destination focus.

Resulting from the above issues, there are many different co-existing perspectives on sustainable tourism in research and how the limits to growth are or could be defined and set in sustainable tourism development (Saarinen, 2006). Table 18.1 summarises these different perspectives that can be labelled as resource-based, industry-based, community-based and system-based limits to growth or sustainability in tourism development (see Saarinen & Gill, 2019). These are ideal types that may not exist 1:1 in actual development situations in different contexts. With the exception of system-based limits, the other terms emphasise the different basic elements of ecological, economic and social aspects of sustainable development on a destination scale. The system-based approach highlights the need to involve and integrate all these elements and to understand destinations as parts of a wider structure that involves tourist-generating regions, related routes and mobilities, and (impacted) societies. From the system-based perspective there is a need to have strong institutions and a regulatory framework if the aim of sustainable tourism is to be realised (see Gössling et al., 2020; Saarinen, 2021a). Furthermore, recent research has called for a need to reset the current neoliberal economy (Bianchi, 2018; Saarinen, 2021b) or replace it by alternative development models that are based on, for example, degrowth or a post-growth economy (Blazquez et al., Chapter 5 of this volume; Fletcher et al., 2020; Hall, 2009; Higgins-Desbiolles et al., 2019; Saarinen, 2018).

Resource-based limits for tourism originate from the conservation movement and highlight the importance of ecological resources and their functionality. With respect to destination management and governance, resource-based limits rely on a regulatory approach in safeguarding the integrity of ecosystems. In this respect, the idea of defining the limits of carrying capacity is central (Getz, 1983). As Butler (1998) stated, carrying capacity should be understood as an integral part of the history of sustainability thinking. He further stated that the operationalisation of carrying capacity is an essential measure if sustainability is to be achieved in future tourism (Butler, 1996).

Industry-based limits are linked with the historical wise-use and sustainable yield thinking in resource management, emphasising the needs of tourism as a form of economy (Burns, 1999; Saarinen, 2006). This is in line with neoliberal market-driven approaches to destination governance, highlighting the need to reduce external regulations and the role of governmental institutions in tourism development (Saarinen & Gill, 2019; see Jessop, 2004). Instead, the industry-based perspective stresses the (sustainable) growth and self-regulatory needs of tourism development (Mosedale, 2014; Saarinen, 2021b). This aspect is even included in the Paris Agreement, which was adopted by 196 countries (United Nations, 2015b). The key goal of the Agreement is to limit global warming to below 2 °C (preferably to 1.5 °C), compared to pre-industrial levels, which would reduce the impacts of climate change by the end of the twenty-first century (Scott et al., 2016a). The Paris Agreement involves the need for economy-wide emissions reductions, but international air travel, a major element in global tourism, is not included in the Agreement. Rather, it is based on a self-regulatory approach (Scott et al., 2016b).

Community-based limits are defined by participatory methods and collaborative governance approaches with local stakeholders (see Holden, 2006; Saarinen, 2019). In some respects, it represents an approach to overcome the problematic relationships between protection (resource-based) and use values (industry-based) in destination governance by utilising participatory planning (Scheyvens, 1999). This creates a broader stakeholder context for sustainable tourism planning and development in which suitable limits to growth are defined collectively based on negotiations and participation. In the past decade, community-based limits to growth have been studied in the context of overtourism, which is said to reflect a crisis of how the visitor flows and host-guest relations have been structured in the pre-COVID-19 time (Milano, Novelli and Cheer, 2019). Although it partly refers to 1960s and 1970s social carrying capacity studies, the issue is not simply about the scale of tourism mobilities or overcrowding. In addition to having many visitors in certain sites and destinations at the same time, overtourism involves the ways in which the 'industry' works by utilising the sharing economy and how that process changes urban spaces, housing markets and neighbourhoods where local people live their everyday lives.

Recently, the community-based approach has evolved towards inclusive tourism approaches that refer to "growth coupled with equal opportunities" (Rauniyar & Kanbur, 2010: 457). Jeyacheya and Hampton (2020: 2) have defined inclusive growth as a (sustainable) development process that aims "to broaden access, rights and participation of the majority of the population to equally prosper from, and contribute to economic growth through productive employment, rising incomes and living standards." Currently, these kinds of inclusive and transformative aims are very evident in research on the relationship between tourism and the SDGs.

Tourism and Sustainable Development Goals

In recent years, the capacity of tourism to work for sustainable development has been highlighted in connection with the United Nations Sustainable Development Goals (SDGs) of the 2030 Agenda

Table 18.2 Sustainable Development Goals (SDGs) (United Nations, 2015a)

Goal 1: No Poverty: Economic growth must be inclusive to provide sustainable jobs and promote equality.

Goal 2: Zero Hunger: The food and agriculture sectors offer key solutions for development, and are central for hunger and poverty eradication.

Goal 3: Good Health and Well-being: Ensuring healthy lives and promoting well-being for all at all ages is essential to sustainable development.

Goal 4: Quality Education: Obtaining a quality education is the foundation to improving people's lives and sustainable development.

Goal 5: Gender Equality: Gender equality is not only a fundamental human right, but a necessary foundation for a peaceful, prosperous and sustainable world.

Goal 6: Clean Water and Sanitation: Clean, accessible water for all is an essential part of the world we want to live in.

Goal 7: Affordable and Clean Energy: Energy is central to nearly every major challenge and opportunity.

Goal 8: Decent Work and Economic Growth: Sustainable economic growth will require societies to create the conditions that allow people to have quality jobs.

Goal 9: Industry, Innovation and Infrastructure: Investments in infrastructure are crucial to achieving sustainable development.

Goal 10: Reduced Inequality: To reduce inequalities, policies should be universal in principle, paying attention to the needs of disadvantaged and marginalised populations.

Goal 11: Sustainable Cities and Communities: There needs to be a future in which cities provide opportunities for all, with access to basic services, energy, housing, transportation and more.

Goal 12: Responsible Consumption and Production: There needs to be responsible production and consumption.

Goal 13: Climate Action: Climate change is a global challenge that affects everyone, everywhere.

Goal 14: Life Below Water: Careful management of this essential global resource is a key feature of a sustainable future.

Goal 15: Life on Land: Sustainably manage forests, combat desertification, halt and reverse land degradation, halt biodiversity loss.

Goal 16: Peace and Justice Strong Institutions: Access to justice for all, and building effective, accountable institutions at all levels.

Goal 17: Partnerships to Achieve the Goal: Revitalise the global partnership for sustainable development.

---

for Sustainable Development (see Hughes & Scheyvens, 2016). This connection underlines the responsibility of tourism, as one of the world's largest industries, to contribute to human development and sustainability on a global scale (Saarinen, 2020). The SDGs were ratified in 2015 (United Nations, 2015a) and they define the agenda for global development to 2030 by addressing the key challenges faced by humanity such as poverty, inequality, climate change, environmental degradation, prosperity, peace and justice. There are 17 goals (Table 18.2) and 169 specific targets, and tourism is explicitly mentioned in relation to SDG8, SDG12 and SDG14. However, tourism should have the capacity to contribute to the SDGs more widely (Scheyvens, 2018): as the UNWTO (2017: 2) has stated:

> tourism is a vital instrument for the achievement of the 17 SDGs and beyond as it can stimulate inclusive economic growth, create jobs, attract investment, fight poverty, enhance the livelihood of local communities, promote the empowerment of women and youth, protect cultural heritage, preserve terrestrial and marine ecosystems and biodiversity, support the fight against climate change, and ultimately contribute to the necessary transition of societies towards greater sustainability.

Indeed, the SDGs and their targets provide a variety of research topics and issues for tourism geographies. Based on the SDGs, the World Bank Group (2017) has developed a list of 20 reasons why tourism works for sustainable development. The long list involves five core pillars: sustainable economic growth; social inclusiveness, employment and poverty reduction; resource efficiency, environmental protection and climate; cultural values, diversity and heritage; and mutual understanding, peace and security. Each pillar consists of several specific reasons why tourism works for sustainable development. For example, the first sustainable economic growth pillar includes reasonings that tourism

(1) stimulates GDP growth;
(2) increases international trade;
(3) boosts international investment;
(4) drives infrastructure development;
(5) supports low-income economies.

In the context of sustainable development thinking beyond the industry-based approach, however, this list of reasoning becomes problematic for research on tourism geographies focusing on the costs and benefits of tourism (Saarinen, 2019). Indeed, most of these reasons are based on the emphasis on growth that may be in conflict with some of the SDGs (e.g., SDG13: climate action). Because of these kinds of contradictions, the research community has been somewhat critical of the prospects of the tourism industry and its potential role in the wider framework of sustainable development that also involves socio-ecological resources, and ultimately the system level (see Bianchi, 2018). Indeed, the growth imperative of the industry, referring to GDP, international trade, investments and infrastructure, for example, will not necessarily lead to sustainable development for environments and communities where tourism takes place (Schilcher, 2007).

Conclusions: A Way Ahead

Sustainable development has become a leading paradigm in tourism policy and planning discourses. Sustainable tourism development aims to exclude practices and processes that are undesirable ones for destinations in relation to the use of natural and cultural resources. In the past three decades, this approach has had a great impact on geographical research on tourism. Growth may have gradually replaced development in sustainable tourism thinking, emphasising the industry-based perspective in sustaining tourism (see Burns, 1999). Currently, however, the SDGs provide a guiding framework for a revival of sustainable tourism geographies that would involve the original core of sustainable development referring to human wellbeing (Saarinen, 2020, 2021a). Related to this, Scheyvens (2018: 341) has called for tourism geographers "to consider how we might utilise the SDGs to analyse the linkages between tourism and sustainable development in a wide range of contexts and at different scales." Although empirical research on the SDGs is still emerging, a key current and future task for research in tourism geographies is a need to find solutions for sustainable tourism development in different places and on different scales.

Therefore, it is crucial for tourism geographers to influence tourism development policies and models with their research, so that the policies would genuinely meet the needs of sustainable development in practice. In the context of climate change policies, related regulations and carbon neutrality targets in the coming decades, this highlights a need for more limited focus on growth aspects, such as international tourist numbers, foreign direct investments or numbers of jobs in tourism development. While tourism has a great capacity to provide employment, for example, we should also analyse what kinds of jobs tourism creates and for whom (SDGs 8 and 10), and who is included or excluded in development based on expanding tourism operations and foreign investments (SDGs 5 and 10) (Saarinen, 2020).

This calls for sustainable tourism geographies that are both critical and context sensitive. While critical acknowledgement of the past and current vagueness of the sustainable development concept for academic research is essential, there is also a need to realise that sustainability thinking is very widely used and incorporated into tourism policy-making, planning and development models on various scales. As these models increasingly impact local and global realities, we need tourism geographies, actions and practices that contribute to the development of sustainable tourism thinking by leaning on critical structural and realist analysis and transformative tourism thinking. This would stimulate sustainable tourism geographies and development that has a capacity to contribute positively to the quality of life, well-being of people and the environment in future.

References

Bianchi, R. (2018). The political economy of tourism development: A critical review. *Annals of Tourism Research, 70*, 88–102.

Burns, P. (1999). Paradoxes in planning tourism elitism or brutalism? *Annals of Tourism Research, 26*(2), 329–348.

Butler, R. W. (1991). Tourism, environment, and sustainable development. *Environmental Conservation, 18*(3), 201–209.

Butler, R. W. (1996). The concept of carrying capacity for tourism destinations: Dead or merely buried? *Progress in Tourism and Hospitality Research, 2*(3-4), 283–293.

Butler, R. W. (1998). Sustainable tourism-looking backwards in order to progress?. In C. M. Hall, & A. A. Lew (Eds.), *Sustainable Tourism: A Geographical Perspective* (pp. 25–3). Longman.

Butler, R. W. (1999). Sustainable tourism: A state-of-the-art review. *Tourism Geographies, 1*(1), 7–25.

Butler, R. W. (2010). Carrying capacity in tourism. In D. Pearce, & R. Butler (Eds.), *Tourism Research: A 20–20 Vision* (pp. 53–64). Goodfellow Publishers.

Clarke, J. (1997). A framework of approaches to sustainable tourism. *Journal of Sustainable Tourism, 5*(3), 224–233.

Crober, U. (2015). The discovery of sustainability: A genealogy of a term. In J. C. Enders & M. Remig (Eds.), *Theories of sustainable development* (pp. 6–15). Routledge.

Daly, H. E. (1990). Sustainable growth: An impossibility theorem. *Development, 3/4*,

Enders, J. C., & Remig, M. (2015). Theories of sustainable development: An introduction. In J. Enders, & M. Remig (Eds.), *Theories of Sustainable Development* (pp. 1–5). Routledge.

Fennell, D. A. (2022). The tourism knowledge translation framework: Bridging the canyon between theory and practice. *Current Issues in Tourism, 25*(5), 674–691.

Fletcher, R., Murray, I., Blanco R. A., & Blázquez-Salom, M. (Eds.) (2020). *Tourism and Degrowth: Towards a Truly Sustainable Tourism.* Routledge.

Getz, D. (1983). Capacity to absorb tourism: Concepts and implications for strategic planning. *Annals of Tourism Research, 10*(2), 239–263.

Gössling, S., Scott, D., & Hall, C. M. (2020). Pandemics, tourism and global change: A rapid assessment of COVID-19. *Journal of Sustainable Tourism, 29*(1), 1–20.

Hall, C. M. (1998). Historical antecedents of sustainable development and ecotourism: New labels on old bottles? In C. M. Hall, & A. A. Lew (Eds.), *Sustainable Tourism: A Geographical Perspective* (pp. 13–24). Longman.

Hall, C. M. (2009). Degrowing tourism: Décroissance, sustainable consumption and steady-state tourism. *Anatolia, 20*(1), 46–61.

Hall, C. M. (2011). Policy learning and policy failure in sustainable tourism governance: From first- and second-order to third-order change? *Journal of Sustainable Tourism, 19*(4–5), 649–671. https://doi.org/10.1080/09669582.2011.555555

Hall, C. M., & Boyd, S. (Eds.) (2005). *Nature-based Tourism in Peripheral Areas: Development or Disaster?,* Channelview Publications.

Hall, C. M., & Lew, A. A. (Eds.) (1998). *Sustainable Tourism: A Geographical Perspective*. Longman.

Higgins-Desbiolles, F., Carnicelli, S., Krolikowski, C., Wijesinghe, G., & Boluk, K. (2019). Degrowing tourism: Rethinking tourism. *Journal of Sustainable Tourism, 27*(12), 1926–1944.

Holden, A. (2006). *Environment and Tourism*. Routledge.

Hughes, E., & Scheyvens, R. (2016). Corporate social responsibility in tourism post-2015: A development first approach. *Tourism Geographies, 18*(5), 469–482.

IUCN (International Union for Conserving Nature) (1980). *World Conservation Strategy*. IUCN.

Jessop, B. (2004). Hollowing out the 'nation-state'and multi-level governance. In P. Kenneth (Ed.), *A Handbook of Comparative Social Policy* (pp. 11–25). Edward Elgar Publishing.

Jeyacheya, J., & Hampton, M. P. (2020). Wishful thinking or wise policy? Theorising tourism-led inclusive growth: Supply chains and host communities. *World Development, 131*, 104960.

Kasim, A., Gursoy, D., Okumus, F., & Wong, A. (2014). The importance of water management in hotels: A framework for sustainability through innovation. *Journal of Sustainable Tourism, 22*(7), 1090–1107.

Lucas, R. C. (1964). Wilderness perception and use: The example of the Boundary Waters Canoe Area. *Natural Resources Journal, 3*(3), 394–411.

McCarthy, J., & Hague, E. (2004). Race, nation, and nature: The cultural politics of "Celtic" identification in the American West. *Annals of the Association of American Geographers, 94*(2), 387–408.

McCool, S. F., Moisey, R. N., & Nickerson, N. P. (2001). What should tourism sustain? The disconnect with industry perceptions of useful indicators. *Journal of Travel research, 40*(2), 124–131.

McMurry, K. C. (1930). The use of land for recreation. *Annals of the Association of American Geographers, 20*(1), 7–20.

Meadows, D., H., Meadows, D. L., Randers, J., & Behrens, W. W. (1972). *The Limits to Growth: A Report for the Club Rome's Project on the Predicament of Mankind*. Earth Island Limited.

Milano, C., Novelli, M., & Cheer, J. M. (2019). Overtourism and tourismphobia: A journey through four decades of tourism development, planning and local concerns. *Tourism Planning & Development, 16*(4), 353–357.

Mosedale, J. (2014). Political economy of tourism. In A. A. Lew, C. M. Hall, & A. M. Williams (Eds.), *The Wiley Blackwell Companion to Tourism* (pp. 55–65). Wiley.

Nash, R. (1967). *Wilderness and the American Mind*. Yale University Press.

Nepal, S. K. (2000). Tourism in protected areas: The Nepalese Himalaya. *Annals of Tourism Research, 27*(3), 661–681.

Rauniyar, G., & Kanbur, R. (2010). Inclusive growth and inclusive development: A review and synthesis of Asian Development Bank literature. *Journal of the Asia Pacific Economy, 15*(4), 455–469.

Redclift, M. (1992). The meaning of sustainable development. *Geoforum, 23*(3), 395–403.

Redclift, M. (2005). Sustainable development (1987–2005): An oxymoron comes of age. *Sustainable Development, 13*(4), 212–227.

Ruggerio, C. A. (2021). Sustainability and sustainable development: A review of principles and definitions. *Science of the Total Environment, 786*, 147481.

Saarinen, J. (2006). Traditions of sustainability in tourism studies. *Annals of Tourism Research, 33*(4), 1121–1140.

Saarinen, J. (2018). Beyond growth thinking: The need to revisit sustainable development in tourism. *Tourism Geographies, 20*(2), 337–340.

Saarinen, J. (2019). Communities and sustainable tourism development: Community impacts and local benefit creation in tourism. In S. F. McCool, & K. Bosak (Eds.), *A Research Agenda for Sustainable Tourism* (pp. 206–222). Edward Elgar Publishing.

Saarinen, J. (2020). Tourism and sustainable development goals: Research on sustainable tourism geographies. In J. Saarinen (Ed.), *Tourism and Sustainable Development Goals* (pp. 1–10). Routledge.

Saarinen, J. (2021a). Is being responsible sustainable in tourism? Connections and critical differences. *Sustainability, 13*(12), 6599.

Saarinen, J. (2021b). Sustainable growth in tourism?: Rethinking and resetting sustainable tourism for development. In C. M. Hall, L. Lundmark, & J. J. Zhang (Eds.), *Degrowth and Tourism: New Perspectives on Tourism Entrepreneurship, Destinations and Policy* (pp. 135–151). Routledge.

Saarinen, J., & Gill, A. M. (2019). Tourism, resilience, and governance strategies in the transition towards sustainability. In J. Saarinen, & A. M. Gill (Eds.), *Resilient Destinations: Governance Strategies in the Transition Towards Sustainability in Tourism* (pp. 15–33). Routledge.

Saarinen, J., & Rogerson, C. M. (2014). Tourism and the Millennium development goals: Perspectives beyond 2015. *Tourism Geographies, 16*(1), 23–30.

Scheyvens, R. (1999). Ecotourism and the empowerment of local communities. *Tourism Management, 20*(2), 245–249.

Scheyvens, R. (2018). Linking tourism to the sustainable development goals: A geographical perspective. *Tourism Geographies, 20*(2), 341–342.

Schilcher, D. (2007). Growth versus equity: The continuum of pro-poor tourism and neoliberal governance. *Current Issues in Tourism, 10*(2–3), 166–193.

Scott, D., Hall, C. M., & Gössling, S. (2016a). A review of the IPCC Fifth Assessment and implications for tourism sector climate resilience and decarbonization. *Journal of Sustainable Tourism, 24*(1), 8–30.

Scott, D., Hall, C. M., & Gössling, S. (2016b). A report on the Paris Climate Change Agreement and its implications for tourism: Why we will always have Paris. *Journal of Sustainable Tourism, 24*(7), 933–948.

Sharpley, R. (2009). *Tourism Development and the Environment: Beyond Sustainability.* Earthscan.

Swarbrooke, J. (1999). *Sustainable Tourism Management.* CAB International.

UNEP (2005). *Making Tourism More Sustainable – A Guide for Policy Makers.* United Nations Environment Programme.

United Nations (2015a). *Transforming our World: The 2030 Agenda for Sustainable Development.* Resolution adopted by the General Assembly on 25 September 2015. United Nations.

United Nations (2015b). Paris Agreement. United Nations. Available at: https://unfccc.int/sites/default/files/english_paris_agreement.pdf.

UNWTO (1993). *Sustainable Tourism Development: Guide for Local Planners.* World Tourism Organization.

UNWTO (2006). *UNWTO's Declaration on Tourism and the Millennium Goals: Harnessing Tourism for the Millennium Development Goals.* World Tourism Organization.

UNWTO (2012). Towards Inclusive & Sustainable Growth & Development: How can the Tourism Sector Contribute? World Tourism Organization. Available at: www.unwto.org/archive/global/event/towards-inclusive-sustainable-growth-development-how-can-tourism-sector-contribute.

UNWTO (2017). Chengdu declaration on 'tourism and the sustainable development goals. *UNWTO General Assembly Documents 1*, 1–4.

WCED (World Commission on Environment and Development) (1987). *Our Common Future.* Oxford University Press.

Weaver, D. (2011). Small can be beautiful, but big can be beautiful too—and complementary: Towards mass/alternative tourism synergy. *Tourism Recreation Research, 36*(2), 186–189.

World Bank Group (2017). Tourism for Development: 20 Reasons Sustainable Tourism Counts for Development. World Bank Group Knowledge Series. https://documents1.worldbank.org/curated/en/558121506324624240/pdf/119954-WP-PUBLIC-SustainableTourismDevelopment.pdf (accessed June 2024).

19
GEOGRAPHIES OF TOURISM AND THE ANTHROPOCENE

Edward H. Huijbens

Introduction

In May 2022 atmospheric concentrations of carbon dioxide (CO_2) reached 420.99 parts per million (ppm) as measured at the Mauna Loa Observatory in Hawai'i.[1] This figure on its own does not say much. However, to grasp the significance of it one needs to place it in the context of time. Only 70 years ago this figure stood at 300 ppm and has not been above that 300 for the preceding 800,000 years (Lindsey, 2020). In effect the almost 421 ppm represents an unprecedented jump in CO_2 concentrations in all of the history of humanity in the span of a few decades. Where our climate is heading with such rapid and unprecedented increase in atmospheric CO_2 is altogether uncertain and the effects this may have on our lives and the lives of other living beings is uncharted terrain (Wakefield, 2020). In this context, it is imperative to interrogate the 'promise of tourism' as it manifests now and how it could look into the future.

The rapid post-war increase of CO_2 in our atmosphere is but one 'Earth System' indicator of many, which have assumed a 'hockey stick' shape since the 1950s. Steffen, et al. (2015) show how these correlate with post-war socio-economic trends whereby, among other things, international tourism, population, GDP, water and energy use have all assumed the 'hockey stick' shape of exponential increase. This they label the 'Great Acceleration' characterising global post-war economies. The graph they generated explicitly shows how these correlate with Earth System trends, which are being decidedly thrown out of kilter at our behest. This correlation has prompted a proposal that a new geological epoch has started, superseding the Holocene, the name of the geological epoch since the end of the last massive glaciation of the Ice Age. The uncharted terrain humanity is entering has been labelled after humankind itself, the Anthropos. This is the 'Anthropocene' representing a move beyond the geologic conditions characteristic of the 10,000 to 12,000 years since the end of the last glacial epoch, and the shift into a novel and perilous epoch whose signature is irreversible human impact on Earth system and all life processes.

This title for a 'geology of mankind' was popularised by the Nobel Prize-winning scientist Paul Crutzen (2002) and originally presented in a paper entitled 'The Anthropocene' (Crutzen & Stoermer, 2000). Many have further speculated on its meaning, but more importantly on which pivotal event this human epoch in geology can be hinged upon. The proposals are numerous in a geological sense (see Biello, 2016: 44–46; Chwałczyk, 2020; Faber, 2018: 52–54) but also

DOI: 10.4324/9781003286301-23

conceptually (see Castree, 2014). Not wanting to address here geological debates around stratigraphic markers, the most appealing of the conceptual proposals is that of Lewis and Maslin (2018). According to them, humankind's environmental impact reached an observable and measurable level at the global scale when Europeans colonised the Americas. More specifically, they pin the onset of the Anthropocene at what they call the 'Orbis spike' in 1610. This spike refers to a dramatic dip in atmospheric carbon dioxide levels in core samples of Antarctic ice. This was a measured dip of 7 ppm, incidentally on a par with the measured increase in CO_2 concentrations from May 2019 (414 ppm) to May 2022 (421 ppm). Lewis and Maslin credit the Orbis spike with being a contributing factor to the onset of the 'Little Ice Age' of the late medieval period. Although not recognised geologically as an Earth System shift (see Zalasiewicz et al. 2019), this Orbis spike came about when indigenous agricultural land reverted to forest in the Americas as the sedentary societies of the indigenous populations were decimated in the wake of the European colonisation, starting in 1492.

The colonial enterprise and the way it set in motion our modern-day economic system, global relations and entrenched inequalities, make this 1610 marker particularly appealing to the study of tourism geographies. Tourism as a global industry is largely premised on the power geometries, transport systems and wealth distribution that this era brought about (Britton, 1991). As a counterpoint, Paul Crutzen (2006) initially assumed that the Anthropocene started with the invention and deployment of Watt's steam engine in 1784 and thereby the emergence of the Industrial Revolution (see also Morton, 2013: 4 and Young, 2017: 85). But with this marker, a long story explaining why the steam engine was deemed necessary is glossed over. The fact that the well-known propelling power of steam only caught on in late eighteenth century Britain indeed ties back to the colonial enterprise, the triangular sugar-slave trade, enclosures and privatisation of land and a host of other relational transformations that led to the necessity for increased labour productivity there and then, to which the steam engine lent itself neatly (Malm & Hornburg 2014). Similarly, a recognition of the Anthropocene in tourism geographies will eventually have to find its truth-value in terms of its usefulness in negotiating and navigating tourism futures in the context in which they have been conditioned by the present. By now, and in the Anthropocene, the shifting Earth System of a becoming Earth needs to be incorporated into the power geometries.

All in all, this chapter builds on the premise that humanity is entering a new climatic regime (Latour, 2018) and simultaneously precipitating an emerging sixth mass extinction of rampant biodiversity loss (Kolbert, 2014). This shift profoundly challenges the ways in which things have been done. As explained by Zalasiewicz et al. (2019):

> Throughout recent history, an underlying stable condition of the Earth System has been taken as a given. This is the premise upon which our legal and political structures have been created over the past several centuries. [...] [T]here has been an implicit assumption that current conditions form an objective and unchanging reality that has surrounded us since time immemorial.

> (p. 36).

In the Anthropocene, the current mode of economy and society is wholly untenable, and the future is profoundly uncertain in the hands of a becoming Earth. As the timescales of geological and human history coalesce in a geo-story, "[w]ith the arrival of the Anthropocene we must now be suspicious of all ideas developed in the last 10,000 years" (Hamilton, 2014: 1). These are indeed times of climate and biodiversity emergency, as made abundantly clear in the sixth report of the Intergovernmental Panel on Climate Change (IPCC, 2021). Yet "[t]hrough the term 'crisis,' the singularity of events is abstracted by a generic logic, making crisis a term that seems self-explanatory" (Roitman, 2014: 3). That, it is most certainly not. The climate and ecological emergency of the Anthropocene is a crisis of our making, but not all of humanity can be held equally accountable. This planetary crisis manifests in particular spaces and places in particular ways and thereby a geography of tourism can contribute to the unravelling of one particular element of the current predicament resulting from the geological force of humans. The focus of the chapter will be on how, what and where current ideas mean for tourism and the current mode it operates. The chapter concludes with some indications of a tourism research agenda which can live up to the 'promise of tourism' (Huijbens, 2021a), not contributing to the planetary climatic and biodiversity crisis that is now unfolding.

The Madness of Our Economic Reason

To most, the 'promise of tourism' entails growth, prosperity and opportunity; at least this is how tourism is invariably promoted in policy and rhetoric worldwide. This amounts to a profoundly anthropocentric take on tourism as explained by Holden (2012) in the first edition of this Handbook. In the global scheme of things, this is neatly captured in the United Nations (UN) 2030 Agenda for Sustainable Development and the associated goals. For tourism the particular goals of economic growth, responsible production and consumption and life below water have been highlighted, but in actual fact, according to the World Tourism Organization and UN Development Programme, "[t]ourism industries play a vital role in all 17 SDGs [Sustainable Development Goals]" (2017: 49). Similarly the World Bank has jumped on the bandwagon celebrating tourism as a tool for development highlighting how it contributes to economic growth, employment and poverty reduction, efficient use of resources, valuing diversity and contributing to peace building. Tourism is indeed high on the agenda, if only allowed to grow and conform to market logic (Bianchi & de Man, 2021). At the same time tourism has been taken up by the eminent Marxist geographer David Harvey (2017, quoting Derrida), as a prime example of the 'madness of our economic reasoning'. This madness is premised on an economic system of perpetual growth and expansion, spiralling at an ever accelerating rate to sustain the surplus value necessary to keep the economic engine running. Therein tourism represents to Harvey the idea of 'planned obsolescence' gone instantaneous. Planned obsolescence is generally about designing things that either quickly break, are single use and/or require constant maintenance and updates. In tourism once the experience is consumed, it is gone and the need for more can be immediately realised. Tourism thus represents the ultimate way to accelerate the turnover time of capital. In this sense, tourism contributes to sustaining surplus-value creation through two means. One contributes to the 'spatial fix' of ever growing capital formation. This can be seen manifesting in more places assuming the guise of a destination and the enclosures of landscapes and cultures for consumption. More profoundly, this cultural logic of consuming spaces and places leads to their abstraction and compartmentalisation into consumable bits, detaching people from their environments and each other. Tourism is hereby part of the spectacularisation of life and the ways in which it is becoming a 'fictitious commodity', as consumption emerges as the privileged site for the fabrication of self and society (Rossi, 2013: 1069) accelerated today by nigh on spatially ubiquitous connectivity.

The growth engine has indeed appropriated life itself, its blueprints as in DNA and genetic modification, our time; as in the need to create value from every moment, our home or couch; as in Airbnb and other 'disruptive innovations' of the Internet economy. Not that sharing is bad; it is the monetising of every aspect of everyday life that is the challenge. And as capital penetrates deeper into the cultural logic of our societies, grand-scale schemes powered by the engine of capital and growth engulf global relations. Our economic logic, global policy making and the very way individuals aspire to wellbeing in the everyday embraces the core element of capitalist functioning; growth. And year on year, tourism grows. Every year more people travel, ever further, ever more frequently, so that tourism has become emblematic of the growth engine and a celebrated key global industry. International inbound tourism is indeed a depicted 'hockey stick' shaped curve in the Great Acceleration graph. And, there is nothing to indicate that COVID-19 pandemic will change this. Signs of recovery indicate how, as in preceding tourism crisis, temporary dips can be recorded in retrospect (Gil-Alana & Huijbens, 2018), but growth will resume with a vengeance not least because of non-discriminatory blanket state subsidies to the industry during COVID – for growth's sake.

This is not the first piece of writing pointing out the correlation between Earth System impacts and the growth madness of our economic reasoning. In the edited volume on *Tourism and the Anthropocene,* Gren and Huijbens (2016) show how tourism and social science have portrayed the Earth as a passive backdrop for social life, a stage where the big actors of nature and society/culture could play out. The question posed here is: How does the new, 'becoming', Earth of the Anthropocene speak back? The Earth obviously does not speak as such. It is not your standard subject of correspondence or any type of inquiry. So reframed and drawing on the work of Gren and Huijbens (2012): How does the Earth become a dynamic and foregrounded matter of fact and concern in the Anthropocene? And how can we develop Earth-led priorities and perspectives under these conditions?

Moving from Madness

Here, geographies of tourism can provide more nuance, lest the grand standing of global agendas is to be replicated. The Earth, as this lonely planet in space, humanity's only known planet that can sustain us, is not doing so in its entirety and life will go on. Thereby there is a need to move away from the hubris of all-encompassing solutions and concepts. Doing away with what Erik Swyngedouw (2019) calls

> a global intellectual and professional technocracy [which] has spurred a frantic search for 'smart', 'sustainable', 'resilient', and/or 'adaptive' socio-ecological management and seeks out the socioecological qualities of eco-development, retrofitting, inclusive governance, the making of new inter-species eco-topes, geoengineering, and technologically innovative – but fundamentally market conforming – eco-design in the making of a "good".
>
> Anthropocene (253)

Premised on growth and fundamentally market conforming, but now just 'green'; Swyngedouw sees an emerging Anthropo(obs)cene. Alternatively the hockey sticks of the Great Acceleration need to be broken. It is in this context that Bruno Latour (2018) tries to call us 'Down to Earth' "while observing, with anguish, that the universal condition today entails living in the ruins of modernization, groping for a dwelling place" (p. 106).

Indeed, ours is merely an existence on a thin crust of geosciences referred to as 'the Critical Zone' (CZ); a term coined by the US National Research Council (2001). The CZ puts the focus squarely on the conditions of life and is recognised as an entity composed of co-evolving systems that create the structured dynamic skin of the Earth (Brantley et al. 2017: 852, 856). A few kilometres down and a few up, relative to the Earth's surface at median sea level, is where all terrestrial life exists, including humanity and tourism. It is in this thin varnish, the skin of the Earth, where life produces and maintains itself. It is in these skin parts of humanity have scarred with the ruins of modernisation and growth premised on the madness of capitalist economic reasoning.

The point is that the CZ offers one alternative to big concepts like Earth System, the Anthropocene and Humanity. As necessary as it may be to try to 'save the planet', to keep the 'Earth System' in a 'safe operating space for humanity', or to focus on the 'planetary scale', these are still operating on a level that is too remote from earthly human geographical practice. It is arguably in the CZ where the Earth can possibly be discerned and denoted. It follows that the CZ becomes an earthly entity that can and needs to be politically and spatially re-composed.

Groping for dwelling amidst the ruins of the post-war dreams and aspirations of growth and progress calls for facing the ruins made, call them for what they are and start looking for life and a future in these ruins as the CZ is recomposed. In concluding her book on the *Mushroom at the End of the World* Anna Tsing (2015) explains:

> The ruin glares at us with the horror of its abandonment. It's not easy to know how to make a life, much less avert planetary destruction. Luckily there is still company, human and not human. We can still explore the overgrown verges of our blasted landscapes – the edges of capitalist discipline, scalability, and abandoned resource plantations. We can still catch the scent of the latent commons – and the elusive autumn aroma.
>
> (p. 282)

The 'latent commons' and the 'elusive aroma' that hold future potential can help in looking beyond the madness. The future can thereby still be gleaned through the ruin and images of catastrophe. Although it is not a pretty sight, it is one that needs to be reckoned with. These are 'geo-social formations' as presented by Clark and Yusoff (2017) and can be recognised as

> becomings of earth and society together [which] might help us probe the richly layered formations we have inherited for the overlooked, marginalised or as yet unactualized geosocial possibilities murmuring within them.
>
> (p. 6)

Grounding Tourism

People impact the Earth System and it impacts in return, in particular ways in particular places at particular times. Scaled up, or zoomed out, it may seem that humanity as an entity is driving up atmospheric CO_2 concentrations, but zoom in and who is actually contributing, how and where can be revealed; with wildly different effects the world over. So there is an imperative to zoom in in order to know what to do, and how to act (Gren, 2017) and therein the murmurings of unactualised geosocial possibilities can be discerned. Going all the way 'down to Earth' there is namely good company, as Anna Tsing reminds us. So many things taken for granted and assumed inert have a role to play. My personal favourite are rocks. The geologist Marcia Bjornerud (2018: 8) described them as "not nouns but verbs – visible evidence of processes", A focus on deep geologic time responsible for rock formation helps in appreciating just how the Earth's temporal flows come together. Rocks hold 'deep time', affording a connection deep into the mist of time and at the same time, deep into the crust of the Earth as explored in a tourism context by Smith, Speiran and Graham (2021). Indeed Bjornerud wants to take time into account when thinking about the Earth and its systems in crisis. Adopting 'timefulness' to appreciate how all things come to pass at their own time, in their own rhythm.

So what about tourism? That instant gratification through experience which Harvey sees as one of the prime motors of the global economy at present and a prime example of the madness of our economic reasoning? Rocks the world over have been madly spectacularised as a view to be snapped, uploaded and shared with the tagline 'been there, done that'. They have generated many a selfie moment. A quick sweep of social media posts will mostly confirm that, but there is more. As opposed to seeing them as static entities to be inscribed with consumptive meaning, much like the becoming Earth of the Anthropocene, rocks can be conceived as so much more than spectacularised ends. As Barua (2016) argues, they can seen as

> becomings, nuptials. They are movements, lines, flows of differing speeds and durations. An encounter [with these] poses problems; it reconfigures identities, space, political economies.
>
> (p. 265)

The rhythms of rock formation are so different, yet part and parcel of this Earth System that makes for us as we make for it. Encountering rocks holds a particular transformative allure which has most certainly been made use of throughout history. Rocks

> invite speculation – demand it, even – but they offer few answers to the questions they pose. [...] Only material remains, summoning archaeopoetics from archaeologists, just as it summons geopoetics from geologists, and poetic poetics from poets, all gathering up the millenia to apprehend life from these and other stones.
>
> (Raffles, 2020: 57–58, partially citing Papapretos, 2012)

Apprehending life from these and other rocks, Raffles says. Indeed, rocks do not speak, no more than the Earth does. They do not provide answers. But as foregrounded matters of fact and concern, these will always be with you as you relate; and in that relating

> a virtual differential field of bioneuro-cultural processes insisting in different existing actual assemblages of prosocial politically-inflected affective cognition. Each person is an assemblage differently incarnating that pattern.
>
> (Protevi, 2019: 134)

Much like the Earth System, we are a system in relation and certain equilibrium. We mostly know what can upset our delicate balance, but we are only fumbling to understand how that extends to the Earth System as a whole.

But the recognition of our extended being is not transcendental. We can and should assign agency to all things material and immaterial and see ourselves as nodes in the endless actor-network of being, but at the end of the day it is us, through our deliberations, stories we tell, myths we create and meaning we make that are making for the Earth as it makes us. Büscher (2022) reminds us that acknowledging certain forms of human exceptionalism is no impediment to convivial more-than-human relations. Regardless of what he labels 'the non-human turn',

> we must place ourselves firmly within the dialectical tension that the co-constitution of nature and society represents, which is always an epistemological, analytical, and political balancing act that responds to forces of power and other (inter)relationships.
>
> (p. 5)

In other words, precisely because we make for the Earth and it makes for us, we need to distinguish between the different elements at play and meaningfully understand the relations that constitute their interrelationship. Foregrounding the Earth, the key questions to emerge then are how, why and where we are attached to this becoming Earth of the Anthropocene?

And here is where I see the future possibilities of tourism and ways in which we can reshape it from growth and grand schemes of fixing things. Through travel and the encounter we can enact the responsibility Robyn Wall-Kimmerer (2013) calls for, and

> find ways to enter into reciprocity with the more-than-human world. We can do it through gratitude, through ceremony, through land stewardship, science, art, and in everyday acts of practical reverence.
>
> (p. 190)

This is a responsibility, not only to the more-than-human world, but also to that of others. This is the true 'promise of tourism' in my mind. An idea of conviviality resting on "the gift and surprise that is the Other can only wander in when that space is open" (Farage, 2013: 46). In an earthly context and in times of the Anthropocene, tourism geographies are about exploring the spatialities of this reciprocity and openness to the other. It is about

> a nuanced "geo-logisation" of hospitality ethics which transgresses issues of equity (e.g. of access and participation), professionalism, or other human centred ethics approaches (Fennell, 2006; 2012; Lovelock & Lovelock, 2013). The hospitality geo-ethics we envision would incorporate geo-resources at its core, their origins and usage, and their co-option into value creation in tourism as geo-ethical decisions about tourism activities in one "EarthWorld".
>
> (Gren & Huijbens, 2016: 194)

Developing responsibilities, attuning to more than human rhythms of life and engaging with affordances and encounters with the earthly elements help us 'narrate the mesh' (Caracciolo, 2021), informing a minimal geo-ethics for the Anthropocene (Zylinska, 2014).

Conclusion

A new Earth politics emerges from our earthly attachments. Faced with the becoming Earth of the Anthropocene and the climate and ecological emergency it entails, our primary task is to listen and comprehend these attachments and allow them to emerge as foregrounded matters of fact and concern in the making of spaces and places. The Anthropocene is not the time for idealised progress and hope for grand fixes to a 'broken' Earth System conceived by an abstracted humanity, or inversely the doom and gloom of the end of times.

For tourism geographies, foreclosing tourism and destinations as spectacular experiences to be ticked off the proverbial bucket list while feeding the hubris of growth cannot go on. We need to recognise ourselves as of this Earth and spawned of the relations forged through encounters with others and the more-than-human in specific corners of the critical zone of a dynamic becoming Earth. We need to wrest the image of tourism from the hands of modernising grand standers most recently manifest in the Sustainable Development Goals – meant as a generic recipe for human emancipation on a static Earth surface – and think through how tourism can become a vehicle for conviviality whereby through the individual rediscovery of everyday life and tools, we begin to imagine convivial commonwealth alternatives to industrialism, cultivated and vitalized as social challenges to industrial forms of life.

(Atasay, 2013: 63)

Our modern day economy is premised on the fictitious commodity of life and the mediating of spectacularised experiences – and tourism is emblematic thereof. We can produce endless ideas of how to remedy the global industry with blanket measures such as the SDGs, and yes, we should try doing so. But at the same time I want to highlight the stories rocks can tell us, how they can help us form meaningful attachments with a becoming Earth and thereby re-write the story of tourism as part of the critical zone of life. How can rocks, or trees, or bees help us create stories of other ways of being and doing that help us understand the becoming Earth of the Anthropocene? How can they make the everyday a spectacular experience we all have access to, even in our backyard? The stories we tell matter; the stories of growth and progress have transformed the Earth to the extent that it is talking back, and we finally recognise it as becoming one with us. What that might mean we will only be able to appreciate in a future we make for now.

So, in terms of tourism and the Anthropocene, research needs to focus on the potential spaces of conviviality, fostered and created in a range of critical zones, forming attachments to a becoming Earth. Therein tourism is about radical hospitality as opposed to formalised commodified service. How can we understand being open to the other, and also the more-than-human-other, and how do these understandings frame the stories making for Earth led priorities and perspectives? What is needed to move beyond our consumptive aspirations and ideas? How can we admit powerlessness as we shift away from the enslaving and debilitating perfectionism of trying to save the planet or humanity en masse? What shape and contours would the creative energies, desires and experimental form of everydayness take, where we allow for a multiplicity of practices, cultivating the art of paying attention to the possible, honouring divergence and alloy differences based on what people make matter (Stengers, 2015)?

We are relational beings and we need to allow for that; spawned in the context of the critical zone on and off this becoming Earth of the Anthropocene. Being open to being relational however cannot come at the cost of taking responsibility for the forcefulness of our being. Researching and understanding our relational responsibility brings us far in order to come to terms with the legacy of colonialism that spawned the Anthropocene (Lewis & Maslin, 2018). Future research into the stories we tell, and how we make sense of the world, most certainly matter. I see therefore exciting new research directions into ideas around radical hospitality, conviviality, the tourism encounter and how to move tourism beyond the growth paradigm through developing earthly attachments in the Anthropocene (Huijbens, 2021b).

Note

1   https://gml.noaa.gov/ccgg/trends/data.html

References

Atasay, E. (2013). Ivan Illich and the study of everyday life. *The International Journal of Illich Studies*, *3*(1), 56–77.
Barua, M. (2016). Encounter. *Environmental Humanities*, 7(1), 265–270.

Bianchi, R. V., & de Man, F. (2021). Tourism, inclusive growth and decent work: A political economy critique. In T. Jamal & J. Higham (Eds.) *Justice and Tourism* (pp. 220–238). Routledge.

Biello, D. (2016). *The Unnatural World: The Race to Remake Civilization in Earth's Newest Age*. Scribner.

Bjornerud, M. (2018). *Timefulness: How Thinking like a Geologist Can Help Save the World*. Princeton University Press.

Brantley, S. L., McDowell, W. H., Dietrich, W. E., White, T. S., Kumar, P., Anderson, S. P., … & Gaillardet, J. (2017). Designing a network of critical zone observatories to explore the living skin of the terrestrial Earth. *Earth Surface Dynamics*, *5*(4), 841–860.

Britton, S. (1991). Tourism, capital, and place: towards a critical geography of tourism. *Environment and planning D: society and space*, *9*(4), 451–478.

Büscher, B. (2022). The nonhuman turn: Critical reflections on alienation, entanglement and nature under capitalism. *Dialogues in Human Geography*, *12*(1), 54–73.

Chwałczyk, F. (2020). Around the Anthropocene in eighty names—Considering the Urbanocene proposition. *Sustainability*, *12*(11), 4458.

Caracciolo, M. (2021). *Narrating the Mesh: Form and Story in the Anthropocene*. University of Virginia Press.

Castree, N. (2014). The Anthropocene and geography III: Future directions. *Geography Compass*, *8*(7), 464–476.

Clark, N., & Yusoff, K. (2017). Geosocial formations and the Anthropocene. *Theory, Culture & Society*, *34*(2–3), 3–23.

Crutzen, P. J. (2002). Geology of mankind. *Nature*, *415*, 23.

Crutzen, P. J. (2006). The "Anthropocene". In E. Ehlers, & T. Krafft (Eds.), *Earth System Science in the Anthropocene* (pp. 13–18). Springer.

Crutzen, P. J., & Stoermer, E. F. (2000). The "Anthropocene". *Global Change Newsletter*, *41*, 17.

Faber, A. (2018). *De Gemaakte Planeet: Leven in het Antropoceen*. Amsterdam University Press.

Farage, S. (2013). A flame in the dark. *The International Journal of Illich Studies*, *3*(2), 45–51.

Fennell, D. A. (2006). *Tourism Ethics*. Channel View.

Fennell, D. A. (2012). Tourism ethics needs more than a surface approach. In T.V. Singh (Ed.) *Critical Debates in Tourism* (pp. 188–198). Channel View.

Gil-Alana, L. A., & Huijbens, E. H. (2018). Tourism in Iceland: Persistence and seasonality. *Annals of Tourism Research*, *68*, 20–29.

Gren, M. (2017). Mayday – a letter from the Earth. In M. Nieuwenhuis, & D. Crouch (Eds.), *The Question of Space: Interrogating the Spatial Turn between Disciplines* (pp. 167–178). Rowman & Littlefield.

Gren, M., & Huijbens, E. H. (2012). Tourism theory and the earth. *Annals of Tourism Research*, *39*(1), 155–170.

Gren, M., & Huijbens, E. (Eds.) (2016). *Tourism and the Anthropocene*. Routledge.

Hamilton, C. (2014). When Earth Juts Through World. A contribution to "The Situation Facing the Moderns after the Intrusion of Gaïa: A Philosophical Simulation" – An Inquiry into Modes of Existence (AIME) Conference, Sciences Po, Paris, July 28–29, 2014.

Harvey, D. (2017). *Marx, Capital and the Madness of Economic Reason*. Profile Book.

Holden, A. (2012). Environmental discourses and tourism. In J. WIlson (Ed.), *The Routledge Handbook of Tourism Geographies* (pp. 194–200). Routledge.

Huijbens, E. H. (2021a). Forum: Sociale afstand en de belofte van toerisme. *Vrijetijdstudies*, 39(1), 33–42.

Huijbens, E. H. (2021b). *Developing Earthly Attachments in the Anthropocene*. Routledge.

IPCC (2021). *Climate Change 2021: The Sixth Assessment Report of the Intergovernmental Panel on Climate Change*. Cambridge University Press.

Kolbert, E. (2014). *The Sixth Extinction: An Unnatural History*. Henry Holt and Company.

Latour, B. (2018). *Down to Earth: Politics in the New Climatic Regime*. Polity.

Lindsey, R. (2020). Climate Change: Atmospheric Carbon Dioxide, 2020. NOAA Climate.gov; www.climate.gov/news-features/understanding-climate/climate-change-atmospheric-carbon-dioxide, viewed 13 July 2021.

Lewis, S. L., & Maslin, M. (2018). *The Human Planet: How We Created the Anthropocene*. Yale University Press.

Lovelock, B. & Lovelock, K.M. (2013). *The Ethics of Tourism. Critical and Applied Perspectives*. Routledge.

Malm, A., & Hornborg, A. (2014). The geology of mankind? A critique of the Anthropocene narrative. *The Anthropocene Review*, *1*(1), 62–69.

Morton, T. (2013). *Hyperobjects: Philosophy and Ecology after the End of the World*. University of Minnesota Press.

National Research Council (2001). *Basic Research Opportunities in Earth Sciences*. National Academy Press.

Protevi, J. (2019). States of nature: Geographical aspects of current theories of human evolution. *Political Geography, 70*, 127–136.

Raffles, H. (2020). *The Book of Unconformities: Speculations on Lost Time*. Pantheon Books.

Roitman, J. (2014). *Anti Crisis*. Duke University Press.

Rossi, U. (2013). On life as a fictitious commodity: Cities and the biopolitics of late neoliberalism. *International Journal of Urban and Regional Research, 37*(3), 1067–1074.

Smith, M., Speiran, S., & Graham, P. (2021). Megaliths, material engagement, and the atmospherics of neo-lithic ethics: Presage for the end (s) of tourism. *Journal of Sustainable Tourism, 29*(2/3), 337–352.

Steffen, W., Broadgate, W., Deutsch, L., Gaffney, O., & Ludwig, C. (2015). The trajectory of the Anthropocene: The great acceleration. *The Anthropocene Review, 2*(1), 81–98.

Stengers, I. (2015). *In Catastrophic Times. Resisting the Coming Barbarism*. Open Humanities Press.

Swyngedouw, E. (2019). "The Anthropo(Obs)cene." In Antipode Editorial Collective (Ed.), *Keywords in Radical Geography: Antipode at 50* (pp. 253–258). Wiley Blackwell.

Tsing, A. L. (2015). *The Mushroom at the End of the World: On the Possibility of Life in Capitalist Ruins*. Princeton University Press.

Wakefield, S. (2020). *Anthropocene Back Loop: Experimentation in Unsafe Operating Space*. Open Humanities Press.

Wall-Kimmerer, R. (2013). *Braiding Sweetgrass: Indigenous Wisdom, Scientific Knowledge, and the Teaching of Plants*. Milkweed Editions.

Young, O. R. (2017). *Governing Complex Systems: Social Capital for the Anthropocene*. The MIT Press.

Zalasiewicz, J, Waters, C. N., Williams, M., & Summerhayes, C. P. (2019). *The Anthropocene as a Geological Time Unit*. Cambridge University Press.

Zylinska, J. (2014). *Minimal Ethics for the Anthropocene*. Open Humanities Press.

20

TOURISM AND SOCIO-ECOLOGICAL SYSTEMS

A Geographical Approach

Rannveig Ólafsdóttir and Anna Dóra Sæþórsdóttir

Introduction

Over the past three decades, there has been a growing emphasis on sustainable development as a guiding principle in tourism. Accelerated climate change and greater public awareness of the environment have further contributed to the significance of sustainability. However, there are many indications that tourism is less sustainable today than ever, and that a disproportionate focus on economic considerations largely influences the development of tourism worldwide.

Tourism is a complex system, composed of multiple elements with intertwining relationships, working on different scales. As such, the tourism system includes a wide range of stakeholders with varying economic, social, and environmental interests. Being one of the world's largest economic sectors, the tourism industry causes enormous socio-economic as well as environmental impacts and the nature and particular characteristics of tourism mean its spatial impacts are much greater than most other industries. In view of its complexity, tourism has increasingly been studied through the social-ecological system concept (SES). There is also growing recognition that understanding ecosystems and the benefits they provide to people, that is, their service to human societies, is the foundation of community wellbeing, as the three value-domains of ecosystems are ecological, socio-cultural and economic.

The underlying assumption of both social-ecological systems (SES) and ecosystem services (ES) is that analysis must be conducted from a holistic and systemic perspective, which in turn enables an integrated assessment of socioeconomic and ecological factors, their interaction and feedback. However, understanding the spatial dynamics of the tourism system, as well as gaining holistic understanding of the causal relations between the different impact factors, is still a challenging task. Some of the challenges in using SES are its limitations in dealing with important geographical factors, such as the varying range of scales, defining system boundaries and the complexities of social systems in general. The analysis of tourism in SES therefore requires a distinct, innovative approach to better deal with the challenges of sustainability.

The emphasis on space and time has long been the fundamental characteristic of geography while abstraction from complex reality in a manner which makes it more understandable is one of its core features. Hence, a geographical approach stresses the importance of being able to identify the relevant elements of a system from which an analogue is being drawn. However, excessive

DOI: 10.4324/9781003286301-24

abstraction can lead to eliminating important influencing spatial and temporal variables, thus altering the functioning of the system. The concept of holistic pluralism encompasses the complex reality of both time and space and might therefore be an ideal approach to capture the dynamic complexity characterising the tourism system. This chapter aims to discuss tourism as an ecosystem service in SES, and explore its potential for sustainable tourism management through the lens of holistic pluralism and geographical approaches that allow for the complexity characterising the tourism system/s.

Tourism Geographies and Socio-ecological Systems

Traditionally, geography is the branch of science that bridges the gap between the natural and social sciences, with the aim of increasing understanding of the complex spatial relationship between humans and nature. Ecological research, on the other hand, has often tended to exclude humans from the system while social research has similarly ignored the consequences of human activities on ecosystems. Many scholars (e.g., Folke, et al., 2005; Hall & Page, 2006; Heslinga, et al., 2017) point out that this narrow-mindedness has long characterised research into tourism, especially nature-based tourism, as the tourism industry has long been seen as a tool to promote regional economic growth and a driving force for rural development. Moreover, Hall & Page (2006) state that this narrow-mindedness within tourism research has led to an ignorance of both the environmental impact and environmental cost of tourism. Thus, tourism research has failed to understand the overall picture.

SES focus on the complexity of the human-nature relationship, by diving into the interconnectedness of social and ecological dilemmas (Everard, 2020; Ostrom 1990). The term originated around 1990, most likely as an attempt to better capture the holistic dimension of sustainable development that had emerged a few years earlier. Many definitions of the term have been put forward. Most of them reflect the interconnection between social- and ecosystems and emphasise the dynamic interdependencies among the factors in these systems (e.g., Everard, 2020; Folke, et al., 2007; Heslinga et al., 2017; Redman et al., 2004).

However, there is a need for a better understanding of these interconnections and interdependencies that span a diverse range of levels and scales. This is emphasised by many scholars (e.g., Avriel-Avnia & Dick, 2019; Everard, 2020; Fabinyi et al., 2014; Heslinga et al., 2017; Olsson & Jerneck, 2018; Partelow & Winkler, 2016; Terkenli, 2004), who point out that the interdisciplinary and systemic approach taken in SES often underestimates different values and power relations in social diversity. Furthermore, if the intention is to achieve overall system resilience, then the different views and interests of the diverse stakeholders must be taken into account. SES also have difficulty coping with important geographical factors, such as different scales, and struggle to define system boundaries and complex social systems in general.

The systemic nature of the SES approach is nevertheless important in understanding and managing complex systems (Everard, 2020). Reality is, after all, a complex system. A systemic approach usually means abstracting from complex realities to make them easier to understand. Thus, an integrated assessment of the multiple different aspects of sustainable development and the complex interrelations within them favours a systemic approach. Managing tourism in natural areas in a more integrated way is, however, not an entirely new idea. Sæþórsdóttir (2011) points out that various scholars began using systemic approaches in tourism in the 1970s, resulting in a number of systems theories of tourism. The previously and widely used mechanistic approach was recognised at the time as being problematic because it advocated analysis of the constituent elements in isolation. Systems thinking partly overcame this constraint by encouraging the analysis of connectivity and causal relationship between elements (Hall & Lew, 2009). However, as stressed by Baggio (2008), the vast complexity of the tourism system means that simple cause and effect relationships between the different components of the system hardly exist. Instead, a small stimulus can cause large and unforeseen effects or have no effect at all. Any simplification of the tourism system is thus likely to leave out important influencing factors.

To capture the complexities of SES more accurately and render it easier to use, the focus over the past decade has shifted towards meeting the need for designing a common framework, a so-called Social Ecological System Framework (SESF). Ostrom (2009) points out that from the start, the fundamental principle of SES has been to increase understanding of the processes that lead to improvements in, or deterioration of, natural resources. In 2009 she demonstrated the importance of fostering a mutual discussion among the different disciplines. Harmonising their language and knowledge would promote a better understanding of the complex relationships between the multiple SES subsystems and their internal variables. Accordingly, she placed great emphasis on the need for a harmonised classification system to facilitate such multidisciplinary efforts, stating that "without a common framework to organise findings, isolated knowledge does not cumulate" (Ostrom, 2009, 419). At the same time, Ostrom also presented a set of potentially relevant variables and sub-variables for use in data collection preparation, fieldwork and analysis of findings related to the sustainability of complex SESs.

In recent years, considerable improvements have been made to the SESF (McGinnis & Ostrom, 2014; Partelow, 2018), in response to the various voices of criticism and to make it more usable for policy settings beyond natural resources. To develop such a universal analytical tool is a critical point of departure for scholars and others seeking to understand the impact of sustainability in a complex SES. However, to the best of our knowledge, there is still a lack of a common classification system and analytical tools aimed at a deeper understanding of the complexity in tourism systems, as well as the specific roles of tourism in SES.

Tourism as an Ecosystem Service

In the wake of sustainable development and SES, the concept of ecological economics and later ES came to the fore in response to demands from both the scientific community and politicians, who stated that world ecosystems could not be separated from the economic debate in the political arena (Costanza, et al., 1991; 2017). Like SES, ES also focuses on the relationship between humans and nature. However, in order to gain a wider political audience and support, the principal focus of ES is on the value of ecosystems for human life or how ecosystems benefit humans the most (Millennium Ecosystem Assessment, 2005).

It is this emphasis in ES that has been criticised as biased towards human needs, arguing that ES promotes the exploitation of ecosystems and not the balance between exploitation and conservation (e.g., Heslinga, 2017; Potschin & Haines-Young, 2011; Schroter, et al., 2014). However, Constanza et al. (2017) disagree with this interpretation of the concept and stress that the fundamental idea of ES implies recognising that human survival and wellbeing are dependent on nature, and that the human being is an integral part of the biosphere. Accordingly, they consider that the essence of the concept clearly revolves around the importance of holistic understanding.

Millennium Ecosystem Assessment (2005) identifies four major categories of ecosystem services: *supporting services*, such as primary production, nutrient cycle and soil formation that protect human life, thus supporting all other services; *provisioning services*, such as food, freshwater, timber and fuel, which are products obtained directly from the ecosystems; *regulating services*, which are the benefits of managing ecosystem processes, such as climate management, natural hazard regulations, water purification and waste management; and *cultural services*, which include the intangible benefits people obtain from ecosystems, such as aesthetic values, spiritual enrichment, intellectual development and recreation.

Tourism may thus be seen as a valuable cultural ecosystem service (CES), the biophysical environment and ecosystem functions being the basis of numerous direct and indirect resources for tourism (Simmons, 2013). Nevertheless, many scholars (e.g., Kosanic & Petzold, 2020; Kulczyk et al., 2018; Palomo, et al., 2018) argue that the distinction between recreation, tourism and other CESs remains blurred, because, for example, the aesthetic experience is often intertwined with recreational ES, merged with elements related to the physical use of the natural environment. Consequently, the role of the CES is not yet adequately understood, especially not its role in human wellbeing, nor how the potential effects of environmental and climate change will affect different communities. However, in light of the growth of global tourism, its value as a CES is gradually increasing.

According to Partelow and Winkler (2016) all attempts to merge SESF and ES have so far been unsuccessful, due to the fact that there is still a lack of mutual learning and strategic direction in knowledge in these systems. In order to obtain a sensible and robust value for the contribution of natural resources to human and social wellbeing, Potschin and Haines-Young (2011) point out the ongoing need for a much deeper understanding of how the driving forces of change affect the services ecosystems have at their disposal. Accordingly, they stress the importance of integrating a geographical approach into the analysis of ES to better understand the mechanism that connects ecosystems to human wellbeing. They further argue that a place-based perspective is critical and that the geographical approach can aid in identifying appropriate spatial units of analysis, and likewise find ways of characterising the 'significant' function and services they deliver. Any attempt to define a set of supporting ES is therefore likely to end up oversimplifying the biophysical complexity they are based on (Partelow & Winkler, 2016; Potschin & Haines-Young, 2011).

Hence, the analytical approaches used must be interdisciplinary and take into consideration the whole range of the dynamic interaction and overlapping functions and processes that characterise the supporting ecosystem services. No single factor or process should be looked at in isolation. This is supported by Costanza et al. (2017), who suggest the use of the pluralist approach to appraise the interconnections of ES, as the relationship between processes and the functioning of ecosystems and human well being is highly complex and its various pathways are still not well understood.

Geographical and Holistic Pluralism Approach

The basic idea of both SES and ES, that is, to create a simplified social-natural system capable of providing a holistic understanding of human-environmental relations and interrelations, seems to be more problematic than initially assumed. This idea is supported by Olsson and Jerneck (2018), who state that despite decades of trying to integrate social and natural sciences, it is still deeply problematic and cannot be taken for granted. They argue that integrating ecosystems and social systems into SES creates a risk of holistic reductionism by clustering elements with opposing points of view, which otherwise would require different approaches and solutions.

To overcome the inconsistencies between the social and natural disciplines, Olsson and Jerneck (2018) propose an approach that allows for complexity: namely, holistic pluralism, originally introduced by Peterson (1979). They believe that pluralism is the best way forward for dealing with sustainability challenges, such as climate change, biodiversity loss, changes to land use and social conflicts. As a concept, pluralism denotes diversity or multiplicity, and as such reflects the view that there is more than one correct logic, supporting a system in which two or more states, groups, principles, sources of authority, etc., coexist (Cambridge English Dictionary, 2008). Hence, by placing diversity and interdependencies at its core, holistic pluralism empirically takes into account issues, historically longer-term views and theoretically broader perspectives (Olsson & Jerneck, 2018; Peterson 1979), and is therefore ideally suited to tourism.

Due to its complexity, it may be stated that tourism is truly a pluralistic phenomenon. In addition to basing its existence on the exploitation of natural and cultural resources, multi-scale settings, crossing different borders, and a multitude of diverse stakeholders, tourism destinations undergo continuous change in response to a wide range of disturbances. However, they often possess an inherent resilience to cope with these changes and reorganise themselves in ways that allow them to maintain their core functions (Becken, 2013; Farrell & Twining-Ward, 2004). Briassoulis (2017; see also Chapter 7, this volume) discusses these multiplicities of tourism destinations as assemblage and suggests three alternative conceptions based on their shifts over time and increasing complexities as regards scales, extent, boundaries, content, identity, perception of destinations and dynamics in general. The first conception is based on positivist traditions of looking at tourism destinations as containers, drawing on a geometric view of space. As such, tourism destinations are considered geographic areas with a multitude of services and tourist attractions. The second conception views tourism destinations as a unitary whole or a system.

This conceptualisation is still the most prevalent within tourism research. Similar to the first conception, it recognises the multiple elements that make up tourism destinations, but also adopts a relative view of space by discussing the relations between different elements, including both tangible and intangible elements. In the third and last conception, Briassoulis (2017) defines tourism destinations as relational places. At the heart of this conception is the notion that tourism destinations are assemblages shaped by processes and performances by both human and non-human actors. Tourism destinations thus reflect dynamic pluralism. Holistic pluralism has proven to be a good approach to raise understanding of social relationships and how they relate to different policy areas and regions (Olsson & Jerneck, 2018). Such an understanding is crucial to capturing the dynamics of tourism. Furthermore, by employing holistic pluralism as an analytical framework for tourism, Wardana (2015) points out that the associated sensitivity to various social and cultural aspects makes discussions about spatial governance both broader and more subjective, greatly reducing conflicts of interest and therefore beneficial to all.

The tourism industry faces many and constant challenges worldwide. The most recent challenge is evidently the COVID-19 pandemic, which has affected tourism from various angles. Hall, et al. (2020) outline how COVID-19 has impacted on tourism destinations, which now face challenges with regard to how to respond to the pandemic in the form of legislative and voluntary actions, as well as how to recover from it, for instance by developing promotional strategies. At the same time, the pandemic has also affected tourism demand and supply. They further argue that destinations' response to COVID-19 will in the long-run largely be shaped by the approach undertaken by competing destinations (Hall et al., 2020), demonstrating the dynamic pluralism characterising tourism destinations. Some may thus choose to use the pandemic as an opportunity to transform themselves and move towards more sustainable forms of tourism, whereas others might continue on the path of deregulation and away from sustainability.

Similarly, some destinations may again face familiar challenges, such as overtourism. Prior to the pandemic, destinations in a state of overtourism suffered from impacts affecting multiple elements, including residents, tourists, infrastructure and social and physical characteristics, thus escalating various conflicts (e.g., Jóhannesson, et al., 2022; Ólafsdóttir, et al., 2020; Sæþórsdóttir, et al. 2020). In this regard, Prayag (2020, 181) emphasises that "understanding the impacts of change on different systems is important but equally important is the capacity to manage impacts and outcomes of change".

The Long and Winding Road towards Sustainable Tourism Destinations

Despite increased emphasis on analysing tourism in SES and ES, the outcome is still far from providing tourism with a sustainable future. Both systems describe how ecosystems and human welfare are linked through some form of demand-supply relationship in a complex system with a myriad of influencing variables. A large part of the challenge seems to be that the systemic approach applied often underestimates different values and relations in social diversity. To achieve the sustainability and resilience of a system, more and more scholars now emphasise the importance of examining different views and interests among the diverse stakeholders (e.g., Avriel-Avnia & Dick, 2019; Everard, 2020; Fabinyi et al., 2014; Olsson & Jerneck, 2018; Partelow & Winkler, 2016). This is particularly important in the tourism system, where stakeholders are numerous and likewise diverse.

Equally important for tourism is a consideration of various spatial variables, but SES is still struggling with important geographical factors, such as different scales and the definition of system boundaries (e.g., Heslinga et al., 2017; Olsson & Jerneck, 2018; Terkenli, 2004). Needless to say, tourism occurs and takes place in space, meaning it requires an area in which to take place (Urry, 1990). In this respect, tourism is primarily a geographical phenomenon, and a geographical approach can thus make a major contribution to tourism research, as stressed by Hall and Page (2006). Furthermore, social constructions of space and time are shaped out of the various spatial and temporal forms (Harvey 1996; Saarinen 2004). It is therefore the form of development that is important in structuring social-natural relations (Massey, 1984).

Accordingly, to understand the overall picture, it is of vital importance to consider all the influencing factors in the tourism system and realise how they are interconnected both in time and space, as any simplification of the system is likely to leave out important influencing factors. Hence, a geographical approach based on holistic pluralism that makes all natural and social elements transparent is required to prevent the risk of holistic reductionism by classifying influencing factors and variables with opposing perspectives. Such transparency is critical to raising understanding of and analysing causal relations while also identifying key driving forces in tourism systems. A geographical approach captures both the spatial and temporal forms needed to understand and distinguish between the myriad of different causal factors characterising the complex evolution of changes in the touristscape, adding significant value to both SES and ES. Tourism geographers can moreover construe how tourism destinations evolve throughout time and space, which in turn has implications on how to best manage them in the future.

The immense growth of global tourism has created many challenges, in particular for natural areas, whose ecosystems are experiencing both social and economic strain. Yet tourism planning and development seems to be driven by neoliberal and growth-centric perspectives (Saarinen, et al., 2017). This idea is supported by many scholars (e.g., Hall, 2019; Ólafsdóttir, 2021; Sharpley, 2020), who point out that despite growing international effort towards sustainable tourism over the past 35 years, the tourism sector is still no closer to being sustainable.

This demonstrates the need for planning of tourism destinations that does not focus exclusively on the tourism industry as an economic issue, but also considers its social and environmental ramifications (Saarinen, et al., 2017) and supports the use of a geographic and holistic pluralism approach. Evidently, tourism planners and managers must abandon the status quo and instead adopt practices that take the overall picture into consideration. This is supported by Pueyo-Ros (2018), who emphasises that the understanding of tourism systems must go beyond conceptualising their activities as cultural ecosystem services and instead discuss tourism's often extractive consumption of natural and cultural resources from a holistic angle.

As a part of such holistic understanding, Saarinen, et al. (2017) also stress the importance of investigating the historical context of contemporary tourism (see also Timothy, Chapter 17). Whilst tourism research has focused predominantly on the present, historical analysis has the potential to shed light on the transformation of destinations, which in turn could provide insight into how current issues came into existence. This is supported by Mitchell and Murphy (1991), who demonstrate the importance of viewing the development of tourism destinations within an evolutionary framework, as decisions made early in the life cycle of a destination undoubtedly influence its later growth options. This is also in line with the growing body of researchers focusing on the evolutionary perspective of tourism and geography (e.g., Brouder, 2014; Brouder & Eriksson, 2013; Haraldsson & Ólafsdóttir, 2018; Ma & Hassink, 2013; Papatheodorou, 2004; Russell & Faulkner, 2004), highlighting the evolutionary process of tourism sectors, their multitude of tourism products and numerous institutions and stakeholders that interact at many levels within a destination, resulting in complex multi-level tourism co-evolution.

Tourism is indeed a complex system. But by looking at the big picture and focusing on understanding the dynamic interconnection of all the diverse spatial and temporal variables affecting the system, a better knowledge of the system's drivers is obtained. This will reduce conflicts and move tourism forward on its winding road towards sustainability.

References

Avriel-Avni, N., & Dick, J. (2019). Differing perceptions of socio-ecological systems: Insights for future transdisciplinary research. *Advances in Ecological Research, 60,* 153–190.

Baggio, R. (2008). Symptoms of complexity in a tourism system. *Tourism Analysis, 13,* 1–20.

Becken, S. (2013). Developing a framework for assessing resilience of tourism sub-systems to climatic factors. *Annals of Tourism Research, 43,* 506–528.

Briassoulis, H. (2017). Response assemblages and their socioecological fit: Conceptualizing human responses to environmental degradation. *Dialogues in Human Geography,* 7(2), 166–185. https://doi.org/10.1177/2043820617720079

Brouder, P., & Eriksson, R.E. (2013). Tourism evolution: On the synergies of tourism studies and evolutionary economic geography. *Annals of Tourism Research, 43,* 370–389.

Brouder, P. (2014). Evolutionary economic geography and tourism studies: Extant studies and future research directions. *Tourism Geographies, 6*(4), 540–545,

Cambridge English Dictionary (2008). *Cambridge Advanced Learner's Dictionary* (3rd Ed.). Cambridge University Press.

Costanza, R., Daly, H. E., & Bartholomew, J. A. (1991). Goals, agenda and policy recommendations for ecological economics. In R. Costanza (Ed.), *Ecological Economics: The Science and Management of Sustainability* (pp. 1–20). Columbia University Press.

Costanza, R., de Groot, R., Braat, L., Kubiszewski, I., Fioramonti, L., Sutton, P., Farber, S., & Grasso, M. (2017). Twenty years of ecosystem services: How far have we come and how far do we still need to go? *Ecosystem Services, 28,* 1–16.

Everard, M. (2020). Managing socio-ecological systems: Who, what and how much? The case of the Banas river, Rajasthan, India. *Current Opinion in Environmental Sustainability, 44,* 16–25.

Fabinyi, M., Evans, L., & Foale, S. J. (2014). Social-ecological systems, social diversity, and power: Insights from anthropology and political ecology. Ecology and Society, *19*(4), 28.

Farrell, B. H., & Twining-Ward, L. (2004). Reconceptualizing tourism. *Annals of Tourism Research, 31*(2), 274–295.

Folke, C., Hahn, T., Olsson, P., & Norberg, J. (2005). Adaptive governance of social-ecological system. *Annual Review of Environment and Resources, 30*(1), 441–473.

Folke, C., Pritchard, L., Berkes, F., Colding, J., & Svedin, U. (2007). The problem of fit between ecosystems and institutions: Ten years later. *Ecology and Society,* 12(1). www.jstor.org/stable/26267849

Hall, C. M. (2019). Constructing sustainable tourism development: The 2030 agenda and the managerial ecology of sustainable tourism. *Journal of Sustainable Tourism, 7,* 1044–1060.

Hall, C. M., & Lew, A. A. (2009). *Understanding and Managing Tourism Impacts: An Integrated Approach.* Routledge.

Hall, C. M., & Page, S. (2006). *The Geography of Tourism and Recreation: Environment, Place and Space.* 4th ed. Routledge.

Hall, C. M., Scott, D., & Gössling, S. (2020). Pandemics, transformations and tourism: Be careful what you wish for. *Tourism Geographies, 22*(3), 577–598.

Haraldsson, H. V., & Ólafsdóttir, R. (2018). Evolution of tourism in natural destinations and dynamic sustainable thresholds over time. *Sustainability, 10,* 4788. https://doi.org/10.3390/su10124788

Harvey, D. (1996). *Justice, Nature & the Geography of Difference.* Blackwell Publishers.

Heslinga, J. H., Groote, P., & Vanclay, F. (2017). Using a social-ecological systems perspective to understand tourism and landscape interactions in coastal areas. *Journal of Tourism Futures, 3*(1), 23–36.

Jóhannesson, G. Þ., Welling, J., Müller, D. K., Lundmark, L., Nilsson, R. O., de la Barre, S., Granås, B., Kvidal-Røvik, T., Rantala, O., Tervo-Kankare, K., & Maher, P. (2022). *Arctic Tourism in Times of Change: Uncertain Futures – From Overtourism to Re-starting Tourism.* Nordic Council of Ministers.

Kosanic, A., & Petzold, J. (2020). A systematic review of cultural ecosystem services and human wellbeing. *Ecosystem Services, 45,* 101168.

Kulczyk, S., Woźniak, E., & Derek, M. (2018). Landscape, facilities and visitors: An integrated model of recreational ecosystem services. *Ecosystem services, 31,* 491–501.

Ma, M., & Hassink, R. (2013). An evolutionary perspective on tourism area development. *Annals of Tourism Research, 41,* 89–109.

Massey, D. (1984). Introduction: Geography and society. In D. Massey, & J. Allen (Eds.), *Geography Matters!: A Reader* (pp. 161–165). Cambridge University Press.

McGinnis, M. D., & Ostrom, E. (2014). Social-ecological system framework: Initial changes and continuing challenges. *Ecology and Society, 19*(2), 30.

Millennium Ecosystem Assessment (2005). *Ecosystems and Human Well-Being: Synthesis.* Island Press.

Mitchell, L. S., & Murphy, P. E. (1991). Geography and tourism. *Annals of Tourism Research, 18,* 57–70.

Ólafsdóttir, R. (2021). The role of public participation for determining sustainability indicators for Arctic tourism. *Sustainability, 13*(1), 295.

Ólafsdóttir, R., Tuulentie, S., Hovgaard, G., Zinglersen, K. B., Svartá, M., Poulsen, H. H. & Søndergaard, M. (2020). The contradictory role of tourism in the northern peripheries: Overcrowding, overtourism and the importance of tourism for rural development. In J. McDonagh, & S. Tuulentie (Eds.), *Sharing Knowledge for Land Use Management: Decision-Making and Expertise in Europe's Northern Periphery* (pp. 86–99). Edward Elgar Publishing.

Olsson, L., & Jerneck, A. (2018). Social fields and natural system: Integrating knowledge about society and nature. *Ecology and Society, 23*(3), 26.

Ostrom, E. (1990). *Governing the Commons: The Evolution of Institutions for Collective Action.* Cambridge University Press.

Ostrom, E. (2009). A general framework for analyzing sustainability of social-ecological systems. *Science,* 325(5939), 419–422. DOI:10.1126/science.1172133

Palomo, I., Willemen, L., Drakou, E., Burkhard, B., Crossman, N., Bellamy, C., ... & Verweij, P. (2018). Practical solutions for bottlenecks in ecosystem services mapping. *One Ecosystem, 3,* e20713.

Papatheodorou, A. (2004). Exploring the evolution of tourism resorts. *Annals of Tourism Research, 31,* 219–237.

Partelow, S. (2018). A review of the social-ecological systems framework: applications, methods, modifications, and challenges. *Ecology and Society, 23*(4), 36.

Partelow, S., & Winkler, K. J. (2016). Interlinking ecosystem services and Ostrom's framework through orientation in sustainability research. *Ecology and Society, 21*(3), 27.

Peterson, R. W. (1979). Holistic pluralism and the environmental movement. *BioScience, 29,* 399.

Potschin, M. B., & Haines-Young, R. H. (2011). Ecosystem services: Exploring a geographical perspective. *Progress in Physical Geography: Earth and Environment, 35*(5), 575–594.

Prayag, G. (2020). Time for reset? Covid-19 and tourism resilience. *Tourism Review International, 24*(2), 179–184.

Pueyo-Ros, J. (2018). The role of tourism in the ecosystem services framework. *Land, 7*(3), 111.

Redman, C., Grove, J. M., & Kuby, L. H. (2004). Integrating social science into the Long-term Ecological Research (LTER) Network: Social dimensions of ecological change and ecological dimensions of social change. *Ecosystems, 7*(2), 161–171.

Russell, R., & Faulkner, B. (2004). Entrepreneurship, chaos and the tourism area lifecycle. *Annals of Tourism Research, 31*, 556–579.

Saarinen, J. (2004). 'Destinations in change': The transformation process of tourist destinations. *Tourist Studies, 4*(2), 161–179.

Saarinen, J., Rogerson, C. M., & Hall, C. M. (2017). Geographies of tourism development and planning. *Tourism Geographies, 19*(3), 307–317.

Schroter, M., van der Zanden, E. H., van Oudenhoven, A. P. E. Remme, R. P., Serna-Chaves, H. M., de Groot, R. S., & Opdam, P. (2014). Ecosystem services as a contested concept: A synthesis of critique and counter-arguments. *Conservation Letters, 7*(6), 514–523.

Sharpley, R. (2020). Tourism, sustainable development and the theoretical divide: 20 years on. *Journal of Sustainable Tourism, 28*, 1932–1946.

Simmons, D. G. (2013). Tourism and ecosystem services in New Zealand. In J. R. Dymond (Ed.), *Ecosystem Services in New Zealand – Conditions and Trends*. Manaaki Whenua Press.

Sæþórsdóttir, A. D. (2011). *Wilderness Tourism in Iceland – Land Use and Conflicts with Power Production*. The Geographical Society of Northern Finland and The Department of Geography, University of Oulu.

Sæþórsdóttir, A. D., Hall, C. M., & Wendt, M. (2020). From boiling to frozen? The rise and fall of international tourism to Iceland in the era of overtourism. *Environments, 7*(8), 59.

Terkenli, T. S. (2004). Tourism and landscape. In A. A. Lew, C. M. Hall, & A. M. Williams (Eds.), *A Companion to Tourism* (pp. 339–348). Blackwell.

Urry, J. (1990). The 'Consumption' of Tourism. *Sociology, 24*(1), 23–35.

Wardana, A. (2015). Debating spatial governance in the pluralistic institutional and legal setting of Bali. *Asia Pacific Journal of Anthropology, 16*(2), 106–122.

21

THE CHANGING GEOGRAPHY OF TOURISM IN A CLIMATE-DISRUPTED WORLD

Natalie L.B. Knowles and Daniel Scott

Introduction

Climate change is considered by the United Nations as the defining challenge of the twenty-first century and by the United Nations World Tourism Organisation (UNWTO) and other leading international tourism organisations to be the greatest threat to sustainable tourism in the twenty-first century (UNFCCC, 2020; UNWTO, 2019; World Travel and Tourism Council (WTTC), 2015; WTTC, UNWTO & Global Travel Association Coalition (GTAC), 2016). As it becomes increasingly apparent that the solutions of global environmental change cannot be achieved in silos, the need for interdisciplinary research on bridging human and natural systems increases (Biesbroek et al., 2009, Klein, 2011). The interdisciplinary nature of geography, including connections between physical and social sciences and socio-ecological systems thinking across scale, have enabled geographers to make major contributions to the study of climate change and the implications for ecosystems and society (Aspinall, 2010; Hulme, 2007; Moser, 2010). Geographers have also contributed foundational and leading work on climate change and tourism (Wall, 1986, Scott & Gössling, 2022). (Bjurstrom & Polk, 2011). This chapter builds on these disciplinary contributions to examine the future of tourism in a decarbonising and increasingly climate-disrupted world.

Our Warming World

Anthropogenic greenhouse gas emissions have already warmed the global average temperature 1.1 °C in the last 150 years according to the Intergovernmental Panel on Climate Change (IPCC) 2021. One degree does not sound like much – tourists love the heat, right? Heatwaves across Europe, North America, Australia and India have turned increasingly deadly (Mora et al., 2017) and are increasingly only possible because of anthropogenic climate change (Philip et al., 2021). Heatwaves and droughts are impacting wildfire regimes from the Mediterranean to Australia to Siberia and western North America. In the last decade, wildfires in North America have burned twice the area than in the 1990s (Congressional Research Service, 2021), also impacting air quality in many destinations. Our oceans absorb 90 per cent of the excess heat from anthropogenic global warming and are warming to greater depths and all the way to the poles, contributing to accelerating sea level rise (IPCC, 2021).

DOI: 10.4324/9781003286301-25

The global average temperature is rising at unprecedented rates (10 times faster than at the end of the last Ice Age – IPCC, 2021, Hickel, 2020), which is leading to multi-system environmental impacts that have direct consequences for tourism (see the global distribution of Climate Change Vulnerability Index for Tourism (CVIT) scores in Figure 21.1). As reported by the media with increased regularity, these impacts are increasingly part of our lived experience in tourism and as tourists. Seeking reefs teeming with colour and life? Divers are increasingly returning home from the Great Barrier Reef and other dive destinations with images of coral bleached white (Curnock et al., 2019), while marine species are going extinct at twice the rate of land animals (Albouy et al., 2020). Looking for fresh powder? Skiers visiting Europe's Alps and other ski tourism destinations around the world are already finding shorter seasons and less snow cover (Knowles, 2019a; Steiger & Scott, 2020; Scott et al., 2021). Glaciers in the Himalayas, Greenland and the polar ice caps are all melting faster than in decades past, contributing to accelerating sea level rise (DeConto & Pollard, 2016; IPCC, 2018). Looking forward to a wine tour in France?

High temperatures during grape-growing season contributes to fruit with less aromas and pigments, and lower-quality wine; some veritable varieties will no longer be able to grow in the regions they are known for (Cardell et al., 2018; Sottini et al., 2021). Although we have only experienced the early onset of climate change and socio-ecological disruption, the landscapes, biodiversity, cultures, foods, and activities that the global tourism industry depends on are increasingly threatened. Accelerating climate impacts and our societal response to climate change will transform travel demand patterns, tourism operations, planning, and investment. The climate change imperative demands much new research to better understand the implications of future changes in climate and develop effective and just adaptation and mitigation responses to advance climate resilient tourism (Scott & Gössling, 2022).

The international community is very concerned about the future implications of climate change for biodiversity, food and water security, human health, security, and the economy (IPCC, 2018; 2021). Some even consider climate change an existential risk (Kumar et al., 2021; Ripple et al., 2021). To avoid the worst consequences of climate change, the world came together and ratified the Paris Climate Agreement (2016), which has the goal to limit global warming to less than 2° C above pre-industrial times. Despite the Glasgow Climate Pact turning the 2020s into a decisive decade of climate action, strengthening emission reduction ambitions, with more than 120 countries pledging to achieve net-zero emissions by 2050 (United Nations Climate Change, 2021; United Nations Environment Programme (UNEP) 2020), most analyses conclude the world is unlikely to achieve the Paris Climate Agreement's highly ambitious emission reduction targets (as low as 5% change – Lui & Raftery, 2021). Even if all current emission reduction pledges are achieved, we can expect 1.5° C in the next 20 years (IPCC, 2021) and over 3° C warming by the end of the century (UNEP 2020) with other models suggesting we should prepare for additional warming (Brown & Caldeira, 2017). Our understanding of what a post +3° C world means for tourism remains very limited (Scott, 2021).

As the social movement for climate action grows internationally, tourism leaders increasingly recognise that climate change risk comes not only from physical disruption to destinations and travel patterns, but also from the low-carbon transition (as well as public concern for climate responsibility influencing financial investment, consumer behaviour and employee satisfaction – Scott & Gössling, 2022). How tourism responds to the climate crisis is critical to the sector's future viability (Scott 2021). Tourism is responsible for creating substantial emissions at the global scale and recognises its responsibility to reduce emissions consistent with science-based targets within the Paris Agreement. Despite international tourism organisations' commitment to "a meaningful and long-term contribution to fighting climate change" (WTTC, UNWTO & GTAC, 2016), tourism accounts for 5–8 per cent of total global emissions (Scott et al., 2010; Lenzen et al., 2018) and there is currently no credible plan whereby tourism emissions could be reduced to be compatible with the net-zero economy of mid-century (Scott & Gössling, 2022). The absence of detailed climate change response strategies, both adaptation and mitigation, raises fundamental questions on whether tourism truly can be one of the economic pillars in a decarbonised and climate resilient economy (Gössling & Scott, 2018; Scott et al., 2016; Becken et al., 2020; Scott, 2021).

Some scientists have termed this complex new era of global scale changes to multiple environmental systems as 'the Anthropocene' – a geological era marked dominantly by human activity (see Huijbens, Chapter 19 of this volume). As we journey into this new era, we must now navigate increasingly uncertain futures wherein tourism must find novel ways to prepare not only for environmental change, but significant political, economic, and social transformations. The recent COVID-19 pandemic, revisions to international trade agreements, social justice movements, and political instability demonstrate the interconnectedness of our social-political-economic-ecological systems within which tourism exists. These complex interactions in many cases will exacerbate the human and environmental impacts of climate change. The global response, and thus tourism's response, to climate change is and will continue to be fraught with trade-offs and synergies with other sustainability objectives, resulting in both new opportunities and greater challenges. While devastating for millions of people and the tourism sector globally (Gössling et al., 2021), the COVID-19 pandemic provides a unique opportunity to reimagine tourism (Benjamin et al. 2020, Stankov et al. 2020) and its role in a decarbonised and more sustainable economy. To support this transition, the tourism academy must also restructure and reprioritise research in a way that reduces barriers between disciplines, sectors, and stakeholders to create viable climate change responses strategies that also foster wider sustainability transitions as the shift to post-pandemic travel begins across the world.

Conceptual Underpinnings

As a pressing issue facing the sustainability of tourism worldwide, scientific research on climate and carbon risks to tourism has developed rapidly over the past 30 years, with the number of total publications on the subject doubling in the last 5 years, although this only represents 3.5 per cent of the publications in the top 5 tourism journals (Scott & Gössling, 2022). Climate and tourism research trends based on bibliometric rankings by Fang et al. (2018) find Geography to be the fifth most utilised discipline behind Environmental Science, Tourism & Leisure, Atmospheric Sciences, and Ecology. A more recent bibliometric analysis of the climate change and tourism literature (Scott & Gössling, 2022) found that half of the top 15 contributors to the field (based on number of author contributions) are Geographers or have a PhD in Geography. Geography or tourism geography focused journals do not rank in the top five publications by Fang et al. (2018) and listed in the lower 20 per cent by Scott & Gössling (2022) in conjunction with transportation, economics, or other interdisciplinary journals. Looking into tourism research more broadly, while Haarstad (2014) feels that geographers have been leaders contributing to past, present, and future interdisciplinary understanding of climate change systems and human responses, Corbera et al. (2016) identify that only one geographer (and only 3 social scientists) was a coordinating lead author of AR5 Working Group III, while only 7 out of 273 authors have been trained in geography (22 per cent of authors were social scientists). Despite a lack of formal integration of the geographic discipline, research across the diverse and multidisciplinary tourism and climate field frequently rely on and utilise geographic and other space, and place-based perspectives to relate and understand both climate change and tourism within the wider socio-economic systems it exists in.

Tourism is a significant sector in the extant growth-based patterns of global production, trade, finance, and consumption (Higgins-Desbiolles, 2018, WTTC, 2020), which are some of the primary drivers of sustainability challenges (climate change, biodiversity loss, inequity) (Hüttel et al., 2020; Intergovernmental Science-Policy Platform on Biodiversity and Ecosystem Services (IPBES), 2019; IPCC 2014b; Wiedmann et al., 2020). Already one of the largest (generating roughly 10% of global GDP and supporting 313 million jobs (WTTC, 2020) and pre-pandemic fastest growing (+30% between 2005–2035) economic sectors in the world (UNWTO, 2019), 1.5 billion international tourists in 2019 (+4% over 2018) generated $1.7 trillion USD of exports represented "the ninth consecutive year of sustained growth in international tourism" (UNWTO, 2019). While COVID-19 has reversed tourism growth in the short term, with the recovery offered by effective vaccines, some remind us that relative to the longer impacts of climate change, this will be a 'very painful, tragic footnote' (Schaal, 2020). As the COVID-19 economic and travel recovery begins there are opportunities to reimagine tourism and avoid many of the issues associated with over-tourism and the impact of tourism's continued growth (Cañada & Murray, 2019; 2021; Koh, 2020). Whether the tourism sector will take advantage of this opportunity remains a salient question (Hall et al., 2020).

Tourism Growth and Climate Ambitions: Are They Compatible?

Becken (2019) has argued that international tourism organisations support policies that reduce emissions, but not at the expense of tourism growth (measured in terms of arrivals). Gössling & Scott (2018) heard similar sentiments in their interviews with global leaders in the tourism sector. The WTTC, for instance, aims to reduce global tourism emissions 50 per cent by 2035 from 2005 levels (2015: 5), while increasing tourist arrivals by 25 per cent over the same time. Many researchers (Hickel, 2020; Higgins-Desbiolles, 2010; 2018; Hall, 2009; Hall et al., 2015; Budeanu et al., 2015) highlight this fundamental incongruity between concepts of perpetual economic growth and the responses needed to mitigate and adapt to climate change within both tourism and broader economic contexts. By 2030 the World Economic Forum's (WEF) *Global Agenda Council for the Future of Travel and Tourism* aims to reduce visa requirements to create "frictionless" travel for 2 billion international tourism arrivals (Budeanu et al., 2015; WEF, 2020; WTTC, 2020). Currently, the WTTC top priority is "ensuring that this *growth is sustainable*" (WTTC, 2020). At current growth rates, tourism emissions are projected to triple by 2035 (Scott et al., 2010) which is incompatible with global climate goals (Gössling et al., 2013, Gössling & Lyle, 2021; Scott & Gössling, 2022). To achieve the Paris Agreement objectives and minimise climate impacts, the IPCC (2018; 2021) estimates that CO_2 emissions must be reduced by 50 per cent by approximately 2030 and reach net-zero by 2050. International shipping and aviation are not covered by the nationally determined contributions each country submits to the Paris Agreement framework and if left unrestricted could emit up to 220 per cent of the allowable global carbon by 2050 under the 1.5° C scenario (UNEP, 2020).

George Monbiot (2012: online) noted the transition of language and logic from 'sustainability' to 'sustainable development' to 'sustainable growth' to 'sustained growth', finding that 'if sustainability means anything it is surely the opposite of sustained growth. Sustained growth on a finite planet is the essence of unsustainability. Despite the commitment of leading international tourism organisations to become 'climate neutral' by mid-century (see the 2021 *Glasgow Declaration: A Commitment to a Decade of Tourism Climate Action,* UNWTO, 2021), Monbiot's argument can be clearly seen within the WTTC's ambition for sustained growth. Sustainable tourism and climate change research now numbers over 5,000 articles (Centre International de Recherches et d'Etudes Touristiques [CIRET] 2012; Sharpley 2020), with research on tourism's climate risks developing rapidly since 2010 (Fang et al., 2018; Scott & Gössling, 2022). Despite researchers developing foundational climate-tourism knowledge highlighting the need to address climate and carbon risks urgently and the growing international consensus on the climate crisis, the sector has responded by "advocating the *need* for responsible growth" to help fulfill the 2030 Sustainable Development Goals (UNWTO, 2019: online). UNWTO Secretary General Pololikashvili stresses that with recent "downgraded global economic perspectives, international trade tensions, social unrest and geopolitical uncertainty" and now pandemic crisis, "our sector keeps outpacing the world economy and calling upon us to *not only grow but to grow better*" (UNWTO 2019: online). Sharpley's (2020) review of two decades of research on sustainable development through tourism concludes that while academic and policy circles continue to advocate for sustainability as a priority, there is little evidence of this happening in practice and a growing uncertainty on whether sacrifice-free decarbonisation and sustainability solutions exist for tourism.

Physical Climate Impacts and Carbon Transitions Risks: Are the Risks to Tourism Understood?

As Westley et al. (2011: 764) stated,

> the private sector is most likely to produce innovative technical response to environmental threats, but it constitutes as the engine of economic growth and is therefore unlikely to place that innovative capacity at the service of greater sustainability unless broad institutional shifts occur to encourage such reinforcement.

Like many economic sectors, tourism's entrenchment in the paradigm of unlimited economic growth coupled with a heavy reliance on fossil fuels and weak accountability on emission reductions, means the sector is not currently positioned well to respond to the increasing climate and carbon risks (Gillard et al., 2016; Scott, 2021; Sharpley, 2020). Industry engagement in sustainability and climate action has thus far been primarily aligned to corporate social responsibility activities, focused on increased energy efficiencies, and cost reductions, without addressing systemic challenges of decarbonising air travel or how deep decarbonisation could be achieved in the sector (Gössling & Scott, 2018; Spector, 2017; Lyle, 2018; Ryley et al., 2020). Some consider the sector's approach to emission reduction thus far as largely greenwash, with some poignantly asking if international tourism leadership is "*capable* of doing *anything other than* cheerleading for the travel industry's growth, particularly mass tourism," (Spinks, 2020). Others also highlight the significant implications of net-zero transition strategies that the sector has yet to come to terms with (Scott & Gössling, 2022).

The private sector is increasingly concerned about the risks associated with a warmer and decarbonised future (Carney, 2015; 2018; Cohen & Munoz, 2017; Espiner et al., 2017), especially as governments and major stock markets are beginning to demand climate change risk reporting (Mazzacurati et al., 2018; Task Force on Climate-Related Financial Disclosures (TFCRFD) 2017). The IPCC (2014b) 5th Assessment Report conceptual framework considers climate change an 'iterative risk management process', centred around reducing *climate risks* in human or natural systems. Climate risk, in this framework, stems from the interaction of *hazards* (natural or human-induced events that cause damage to humans, infrastructure or the environment), *exposure* (presence of people, livelihoods, species, natural resources or cultural assets that could be adversely affected), and *vulnerability* (sensitivity to harm and capacity to cope or adapt) (IPCC, 2014b). This conceptual framework can be applied to the tourism sector by assessing *physical climate risks* to infrastructure, operations, supply chains, natural or cultural heritage assets, *liability risks* of new regulations, business reputation and *transition risks* from altered product demands, adjustment to low-carbon economy (Agrawala et al., 2011; Carney, 2018) with the aim of identifying 'no (or low) regrets strategies' to reduce vulnerability and exposure.

Although risk-based language resonates with the business or financial case for climate action, by using timelines and language applicable to preserving tourism business-as-usual growth trajectory (Bisaro & Hinkel, 2018; Birkmann & Mechler, 2015; Buchner et al., 2014; Crick et al., 2017), the IPCC 5th Assessment Report risk framework (IPCC, 2014b; 2014c) focuses on negative impacts, which as a singular framing of climate change is problematic for tourism research and response. It has become increasingly clear that, 'the persuasive tactic of arousing fear in order to promote precautionary motivation and behavior is neither effective nor appropriate in the context of climate change communication and engagement' (Reser & Bradley, 2017: 1). The tourism industry is highly image-sensitive and carefully manages destination reputation. Research and resulting media content that promotes a predominantly negative future is a barrier to industry acceptance of credible science and has likely hindered climate responsiveness particularly at the destination or business scale (Knowles & Scott, 2021). Discussions on the incompatibility between continued growth and a liveable climate (Becken, 2019; Hickel, 2020; Higgins-Desbiolles et al., 2019), while scientifically valid, are also not likely to foster enthusiastic engagement from a sector that primarily measures its success via growing tourist arrivals and spending. New strategies to address the massive knowledge needs and collaboration are needed to accelerate the sector's climate response.

Value and Vulnerabilities

Like most sectors, economic valuation dominates tourism policy (e.g., spending, jobs, GDP contribution). This perspective considers economic indicators to be a neutral way to assign value, able to recognise, and capture the value of environmental and social integrity for human well-being and use the market to solve the climate crisis. The IPCC 2nd Assessment Report (1995: 10) warned of non-monetary value and impacts of climate change, stating "monetary valuation should not obscure the human consequences of anthropogenic climate change damages because the value of life has meaning beyond monetary valuation". In pulling exposure out from under the previous conceptualisation of vulnerability, IPCC 5th Assessment Report (2014a: 127) risk framework evaluates exposure as whether "something of value is at stake, recognising the diversity of values". An economic-only conceptualisation of value (and thus risk) will likely exacerbate inequities in the tourism sector by prioritising investment in risk reduction within areas already holding significant resources and capacity. There are persistent regional information gaps where the global south, including key emerging markets like Asia and South America, lack climate change and tourism research (Scott et al., 2016), thus hindering an understanding of climate risk and the development of equitable response strategies.

Furthermore, this economic-only perspective fails to capture tourism's reliance on surrounding natural ecosystems, regional economies, cultural heritage, and local communities. Tourism destinations derive direct, yet qualitative or non-monetary, value and cannot be sustained independently of their socio-ecological system (Bosak, 2016; Hall, 2011; McCool, 2015). This may create challenges assessing, quantifying, and responding to climate or carbon vulnerability and exposure across and between various elements of tourism destinations and regions for example, species versus infrastructure versus livelihood versus cultural asset, which are prevalent in tourism systems (Folkersen, 2018; Van den Bergh & Botzen, 2015). How climate change impacts will interact within these integrated socio-ecological systems and macro-scale drivers of tourism development is largely unknown and will vary by market segment and tourism region (Buckley et al., 2015). Additional governance challenges ensue when the multifaceted and scaled vulnerabilities of the system fall outside or between the scope of one organisation or government mandate, as is the case of many of the direct and indirect hazards of climate change (e.g., droughts, famines, biodiversity loss, invasive species, wildfires) (Klein, 2011). Tourism's deep foundation in local and global socio-economic systems suggests that pursuing an integrated sustainability transformation (rather than a narrow focus on climate change) has the potential to yield benefits for both climate adaptation and mitigation (Burch et al., 2019; Dale et al., 2019; O'Neill et al., 2014).

Transformation and Thinking Differently: Time for Critical Reflection?

Sustainability transformations emphasise the importance of creating spaces in research, policy, and practice to integrate multiple values for and of sustainability into economic narratives, including those aspects of human and natural systems which are not amenable to pricing, valuation, or markets (Krauss, 2020; Higgins-Desbiolles, 2018). Transformative change seeks to shift the underlying values and perceptions of what tourism's role is in society and aims to redefine its success relative to ecological sustainability and ethical principles. In broader economic contexts the beginnings of sustainable transformations can be seen as concepts and movements such as degrowth, circular economy, regenerative economy and post-consumerism gain traction and happiness or wellness indices complement or take the place of GDP as key indicators of economic success (Hickel, 2020; Raworth, 2017; Fath et al., 2019). The tourism sector has potential to transform perceptions of its 'success' via mobilising large-scale collective action from diverse stakeholders to mainstream the qualitative value tourism derives and provides socio-ecological systems, such as health and well-being, infrastructure, ecological conservation, social cohesion, cross-cultural exchange, education (Krauss, 2020; Higgins-Desbiolles, 2018).

Understanding these opportunities, tourism research has begun to move beyond sustainable tourism to innovative theoretical approaches that redefine and incorporate new values into tourism to achieve climate and sustainability objectives. *Steady-state tourism* was advocated as early as 2009 by Hall (2009: 1) who encouraged "qualitative development but not aggregate quantitative growth to the detriment of natural capital". Building on this concept, Fletcher et al. (2019) defined *degrowth tourism* as the drastic transformation of tourism necessary to meet the scale of global environmental challenges with further research promoting a radical transition from private tourism activity to that founded in and contributing to the common (Fletcher et al., 2019; Cañada & Murray, 2019; 2021; see also Blazquez et al., Chapter 5 of this volume). Higgins-Desbiolles et al., (2018) similarly uses degrowth thinking to reimagine the objectives of tourism by placing the rights of the local community above the rights of tourists for holidays and the rights of tourism corporations to make profits. Cave and Dredge (2020: 4) take a more pluralistic economic framework that envisages the "coexistence of capitalist, alternative capitalist and non-capitalist practices" in conceptualising *regenerative tourism* (see also Bellato et al., Chapter 23 of this volume). Others suggest alternative approaches such as *slow tourism* (see Moira et al., 2017), *inclusive growth* and *pro-poor tourism* (Bakker & Messerli, 2017) as ways to transform how and why we travel and with the aim to achieve global social, ecological, and economic objectives through this industry. This framing of climate responsiveness as innovations that enhance the industry and its future, may encourage tourism researchers and decision-makers to see social and ecological tension and disruption as opportunities for thinking and acting differently (Gillard et al., 2016).

The State of Existing Climate Change and Tourism Research

In looking forward to advancing tourism management, policy, and research for a warmer and decarbonised future, let's first reflect on where tourism as a sector has come from and the current state of knowledge. The IPCC (2014b) considers tourism to be a sector most likely to be impacted by climate change, yet despite 30 years of climate and tourism research, this topic remains a priority knowledge gap (Scott, 2021). Scott and Gössling (2022: 10) emphasise that "climate change does not appear to be a high priority on the tourism academy's research agenda" ... "nor is there evidence that the tourism industry makes widespread use of the information base that is available". There is an important role for geographers to support tourism researchers, industry stakeholders, and destination communities address the urgent and wicked climate problem and enhance climate and tourism research by adding a natural and social sciences scope beyond the capacity of the tourism academy.

In particular, the global tourism sector remains hindered by a lack of research in key geographic locations (particularly Asia and the global south which are simultaneously the regions expected to experience the greatest future growth) (Fang et al., 2018; Scott et al., 2016; Scott & Gössling, 2022), limited capacity to monitor its emissions and report on progress towards climate ambitions (Scott & Gössling, 2022), uncertainty regarding tourist responses to climate change (Gössling et al., 2013; Scott et al., 2016; Weaver, 2011), the carbon intensive tourist mobility system (Higham et al., 2016), reliance on technology efficiency versus behaviour change (Becken, 2019), limited understanding of the integrated impacts of climate and carbon transition risks at the destination scale (Scott et al., 2016), and the disconnect between researchers and practitioners (Gössling & Scott, 2018; Knowles & Scott, 2021). Beyond increasing the prioritisation of research (currently only 1–3 per cent of research output, as measured by the content of high-ranking tourism journals) (Scott, 2021), we must ensure the methods by which we conduct and communicate research meet the scope and scale of this industry's unique climate challenges and are increasingly decision relevant.

Despite the nuanced, contextual, and interconnected nature of tourism, science-based climate risk discourse traditionally aims to define, synthesise, simplify, and reduce complexity (Krauss, 2020); often aiming to isolate the impacts of climate change (Mora et al., 2018) and prioritising technological innovation over social solutions (Cinner et al., 2018). The isolation of climate from other factors creates an incomplete picture and potentially misleading conclusions (Mora et al., 2018). Instead, given climate change and tourism both incorporate a complex landscape of actors and exist beyond the scope of single regional or national government's ability to manage, geography's interdisciplinary lens, relational concept of scale, and process-based thinking (Gregory et al., 2009) may be relevant to assessing tourism's climate and carbon risk. Even when armed with robust knowledge of emission reduction levels required to keep the global climate stable, and availability of technologies to limit or sequester emissions (Falkner, 2016; IPCC 2014a; 2018), mitigation efforts face significant challenges in governance and social acceptance (Griscom et al., 2020) and can cause impacts inconsistent with climate justice (Jakob & Steckel, 2013). Geography represents an underutilised avenue to bridge human and natural systems, work within and across political borders and embrace rather than reduce complexity of climate change, tourism, and the world within which they exist. Still, despite the interdisciplinary tendencies of geographers, translation of concepts, terminology, and logic within geography and across disciplines remains a challenge, especially as new interdisciplinary fields such as hazards research, political or cultural ecology, and sustainable development emerge (Gregory et al., 2009; Turner, 2002, Bracken & Oughton, 2006).

Similarly, climate adaptation literature has thus far been primarily what Klein (2017) calls research *for* adaptation, focused on the effects and management of weather-related impacts on human and natural systems, while putting less effort into studying the adaptation process and outcomes (the study *of* adaptation) (Klein & Juhola, 2014; Juhola et al., 2016), despite findings that key challenges to building support for climate adaptation are frequently political, cognitive, and social (Cinner et al., 2018). Within the tourism sector leaders have confirmed that substantive scientific literature on climate and carbon risks is available to inform policy and action (Gössling & Scott, 2018). The availability of information and simultaneous lack of tourism policy that is consistent with national or international climate commitments (Becken et al., 2020) or strategies that position tourism to achieve the deep emission reductions needed to exist in a decarbonised economy by mid-century, suggests the need to strengthen collaborations among government, business, non-governmental organisations, and universities (Scott, 2021).

Agenda for Advancing the Future

As the urgency and complexity of the climate crisis grows ever more apparent, there are less than 30 years to fundamentally transform global tourism into a decarbonised and climate resilient sector. The timelines to achieve the aims of the Paris Agreement serve as a striking reminder of the responsibility of all parties in creating and implementing a more sustainable pathway forward. To mobilise existing knowledge, create new information and foster shared learning throughout the sector we highlight three priorities: (1) ensure local tourism participants, policymakers and planners on the ground are engaged in the research and climate action process, (2) foster urgency, leadership and vision within the industry and governments, particularly international tourism organisations (e.g., UNWTO (United Nations World Tourism Organisation), WTTC (World Travel and Tourism Council), European Travel Commission, Pacific Asia Travel Association), and (3) produce and communicate research that is accessible and relevant to decision-makers.

Recognising the messy reality of global tourism, the wide-ranging stakeholders (tourists, businesses, policymakers, local communities), and unique situated context of each destination, means climate and carbon risks and responses may not be comparable or applicable across businesses and destinations. While "decision-makers don't want stories. They want data to go into their financial risk models" (Krauss 2020: 3), in real world contexts, climate actions are rarely taken in response to climate change alone and a clearly identifiable and a rational decision-maker with the mandate to make decisions rarely exists (Smit & Wandel, 2006; Wise et al., 2014). Incorporating industry stakeholders' cultural values, orientations, priorities, and alternative forms of understanding can contribute to deeper insight on climate change risks and responses (Krauss, 2020). Decision frameworks such as adaptation pathways (Burnham et al., 2018; Fazey et al., 2016; Klein, 2017), transition management (Gillard et al. 2016), sustainability transformations (Burch, 2019), stakeholder narratives (Bosomworth et al., 2015; O'Neill et al., 2014), political ecology (Douglas, 2014; Knowles, 2019b; Saarinen & Nepal, 2016), and other decision-centred approaches (Wise et al., 2014) may be helpful. These approaches represent inclusive ways to harness representative and broad input, mobilising knowledge between and within the scientific community, political institutions, private sector, and individuals and create buy-in to climate research and engagement in risk response strategies that suit their unique situation.

In terms of leadership and vision, Scott (2021) suggested the UNWTO and sector leaders create a task force to rapidly develop an industry-wide roadmap to decarbonisation, including a just transition and climate resiliency building, similar to the Global Tourism Crisis Committee established to develop a tourism wide response to COVID-19. A key element of this roadmap would be to address the contradiction between climate and reliance on carbon-intensive air and ground transportation networks and the implications of related emission reduction strategies (Gössling & Scott, 2018; 2021; Spector, 2017; Stoddart, 2011). Aviation emissions are currently considered 'extra-national' and thus outside the scope of any nation's responsibility within the Paris Agreement targets, and the International Civil Aviation Organisation (ICAO) has thus far been ineffective at reducing sector emissions despite proposing the Carbon Offsetting and Reduction Scheme for International Aviation (Lyle, 2018; Ryley et al., 2020; Gössling & Lyle, 2021). Establishing relationships with other governing bodies including the ICAO, the IPCC, and federal governments to identify responsibility and develop effective strategies to reduce these international travel emissions is essential to achieving global climate targets. This may also help the tourism sector address concerns of credibility and greenwashing in their broader pursuit of decarbonisation and sustainability.

Finally, if we are asking the tourism sector to transform itself, we, as tourism scholars and professionals, must be reflective on our own role and look for ways to personally be part of the solution. First, if we want, as Robin Wall Kimmerer (2013: 123) eloquently puts it, "to be heard, you must speak the language of the one you want to listen". Individually, this means reflecting on our privilege, monitoring our own impacts, and living the behaviour change and theoretical solutions we purport in our daily lives and travels (Higham & Font, 2020). Institutionally, it seems as long scientific articles behind the paywalls of academic journals, filled with jargon, are a foreign tongue to many in the tourism sector, and it should perhaps come as no surprise that industry has not internalised the challenges we document, nor put our messages to action (Sharpley, 2020). In an age of rapid communication, whether by text or Tiktok, blog or podcast, what are the best mediums for climate and tourism scientists to mobilise research findings and communicate with action-oriented stakeholders? Fostering relationships and engagement within the destinations and local communities we study are of key importance (Scott, 2021). As tourism scholars we already have these relationships in diverse destinations around the world, and now we need to build on this trust to support these destinations in the journey to respond to climate change. These relationships will likely look and require engagement that is very different from that needed from our policymakers, international organisations, or business audiences. Rather than research for research's sake, we must research with a shared purpose in mind: Re-envisioning and operationalising tourism that is compatible with and fosters the decarbonised, resilient, and sustainable future the world needs. Beyond sharing our knowledge, we must listen to the sector to make sure we are asking the right questions to begin with. Instead of siloed academic research and ownership of information, we need to incorporate local knowledge and share transdisciplinary understanding to create decision-relevant information. Rather than prioritising theoretical arguments and coveting the prestige of publication metrics, the climate crisis, perhaps like no other crisis tourism has faced, demands we measure the success of our research by societal impact (Ravenscroft et al., 2017) such as on-the-ground implementation and policy and planning outcomes.

Conclusion

Climate change and societal responses will redraw the global geography of tourism in the decades ahead (Scott & Gössling, 2022). As the WTTC (2015: 5) concluded, "the next 20 years will be characterised by our sector fully integrating climate change and related issues into business strategy, supporting the global transition to a low carbon economy, strengthening resilience at a local level against climate risks". As tourism is both highly impacted by the impacts of climate change on environmental and economic systems and a large contributor of emissions with important transition risks, failure to do so would be to the substantial detriment of global tourism, sustainable tourism, and sustainable development more broadly (Scott, 2011; 2021). As we emerge from the unprecedented challenges of COVID-19, it serves to remember that while devastating, this three-to-five-year disruption would be dwarfed by the impact of a +3° C or warmer world. The disruption of the pandemic offers a once in a generation opportunity to rethink the status of tourism, re-envision its place in a carbon-constrained and warmer world, as well as offers the sector lessons on how to set out on a new trajectory with hope, commitment, and unity.

With its massive economic size, diverse stakeholders, local to global reach, transnational governance, and integration into nearly all industries from aviation to agriculture, the global tourism industry has the potential to be an international catalyst for change. Whether a tourist, a tourism operator, a policymaker, an investor, or scholar, in the tourism and climate change challenge there is no space for competition only collaboration. Across geographic regions, economic sectors, or academic disciplines, we either all win or we all lose the climate crisis. While one market or region may face relative gains in the short-term, extending outside the scope of any of our planetary boundaries will create cascading and unknown consequences that will resonate across the international tourism community and generations. We understand the urgency and have the information and technology to act (Gössling & Scott, 2018; Griscom et al., 2020; Jakob & Steckel, 2013; Klein, 2017), and thus must focus on creating the social capital and will to change our underlying values and metrics of success, share knowledge and mobilise diverse stakeholders to implement behaviour change, management decisions, and regulations within tourism that align with a decarbonised and climate resilient future.

References

Agrawala, S., Carraro, M., Kingsmill, N., Lanzi, E., Mullan, M., & Prudent-Richard, G. (2011). *Private Sector Engagement in Adaptation to Climate Change: Approaches to Managing Climate Risks* (OECD Environment Working Papers No. 39). Paris, France: OECD.

Albouy, C., Delattre, V., Donati, G., Frolicher, T., Albouy-Boyer, S., Rufino, M., Pellissier, L., Mouillot, D., & Leprieur, F. (2020). Global vulnerability of marine mammals to global warming. *Nature Scientific Reports, 10,* 548.

Aspinall, R. (2010). Introduction: Geographical perspectives on climate change. *Annals of the Association of American Geographers, 100*(4), 715–718.

Bakker, M., & Messerli, H. (2017). Inclusive growth versus pro-poor growth: Implications for tourism development. *Tourism and Hospitality Research, 17(4)*384–391.

Becken, S. (2019). Decarbonising tourism: Mission impossible? *Tourism Recreation Research, 44*(4), 419–433.

Becken, S., Whittlesea, E., Loehr, J., & Scott, D. (2020). Tourism and climate change: Evaluating the extent of policy integration. *Journal of Sustainable Tourism, 28*(10), 1603–1624.

Benjamin, S., Dillette, A., & Alderman, D. H. (2020). "We can't return to normal": Committing to tourism equity in the post-pandemic age. *Tourism Geographies, 22*(3), 476–483.

Biesbroek, R., Swart, R. J.. & van der Knaap, W. G. M. (2009). The mitigation – adaptation dichotomy and the role of spatial planning. *Habitat International, 33*(3), pp.230–237.

Birkmann, J., & Mechler, R. (2015). Advancing climate adaptation and risk management: New insights concepts and approaches: What have we learned from the SREX and the AR5 process? *Climatic Change, 133,* 1–6.

Bisaro, A., & Hinkel, J. (2018). Mobilizing private finance for coastal adaptation: A literature review. *WIREs Climate Change, 9,* 1–15.

Bjurström, A., & Polk, M (2011). Climate change and interdisciplinarity: A co-citation analysis of IPCC Third Assessment Report. *Scientometrics, 87,* 525–550.

Bracken, L. J., & Oughton, E. A. (2006) 'What do you mean?' The importance of language in developing interdisciplinary research. *Transactions of the Institute of British Geographers, 31*, 371–82.

Brown, P., & Caldeira, K. (2017). Greater future global warming inferred from Earth's recent energy budget. *Nature, 552*, 45–50.

Bosak, K. (2016). Tourism, development and sustainability. In S. McCool, & K. Bosak (Eds.), *Reframing Sustainable Tourism* (pp. 33–44). Springer.

Bosomwoth, K., Harwood, A., Leith, P., & Wallis, P. (2015). *Adaptation Pathways: A Playbook for Developing Robust Options for Climate Change Adaptation in Natural Resource Management.* Southern Slopes Climate Change Adaptation Research Partnership: RMIT University, University of Tasmania, Monash University.

Buchner, B., Stadelmann, M., Wilkinson, J., Mazza, F., Rosenberg, A., & Abramskiehn, D. (2014). The global landscape of climate finance 2014. *Climate Policy Initiative.* https://climatepolicyinitiative.org/wp-cont ent/uploads/2014/11/The-Global-Landscape-of-Climate-Finance-2014.pdf

Buckley, R., Gretzel, U., Scott, D., Weaver, D., & Becken, S. (2015) Tourism megatrends. *Tourism Recreation Research, 40*, 59–70.

Budeanu, A., Miller, G., Moscardo, G., & Ooi, C.-S. (2015). Sustainable tourism, progress, challenges and opportunities. *Journal of Cleaner Production, 111*, 285–294.

Burch, S. (2019). *Interjecting Politics into Business-led Sustainability Innovation: New Data from Small Businesses in Canada.* Centre for International Governance Innovation.

Burnham, M., Rasmussen, L., & Ma, Z. (2018). Climate change adaptation pathways: Synergies, contradictions and trade offs across scales. *World Development, 108*, 231–234.

Cañada, E., & Murray, I. (Eds.) (2019). *Turistificacion Global: Perspectivas Críticas en Turismo.* Icaria.

Cañada, E., & Murray, I. (Eds.) (2021). *#TourismPostCOVID19 Turistificación confinada.* Alba Sud.

Cardell, M., Amengual, A., & Romero, R. (2018). Future effects of climate change on the suitability of wine grape production across Europe. *Regional Environmental Change, 19*, 2299–2310.

Carney, M. (2015). Breaking the Tragedy of the Horizon – Climate Change and Financial Stability. Speech given at Lloyd's of London, 29 September, www.bankofengland.co.uk/publications/Documents/speeches/2015/speech844.pdf

Carney, M. (2018). *A Transition in Thinking and Action.* International Climate Risk Conference for Supervisors, De Nederlandsche Bank, Amsterdam (April 6).

Cave, J., & Dredge, D. (2020). Regenerative tourism needs diverse economic practices. *Tourism Geographies, 22*(3), 503–513.

Cinner, J. E., Adger, W. N., Allison, E. H., Barnes, M. L., Brown, K., Cohen, P. J., Gelcich, S., Hicks, C. C., Hughes, T. P., Lau, J., Marshall, N. A., & Morrison, T. H. (2018). Building adaptive capacity to climate change in tropical coastal communities. *Nature Climate Change, 8*, 117–123.

CIRET (2012). *Encyclopedia de la recherche touristique mondiale.* Center Internationale de Recherches et d'Etudes Touristiques.

Cohen, B., & Munoz, P. (2017). Entering conscious consumer markets: Towards a new generation of sustainability strategies. *California Management Review, 59*(4), 23–48.

Congressional Research Service (2021). Wildfire Statistics: In Focus. Updated Aug 6 2021. Accessible at: https://crsreports.congress.gov/product/pdf/IF/IF10244

Corbera, E. Calvet-Mir, L., Hughes, H., & Paterson, M. (2016). Patterns of authorship in the IPCC Working Group III report. *Nature Climate Change, 6*, 94–99.

Crick, F., Gannon, K., Diop, M., & Snow, M. (2017). Enabling private sector adaptation to climate change in sub-Saharan Africa. *WIREs Climate Change, 9*, 1–17.

Curnock, M., Marshall, N., Thiault, L., Heron, S., Hoey, J., Williams, G., & Goldberg, J. (2019). Shifts in tourists' sentiments and climate risk perceptions following mass coral bleaching of the Great Barrier Reef. *Nature Climate Change, 9*, 535–541.

Dale, A., Robinson, J., King, L., Burch, S., Newell, R., Shaw, A., & Jost, F. (2019). Meeting the climate change challenge: Local government climate action in British Columbia, Canada. *Climate Policy, 20*(7), 866–880.

DeConto, R., & Pollard, D. (2016). Contribution of Antarctica to past and future sea level rise. *Nature, 531*(7596), 591–597.

Douglas, J. (2014). What's political ecology got to do with tourism? *Tourism Geographies, 16*(1), 8–13.

Espiner, C., Orchiston C., & Higham J. (2017). Resilience and sustainability: A complementary relationship? Towards a practical conceptual model for the sustainability–resilience nexus in tourism. *Journal of Sustainable Tourism, 25*(10), 1385–1400.

Falkner, R. (2016). The Paris agreement and the new logic of international climate politics. *International Affairs, 92*(5), 1107–1125.

Fang, Y., Yin, J., & Wu, B. (2018). Climate change and tourism: A scientometric analysis using CiteSpace. *Journal of Sustainable Tourism, 26*(1), 108–126.

Fath, B., Fiscus, D., Goerner, S., Berea, A., & Ulanowicz, R. (2019). Measuring regenerative economics: 10 principles and measures undergirding systemic economic health. *Global Transitions, 1*, 15–27.

Fazey, I., Wise, R., Lyon, C., Campeanu, C., Moug, P., & Davies, T. (2016). Past and future adaptation pathways. *Climate and Development, 8*(1), 26–44.

Fletcher, R., Murray Mas, I., Blanco-Romero, A., & Blázquez-Salom, M. (2019). Tourism and degrowth: An emerging agenda for research and praxis. *Journal of Sustainable Tourism, 27*(12), 1745–1763.

Folkersen, M. (2018). Ecosystem valuation: Changing discourse in a time of climate change. *Ecosystem Services, 29*(A), 1–12.

Gillard, R., Gouldson, A., Paavola, J., & Van Alstine, J. (2016). Transformational responses to climate change: Beyond a systems perspective of social change in mitigation and adaptation. *Wiley Interdisciplinary Reviews: Climate Change, 7(2)*, 251–265.

Gössling, S., & Lyle, C. (2021). Transition policies for climatically sustainable aviation. *Transport Reviews, 41*(5), 643–658.

Gössling, S., & Scott, D. (2018). The decarbonisation impasse: Global tourism leaders' views on climate change mitigation. *Journal of Sustainable Tourism, 26*(12), 2071–2086.

Gössling, S., Scott, D., & Hall, C. M. (2013). Challenges of tourism in a low-carbon economy. *Wiley Interdisciplinary Reviews Climate Change, 4*(6), 525538.

Gössling, S., Scott, D., & Hall, C. M. (2021). Pandemics, tourism and global change: A rapid assessment of COVID-19. *Journal of Sustainable Tourism, 29*(1), 1–20.

Gregory, D., Johnson, R., Pratt, G., Watts, M., & Whatmore, S. (Eds.) (2009). *The Dictionary of Human Geography*. Blackwell Publishing.

Griscom, B. W., Busch, J., Cook-Patton, S. C., Ellis, P. W., Funk, J., Leavitt, S. M., … & Worthington, T. (2020). National mitigation potential from natural climate solutions in the tropics. *Philosophical Transactions of the Royal Society B, 375*(1794), 20190126.

Haarstad, H. (2014). Climate change, environmental governance and the scale problem. *Geography Compass, 8*(2) 87–97.

Hall, C. M. (2009). Degrowing tourism: Decroissance, sustainable consumption and steady-state tourism. *Anatolia: An International Journal of Tourism and Hospitality Research, 20*(1), 46–61.

Hall, C. M. (2011). Policy learning and policy failure in sustainable tourism governance: From first-and second-order to third-order change? *Journal of Sustainable Tourism, 19*(4–5), 649–671.

Hall, C. M., Amelung, B., Cohen, S., Eijgelaar, E., Gössling, S., Higham, J., … & Scott, D. (2015). On climate change skepticism and denial in tourism. *Journal of Sustainable Tourism, 23*(1), 4–25.

Hall, C. M., Lundmark, L. & Zhang, J. J. (2020) *Degrowth and Tourism: New Perspectives on Tourism Entrepreneurship, Destinations and Policy*. Routledge.

Hickel, J. (2020). *Less is More*. Heinemann.

Higgins-Desbiolles, F. (2010). The elusiveness of sustainability in tourism: The culture-ideology of consumerism and its implications. *Tourism and Hospitality Research, 10*(2), 116–129.

Higgins-Desbiolles, F. (2018). Sustainable tourism: Sustaining tourism or something more? *Tourism Management Perspectives, 25*, 157–160.

Higgins-Desbiolles, F., Carnicelli, S., Krolikowski, C., Wijesinghe, G., & Boluk, K. (2019). Degrowing tourism: Rethinking tourism. *Journal of Sustainable Tourism, 27*(12), 1926–1944.

Higham, J., Cohen, S. A., Cavaliere, C. T., Reis, A., & Finkler, W. (2016). Climate change, tourist air travel and radical emissions reduction. *Journal of Cleaner Production, 111*, 336–347.

Higham, J., & Font, X. (2020). Decarbonising academia: Confronting our climate hypocrisy. *Journal of Sustainable Tourism, 28*(1), 1–9.

Hulme, M. (2007). Geographical work at the boundaries of climate change. *Transactions of the Institute of British Geographers, 33*(1), 5–11.

Hüttel, A., Balderjahn, I., & Hoffmann, S. (2020). Welfare beyond consumption: The benefits of having less. *Ecological Economics, 178*, 106719.

Intergovernmental Panel on Climate Change (IPCC) (1995). *Climate Change 1995. Synthesis Report, A report of the Intergovernmental Panel on Climate Change*.

IPCC (2014a).*Climate Change 2014: Annex II Glossary*. Cambridge University Press.

IPCC (2014b). *Climate Change 2014: Impacts, Adaptation, and Vulnerability. Part A: Global and Sectoral Aspects: Contribution of Working Group II to the Fifth Assessment Report of the Intergovernmental Panel on Climate Change*. Cambridge University Press.

IPCC (2014c). *Climate Change 2014: Mitigation of Climate Change: Working Group III Contribution to the Fifth Assessment Report of the Intergovernmental Panel on T. Z. and J. C. M.* Cambridge University Press.

IPCC (2018). Summary for policymakers. In M. Allen, O. P. Dube, W. Solecki, F. Aragón-Durand, W. Cramer, S. Humphreys … & K. Zickfeld (Eds.), *Global Warming of 1.5 C*. An IPCC Special Report on the Impacts of Global Warming of 1.5 C above Pre-Industrial Levels and Related Global Greenhouse Gas Emission Pathways, in the Context of Strengthening the Global Response to the Threat of Climate Change, Sustainable Development, and Efforts to Eradicate Poverty. IPCC.

IPCC (2021). Climate change widespread, rapid, intensifying. Press Release Aug 9 2021, accessible at: www.ipcc.ch/site/assets/uploads/2021/08/IPCC_WGI-AR6-Press-Release_en.pdf

Intergovernmental Science Policy Platform on Biodiversity and Ecosystem Services (IPBES) (2019). *Summary for Policymakers of the Global Assessment on Biodiversity and Ecosystem Services*. https://ipbes.net/news/ipbes-global-assessment-summary-policymakers-pdf

Jakob, M., & Steckel, J. C. (2013). How climate change mitigation could harm development in poor countries. *Wiley Interdisciplinary Reviews: Climate Change*, 5(2), 161–168.

Juhola, S., Glaas, E., Linnér, B., & Neset, T. S. (2016). Redefining maladaptation. *Environmental Science and Policy*, 55(1), 135–140.

Kimmerer, R. W. (2013). *Braiding Sweetgrass: Indigenous Wisdom, Scientific Knowledge and the Teachings of Plants*. Milkweed editions.

Klein, R. (2011). Adaptation to climate change: More than technology. In I. Linkov, & T. Bridges (Eds.), *Climate: Global Change and Local Adaptation* (pp. 157–168). NATO Science for Peace and Security Series.

Klein, R. (2017). Both research for adaptation and research on adaptation are needed to inform society's response to climate change impacts. Adaptation Futures 2016 – Discussion Brief *1*. https://mediamanager.sei.org/documents/Publications/Climate/AF-DB-2017-Research-for-and-on-adaptation.pdf

Klein, R. J., & Juhola, S. (2014). A framework for Nordic actor-oriented climate adaptation research. *Environmental Science & Policy*, 40, 101–115.

Knowles, N. L. B. (2019a). Can the North American ski industry attain climate resiliency? A modified Delphi survey on transformations towards sustainable tourism. *Journal of Sustainable Tourism*, 27(3), 380–397.

Knowles, N. L. B. (2019b). Targeting sustainable outcomes with adventure tourism: A political ecology approach. *Annals of Tourism Research*, 79, 102809.

Knowles, N. L., & Scott, D. (2021). Media representations of climate change risk to ski tourism: A barrier to climate action? *Current Issues in Tourism*, 24(2), 149–156.

Koh, E. (2020). The end of over-tourism? Opportunities in a post-Covid-19 world. *International Journal of Tourism Cities*, 6(4), 1015–1023.

Krauß, W. (2020). Narratives of change and the co-development of climate services for action. *Climate Risk Management*, 28, 100217.

Kumar, A., Nagar, S., & Anand, S. (2021). Climate change and existential threats. *Global Climate Change*, 2021, 1–31.

Lenzen, M., Sun, Y. Y., Faturay, F. et al. (2018). The carbon footprint of global tourism. *Nature Climate Change*, 8, 522–528. https://doi.org/10.1038/s41558-018-0141-x

Liu, P., & Raftery, A. (2021). Country-based rate of emissions reductions should increase by 80% beyond nationally determined contributions to meet the 2 °C target. *Communications Earth & Environment*, 2, 29.

Lyle, C. (2018). Beyond the ICAO's CORSIA: Towards a more climatically effective strategy for mitigation of civil-aviation emissions. *Climate Law*, 8, 104–127.

Mazzacurati, E., Firth, J., & Venturini, S. (2018). *Advancing Task Force on Climate-related Financial Disclosures Guidance on Physical Climate Risks and Opportunities*. European Bank for Reconstruction and Development.

McCool, S. (2015). Sustainable tourism: Guiding fiction, social trap or path to resilience? In T. V. Singh (Ed.), *Challenges in Tourism Research* (pp. 224–234). Channel View Publications.

Moira, P., Mylonopoulos, D., & Kondoudaki, A. (2017). The application of slow movement to tourism: Is slow tourism a new paradigm. *Journal of Tourism and Leisure Studies*, 2(2), 1–10.

Mora, C., Dousset, B., Caldwell, I. R., Powell, F. E., Geronimo, R. C., Bielecki, C. R., Counsell, C. W. W., Dietrich, B. S., Johnston, E. T., Louis, L. V., Lucas, M. P., McKenzie, M. M., Shea, A. G., Tseng, H., Giambelluca, T. W., Leon, L. R., Hawkins, E., & Trauernicht, C. (2017). Global risk of deadly heat. *Nature Climate Change, 7*(7), 501–506.

Monbiot, G. (2012). *How Sustainability Became "Sustained Growth"*. Accessed: 3 June 2016 from www.monbiot.com/2012/06/22/how-sustainability-became-sustained-growth/

Mora, C., Spirandelli, D., Franklin, E. C., Lynham, J., Kantar, M. B., Miles, W., … & Hunter, C. L. (2018). Broad threat to humanity from cumulative climate hazards intensified by greenhouse gas emissions. *Nature Climate Change, 8*(12), 1062–1071.

Moser, S. (2010). Now more than ever: The need for more societally relevant research on vulnerability and adaptation to climate change. *Applied Geography, 30,* 464–474.

O'Neill, B., Kriegler, E., Riahi, K., Ebi, K., Hallegatte, S.,Carter, T., … & van Vuuren, D. (2014). A new scenario framework for climate change research: the concept of shared socioeconomic pathways. *Climatic Change, 122,* 387–400.

Philip, S., Kew, S., van Oldenborgh, G., Yang, W., Vecchi, G., Anslow, F … & Otto, F. (2021). Rapid attribution analysis of the extraordinary heatwave on the Pacific Coast of the US and Canada June 2021. Accessible at: www.worldweatherattribution.org/western-north-american-extreme-heat-virtually-impossible-without-human-caused-climate-change/

Ravenscroft, J., Liakata, M., Clare, A., & Duma, D. (2017). Measuring scientific impact beyond academia: An assessment of existing impact metrics and proposed improvements. *PloS one, 12*(3), e0173152.

Raworth, K. (2017). *Doughnut Economics*. Chelsea Green Publishing.

Reser, J. P., & Bradley, G. L. (2017). Fear appeals in climate change communication. In *Oxford Research Encyclopedia of Climate Science*. https://doi.org/10.1093/acrefore/9780190228620.013.386

Ripple, W., Wolf, C., Newsome, T., Gregg, J., Lenton, T., Palomo, I., Eikelboom, J., Law, B., Huq, S., Duffy, P., & Rockström, J. (2021). World scientists' warning of a climate emergency. *BioScience, 71*(9), 894–898.

Ryley, T., Baumeister, S., & Coulter, L. (2020). Climate change influences on aviation: A literature review. *Transport Policy, 92,* 55–64.

Saarinen, J., & Nepal, S. (Eds.) (2016). *Political Ecology and Tourism*. Routledge.

Schaal, D. (2020). *Escaping Climate Change for the Travel Industry even When Covid-19 Becomes a Distant Memory*. Skift Research.

Scott, D. (2011). Why sustainable tourism must address climate change. *Journal of Sustainable Tourism, 19*(1), 17–34. https://doi.org/10.1080/09669582.2010.539694

Scott, D. (2021). Sustainable tourism and the grand challenge of climate change. *Sustainability, 13*(4), 1966.

Scott, D., & Gössling, S. (2022). A review of research into tourism and climate change-Launching the annals of tourism research curated collection on tourism and climate change. *Annals of Tourism Research, 95,* 103409.

Scott, D., Hall, C. M., & Gössling, S. (2016). A report on the Paris climate change agreement and its implications for tourism: Why we will always have Paris. *Journal of Sustainable Tourism, 24*(7), 933–948.

Scott, D., Peeters, P., & Gössling, S. (2010). Can tourism deliver its "aspirational" greenhouse gas emission reduction targets? *Journal of Sustainable Tourism,* 18(3), 393–408. https://doi.org/10.1080/0966958100 3653542

Sharpley, R. (2020). Tourism, sustainable development and the theoretical divide: 20 years on. *Journal of Sustainable Tourism, 28*(11), 1932–1946.

Smit, B., & Wandel, J. (2006). Adaptation, adaptive capacity and vulnerability. *Global Environmental Change, 16,* 282–292.

Sottini, V., Barbierato, E., Bernetti, I., & Capecchi, I. (2021). Impact of climate change on wine tourism: An approach through social media data. *Sustainability, 13,* 7489.

Spector, S. (2017). Environmental communications in New Zealand's skiing industry: Building social legitimacy with-out addressing non-local transport. *Journal of Sport & Tourism, 21*(3), 159–177.

Spinks, R. (2020). It's time to rethink travel's global leadership—Starting with WTTC. Skift (accessed June 2024). https://skift.com/2020/09/28/its-time-to-rethink-travels-global-leadership/

Stankov, U., Filimonau, V., & Vujičić, M. D. (2020). A mindful shift: An opportunity for mindfulness-driven tourism in a post-pandemic world. *Tourism Geographies, 22*(3), 703–712.

Steiger, R., & Scott, D. (2020). Ski Tourism in a warmer world: Increased adaptation and regional economic impacts in Austria. *Tourism Management, 77,* 104032.

Stoddart, M. (2011). "If we wanted to be environmentally sustainable, we'd take the bus": Skiing, mobility and the irony of climate change. *Human Ecology Review*, *18*(1), 19–29.

Task Force on Climate-Related Financial Disclosures (TFCRFD) (2017). *Final Report: Recommendations of the Task Force on Climate-related Financial Disclosures*. Retrieved from: www.fsb-tcfd.org

Turner, B. L. (2002). Contested identities: Human-environment geography and disciplinary implications in a restructuring academy. *Annals of the Association of American Geographers*, *92*(1), 52–74.

United Nations Climate Change (2021). The Glasgow Climate Pact – Key Outcomes from COP26. United Nations Framework Convention on Climate Change (UNFCCC). Accessible at: https://unfccc.int/process-and-meetings/the-paris-agreement/the-glasgow-climate-pact-key-outcomes-from-cop26

United Nations Environment Programme (UNEP) (2020). Emissions Gap Report 2020. Available online: www.unenvironment.org/emissions-gap-report-2020

United Nations Framework Convention on Climate Change (UNFCCC) (2020). Annual Report 2020. Accessible at: https://unfccc.int/sites/default/files/resource/UNFCCC_Annual_Report_2020.pdf

UNTWO (2021). The Glasgow Declaration on Climate Action in Tourism (accessed June 2024). www.unwto.org/the-glasgow-declaration-on-climate-action-in-tourism

UNWTO (2019). World Tourism Barometer, 18, January 2020.

Van den Bergh, J., & Botzen, W. (2015). Monetary valuation of the social cost of CO2 emissions: A critical survey. *Ecological Economics*, *114*, 33–46.

Wall, G. (1986). Climate. change and downhill skiing in Ontario. *Ontario Geographer*, *28*, 51–68.

Weaver, D. (2011). Can sustainable tourism survive climate change? *Journal of Sustainable Tourism*, *19*(1), 5–15.

Westley, F., Olsson, P., Folke, C., Homer-Dixon, T., Vredenburg, H., Loorbach, D., Thompson, J., Nilsson, M., Lambin, E., Sendzimir, J., Banerjee, B., Galaz, V., & van der Leeuw, S. (2011). Tipping toward sustainability: Emerging pathways of transformation. *AMBIO*, *40*, 762–780.

Wiedmann, T., Lenzen, M., Keyßer, L., & Steinberger, J. (2020). Scientists' warning on affluence. *Nature Communications*, *11*, 3107.

Wise, R., Fazey, I., Stafford Smith, M., Park, S., Eakin, H., Van Garderen, E., & Campbell, B. (2014). Reconceptualising adaptation to climate change as part of pathways of change and response. *Global Environmental Change*, *28*, 325–336.

World Economic Forum (WEF) (2020). *The Global Risks Report 2020*. (Accessed June 2024). www.weforum.org/publications/the-global-risks-report-2020/

World Travel & Tourism Coalition (WTTC), UNWTO, Global Tourism Association Coalition (GTAC) (2016). Travel and Tourism 2015: Connecting Global Climate Action. Accessible at: www.wttc.org/research/policy-research/travel-and-tourism-2015-connecting-global-climate-action

WTTC (2015). *Travel & Tourism 2015; Connecting Global Climate Action*. Accessible at www.wttc.org/-/media/files/reports/policy-research/tt-2015--connecting-global-climate-action-a4-28pp-web.pdf

WTTC (2020). Sustainable Growth. www.wttc.org/priorities/sustainable-growth/

22

THE FOOTPRINT OF RESPONSIBILITY IN TOURISM DESTINATIONS

The Role of Governance

Theodora Papatheochari, Spyros Niavis and Harry Coccossis

Introduction

Tourism is regarded as a force that can either discover ('make') destinations and/or pressure them to a point beyond their capacity ('unmake') (Chang, 2012 and Chapter 13 of this volume). As a sector, it is also highly dependent on places, people and the interactions between them and, therefore, it is related to the physical and social conditions of a destination. At the same time, it is sensitive to the availability of resources and the market (Musavengane, 2019).

A behavioural change could potentially alter the geography and dynamics of tourism destinations through a shift of tourism demand and preferences towards new types of tourism products offered by destinations. Recent literature has examined such changes through research on tourism responsibility. Responsibility in tourism has often been confused with tourism sustainability. According to Mihalic (2016), the concept of responsibility is related to behaviour and action, in contrast to sustainability, which is the goal, presupposing responsibility. This approach is enhanced by the three dimensions of responsibility proposed by Goodwin (2011, cited in Mihalic, 2016: 465): Accountability, capacity to act, and capacity to respond.

Capacity to respond is directly linked to tourism behaviour and suggests cooperation in effectively addressing tourism and leading to the creation of actions, which leads to the dimension 'capacity to act'. Both dimensions have their origins, of course, in the fact that people should be held accountable for their actions ('accountability') (Mihalic, 2016). In the literature, responsible tourism is considered as behaviour-based and action-based and reflects respect for others and their environment. In this sense, responsible tourism could be regarded as a notion that links all tourism-related stakeholders with taking action towards sustainability, placing local priorities at its core. It aims to create tourism business opportunities while enhancing tourist experiences, improving the residents' quality of life, creating socio-economic benefits and protecting the natural and cultural resources of the destination (Juvan & Dolnicar, 2016; Mihalic, 2016: Mathew & Sreejesh, 2017).

There are several terms used to address responsibility. On the demand side, there is environmentally responsible behaviour (ERB) or pro-environmental behaviour (PEB), also termed as complying (CERB), based on relevant regulations, and promoting (PERB), referring to the

DOI: 10.4324/9781003286301-26

promotion of responsible behaviour at the destination. Focusing on the business side, there is corporate social responsibility (CSR), while, at the destination level, there is the concept of destination social responsibility (DSR)[1] (Lee et al., 2013; He et al., 2018; Li & Wu, 2019; Luo et al., 2020). The four main groups of stakeholders involved include tourists, entrepreneurs (businesses and/or freelancers either directly or indirectly related to tourism activities), the public administration and residents (Su et al., 2018).

Tourists seem to be gradually shifting towards a more responsible way of travelling. Therefore, this chapter examines tourist motivations that lead them to adopt responsible behaviours, reflecting the demand side, as well as the reaction of the tourism industry and destination, constituting the supply side.

The Demand Side of Tourism Responsibility

Tourism does not only influence the development of places but also affects how people perceive the geographies of destinations, including places and their population (Palomino-Schalscha, 2012). These perceptions are reflected in tourist behaviour by shaping their motivations, reactions and attitudes. According to Lee et al. (2013), tourists affect the destinations they visit either intentionally or unintentionally through their activity. In order to mitigate the negative impacts of tourists it is essential to understand their behaviour, motivations, perceptions and attitudes.

Two theoretical approaches in social psychology have been used to address tourism behaviour, one based on morality, influenced by personal benefits and costs, and one on rationality, influenced by social motives (Li & Wu, 2019). In addition, a series of studies have attempted to 'segment' tourists in terms of a specific source market, a specific destination, a specific product within a destination, or a combination of the three (Mody et al., 2014).

Tourist motivations can be divided into push and pull factors. Push factors are related to the travellers and their emotions, values and desires, and affect their decisions. Pull factors relate to the factors engaging them to travel, the destination, its distinctive characteristics, services, experiences and benefits that influence tourists' choice of a destination (Dann, 1977; Kozak, 2002; Mody et al., 2014; Villamediana-Pedrosa et al., 2020; Bayih & Singh, 2020). The analysis of push and pull factors is considered essential in order to segment markets, develop products and design packages and promotional plans (Villamediana- Pedrosa et al., 2020). Bayih & Singh (2020) examine the factors influencing tourist behavioural intentions partially through the analysis of push (knowledge or pursuit of knowledge and family and togetherness) and pull factors (cultural heritage, events and festivals and natural heritage).

Similarly, Mody et al. (2014) define nine tourist motivations and divide them into push and pull factors. Out of the nine motivations – namely nature, responsible operator, rurality, responsibility, escape, personal development, family, socialisation and travel bragging – the two responsibility-related motivations (responsibility and responsible operator) rank highly. This means that tourists seem to be pushed by an inner motivation towards more responsible experiences while, at the same time, they seem to be pulled by responsible activities in the destinations.

Therefore, one might argue that tourist behaviour might be influenced not only by the different types of motivations but also by the type of destination (Villamediana-Pedrosa et al., 2020). Under this notion, Sirgy & Su (2000 cited in Martin et al., 2013: 690), consider that self-congruence (matching tourists' image of a destination) and functional congruence (matching tourists' needs from a destination) are factors that affect tourists' choice of destination. Furthermore, Zgolli & Zaiem (2018) conclude that social engagement[2] and exemplarity of public power[3] have an important effect on tourists' responsible behaviour. Their results show that tourists who already show an interest in the environment seem to take the environmental and social aspects of the destination into account before choosing the place to visit. Moreover, destinations that engage in sustainable activities manage to encourage tourists to act responsibly.

According to Luo et al. (2020), tourist social responsibility includes two dimensions, one stressing the tourists' personal commitment to following regulations and restricting their behaviour (self-responsibility awareness) and one addressing the tourists' public behaviour (towards other tourists and local communities). In addition, Musavengane (2019) divides the forces that lead to responsible tourism into endogenous forces (altruism, legitimacy, competitiveness) and exogenous ones (pressure to adopt responsibility and conform to regulations, and advantages related to economic benefits). According to Wang et al. (2019a), there are three types of contextual factors affecting a tourist: personal, material (e.g., environmental regulation and interpretation signs) and environmental factors in the tourist destination (e.g., quality of the built environment).

He et al. (2018) summarise the factors affecting environmentally responsible behaviour, namely, attitude toward the environment, emotion, environmental knowledge, subjective norms, environmental sensitivity, concern for the environment, environmental awareness, place attachment, recreational involvement, recreation experience, ascribed responsibility, and injunctive and personal norms. The authors conclude that environmental commitment significantly impacts on tourist ERB.

The relationship established between tourists and the destination seems to affect their behaviour. The higher the commitment towards a destination the more likely tourists are to engage in responsible behaviour. Additionally, the relationship between the actual behaviour of tourists and the intention in their environmental behaviour is positively regulated by the environmental background demonstrated by the tourist destination (Wang et al., 2019a; Panwanitdumrong & Chen, 2021). Lee et al. (2021) found that sustainable intelligence[4], DSR, biospheric value and visit experience affect pro-environmental behaviour, with sustainable intelligence having the highest influence. Gautam (2020) examines the environmentally friendly behaviour of tourists and concludes that greater environmental knowledge of environmentally friendly products leads to a more positive attitude towards these products.

In terms of tourists' perceptions, Caruana et al. (2014) identify a set of narratives that address how tourists perceive their responsible experience. These narratives (instrumental opportunism, mindful minimising, educational empathy and conscious advocating) combine the level of involvement with the goal direction to responsible tourism.

Hosany & Prayag (2013) summarise the variables that affect tourists' interests, motivations and behaviour, namely, lifestyles, personal values, personality traits, novelty seeking, benefits sought and emotional reaction. They also review tourist typologies, starting with their general division into interactional, based on the interactions between the tourists and the destinations, and cognitive-normative, based on travel motivations. The interactional typology provides a 'continuum' that includes the organised mass tourist, the individual mass tourist, the explorer and the drifter. Under the cognitive–normative tourist typology, there is a continuum that includes the allocentric, near-allocentric, mid-centric, near-psychocentric, and psychocentric tourist, based on personality attributes. In addition, François-Lecompte & Prim-Allaz (2011, cited in Zgolli & Zaiem, 2018: 170) identify two types of responsible tourists, the "sustainable adventurer" and the "neo-sustainable" tourist, based on five factors related to tourists' comfort: use of responsible tour operators, protection of local resources and heritage, and travel distance.

The analysis of the existing research presented above indicates that responsible behaviour cannot only be attributed to personal factors, individual values and behaviour but is also inextricably linked to the destination, its characteristics, image and regulations, and reactions on the part of local communities. A list of push and pull factors emerges, formulating the demand side of responsibility in tourism, in the form of motivations and perceptions, types of tourists and actions on the part of the destinations. The push factors identified in recent research provide a new profile of responsible tourists, while pull factors enable destination communities to react accordingly in order to adopt responsible actions and address this new challenge.

Responsible tourists appear to have been part of the demand for some time, to which tourism operators eventually responded (Caruana et al., 2014). It takes two in this relationship between tourist and destination for responsibility to flourish as a practice and show its results and effectiveness. Therefore, it should only be considered under an integrated governance framework and perspective.

Supply Side of Tourism Responsibility

Perceptions and Practices of Social Responsibility

According to Su et al. (2018), tourism destinations are geographical units that offer appropriate infrastructure, services and experiences in order to fulfil tourists' needs. As Gibson (2012) argues, tourism as a production sector does not distance the consumer from the worker, thus creating an inextricable link between them that highlights the complexity of encounters in tourism.

When taking into account tourism-related businesses, the term most usually found is corporate social responsibility (CSR). CSR is related to actions taken by the tourism industry in order to maximise economic benefits while, at the same time, improving the quality of life of the workforce and the conditions of the local community by developing strong connections leading to responsible supply chain management. The term has also evolved to include activities such as stakeholder engagement, community action, ethical leadership and environmental stewardship (Hall & Brown, 2006; Coles et al., 2013; Garay Tamajon & Font i Aulet, 2013).

Taking into account the heterogeneity of tourism-related businesses, the reasons for adopting a CSR approach differ in the case of SMEs and large enterprises. For instance, SMEs may be influenced by the owner's personal values, the need to be a part of and improve their connections among the local community, the desire to improve the quality of their workforce and the need to improve their reputation within and outside the destination. The main reasons for SMEs to adopt a CSR approach seem to be altruistic (environmental protection, commitment, improvement of society) or related to lifestyle. Economic reasons also appear to be important, but of secondary priority, along with image differentiation and legitimacy (Garay Tamajon & Font i Aulet, 2013).

Frey and George (2010) conclude that tourism businesses are in favour and realise the benefits of responsible actions. However, they face numerous constraints in changing their management practices. Such constraints include the high costs of responsible tourism management initiatives, lack of government support and investment in responsible tourism systems, the low level of environmental awareness and lack of knowledge and skills in understanding and establishing responsible mechanisms. Additional issues emerge when adding the small size of a tourism business, such as lack of knowledge of the process to adopt and the benefits of CSR actions, and the long and complex bureaucratic procedures (Frey & George, 2010; Garay Tamajon & Font I Aulet, 2013; Musanvengane, 2019).

The actions most usually found in CSR (for large or small companies) are related to environmental management, community dialogue and staff relationships (Garay Tamajon & Font i Aulet,

2013). However, CSR does not yet reflect the destination as a whole, rather it focuses on businesses and economic benefits. The way that residents perceive tourism could ultimately assist in demonstrating overall the destination through benefits for the residents and satisfaction towards tourism development (Su et al., 2018). Indeed, Mathew & Sreejesh (2017) argue that responsible tourism actions that contribute to the sustainability of a destination enhance the local community's quality of life. Their findings show that four dimensions stand out: community engagement, employment opportunities, skill development programs and public awareness.

In destinations where local communities play an important role in tourism development through their authenticity, distinctive culture and traditions, they become part of the attractions provided by the destinations. Therefore, their incorporation in tourism planning is essential and should include profiling of the residents, their behaviour and commitment, and the development of tools to effectively engage them in decision- making. Such initiatives not only create a sense of confidence, ownership, belonging and acceptance of tourism development but are also related to increased socio-economic benefits (Presenza et al., 2013; Zgolli & Zaiem, 2018).

Su et al. (2020b) argue that not all social responsibility strategies are appropriate for every destination, and each strategy may receive different responses from tourists. In addition, it is considered that social responsibility strategies can be either proactive, designed before nonresponsible behaviour, or reactive, designed after harmful effects. Proactive strategies are found to create stronger intentions and more positive attitudes from tourists, are more related to altruism, less connected to the destination's interests, and reflect the destination's moral obligations and social benefits.

Destinations and Visitors: Situational Factors Affecting Responsibility

In the context of linking destination attributes with tourist behaviour, four types of situational factors that influence tourist behaviour were identified: environmental background, environmental interpretations, behavioural reference and environmentally proactive facilities (Wang et al., 2018; Wang et al., 2019a). In addition, the improvement of laws, regulations and behavioural reference conditions can enhance tourists' responsible behaviour (Wang et al., 2019b). Furthermore, recent literature acknowledges that the reputation of a destination, in terms of environmental protection, security and trustworthiness, plays an important role in tourist behaviour as well as shaping the preferences of tourists towards different destinations (Marinao Artigas et al., 2017; Su et al., 2020a; Hassan & Soliman, 2021; Penagos-Londoño et al., 2021). The relationship between host communities and visitors has also been examined by Lee et al. (2013) and Li et al. (2021), who find that tourists who have strong environmental values, sensitivity and place attachment are more likely to show environmentally responsible behaviour.

Tourist behaviour tends to be more responsible when visiting a destination with high-level environmental conditions and background, security, trustworthiness and values close to the tourists' personal values, while, at the same time, attracting responsible tourists could lead to more effective environmental protection actions (Wang et al., 2019a; Li & Wu, 2019; Penagos-Londoño et al., 2021). Therefore, it is considered imperative to take into account the behaviour of all stakeholders who impact on a destination when addressing tourism development. This will ensure, in turn, a positive influence on the visitors' attitudes and behaviour toward responsible actions (Su & Swanson, 2017).

Overall, it seems that the notion of responsibility, on the supply side, is as complicated as on the demand side. On the one hand, supply-driven proactive actions of the destination could develop specific attributes to reflect responsible actions, attract responsible tourists and stimulate responsible behaviour. In this case, the motives for such actions can be either altruistic (e.g., environmental awareness) or based on socio-economic benefits for both businesses and destinations.

On the other hand, reactions of entrepreneurs and local communities in terms of adopting and addressing the challenge of responsibility, usually draw their inspiration from the emerging needs and preferences of tourists. The latter case could also be regarded on the side of tourists who are already searching for established responsible practices by destinations. This is only one example of the complex interrelationship between the destination and the visitor. These greater complexities and interactions among the groups involved create a necessity to identify and analyse governance structures and schemes that could facilitate the incorporation of responsible actions in tourism development.

The Role of Governance in Tourism Responsibility

According to Mihalic (2016), tourism responsibility basically expresses the fourth pillar of sustainability, namely governance, addressed to all those involved in tourism. The elements that reflect governance issues in tourism responsibility stem from the need for awareness of sustainability and ethics (including a willingness to pay), environmental education and information for all groups of stakeholders involved, representing both the demand and supply sides. In addition, when focusing on the destination, participation, political consensus and leadership are key aspects to achieving both sustainability and responsibility.

Governance in tourism responsibility seems to entail four aspects that affect decision-making, the allocation of resources for responsible practices and the behaviour and actions of the stakeholders involved (Bramwell et al., 2008; Hall, 2012). The first aspect is related to the identification of such stakeholders. As stated above, these can be grouped into four categories, according to their position on the demand or supply side. The second aspect refers to the relationships among these groups of stakeholders. Even in the case of CSR, which is related to increasing the benefits of tourism businesses, one could argue that its definition and basic components promote the development of a system of stakeholders that build interrelationships in order to benefit the destination they are activated in. This notion leads directly to enhancing a governance system that relies on the key dimension of responsibility: Accountability. Here, accountability can be regarded as the allocation and acceptance of taking responsibility for tourism actions and assessing whether these actions of responsibility have actually been met by the network of stakeholders (Musavengane, 2019).

The third aspect addresses the issue of individual responsibility or responsibility reflecting the perceptions and actions of certain groups or the destination as a whole. Actions related to individual environmentally sustainable tourist behaviour include paying carbon offsets, choosing more environmentally friendly means of transportation, choosing accommodation based on environmentally friendly standards, using a provider that follows a sustainable tourism policy, reusing towels and making careful use of water, low electricity consumption, purchasing of green products, minimising waste production (e.g., recycling, reusing products, picking up litter, zero-food waste approach) (Juvan & Dolnicar, 2016; Bahja & Hancer, 2021). Examples of actions reflecting destination responsibility and requiring cooperation and collective action on the part of certain groups of stakeholders include promoting recycling, green transportation, use of renewable resources, consuming more sustainable meals, providing interpretation services, recreational activities and experiences and educating tourists towards environmental behaviour in order to avoid negative effects on the environment, and moving only within designated areas (Lee et al., 2013; John, 2020).

The fourth aspect raises the issue of whether the notion of responsibility should be a part of a local vision or a mass movement. It could be argued that tourism responsibility could be used as a tool to enhance skills and conservation actions at the destination level (Palomino-Schalscha, 2012). The concept of responsibility entails planning, policy and development processes with an emphasis on community involvement, sustainable resource management, equitable distribution of benefits and the minimisation of impacts on local communities (Grimwood et al., 2015). However, at the same time, external factors could call for responsibility in order to deal with a global effect or external factors such as the health crisis that has recently severely affected the sector. The crisis has rendered the existence of dynamic governance structures even more imperative in order to address problems collectively. Destinations can be very sensitive to positive or negative external factors that influence tourism flows, especially those destinations most dependent on tourism, and that generate adaptive or proactive behaviour (Guaita Martínez et al., 2020; Hassan & Soliman, 2021; Almeida et al., 2021).

As Sigala (2020) argues, crises, such as the COVID-19 pandemic, can affect tourist behaviour in psychological terms, changing their values, lifestyles and priorities, reshaping their travel mobilities and perceptions of tourism experiences and improving their understanding of overtourism, their respect for the places they visit. At the same time, destinations are rethinking and redefining tourism, planning for less crowded activities, keeping health and distancing standards, revisiting value chains and incorporating new technologies in order to meet the new demands.

Reflections

This chapter concludes by arguing that integrated actions reflecting the interests and attitudes of both the demand and supply side of responsible tourism are essential in order to ensure that all parties involved are acting in favour of the destination, a key for effective governance of sustainable tourism. In this context, the improvement of governance structures and relationships at the destination level is imperative in order to achieve the desired outcome. Such improvements include the creation and/or improvement of networks among the four basic groups of stakeholders, efforts for integrated and collective responsible actions, and the incorporation of such actions into destination governance frameworks. An integral part of these actions is their extension to spatial dimensions affecting local destinations. It is evident that the incorporation of responsibility in destination tourism development will not necessarily lead to changes from one type of tourism to the other (mass to slow tourism or ecotourism to coastal tourism). However, it will gradually shift the focus towards a more sustainable model which will eventually affect the type of activities and measures undertaken and thus the overall socio-economic performance and environmental conditions of the destination as well as cooperation and interactions among stakeholders.

Tourists, as consumers of tourism products, are considered unpredictable and inconsistent in their behaviour (Caruana et al., 2014), thus further research that links attitudes and behavioural patterns to the destinations could provide more insights into tourists' behaviours and their extension on tourism geography. In addition, this 'ascription of responsibility' that has emerged, especially due to the COVID-19 pandemic, could be further examined in terms of enhancing the sustainability of the sector (Han et al., 2020). In addition, as Hall & Brown (2006) claim, the tourism industry should not only take into account the consumers but also be inspired by them in order to build responsibility and mainstream its path as a viable process.

The increased interest in tourism responsibility, on both the demand and the supply side, and the complexity of the relationships among the stakeholders involved and their motivations call for further research into sustainable tourism governance schemes. The discussion often leads to the questions of who is responsible for initiating responsible actions and how could the groups of stakeholders involved be motivated and supported. It seems that searching for responsibility – on either the demand or the supply side – is simply not enough. Effective cooperation and governance could promote and establish responsible actions in tourism destinations for safeguarding the destinations' tourism potential for sustainable development.

Notes

1 DSR is defined as collective stakeholder obligations and activities that minimise their impacts on and protect and improve the socio-economic and environmental conditions of a destination (Su & Swanson, 2017; Su et al., 2018; Su et al., 2020b).
2 The engagement of a tourist in actions that protect the environment.
3 Here referred to as responsible actions by Public Administrations that benefit local communities, the natural and cultural environment.
4 Sustainable intelligence refers to visitors' awareness of tourism and their ability to demonstrate responsible behaviour (López-Sánchez & Pulido-Fernández, 2016; Lee et al., 2021).

References

Almeida, A., Golpe, A., & Justo, R. (2021). Regional tourist heterogeneity in Spain: A dynamic spatial analysis. *Journal of Destination Marketing & Management, 21*, 100643.

Bahja, F. & Hancer, M. (2021). Eco-guilt in tourism: Do tourists intend to behave environmentally friendly and still revisit? *Journal of Destination Marketing & Management, 20*, 100602.

Bayih, B. E. & Singh, A. (2020). Modeling domestic tourism: Motivations, satisfaction and tourist behavioral intentions. *Heliyon, 6*, e04839.

Bramwell, B., Lane, B., McCabe, S., Mosedale, J., & Scarles, C. (2008). Research perspectives on responsible tourism. *Journal of Sustainable Tourism, 16* (3), 253–257.

Caruana, R., Glozer, S., Crane, A., & McCabe, S. (2014). Tourists' accounts of responsible tourism. *Annals of Tourism Research, 46*, 115–129.

Chang, T. C. (2012). Making and unmaking places in tourism geographies. In J. Wilson (Ed.), *The Routledge Handbook of Tourism Geographies* (pp. 133–138). Routledge.

Coles, T., Fenclova, E., & Dinan, C. (2013). Tourism and corporate social responsibility: A critical review and research agenda. *Tourism Management Perspectives, 6*, 122–141.

Dann, G. M. S. (1977). Anomie, ego-enhancement and tourism. *Annals of Tourism Research, 4* (4), 184–194.

François-Lecompte, A., & Prim-Allaz, I. (2011). Les Français et le tourisme durable: étude des représentations. *Décisions Marketing, 64*, 47–61.

Frey, N., & George, R. (2010). Responsible tourism management: The missing link between business owners' attitudes and behaviour in the Cape Town tourism industry. *Tourism Management, 31*, 621–628.

Garay Tamajón, L., & Font i Aulet, X. (2013). Corporate social responsibility in tourism small and medium enterprises evidence from Europe and Latin America. *Tourism Management Perspectives, 7*, 38–46.

Gautam, V. (2020). Examining environmental friendly behaviors of tourists towards sustainable development. *Journal of Environmental Management, 276*, 111292.

Gibson, C. (2012). Geographies of tourism: Space, ethics and encounter. In J. Wilson (Ed.), *The Routledge Handbook of Tourism Geographies* (pp. 55–60). Routledge.

Goodwin, H. (2011). *Taking Responsibility for Tourism.* Goodfellow Publishers Limited.

Grimwood, B. S. R., Yudina, O., Muldoon, M., & Qiu, J. (2015). Responsibility in tourism: A discursive analysis. *Annals of Tourism Research, 50*, 22–38.

Guaita Martínez, J. M., Martín Martín, J. M., & del Sol Ostos Rey, M. (2020). An analysis of the changes in the seasonal patterns of tourist behavior during a process of economic recovery. *Technological Forecasting & Social Change, 161*, 120280.

Hall, C. M. (2012). Governance and responsible tourism. In D. Leslie (Ed.), *Responsible Tourism. Concepts, Theory and Practice* (pp. 107–118). CABI.

Hall, D., & Brown, F. (2006). *Tourism and Welfare: Ethics, Responsibility and Sustained Well-being.* CABI.

Han, H., Lee, S., Kim, J. J., & Ryu, H. B. (2020). Coronavirus disease (COVID-19), traveller behaviors, and international tourism businesses: Impact of the corporate social responsibility (CSR), knowledge, psychological distress, attitude, and ascribed responsibility. *Sustainability, 12*(20), 8639.

Hassan, S., & Soliman, M. (2021). COVID-19 and repeat visitation: Assessing the role of destination social responsibility, destination reputation, holidaymakers' trust and fear arousal. *Journal of Destination Marketing & Management, 19*, 100495.

He, X., Hu, D., Swanson, S. R., Su, L., & Chen, X. (2018). Destination perceptions, relationship quality, and tourist environmentally responsible behavior. *Tourism Management Perspectives, 28*, 93–104.

Hosany, S., & Prayag, G. (2013). Patterns of tourists' emotional responses, satisfaction, and intention to recommend. *Journal of Business Research, 66*, 730–737.

John, E. (2020). Do tourists behave in a sustainable manner at destinations? Empirical evidence from Munnar. *Materials Today: Proceedings, 33*, 4866–4875.

Juvan, E., & Dolnicar, S. (2016). Measuring environmentally sustainable tourist behaviour. *Annals of Tourism Research, 59*, 30–44.

Kozak, M. (2002). Comparative analysis of tourist motivations by nationality and destinations. *Tourism Management, 23*, 221–232.

Lee, T. H., Jan, F.-H., & Yang, C.-C. (2013). Conceptualizing and measuring environmentally responsible behaviors from the perspective of community-based tourists. *Tourism Management, 36*, 454–468.

Lee, C.-K., Olya, H., Ahmad, M. S., Kim, K. H., & Oh, M.-J. (2021). Sustainable intelligence, destination social responsibility, and pro-environmental behaviour of visitors: Evidence from an eco-tourism site. *Journal of Hospitality and Tourism Management, 47*, 365–376.

Li, Q-C., & Wu, M-Y. (2019). Rationality or morality? A comparative study of pro-environmental intentions of local and nonlocal visitors in nature-based destinations. *Journal of Destination Marketing & Management, 11*, 130–139.

Li, S., Liu, M., & Wei, M. (2021). Host sincerity and tourist environmentally responsible behavior: The mediating role of tourists' emotional solidarity with hosts. *Journal of Destination Marketing & Management, 19*, 100548.

López-Sánchez, Y., & Pulido-Fernández, J. I. (2016). In search of the pro-sustainable tourist: A segmentation based on the tourist "sustainable intelligence". *Tourism Management Perspectives, 17*, 59–71.

Luo, W., Tang, P., Jiang, L., & Su, M. (2020). Influencing mechanism of tourist social responsibility awareness on environmentally responsible behavior. *Journal of Cleaner Production, 271*, 122565.

Marinao Artigas, E., Chasco Yrigoyen, C., Torres Moraga, E., & Barra Villalón, C. (2017). Determinants of trust towards tourist destinations. *Journal of Destination Marketing & Management, 6*, 327–334.

Martin, D., Sirakaya-Turk, E., & Cho, W. (2013). International tourism behavior in turbulent times: Introduction to the special issue. *Journal of Business Research, 66*, 689–691.

Mathew, P. V., & Sreejesh, S. (2017). Impact of responsible tourism on destination sustainability and quality of life of community in tourism destinations. *Journal of Hospitality and Tourism Management, 31*, 83–89.

Mihalic, T. (2016). Sustainable-responsible tourism discourse: Towards 'responsustable' tourism. *Journal of Cleaner Production, 111*, 461–470.

Mody, M., Day, J., Sydnor, S., Jaffe, W., & Lehto, X. (2014). The different shades of responsibility: Examining domestic and international travelers' motivations for responsible tourism in India. *Tourism Management Perspectives, 12*, 113–124.

Musavengane, R. (2019). Small hotels and responsible tourism practice: Hoteliers' perspectives. *Journal of Cleaner Production, 220*, 786–799.

Palomino-Schalscha, M. (2012). Geographies of tourism and development. In J. Wilson (Ed.), *The Routledge Handbook of Tourism Geographies* (pp. 187–193). Routledge.

Panwanitdumrong, K., & Chen, C.-L. (2021). Investigating factors influencing tourists' environmentally responsible behavior with extended theory of planned behavior for coastal tourism in Thailand. *Marine Pollution Bulletin, 169*, 112507.

Penagos-Londoño, G. I., Rodriguez–Sanchez, C. Ruiz-Moreno, F., & Torres, E. (2021). A machine learning approach to segmentation of tourists based on perceived destination sustainability and trustworthiness. *Journal of Destination Marketing & Management, 19*, 100532.

Presenza, A., Del Chiappa, G., & Sheehan, L. (2013). Residents' engagement and local tourism governance in maturing beach destinations. Evidence from an Italian case study. *Journal of Destination Marketing & Management, 2*, 22–30.

Sigala, M. (2020). Tourism and COVID-19: Impacts and implications for advancing and resetting industry and research. *Journal of Business Research, 117*, 213–321.

Sirgy, M. J., & Su, C. (2000). Destination image, self-congruence, and travel behaviour: toward an integrative mode. *Journal of Travel Research,* 38, 340–52.

Su, L., Gong, Q., & Huang, Y. (2020b). How do destination social responsibility strategies affect tourists' intention to visit? An attribution theory perspective. *Journal of Retailing and Consumer Services, 54*, 102023.

Su, L., Hsu, M. K., & Boostrom, R. E. Jr. (2020a). From recreation to responsibility: Increasing environmentally responsible behavior in tourism. *Journal of Business Research, 109*, 557–573.

Su, L., Huang, S., & Pearce, J. (2018). How does destination social responsibility contribute to environmentally responsible behaviour? A destination resident perspective. *Journal of Business Research, 86*, 179–189.

Su, L., & Swanson, S. R. (2017). The effect of destination social responsibility on tourist environmentally responsible behavior: Compared analysis of first time and repeat tourists. *Tourism Management, 60*, 308–321.

Villamediana-Pedrosa, J. D., Vila-López, N., & Küster-Boluda, I. (2020). Predictors of tourist engagement: Travel motives and tourism destination profiles. *Journal of Destination Marketing & Management, 16*, 100412.

Wang, C., Zhang, J., Cao, J., Hu, H., & Yu, P. (2019a). The influence of environmental background on tourists' environmentally responsible behaviour. *Journal of Environmental Management, 231*, 804–810.

Wang, C., Zhang, J., Cao, J., Duan, X., & Hu, Q. (2019b). The impact of behavioral reference on tourists' responsible environmental behaviors. *Science of the Total Environment, 694*, 133698.

Wang, C, Zhang, J., Yu, P., & Hu, H. (2018). The theory of planned behavior as a model for understanding tourists' responsible environmental behaviors: The moderating role of environmental interpretations. *Journal of Cleaner Production, 194*, 425–434.

Zgolli, S., & Zaiem, I. (2018). The responsible behavior of tourist: The role of personnel factors and public power and effect on the choice of destination. *Arab Economic and Business Journal, 13*, 168–178.

23

REGENERATIVE TOURISM IN THE MAKING

Reflections about an Emerging Frontier for Tourism Geographies

Loretta Bellato, Niki Frantzeskaki and Christian A. Nygaard

Introduction

Various tourism scholarship approaches conform to the sustainable development paradigm. These include steady-state, responsible, sustainable, resilient, transformative and hopeful tourism, among others (Ateljevic, 2020; Cheer, 2020). However, many tourism scholars and practitioners criticise the United Nations-led sustainable development agenda for failing to question the underlying growth paradigm. In doing so, they also sustain the drivers of social-ecological destruction and loss (Dwyer, 2018; Higgins-Desbiolles, 2018; Pollock, 2019a; 2019b; Scheyvens & Biddulph, 2018; UNWTO & UNDP, 2017). In contrast to sustainable tourism, regenerative tourism positions tourism activities as interventions that develop the capacities of places, communities and their guests to operate in harmony with interconnected social-ecological systems. As such, regenerative tourism aligns with the regenerative development paradigm despite resembling sustainable development approaches.

Regenerative tourism promotes tourism innovations by embedding tourism practices within local communities and ecological processes that elevate human and non-human wellbeing (Bellato & Cheer, 2021). Evolving from a long conceptual lineage drawing from Indigenous perspectives and knowledges, and Western science, regenerative tourism has emerged as a niche aiming to improve and transform social-ecological systems where tourism practices occur (Hes & Coenen, 2018). The practice-led regenerative development paradigm applies to numerous sectors, including built environments and urban planning (Mang & Haggard, 2016), regenerative agriculture (Haines, 2020) and regenerative economies (Lovins, 2020; Raworth, 2017). The theory and practice of regenerative approaches also address climate change, urbanisation, justice and inequality (Caniglia et al., 2020). However, there is little clarity and agreement on the transformative potential of regenerative tourism or its applications to practice (Cheer & Lew, 2018).

This chapter asks, therefore, 'What is known about the transformative basis of regenerative tourism?' Methodologically, this chapter updates the literature review and findings published in *Tourism Geographies* (Bellato, Frantzeskaki, & Nygaard, 2023) with additional research and insights until October 2022. Regenerative tourism is an emergent concept and theme in the tourism

DOI: 10.4324/9781003286301-27

field. The updated analysis of theory and practice, thus, offers an additional critical reflection on the use and application of the regenerative tourism concept in tourism scholarship.

Specifically, the chapter (1) maps the nature and scope of the existing literature, (2) traces the conceptual and philosophical antecedents of the regenerative concepts and practice, (3) discusses regenerative tourism definitions, (4) analyses the positioning of regenerative tourism in relation to sustainable tourism, (5) discusses regenerative tourism practice principles and conceptual frameworks, and (6) reflects on current advancements of the concept within tourism scholarship and implications for tourism geographies. The analysis is grounded through leading practitioner consultations to identify regenerative tourism practices, their distinctness, and transformative potential.

Materials and Methods

A scoping review method, supplemented by a consultation exercise, was undertaken by non-Indigenous scholars from diverse disciplines using the methodological framework for mapping emerging fields of knowledge outlined by Arksey and O'Malley (2005). Given the niche status of regenerative tourism as praxis, the perspectives, concepts and frameworks shared via the practitioner consultations and grey literature are critical to conceptual development. The authors acknowledge the perspectives, practices, custodianship and their enduring connections with the lands and waterways included in this study by the Traditional Owners and their ancestors. The authors also thank the scoping review consultation participants who shared their wisdom and insights. Figure 23.1 summarises the research process for the study, including its recent update in chronological order.

Stages 1–3: Identifying and Selecting Relevant Papers

The search terms applied in the peer-reviewed searches include: Regenerative tourism, regenerative AND tourism, regenerative travel, conscious travel, conscious tourism, tourism AND regeneration, travel AND regeneration, regenerative development AND tourism, regenerative design AND tourism, *turismo regenerativo* (Spanish). 'Conscious travel' as a search term enabled a review of early concept formations before the niche consolidated around the term 'regenerative tourism'. The following search terms were applied to identify grey literature publications: regenerative tourism, *turismo regenerativo*, conscious travel. Additional search terms derived from an initial review of peer-reviewed papers were searched: Tourism AND regenerative economy OR sustainable futures OR stewardship OR beyond sustainability OR living system OR ecological worldview OR decolon*. The publication search period was limited to 2007–July 2020 with three subsequent relevant publications added during 2020 to map the early formations of the concept. English and Spanish language publications were included, as these languages dominate discussions regarding regenerative tourism.

The search involved applying the search terms to the following data sources: EBSCOhost; Scopus; Web of Science; and hand search. In addition, we included grey literature sourced from practitioner blogs, websites, reports, conference proceedings and books to capture contributions from pioneering practitioners of regenerative tourism and non-peer-reviewed publications by scholars. The study did not include tourism led or owned by Indigenous Peoples, drawing from their own perspectives, knowledge systems and practices. We performed two levels of screening (see Figure 23.1). A total of 84 peer-reviewed publications and 116 grey literature publications were included for review over two iterations. Nine publications were written in Spanish.

Stage 1: Identifying the research question	What is known about regenerative tourism from the existing literature?	
Stage 2: Identifying relevant studies	Database and grey literature search (2007 – July 2020)	Title, Keywords, Abstract: Duplicate and irrelevant publications removed
Stage 3: Study selection	2 levels of screening using Endnote 1. Abstracts: n=1889 2. Full text: n = 359	2. Full text: Publications excluded if inclusion criteria not met
Stage 4: Charting the data	NVivo for charting Peer reviewed: n=59 Grey literature: n=116	Categorisation of literature and coding of themes
Stage 5: Collating, summarising the results	Qualitative synthesis and mapping core concept elements	Analytical framework determined initial analysis
Stage 6: Consultation exercise	PHASE 1: Consultation undertaken with 9 non-Indigenous practitioner experts	Analysis revised - findings validation, gaps, future research goals identification
	PHASE 2: Consultation undertaken with 5 Indigenous practitioner experts	Analysis revised - findings validation, gaps, future research goals identification
Stage 7: Reporting the results	Brief report provided to participants and findings presented to academic audiences	
Stage 8: Updating the results	Stages 3,4,5,7 repeated – 25 peer-reviewed papers published up to October 2022 included	Analysis revised and report updated

Figure 23.1 Scoping review and consultation process.

Stages 4 and 5: Charting and Analysing the Data

Using NVivo, themes were created to code the qualitative data according to the fundamental research question. Initial findings were produced using two analytical categories: Use and core ideas. *Use* relates to how the literature operationalises the concept into practice. Core ideas identify conceptual interpretations of practice. The strengths and weaknesses of the concept were then interpreted against the core elements of a concept (including origins, definition, dimensions, and principles) and how it relates to dominant tourism-related discourses.

Stage 6: Consultation Exercise

A consultation exercise comprising interviews and focus groups with leading practitioners was completed in two phases following the recommendations of Levac et al. (2010). Nine leading regenerative tourism practitioners participated in the consultation exercise in the first phase. Five Indigenous practitioners were interviewed in the second consultation phase, and three participated in a focus group. In keeping with a decolonial approach, including Indigenous perspectives and co-production of knowledges with Indigenous Peoples is essential. Each consultation exercise phase involved presenting preliminary findings outlining the central characteristics of regenerative tourism and its evolution; validating the findings; and identifying gaps and practitioner research priorities. In addition, the consultation exercise phases informed refinements of the working definition, conceptual framework, the evolutionary lineage of regenerative tourism, and future research priorities.

Stage 7: Reporting the Results

The results of this study were published in the *Tourism Geographies* journal and reported at various conferences.

Stage 8: Updating the Results

From the time of conducting the literature review and writing this chapter, 25 published peer-reviewed papers were identified using the same sourcing and analysing methods above (not including the *Tourism Geographies* paper upon which this chapter is based). Additional perspectives were included and the emerging discourse within tourism scholarship was mapped.

Findings

1   Mapping the State-of-the-art

The first review of publications published between 2007 and July 2020 mapped the early formations of the concept by pioneering practitioners and initial adoption by scholars. Evidently, few peer-reviewed papers consider the concept of regenerative tourism. The peer-reviewed literature investigated regenerative approaches or originated from tourism scholars more broadly interested in sustainable tourism. Sixty-six percent of the literature identified in the scoping review was grey literature, reflecting the emergent and practitioner-led evolution of regenerative tourism.

Of the 84 peer-reviewed papers analysed, thirty-three (39%) academic articles that discussed 'regenerative tourism'. Only eight (10%) papers discussing regenerative tourism were published before the initial search. Five peer-reviewed articles discussed conscious travel (forerunner to regenerative tourism), all from the tourism field. None of the most recently published articles discussed conscious travel. The remaining 48 peer-reviewed papers discussed concepts other than but closely related to regenerative tourism. Most scholars come from English-speaking Western countries and write mostly about places in Europe and Oceania. The review update in 2022 identified an additional 25 peer-reviewed papers, either with a direct focus on regenerative tourism or some discussion of the concept. All papers were written in English, with none written in Spanish. Tourism journals published most papers except one ecology journal and the other a land system science journal. Notably, a high proportion of the papers about regenerative tourism have been viewpoint papers proposing various conceptual models or notions claimed to be relevant to regenerative tourism. The regional bases of scholars and case study contexts have widened to places in Asia and North America, with a concentration of papers focusing on Aotearoa New Zealand.

The grey literature included in the review (*n*=116) focused on conscious travel, regenerative tourism or regenerative approaches to tourism. The search identified 33 (28%) grey literature publications using the term regenerative tourism and 37 (32%) publications using conscious travel. The practitioner publications and locations of authors are more diverse than the scholars. Amongst the grey literature, scholars published two conference papers, two books, and one book chapter, practitioners primarily used blogs, books, masters' theses, reports and conference papers. A tourism practitioner first published the term regenerative tourism in 2017 (Araneda, 2017) before gaining wider use by practitioners.

The regenerative tourism-focused publications are mostly by non-Indigenous people drawing from practice, scientific and Indigenous perspectives and knowledges. Only one paper is identified as authored by Indigenous scholars (Matunga et al., 2020). No practitioner authors self-identified as Indigenous persons. The reviewed publications and consultation participants draw upon various Indigenous perspectives, knowledge systems and practices. In some regenerative tourism publications, the Indigenous origins of the knowledge are unclear. In others, the authors identify the following Indigenous communities: Māori (Cave & Dredge, 2020; Matunga et al., 2020; Pollock, 2012c, 2020); Arawak Peoples of the Caribbean (Sheller, 2021); Australian Aboriginal (Owen, 2007a; Pollock, 2012c) perspectives, knowledge systems and practices. Most non-Indigenous consultation participants revealed that their understandings of regeneration and its tourism applications were fundamentally shaped by working with and learning from Indigenous Peoples and nominated Australian Aboriginal, Canadian Aboriginal, Māori, Indigenous Peoples of Central and South America and Native American Peoples.

2 Evolving Regenerative Tourism

Regenerative tourism draws on Indigenous Peoples' continuous living cultures and evolving perspectives, knowledge systems, Western science and practice (Matunga et al., 2020). These influences form the basis of the ecological worldview, which sees the world as a dynamic complex whole with self-organising properties (Mang & Reed, 2012). This section traces the conceptual lineage, key influences, ideas, and events shaping the development of regenerative tourism as identified in the reviewed literature and practitioner consultations. We trace the conceptual origin from pre-Enlightenment worldviews that support living in connection with nature to the distinct separation of humans and nature following the scientific revolution (Pollock, 2012c), before the revival of understanding life as complex living systems (Mang & Reed, 2019; Raworth, 2017), leading to the recent introduction of the regenerative tourism concept.

Pre-Enlightenment

Underpinning regenerative tourism, permaculture and regenerative development draw upon Indigenous Peoples' worldviews, perspectives, knowledge systems, and cultures (Mang & Reed, 2019; Pollock, 2012a). In addition, the concept of flourishing (Cheer, 2020; Pollock, 2019a) used to describe desired outcomes of regenerative tourism was first considered by Aristotle in 340 BC. During this period, travel practices reflected a strong symbiosis between people and nature.

1600s to 1970: Initiation of the Scientific and Industrial Revolutions and Mass Tourism Growth

The scientific revolution, including Newton and Descartes mechanical philosophy, began in the 1600s and was followed by the Industrial Revolution from the 1760s (Teruel, 2018). In the late 1800s, tourism research commenced but focused on land and economics, later expanding to disciplines such as sociology and geography from the 1970s (Butler, 2015). The seminal work introducing the ecological worldview was *A Sand County Almanac* by Leopold in 1949 (Matunga et al., 2020). Boosterism emerged in the 1950s, sparking mass industrial tourism (Butler, 2015). Finally, in the 1960s, tourism scholars began researching supply and demand, carrying capacity, and managing natural areas (Butler, 2015).

1971 to 2004: The Emergence of Sustainability and Regeneration

Approaches to regenerative agriculture and economies, permaculture, regenerative design and development emerged simultaneously as sustainable development and sustainable tourism (Butler, 2015; Mang & Reed, 2012; Pollock, 2019b). Coincidentally, various concepts arose contributing to understanding sustainability and tourism (Butler, 2015). While Zaman et al. (2023) accurately claimed that Anders Arfwedson was the first to use the term 'regenerative tourism', Arfwedson's use of the term mirrors common conceptualisations of sustainable tourism development.

2005 to 2013: The Emergence of Regenerative Tourism and Expansion of Regenerative Approaches

Owen (2007a, 2007b) first introduced the term regenerative tourism regarding the architectural design of ecotourism facilities from a regenerative development perspective. In 2011, drawing from numerous influences, Pollock introduced the 'conscious travel' approach, which applied an ecological worldview to tourism (Pollock, 2012b). In 2012, a peer-reviewed paper on regenerative development included the Playa Viva resort established in 2005 (Mang & Reed, 2012). The interwoven influences of diverse Indigenous perspectives, knowledge systems and Western science on the development of regenerative tourism are evident in the referenced literature from this period.

2014 to 2021: Further Expansion and Development of Regenerative Tourism

Further conceptual and praxis maturation has been evident in recent years. Publications include those by Pollock (2015), Mang and Haggard (2016), and others related to economics (Fullerton, 2015; Raworth, 2017) and business (Sanford, 2019). Consequently, tourism practitioners have adopted regenerative tourism approaches extending beyond the design of facilities towards destination planning, tourism stakeholder and enterprise capacity development. From 2015 tourism scholars, including Dwyer (2018); Zivoder et al. (2015), and Becken (2019) began publishing papers drawing on the 'conscious travel' notion. During the COVID-19 pandemic, calls to rethink tourism saw tourism scholars considering regenerative tourism as an alternative (Ateljevic, 2020; Cave & Dredge, 2020; Cheer, 2020; Duxbury et al., 2021; Sheller, 2021).

2021 and 2022: Emerging Themes Arising from Regenerative Tourism Scholarship

The last two years have seen scholars beginning to examine regenerative tourism and journals hosting special issues with more journal articles, special issues and book chapters in the pipeline. Most recent papers explored or acknowledged the contributions of Indigenous knowledges, perspectives and practices as well as the significance of wellbeing. The special issue theme led by Ateljevic and Sheldon (2022) generated papers that linked the notion of transformational tourism with regenerative tourism. Many papers also explored inclusive tourism development as an element of a regenerative approach. In addition, scholars have begun introducing to the discourse concepts such as a-growth, degrowth, placemaking, harmony, feminist perspectives and nature-based solutions. These conceptual links to regenerative tourism show that it evolves as a boundary concept with the potential to bridge with other agendas that move beyond the growth paradigm and adopt different worldviews. Finally, scholars are also considering how various types of tourism, such as food and yachting, intersect with the regenerative tourism approach.

3 *Progressing Regenerative Tourism Practice*

Regenerative tourism can also be understood as a niche innovation pioneered by three non-Indigenous practitioner groups based in the UK, United States and Chile (Dwyer, 2018; Mang & Reed, 2012; Pollock, 2012b; Teruel, 2018). These niche innovators have developed varied regenerative tourism approaches from diverse place contexts, knowledges and practice bases. However, all broadly draw from regenerative development approaches. Uniquely, the two consultation participants in Chile reported being strongly influenced by Bernard Lievegoed's living organisations work and Steiner and Goethe's ideas about the phenomenology of imaginative consciousness.

Since 2005, these practitioners have advanced the regenerative tourism concept internationally by producing and disseminating publications, consulting enterprises and governments, and delivering practitioner training programs. Their efforts have been instrumental in broadly linking regenerative tourism niche innovations and actors with the more dominant sustainable tourism regime. Consequently, several enterprises are implementing regenerative tourism initiatives at local or regional levels, and two international alliances have formed. The Global Initiative for Regenerative Tourism was established in Latin America in 2015 (Araneda, 2019). The Regenerative Travel Alliance was initiated in 2019 (Regenerative Travel, 2020). Additional applications of regenerative tourism approaches are being developed by practitioners who draw upon non-tourism regenerative development innovations. Among the peer-reviewed papers, several authors crossed the academia–practitioner divide to undertake regenerative tourism work as government employees or social entrepreneurs. Some scholars have undertaken research deeply embedded in the places examined in their research.

4 *Defining Regenerative Tourism*

The review revealed that a universal definition of regenerative tourism is yet to be developed or adopted. Nevertheless, in the literature originating from the practitioner pioneers (Araneda, 2019; Pollock, 2019a; Teruel, 2018), several regenerative tourism attributes can be identified and distilled that form a conceptual core and enable further conceptual development. These eight core definition aspects include: draws from an ecological worldview, tourism is a living system and also is part of living systems, is a transformational approach, increases systems capacity for net positive effects, the purpose of tourism is for ongoing regeneration, design from the potential of the place, develops reciprocal relationships amongst stakeholders, and evolves and varies across places (Bellato et al., 2022: 9).

The peer-reviewed regenerative tourism literature has identified a more varied understanding of the concept, with some scholars emphasising certain aspects more than others. The earliest use of the term identified was applied to ecotourism facilities by Owen (2007a), an architecture scholar. She described regenerative tourism as critically engaging with the place, creating a positive impact, seeing humans as part of nature, and connecting environmentalism with socio-political processes. Cheer (2020) examines the concept of human flourishing underpinned by systems thinking and interconnectedness with nature. He identifies the prioritisation of a net positive benefit, including Indigenous approaches previously displaced by colonisation and inclusive development. Similarly, Matunga et al. (2020) explain regenerative tourism as an additive approach, an interconnected and reciprocal relationship between people and place for mutual benefit. Finally, Duxbury et al. (2021) describe regenerative tourism as systems-based, aligned with cultural and natural patterns, integrated with local development approaches and positions tourism practices as processes of regeneration.

Emphasising economic practices, Sheller (2021) describes regenerative tourism as embracing "alternative non-capitalist forms of ownership, non-monetary exchange and beneficial community-based development" (p. 2). She calls for a departure from colonisation, racial inequity, and extractive neoliberal development towards an alternative collective future. Cave and Dredge (2020) similarly envision regenerative tourism incorporating alternative economic practices to mediate global and local values and create 'well-th' (as defined by Māori), thus a more holistic view of wellbeing.

More recently published peer-reviewed papers show a continuation of the alternative economy discourse strongly influenced by the work of economists such as Raworth (Hartman & Heslinga, 2022), Graham and Gibson (Mathisen et al., 2022) and Fullerton (Sheldon, 2021), the degrowth and circular economy discourses (Tomassini & Cavagnaro, 2022; Blazquez et al., Chapter 5 of this volume). Explorations of the well-being concept open opportunities to redefine wealth and value creation involving tourism. Such conceptual work by Becken and Kaur (2021) may offer a bridge between seeing tourism as an alternative economy and more holistic thinking closely aligned with regenerative development concepts oriented to place. Concepts derived from regenerative approaches include living systems applications to tourism (Bellato et al., 2022), responsible gardener metaphor (Mathisen et al., 2022), capability development principles (Bellato & Cheer, 2021), co-evolutionary and mutually beneficial relationships (Becken & Kaur, 2021) and ecological worldview (Bellato et al., 2022; Dredge, 2022) and worldviews derived from specific place contexts (Becken & Kaur, 2021).

Among the most recently published peer-reviewed literature, some papers sought to offer simple descriptions of regenerative tourism. They tended to use phrases such as 'giving back more than we take' (Ateljevic & Sheldon, 2022) or leaving the destination in a better condition than it was found (Hartman & Heslinga, 2022) to reflect the intention to create net positive effects. Notions of improving tourism communities (Boluk & Panse, 2022) and 'replenishing and restoring what we have lost and building economies and communities that thrive' (Bhalla & Chowdhary, 2022) were also identified to reflect the emphasis on going beyond maintenance of the status quo and the need to heal social-ecological systems. Some papers also implied that regenerative tourism focused on achieving a stabilised end state (Tomassini & Cavagnaro, 2022), while others recognised the ongoing evolution of living systems and regeneration (Becken & Kaur, 2021; Dredge, 2022).

In our attempt to co-construct a working definition with consultation participants, some niche innovators argued that definitions are incongruent with regenerative approaches as they reinforce universalising, mechanistic, and reductive thinking. They instead advocated for inquiry approaches based on pluriversal knowing beyond abstractions and situated within real-world contexts and stories. This echoes scholars of decolonial studies and regenerative development practitioners who argue that knowledge is pluriversal (Chambers & Buzinde, 2015; Kramvig & Forde, 2020; Mang & Reed, 2012). Pluriversality asserts that knowledge cannot be universal due to different cultural contexts and varying impacts of processes such as colonisation and modernity in different places and communities (Chambers & Buzinde, 2015). Rather than republishing the working definition proposed in the (Bellato, Frantzeskaki, & Nygaard, 2023) paper, we invite readers to reflect upon and encourage engagement with the eight aspects above informed by the concept pioneers' literature to use as a basis from which to develop place-sourced definitions of regenerative development in keeping with pluriversality.

5 Distinguishing Regenerative Tourism and Sustainable Tourism

According to its proponents (Araneda, 2019; Howard et al., 2008; Pollock, 2019a), regenerative tourism is a distinct approach originating from the ecological worldview and regenerative paradigm. By comparison, the mechanistic worldview and industrial paradigm dominate sustainable tourism conceptualisations. The emerging academic regenerative tourism literature concurs with the regenerative tourism pioneers regarding the concept as distinct from sustainable tourism, requiring different approaches. Bellato, Frantzeskaki and Nygaard (2023) and Becken and Kaur (2021) have analysed and compared their distinctive characteristics.

6 Applying Tourism Practice Principles to Transformational Regenerative Development Practice Frameworks

No commonly adopted practice principles for regenerative tourism were identified in the literature. Nevertheless, several early publications offer principles for regenerative tourism or hospitality (Howard et al., 2008; Pollock, 2015; Regenerative Travel, 2020; Teruel, 2018). Matunga et al. (2020) offer six Ngāi Tahu values from a Māori worldview for designing regenerative tourism in Aotearoa – New Zealand. The Indigenous consultation participants supported regenerative tourism proponents incorporating Indigenous principles and approaches to benefit all human and non-human beings on the proviso that Indigenous cultural integrity is honoured. Relatedly, regenerative development principles were found in Mang and Haggard (2016) and Sanford (2019). In developing a conceptual framework for regenerative tourism, we drew on these publications and practitioner consultations to identify seven conceptual principles.

These seven principles, outlined in Bellato, Frantzeskaki and Nygaard (2023), are:

(i)   Draw from an ecological worldview,
(ii)   Use living systems thinking,
(iii)   Discover the unique potential of a regenerative tourism place,
(iv)   Leverage the capability of tourism living systems to catalyse transformations,
(v)   Adopt healing approaches that promote cultural revival, returning lands, and privileging of the perspectives, knowledges and practices of Indigenous and marginalised peoples,
(vi)   Create regenerative places and communities, and
(vii)   Collaborate to evolve and enact regenerative tourism approaches.

The principles were categorised under five design dimensions to guide regenerative tourism development transformations. The design dimensions were adapted from the Regenerative Development Framework (Mang & Reed, 2012: 34; 2019: 15). In addition, the tourism living system (TLS) proposed by Bellato and Frantzeskaki (2021) was added to represent the relational stakeholder roles of regenerative tourism. Bellato et al. (2022) paper co-produced with four practitioner pioneers, further developed the TLS concept.

Successful applications of the framework guiding regenerative initiatives are dependent on engagement with the underpinning principles as interdependent relational elements to activate the whole rather than separate parts, and the design dimensions guide the development of practice. The appropriate balance amongst framework dimensions enables tourism stakeholders to design effective, self-sustaining interventions and contribute towards regenerating the system. The framework can be used to investigate the applications of regenerative understandings of place, the roles of tourism change agents and other stakeholders within tourism systems in working towards regeneration, and the use of tourism levers for catalysing systemic transformations.

7 *Proposing Applications of Regeneration Tourism from the Most Recent Peer-reviewed Papers*

In the past two years, various ideas, conceptual models and other applications of regenerative tourism were proposed to guide the concept's practical implementation. These range from elemental integrations of existing concepts for further development in viewpoint papers to more comprehensive frameworks developed from empirical research. Key areas in the literature include

- the use of design thinking and transformational leadership (Cave et al., 2022; Major & Clarke, 2021),
- stakeholder roles of a tourism living system (Bellato et al., 2022),
- shifting paradigms by changing mindsets, systems and practice (Dredge, 2022),
- the 'tourism tree' offering a values-based wellbeing framework (Becken & Kaur, 2021),
- regenerative tourism capacity development framework (Bellato & Cheer, 2021),
- eco-conscious design framework for tourist destinations (Blau & Panagopoulos, 2022),
- education and learning as a strategy (Boluk & Panse, 2022; Cave et al., 2022),
- business and entrepreneur-led innovations (Fusté-Forné & Hussain, 2022; Lupton & Samy, 2022),
- redesign of economic systems (Sheldon, 2021),
- a theoretical framework of circular regenerative process, 'agrowth' and placemaking (Tomassini & Cavagnaro, 2022), and
- the Doughnut Destination (Hartman & Heslinga, 2022).

Reflections on the Regenerative Tourism Discourse to Date

At a fundamental level, regenerative tourism challenges the dominant industrial tourism paradigm that seeks economic growth as an ultimate priority by focusing on regenerating whole systems (Ateljevic, 2020; Pollock, 2015). The study revealed that scholars and practitioners publishing regenerative tourism papers distinguish sustainable from regenerative tourism, indicating that these terms are not interchangeable. Despite misgivings by consultation participants about applying a universal definition that limits pluriversal place-based perspectives, they continue to be sought.

In this chapter, we proposed eight aspects that can inform the development of locally co-constructed definitions of regenerative tourism by practitioners and scholars in diverse contexts. The synthesised aspects were identified primarily from the literature that draws from the regenerative development lineage. The aspects, considered an interconnected whole, could be used as core parameters to advance the concept. In our view, definitions or descriptions that do not honour the lineage of regenerative development or align with its core concepts risk diminishing the integrity of this work and leading towards non-regenerative paths.

One emerging theme arising from the regenerative tourism academic literature relates to the emphasis of regenerative tourism scholars on creating net positive effects, using phrases such as 'leaving the destination in better condition than it was found' without adding the importance of co-evolution and continued capability development with place and communities. Additionally, papers that propose to apply regenerative tourism to goals centred around achieving system stability also leave out the co-evolutionary foundations of regeneration. Importantly, regeneration is not an end goal or state of being, it is an ongoing evolving process creating higher levels of wellbeing, connection and capability for all human and non-human stakeholders.

A second observation relates to key practitioner pioneers' influence on tourism scholarship. While publications by Pollock and regenerative development practitioners such as Mang, Haggard and Reed have significantly shaped the regenerative tourism academic discourse, publications by Spanish-speaking practitioner pioneers highlighted in this study are yet to feature strongly. While this study attempted to identify papers written in Spanish, the search was limited to English language databases and journal searches. Additionally, these pioneers may not have published as much as the others. This observation also relates to the less recognised contributions of the many Indigenous Peoples whose approaches tend to be inherently regenerative and others who have contributed their knowledges, perspectives, and practices toward the developing regenerative tourism concept. Thus, privileging Indigenous Peoples' and others with deep connections and understandings derived from the land, waterways, and all the beings who inhabit those habitats is essential.

The third theme relates to the separation of humans from the rest of nature. Some scholars are embracing the core challenge of the paradigm shift called for by regenerative tourism to reconnect and restore interconnected relationships between humans and nature. Others reflect an ongoing focus on people, communities, and economies. Rather than humans being at life's centre, they are partners with other living beings to promote the wellbeing of the whole social-ecological system. Therefore, we argue that place is the ground from which regenerative tourism approaches must be co-created. While exciting developments are occurring where governments are designing strategies at national and regional scales to provide necessary conditions for regeneration, interventions need to be derived from unique place potential.

The fourth theme relates to the capabilities and transformations of humans in a tourism living system. Scholars have begun exploring the implications and contributions of visitors, tourism professionals and communities. Existing concepts and frameworks previously applied in tourism scholarship, such as leadership and inclusive development, have been examined. Capability development efforts are beginning to be documented so that we can sit in the uncomfortable spaces of wanting to urgently develop answers to the ever-increasing calamities caused by the extraction and exploitation of nature yet needing to take time to reconnect and re-learn how to partner with nature.

Conclusion

We analysed the origins of regenerative tourism and its current state as described by literature and leading practitioners. We developed a conceptual framework to guide regenerative tourism theory and practice. A particular focus was examining the transformational potential of regenerative tourism approaches. Through adopting a regenerative paradigm, regenerative tourism seeks to transform tourism and envisions tourism as processes facilitating encounters between visitors, hosts, communities and places rather than an industry or alternative economy.

Regenerative tourism reorients us towards place as our grounding point and to remembering how travel practices can support the emergence of new life energies as was done for thousands of years. Economies should therefore be considered a flow that serves and supports this process. Tourism living systems facilitate encounters, create connections and develop reciprocal and mutually beneficial relationships through travel practices and experiences, uniquely reflecting tourism places. The regeneration processes of a tourism living system occur mentally, physically, emotionally, spiritually, culturally, socially, environmentally, and economically. Regenerative effects must be demonstrated to transform the sustainable tourism regime and warrant categorisation as a regenerative tourism approach.

We propose a research agenda to advance regenerative tourism from a niche concept to a transformative paradigm for tourism research and practice. Future research on regenerative tourism needs to bring empirical cases to test and strengthen the conceptual writings with evidence on approaches, tools, designs that adhere to regenerative tourism thinking and principles. As regenerative tourism still evolves, it is important to remain open to new models, approaches and designs of regenerative tourism that stem from its main principles but are informed by empirical case studies and comparative or multiple case study research. Simply put, the regenerative tourism concept needs to move from the narrative stage to the research stage to create an empirical base and foundation that will benefit practitioners and researchers and, of course, the communities and places that may adopt this transformative paradigm.

First, we propose employing interdisciplinary and transdisciplinary research methodologies that explore tourism as a phenomenon rather than an industry. These methodologies should be informed by and build upon theories and discourse such as systems theories, living systems thinking, critical theory, and decolonisation. Lessons from fields such as sustainability transitions would advance understanding and appropriate measurement of regenerative change processes. Indigenous science would significantly complement this work through its various methodological approaches and contributions of traditional cultural and ecological knowledges.

Second, we propose that the inclusion and incorporation of non-English, non-Western scientific, Indigenous and other marginalised peoples' perspectives would overcome the limited scope of this review. Their meaningful inclusion would contribute to understanding core practice principles, regeneration processes, contextual factors and the role of Indigenous Peoples in regenerative tourism development. Pivotal to future regenerative tourism research is active engagement and reciprocal partnerships with Indigenous scholars, communities, and decolonial research approaches, thus expanding upon tourism scholarship (Chambers & Buzinde, 2015; Grimwood et al., 2019; Jacobsen, 2020). Co-production of publications, co-creation of initiatives and expanded ways to include the voices of practitioners and scholars in languages other than English is needed.

Third, future regenerative tourism research should explore connections with existing concepts in tourism, such as tourism area life-cycle, carrying capacity, stakeholder inclusion and others highlighted by papers included in this review. Fourth, case studies across diverse contexts will enrich regenerative tourism research, deepen understandings of sustainability in tourism, provide comparisons with alternative tourism approaches and transfer lessons to advance the concept. Fifth, other regeneration spheres may inform regenerative tourism scholarship, especially regenerative leadership, economies and agriculture. Sixth, future research can benefit from applying and testing the proposed transformational frameworks (Bellato et al., 2022; Bellato, Frantzeskaki, & Nygaard, 2023) across varied places and communities.

Tourism geographers are well placed to progress regenerative tourism knowledge co-production due to their central concerns with place, space and environment and their capacity for engagement with plural ontologies and epistemologies. The travel practices and societal and environmental drivers shaping the multifarious development of regenerative tourism approaches remain unexamined. The perspectives of tourism geographers to interrogate the concept critically and support its development are needed, to counter the dominance of business and management perspectives that privilege and centre economies. Accordingly, future regenerative tourism research invites a deep exploration of its ontological and theoretical roots and the co-creation of tourism development approaches towards regenerative futures one place at a time.

References

Araneda, M. (2017). Why Regneration? Available at: http://turismoregenerativo.org/2017/11/por-que-regen eracion/

Araneda, M. (2019). Regenerative Experiences Design. Available at: http://turismoregenerativo.org/2019/09/diseno-de-experiencias-regenerativas/

Arksey, H., & O'Malley, L. (2005). Scoping studies: Towards a methodological framework. *International Journal of Social Research Methodology, 8*(1), 19–32.

Ateljevic, I. (2020). Transforming the (tourism) world for good and (re)generating the potential 'new normal'. *Tourism Geographies, 22*(3), 467–475.

Ateljevic, I., & Sheldon, P. J. (2022). Guest editorial: Transformation and the regenerative future of tourism. *Journal of Tourism Futures, 8*(3), 266–268.

Becken, S. (2019). Decarbonising tourism: Mission impossible? *Tourism Recreation Research, 44*(4), 419–433.

Becken, S., & Kaur, J. (2021). Anchoring "tourism value" within a regenerative tourism paradigm: A government perspective. *Journal of Sustainable Tourism, 30*(1), 52–68.

Bellato, L., & Cheer, J. M. (2021). Inclusive and regenerative urban tourism: Capacity development perspectives. *International Journal of Tourism Cities, 7*(4), 943–961.

Bellato, L., & Frantzeskaki, N. (2021). *Transformative Roles in a Tourism Living System.* CAUTHE 2021 Transformations in uncertain times: Future perfect in tourism, hospitality and events.

Bellato, L., Frantzeskaki, N., Fiebig, C. B., Pollock, A., Dens, E., & Reed, B. (2022). Transformative roles in tourism: Adopting living systems' thinking for regenerative futures. *Journal of Tourism Futures, 8*(3), 312–329.

Bellato, L., Frantzeskaki, N., & Nygaard, C. A. (2023). Regenerative tourism: A conceptual framework leveraging theory and practice. *Tourism Geographies, 25*(4), 1026–1046.

Bhalla, R., & Chowdhary, N. (2022). Green workers of Himalayas: Evidence of transformation induced regeneration. *Journal of Tourism Futures, 8*(3), 380–392.

Blau, M. L., & Panagopoulos, T. (2022). Designing healing destinations: A practical guide for eco-conscious tourism development. *Land, 11*(9), 1595.

Boluk, K. A., & Panse, G. (2022). Recognising the regenerative impacts of Canadian women tourism social entrepreneurs through a feminist ethic of care lens. *Journal of Tourism Futures, 8*(3), 352–366.

Butler, R. (2015). The evolution of tourism and tourism research. *Tourism Recreation Research, 40*(1), 16–27.

Caniglia, B. S., Knott Jr, J. L., & Frank, B. (2020). Conclusion. In B. S. Caniglia (Ed.), *Regenerative Urban Development, Climate Change and the Common Good* (pp. 261–272). Routledge.

Cave, J., & Dredge, D. (2020). Regenerative tourism needs diverse economic practices. *Tourism Geographies, 22*(3), 503–513.

Cave, J., Dredge, D., van't Hullenaar, C., Koens Waddilove, A., Lebski, S., Mathieu, O., Mills, M., Parajuli, P., Pecot, M., Peeters, N., Ricaurte-Quijano, C., Rohl, C., Steele, J., Trauer, B., & Zanet, B. (2022). Regenerative tourism: The challenge of transformational leadership. *Journal of Tourism Futures, 8*(3), 298–311.

Chambers, D., & Buzinde, C. (2015). Tourism and decolonisation: Locating research and self [Article]. *Annals of Tourism Research, 51*, 1–16.

Cheer, J. M. (2020). Human flourishing, tourism transformation and COVID-19: A conceptual touchstone. *Tourism Geographies, 22*(3), 514–524.

Cheer, J. M., & Lew, A. A. (Eds.) (2018). *Tourism, Resilience and Sustainability: Adapting to Social, Political and Economic Change.* Routledge.

Dredge, D. (2022). Regenerative tourism: Transforming mindsets, systems and practices. *Journal of Tourism Futures, 8*(3), 269–281.

Duxbury, N., Bakas, F. E., de Castro, T. V., & Silva, S. (2021). Creative tourism development models towards sustainable and regenerative tourism. *Sustainability, 13*(1), 2.

Dwyer, L. (2018). Saluting while the ship sinks: The necessity for tourism paradigm change. *Journal of Sustainable Tourism, 26*(1), 29–48.

Fullerton, J. (2015). *Regenerative Capitalism: How Universal Principles And Patterns Will Shape Our New Economy?* Capital Institute.

Fusté-Forné, F., & Hussain, A. (2022). Regenerative tourism futures: A case study of Aotearoa New Zealand. *Journal of Tourism Futures, 8*(3), 346–351.

Grimwood, B. S. R., Stinson, M. J., & King, L. J. (2019). A decolonizing settler story. *Annals of Tourism Research*, *79*, 102763.

Haines, A. (2020). The Climate Change Solution Lies in Nature: Regenerative Agriculture at Finca Luna Nueva. Available at: www.regenerativetravel.com/impact/the-climate-change-solution-lies-in-nature-regenerative-agriculture-at-finca-luna-nueva/

Hartman, S., & Heslinga, J. H. (2022). The Doughnut destination: Applying Kate Raworth's Doughnut economy perspective to rethink tourism destination management. *Journal of Tourism Futures*, *9*(2), 279–284.

Hes, D., & Coenen, L. (2018). Regenerative development and transitions thinking. In D. Hes, & J. Bush (Eds.), *Enabling Eco-Cities: Defining, Planning, and Creating a Thriving Future* (pp. 9–20). Springer Singapore.

Higgins-Desbiolles, F. (2018). Sustainable tourism: Sustaining tourism or something more? *Tourism Management Perspectives*, *25*, 157–160.

Howard, P., Hes, D., & Owen, C. (2008). *Exploring Principles of Regenerative Tourism in a Community Driven Ecotourism Development in the Torres Strait Islands.* Proceedings of the World Sustainable Building Conference SB08. Melbourne, Australia.

Jacobsen, D. (2020). The Aboriginalization of inquiry: Tourism research by and for Aboriginal and Torres Strait Islander people. *Tourism Culture & Communication*, *20*(1), 51–56.

Kramvig, B., & Forde, A. (2020). Stories of reconciliation enacted in the everyday lives of Sami tourism entrepreneurs. *Acta Borealia*, *37*(1–2), 27–42.

Levac, D., Colquhoun, H., & O'Brien, K. K. (2010). Scoping studies: Advancing the methodology. *Implementation Science*, *5*(1), 69.

Lovins, L. H. (2020). Regenerative economics. In B. S. Caniglia (Ed.), *Regenerative Urban Development, Climate Change and the Common Good* (pp. 136–155). Routledge.

Lupton, K., & Samy, C. (2022). Restoring the balance between humanity and nature through tourism entrepreneurship: a conceptual framework. *Journal of Tourism Futures*, *8*(3), 367–374.

Major, J., & Clarke, D. (2021). Regenerative tourism in Aotearoa New Zealand–A new paradigm for the VUCA world. *Journal of Tourism Futures*, *8*(2), 194–199.

Mang, P., & Haggard, B. (2016). *Regenerative Development and Design: A Framework for Evolving Sustainability*. Wiley.

Mang, P., & Reed, B. (2012). Designing from place: A regenerative framework and methodology. *Building Research & Information*, *40*(1), 23–38.

Mang, P., & Reed, B. (2019). Regenerative development and design. In R. Meyers (Ed.), *Encyclopedia Sustainability Science & Technology* (pp. 1–44). Springer.

Mathisen, L., Søreng, S. U., & Lyrek, T. (2022). The reciprocity of soil, soul and society: The heart of developing regenerative tourism activities. *Journal of Tourism Futures*, *8*(3), 330–341.

Matunga, H., Matunga, H., & Urlich, S. (2020). From exploitative to regenerative tourism. *MAI Journal*, *9*(3), 295–308.

Owen, C. (2007a). Regenerative tourism: A case study of the resort town Yulara. *Open House International*, *32*(4), 42–53.

Owen, C. (2007b). Regenerative tourism: Re-placing the design of ecotourism facilities. *The International Journal of Environmental, Cultural, Economic and Social Sustainabilty*, *3*(2), 175–181.

Pollock, A. (2012a). Can tourism change its operating model? The necessity and inevitability. Available at: www.slideshare.net/AnnaP/can-tourism-change-its-operating-model-13583914?from_action=save

Pollock, A. (2012b). *Conscious Travel: Signposts towards a New Model for Tourism.* 2nd UNWTO Ethics and Tourism Congress, Quito.

Pollock, A. (2012c). The Role of Indigenous Tourism in Developing Conscious Hosts and Accelerating the Tourism Shift. Available at: www.slideshare.net./AnnaP/the-role-of-indigenous-tourism-in-developing-conscious-hosts

Pollock, A. (2015). *Social Entrepreneurship in Tourism: The Conscious Travel Approach.* TIPSE—Tourism Innovation Partnership for Social Entrepreneurship.

Pollock, A. (2019a). *Flourishing Beyond Sustainability.* ETC Workshop in Krakow, February 6th, 2019.

Pollock, A. (2019b). Regenerative Tourism: The Natural Maturation of Sustainability. Available at: https://medium.com/activate-the-future/regenerative-tourism-the-natural-maturation-of-sustainability-26e6507d0fcb

Pollock, A. (2020). Reimagining a Regenerative Recovery – Why and How. Available at: www.linkedin.com/pulse/reimagining-regenerative-recovery-why-how-anna-pollock/

Raworth, K. (2017). *Doughnut Economics: Seven Ways to Think like a 21st-Century Economist*. Random House Business Books.

Regenerative Travel. (2020). Regenerative Travel Principles for Hospitality. Available at: www.hospitality net.org/file/152008917.pdf

Sanford, C. (2019). The regenerative paradigm: Discerning how we make sense of the world. In B. S. Caniglia (Ed.), *Regenerative Urban Development, Climate Change and the Common Good* (pp. 13–32). Routledge.

Scheyvens, R., & Biddulph, R. (2018). Inclusive tourism development. *Tourism Geographies, 20*(4), 589–609.

Sheldon, P. J. (2021). The coming-of-age of tourism: Embracing new economic models. *Journal of Tourism Futures, 8*(2), 200–207.

Sheller, M. (2021). Reconstructing tourism in the Caribbean: Connecting pandemic recovery, climate resilience and sustainable tourism through mobility justice. *Journal of Sustainable Tourism, 29*(9), 1436–1449.

Teruel, S. A. (2018). *Analysis and Approach to the Definition of Regenerative Tourism Paradigm* Masters thesis, Universidad Para La Cooperación Internacional (UCI)]. Costa Rica.

Tomassini, L., & Cavagnaro, E. (2022). Circular economy, circular regenerative processes, and placemaking for tourism future. *Journal of Tourism Futures, 8*(3), 342–345.

UNWTO & UNDP. (2017). *Tourism and the Sustainable Development Goals – Journey to 2030* (9789284419401). UNWTO.

Zaman, U., Aktan, M., Agrusa, J., & Khwaja, M. G. (2023). Linking regenerative travel and residents' support for tourism development in Kaua'i Island (Hawaii): Moderating-mediating effects of travel-shaming and foreign tourist attractiveness. *Journal of Travel Research, 62*(4), 782–801.

Zivoder, S. B., Ateljevic, I., & Corak, S. (2015). Conscious travel and critical social theory meets destination marketing and management studies: Lessons learned from Croatia. *Journal of Destination Marketing & Management, 4*(1), 68–77.

24

POLITICAL ECOLOGIES OF TOURISM

Key Issues and Research Prospects

Jarkko Saarinen and Sanjay K. Nepal

Introduction

Political ecology as applied to tourism geographies can relate to tourism development in protected areas, ecotourism, community-based tourism, and ethnic and indigenous peoples, among other dimensions. According to Douglas (2014), political ecology has a range of significant implications for developing an understanding of the social and economic relations and power structures often associated with tourism in the Global South and peripheries more generally. Past research has demonstrated that environmental conflicts have emerged through the planning, development and implementation of various forms of tourism ventures (Nepal et al., 2016). These conflicts can refer to displacing local communities (Adams & Hutton, 2007), and appropriating local resources or misallocating benefits in tourism development (Scheyvens, 2011; Scheyvens & Russell, 2012; Spenceley & Meyer, 2012).

As an interdisciplinary approach, political ecology provides fruitful and relevant avenues to analyse and understand how tourism utilises, operates and creates meanings and priorities in natural resource use, conservation and management contexts. Furthermore, the political ecology approach can be used to examine what kind of power issues, inequalities, conflicts and discourses are taking place in tourism-environment-community relations and the changes in them. In tourism, political ecology studies are characterised by approaches that aim to understand the dynamics and transformations taking place in various places and between different spatial scales and stakeholders. From this perspective, there is nothing apolitical about nature and environmental uses in tourism development.

In explicit terms, the political ecology approach has been largely absent from previous tourism studies. However, there is a long tradition in tourism research that focuses on tourism-environment and tourism-community relations, recently with emphases on sustainable and/or responsible approaches in tourism development (see Butler, 1991; 1992; Cole, 2012; Gössling, 2001; Holden, 2015; Lu & Nepal, 2009; Mathieson & Wall, 1982; McMurry, 1930; Murphy, 1985; Saarinen, 2014; 2021).

Indeed, as noted by Susan Stonich (1998: 30) in her seminal paper 'Tourism Ecology': "[over the last two decades] a burgeoning number of studies have dealt with the impacts of tourism development on environmental quality, including effects related to diminishing biodiversity, erosion,

DOI: 10.4324/9781003286301-28

pollution, and degradation of water and other natural resources." The community aspects with links to uses and access to natural resources in tourism have been studied, especially in the context of the political economy of tourism and power relations (see Britton, 1982, 1991; Brohman, 1996; Cheong & Miller, 2000; Dieke, 2000; Gill, 2004; Mosedale, 2011; Lacher & Nepal, 2010; Lenao, 2014; Zapata et al., 2011), which shares common ground with the political ecology approach (Robbins, 2012, 2019; Watts, 2000). Indeed, as Blaikie and Brookfield (1987: 17) have defined, political ecology "combines the concerns of ecology and a broadly defined political economy" by aiming to explain environmental change within global socio-economic forces and by opening up unequal power relations that guide and control benefits from the utilisation of natural and/or cultural resources (see also Bryant & Bailey, 1997).

A good example of the connections between tourism and political ecology without explicitly referencing the latter term is Martin Mowforth and Ian Munt's (1998) book *Tourism and Sustainability: Development, Globalisation and New Tourism in the Third World*. The book covers major themes such as globalisation, sustainability, development, power, class, host-guest relations, governance and poverty issues. All these and many of the socio-cultural and environmental issues, concepts and utilised case studies it raises are highly loaded with the elements that are typical of studies in political ecology. That said, the index of the book does not recognise the term 'political ecology'.

Thus, it seems that the majority of the (implicitly) practised political ecology approaches in tourism studies have been outlined by scholars who do not necessarily identify themselves as doing 'political ecologies'. Obviously, while this can be problematic for the overall development of the political ecology research agenda in tourism studies, it is a rather characteristic situation in political ecology studies per se. As stated by Paul Robbins (2012: 21) much of the work we can label as political ecology "is carried out by people who might never refer to themselves as political ecologists." This is partly due to the nature of political ecology being perhaps less based on a specified body of theory than being what Robbins (2012) calls "a community of practice" (p. 5) "based on the myriad rigorous methods" (p. 87) with a heavy emphasis on the utilisation of versatile empirical materials and texts. On one hand, this has led to a very diverse field of empirical studies and cases, but on the other hand, to highly ideologically driven academic approaches, with elements of action or radical research.

In spite of this complex setting, Robbins (2012) has identified five dominant narratives in political ecologies: the degradation and marginalisation thesis; the conservation and control thesis; the environmental conflict and exclusion thesis; the environmental subjects and identity thesis; and the political objects and actors thesis. In tourism geographies these narratives or broad themes have been related to intertwined research areas, such as communities and livelihoods; class, representation and power; dispossession and displacement; and environmental justice and community empowerment (Saarinen & Nepal, 2016). Against this backcloth, this chapter aims to highlight some of the key issues and potential research avenues that could guide future studies in tourism geographies and political ecologies.

Future Prospects in Political Ecologies of Tourism

Focus on Communities and Livelihood Contexts

Local livelihoods and especially community-based approaches have been typical for tourism studies that focus on people-environment relations in tourism development. The communities and livelihood theme emphasises the necessity to understand communities and their specific natural resource utilisation priorities and needs in a contextual way. In addition to the community views and preferences towards natural resource utilisation and tourism, this calls for an understanding of what the idea of community means as a concept and as a unit of analysis. Basically, the territorial, that is, geographically bound idea of a community, referring to a taken-for-granted definition of community as a fixed and homogenous setting for empirical research, often conceals critical internal issues such as class, ethnicity, and gender (see Ramutsidela, 2014). These elements are also deeply connected to the power, inequality and sustainability issues in tourism development, the perspectives of which will be discussed later. The need to understand the very idea of community has an 'intrinsic' academic value for tourism and political ecology studies. Obviously, this understanding enables us to influence the discourses and practices taking place in various settings. Thus, the way we understand a community and its specific needs, characteristics, history and internal and external dynamics has a great applied value.

Basically, instead of seeing communities solely as territorially or temporally fixed entities, there is a need for relational perspectives to place communities into broader socio-cultural and political economy contexts in analysis (see Lenao et al., 2014; Spiteri & Nepal, 2006; 2011; Stone & Nyaupane, 2014). Thus, many of the 'local' natural resource conflicts and related inequalities in use and access, for example, are typically based on multi-scalar and multi-layered landscapes (Neumann, 2005) in which the spatial fix of global capital operates (Harvey, 2020). Furthermore, in these complex settings of tourism and economic development, communities and natural resource uses, it is crucial to understand other key dimensions of communities than localised spatiality alone, namely shared identity and social interaction (Lehtonen, 1990). Thus, the sense of belonging with social production and reproduction of shared identity and interests are crucial aspects for a relational idea of community (see Stone & Nyaupane, 2014). These dimensions can also shed light on past community issues that may explain current local interests, priorities and divisions, and so forth in natural resource uses and management. This is a valuable approach, as "a problem in tourism studies has been a prevailing present-mindedness [...], refusing deep, grounded or sustained historical analysis" (Walton, 2005: 6).

Together with the territorial dimension of community, the shared identity and interaction dimensions provide opportunities for researchers to understand communities better from below and contextually and, thus, how people construct local social capital, for example, and related norms, trust and interaction in natural resource utilisation and management. Although these partly imagined bounded spaces (see Murphy, 2013) may represent what Anderson (1991) terms 'imagined community' (see Gregory, 1994), they are often very 'real' for people guiding their worldviews and practices, explaining how they use and interact with their environment, and so forth. However, in order to understand the transformation, that is., the changes and pressures taking place in certain localities and communities, the multi-scalar relational approach is crucial. These changes and global-local relations in tourism, communities and natural resource utilisation are highly influenced by power issues producing inequalities and injustices in economic development.

Power: Inequalities and Empowerment

In political ecology, power issues and empowerment have played a major role (Forsyth, 2008; Neumann, 2005; Peet, Robbins & Watts, 2011). Power relates to the control and access of natural resources and basically defines winners and losers or inclusions and exclusions in natural resource utilisation and management (see Carlisle & Jones, 2012; Cheong & Miller, 2000; Church & Coles, 2007; Hall, 1994). These complex multi-scalar and multi-layered power relations, characterised by economic, social and cultural conflicts and marginalisation processes, are often influenced by ethnicity and social class differences between the different stakeholder groups in different scales. There is still a limited body of class focused analyses in tourism (see Hall, 2011, Urry, 1990; Ying et al., 2016), calling for more research.

Interestingly, while tourism operates relationally in the global-local nexus, it often aims to create firmer territoriality in destination community scales. For Sack (1986), territoriality is a form of control that involves organising or classifying by area and creating physical or symbolic boundaries that serve to manage and advertise that classification. Indeed, territories can serve to clarify and powerfully communicate power relations by giving certain spatial units a concrete material and ideological significance and meanings (Murphy, 2013). In tourism, good examples of the territorialisation process are tourism enclaves, which refer to a form of tourism development that is characterised by socio-spatial regulations of host-guest relations, natural resource uses and related mobilities in destination environments (dell'Agnese, 2019; Saarinen & Wall-Reinius, 2019; Weaver, 2019). Critically interpreted contemporary enclavic tourism development often represents a form of 'neo-colonisation' (see Hall & Tucker, 2004; Mbaiwa, 2005). While tourism enclaves and related power issues, empowerment and inequalities require much more scholarly interest in the future, there is one relatively well established subject area of research in tourism and political ecology studies on human territoriality which incorporates organised spatial exclusions and inclusions: nature conservation areas and their uses, meanings, and management.

Conservation: What do We Protect for and from Whom?

Basically, conservation is a political practice focusing on how to organise human-nature relations in a sustainable way. However, there are critical questions on what and whose definition of sustainable development is in a hegemonic position in land use planning and related decision-making, turning nature conservation as a highly topical issue for political ecology approach in tourism geographies. As a result, the role of nature conservation areas has been a major focus of research in tourism and political ecology (Liburd & Becken, 2017; Saarinen, 2019). In addition to national parks, there has been a very specific interest in wilderness areas and how their utilisation is historically and socio-spatially organised and the ways related discourses and representations are constructed (see Cronon, 1998; Neumann, 1998; Ólafsdóttir & Sæþórsdóttir, Chapter 20 of this volume; Sæþórsdóttir et al., 2011; Saarinen, 1998). Wilderness is a highly contested idea, which institutionally is largely based on the United States Wilderness Act that was prescribed just over 50 years ago (Public Law 1964). The idea of the Act – 'man himself is a visitor who doesn't remain' – represents a continuation of Western conservation thinking that originates from the establishment of the first conservation areas, for example, Yellowstone National Park (USA) in 1872 and Banff National Park (Canada) in 1883.

These early conservation areas and their formation processes have served as models of how to create and manage conservation units globally (Hall, 1992), including countries in the Global South. However, this "fortress" model of global conservation thinking, separating wilderness from culture and nature from people, has been increasingly challenged (Nash, 1967; Nelson 2010) by views calling for more people-centred approaches in natural resource management (Agrawal, 2001; Berkes, 2007; Jamal & Stronza, 2009; Ostrom, 1990). One of the key ideas centring on the role of people and communities in wilderness conservation and utilisation has been a community-based natural resource management (CBNRM) approach (Adams & Hulme 2001; Poteete 2009). The CBNRM approach represents a contrast to the fortress wilderness conservation strategy by stating that local communities must have access to and direct control over the uses and benefits of adjacent natural resources. By securing the control and benefits, local communities are assumed to value and manage those resources in a responsible and sustainable way (Blaikie, 2006; Ostrom, 1990; Swatuk, 2005). However, while there are numerous success stories of 'community conservation', there is also a growing criticism concerning the usability of CBNRM or other similar kinds of approaches (see Brockington, 2004; Oates, 1999).

Therefore, a further development of people-centred approaches in nature conservation and natural resource management in general is urgently needed, with support from academic studies at the intersection of tourism and political ecology. There is one emerging critical perspective that requires the combination of tourism and political ecology research interests in the future. This research need is related to neoliberal conservation, which is characterised by market-driven elements in organising spatial and social structures in destination regions and communities. According to Büscher and Dressler (2012: 369), neoliberal conservation refers to "the continuing reconstitution of the relationships between people and between people and "nature" according to the market [...] with a special emphasis on devolved governance that facilitates self-regulation." This devolution (or related rhetoric and marketing) plays a major role in neoliberal conservation and various community-based natural resource management models.

Neoliberal conservation commodifies nature, as it turns local intrinsic or use values into exchange values in nature conservation management. A common vehicle that is used in this process is tourism development: as stated by Büscher (2013: 57), tourism often represents a kind of "Holy Grail" that has a magical power to integrate "all the different goals of contemporary [...] conservation." All this can exclude communities and include some other stakeholders in development (see Ramutsindela, 2007), and the potentially resultant inequalities and exclusions/inclusions in development have been widely discussed topics with regard to devolution strategies in destination governance (see Duffy, 2002; Mowforth & Munt, 1998). Currently, a key element in destination governance is the idea of sustainability, with increasing calls for responsibility in tourism.

Sustainability and Responsibility

Over the past 25 years, sustainability has emerged as a paradigm in tourism planning and development discourses and practices. As stated by Hall, Gössling, and Scott (2015: xiii), "sustainability is one of the most important issues currently facing the tourism sector." Thus, the recent calls referring to the "death of sustainability" research in tourism (see, e.g., Sharpley, 2009) may be a premature yet much-needed awakening for scholars. In this respect, there are urgent research needs in tourism development and political ecology, especially in terms of the aspects of governance (see Siakwah, Musavengane & Leonard (2020). As stated by Bill Bramwell (2012: 51) "destinations wanting to promote sustainable tourism are more likely to be successful when there is effective governance." However, while governance can be seen as a tool by which the industry and destinations adapt to change, it also involves power relations and potential exclusions (Jessop, 2010). Thus, the key questions are who and what is governed by whom? In addition, it is crucial to acknowledge in future studies on the political ecologies of tourism that the industry is not "just an economy" but is also a form of governing localities with implications for local livelihoods, ways of living, social networks, culture, biopolitics, access to resources and the environment, and so on. Therefore, the role of communities should be central to the analyses of the political ecologies of (sustainable) tourism.

Recently sustainability has been linked to ethical tourism consumption (see Seyfi & Hall, 2019), including high-level policy aims to reduce global poverty, for example (see Brickley et al., 2013; Goodwin & Francis, 2003; Gibson, 2010; Saarinen et al., 2013). The issues of poverty and poverty reduction, with references to the achievement of the United Nations Sustainable Development

Goals (SDGs) (Scheyvens & Hughes, 2019; Scheyvens & Laeis, 2021), are highly topical focus areas in current and future political ecology studies and therefore provide fruitful avenues for tourism research to contribute societally relevant, global-scale policy arenas. Therefore, the tourism industry has emerged at various policy levels as a form of responsible production and consumption (see Goodwin, 2009; UNWTO, 2006).

Although the idea of responsible tourism is often used as a specific form of tourism, its principles and guidelines are rather similar to the general aims of sustainable tourism. As Richard Sharpley (2013: 385) has noted "it is difficult, or even impossible, to distinguish responsible tourism from the concept of sustainable tourism." However, while responsibility is partly built on the same grounds as sustainability in tourism, there is a societal framework difference (Saarinen, 2021). Similar to neoliberal conservation, the responsibility discourse in tourism is also a product of neoliberal 'self-organising' modes of new governance with resulting corporate social responsibility (CSR) initiatives and/or the creation of a 'perfect green consumer', who does not consume less but consumes in a responsible way (see Rutherford, 2011). In contrast to that, the origins of sustainability are derived from the United Nations policy arenas and preceding conservation and natural resource management debates (see WCED, 1987). Altogether, the aspects of business and/ or consumer responsibility are critical research perspectives and needs in the political ecologies of tourism of the future.

Conclusion

There are many fruitful -and both academically and societally – relevant intersections between tourism and political ecology studies. It is justified to conclude that political ecology provides fruitful avenues for considering the nature of tourism-community-natural resource relations and related environmental and social changes. Many prospective research themes may go beyond the above-mentioned key areas of communities and livelihood contexts, power issues, conservation and sustainability/responsibility in tourism. As indicated, these areas emerging from previous research are obviously simplified and thematically overlap each other. While the result may look partially 'messy', it also demonstrates the complexity and intertwined nature of the political ecology approach. According to Robbins (2012: 5), "the field is so fragmented that citation in it, as senior political ecologist Piers Blaikie once remarked, 'is largely a random affair'."

The development of the political ecology approach in tourism studies could also provide answers to criticism by Richard Butler (2015: 24), who has stated that "one area that is clearly missing in tourism scholarship is research on the environmental aspects of tourism." Thus, while there is a long and relatively rich tradition focusing on tourism and environmental relations, the factual environmental (e.g., ecological) analyses of tourism impacts have probably been less evident although not completely missing, but mainly published outside the tourism and recreation journals (see, e.g., Hawkings et al., 1999; McClung et al., 2004). However, the same criticism has been targeted at the political ecology approach being more political and textual than ecological in terms of its analysis (see Robbins, 2012; 2019). In spite of this, the political ecology approach could support a better equalisation of the economic, socio-cultural and ecological pillars of a triple bottom line of sustainable development in tourism.

Based on the political ecology approach, it is evident that the changing environment and uses of natural resources by local communities and the emerging tourism industry are highly interlinked. Past research indicates that the expected positive socio-economic impacts of emerging tourism and tourism growth are not always realised but are unevenly distributed, based as they are on structural marginalisation and exclusion, where powerful actors have control over the operational environment of local communities and social groups. This process may further marginalise people in development, indicating that certain groups of people or ways of thinking are potentially excluded from socio-economic development and related decision-making. Obviously, a crucial issue is to understand and reveal the elements and processes that create these exclusions, including addressing issues of knowledge, power, and empowerment. For all this, there needs to be a constructive understanding of what a specific community is, what characterises its interests and dynamics, and how it is related to broader social and political economy settings.

The absence of an explicitly defined political ecology approach in tourism research in the past should not be interpreted as a symptom of a lack of interest in tourism-environment and tourism-community relations. Indeed, despite some exceptions, like Douglas (2014) and Stonich (1998), the research gap has been partly based on the issue of a non-explicit or at least indirect conceptualisation of political ecology in tourism geographies. However, it is equally evident that there could and should be a more explicitly acknowledged cross-fertilisation between political ecology approaches and tourism research in future. With the political ecology approach, tourism geographies are better suited to analyse situations where different actors with asymmetrical positions and power are using and competing for control of and access to natural resources.

References

Adams, William M. & Hulme, David (2001). If Community Conservation is the Answer in Africa, What is the Question? *Oryx* 35, 193–200.

Adams, William M. and Jon, Hutton (2007). People, Parks and Poverty: Political Ecology and Biodiversity Conservation. *Conservation and Society* 5(2), 147–183.

Agrawal, Arul (2001). Common Property Institutions and Sustainable Governance of Resources. *World Development* 29, 1649–1672.

Anderson, Benedict (1991). *Imagined Communities: Reflections on the Origin and Spread of Nationalism*. Verso.

Berkes, Fikret (2007). Community-based Conservation in a Globalized World. *PNAS* 104, 15188–15193.

Blaikie, Piers (2006). Is Small Really Beautiful? Community-Based Natural Resource Management in Malawi and Botswana. *World Development* 34, 1942–1957.

Blaikie, Piers & Brookfield, Harold (1987). *Land Degradation and Society*. Methuen.

Bramwell, B. (2012). Governance, the State and Sustainable Tourism: A Political Economy Approach. In B. Bramwell & B. Lane (Eds.) *Tourism Governance: Critical Perspectives on Governance and Sustainability* (pp. 459–477). Routledge.

Bricker, Kelly, Black, Rosemary & Cottrell, Stuart (2013, Eds.) *Sustainable Tourism and Millennium Development Goals*. Jones & Bartlett Learning.

Britton, Stephen (1982). The Political Economy of Tourism in the Third World. *Annals of Tourism Research* 9, 331–338.

Britton, Stephen G. (1991). Tourism, Capital, and Place: Towards a Critical Geography of Tourism. *Environment and Planning D: Society and Space* 9, 451–478.

Brockington, Dan (2004). Community Conservation, Inequality and Injustice: Myths of Power in Protected Area Management. *Conservation & Society* 2, 411–432.

Brohman, John (1996). New Directions in Tourism for Third World Development. *Annals of Tourism Research* 23, 48–70.

Butler, Richard (1991). Tourism, Environment, and Sustainable Development. *Environmental Conservation* 18, 201–209.

Butler, Richard (1992). Tourism Landscapes: For the Tourist or of the Tourist? *Tourism Recreation Research* 17, 3–9.

Butler, Richard (2015). The Evolution of Tourism and Tourism Research. *Tourism Recreation Research* 40, 14–27.

Bryant, Raymond L. & Bailey, Sinead (1997). *Third World Political Ecology*. Routledge.

Büscher, Bram (2013). *Transforming the Frontier: Peace Parks and the Politics of Neoliberal Conservation in Southern Africa*. Duke University Press.

Büscher, Bram & Dressler, Wolfram (2012). Commodity Conservation. The Restructuring of Community Conservation in South Africa and the Philippines. *Geoforum* 43, 367–376.

Carlisle, Sheena & Jones, Eleri (2012). The Beach Enclave: A Landscape of Power. *Tourism Management Perspectives* 1, 9–16.

Cheong, So-Min & Miller, Marc L. (2000). Power and Tourism: A Foucaultian Observation. *Annals of Tourism Research* 27, 371–390.

Church, Andrew & Coles, Tim (2007, Eds.). *Tourism, Power and Space*. Routledge.

Cole, Stroma (2012). A Political Ecology of Water Equity and Tourism: A Case Study from Bali. *Annals of Tourism Research* 39, 1221–1241.

Cronon, William (1998). The Trouble with Wilderness, or Getting Back to the Wrong Nature. In J. Baird Callicott & Michael J. Nelson (Eds.) *The Great New Wilderness Debate* (pp. 471–499). University of Georgia Press.

dell'Agnese, Elena (2019). 'Islands within Islands?' The Maldivian resort, between Segregation and Integration. *Tourism Geographies* 21, 749–765.

Dieke, Peter U.C. (2000, Ed.). *The Political Economy of Tourism Development in Africa*. Cognizant.

Douglas, Jason A. (2014). What's Political Ecology got to do with Tourism? *Tourism Geographies* 16, 8–13.

Duffy, Rosaleen (2002). *A Trip Too Far: Ecotourism, Politics and Exploitation*. Earthscan.

Forsyth, Tim (2008). Political Ecology and the Epistemology of Social Justice. *Geoforum* 39, 756–764.

Gibson, Chris (2010). Geographies of Tourism: (Un)ethical Encounters. *Progress in Human Geography* 34, 521–527.

Gill, Alison M. (2004). Tourism Communities and Growth Management. In Alan A. Lew, Michael C. Hall and Alan M. Williams (Eds.) *A Companion to Tourism* (pp. 569–583). Blackwell.

Goodwin, Harold (2009). Contemporary Policy Debates: Reflections on 10 Years of Pro-Poor Tourism. *Journal of Policy Research in Tourism, Leisure and Events* 1(1), 90–94.

Goodwin, Harold & Francis, Justin (2003). Ethical and Responsible Tourism: Consumer Trends in the UK. *Journal of Vacation Marketing* 9, 271–284.

Gregory, Derek (1994). *Geographical Imaginations*. Blackwell.

Gössling, Stefan (2001). The Consequences of Tourism for Water use on a Tropical Island: Zanzibar, Tanzania. *Journal of Environmental Management* 61, 179–191.

Hall, Michael C. (1992). *Wasteland to World Heritage: Preserving Australia's Wilderness*. Melbourne University Press.

Hall, Michael C. (1994). *Tourism and Politics: Policy, Power and Place*. Chichester: John Wiley & Sons.

Hall, Michael C., Gössling, Stefan & Scott, Daniel (2015, Eds.). *The Routledge Handbook of Tourism and Sustainability*. Routledge.

Hall, Michael C. & Tucker, Hazel (2004, Eds.) *Tourism and Postcolonialism: Contested Discourses, Identities and Representations*. Routledge.

Hall, Michael C. & Yes, Virginia (2011). There is a Tourism Class: Why Class Still Matters in Tourism Analysis. In Jan Mosedale (Ed.) *Political Economy of Tourism* (pp. 111–125). Routledge.

Harvey, David (2020). *The Anti-Capitalist Chronicles*. Pluto Press.

Hawkings, Julie P., Roberts Callum M., Van't Hof, Tom, De Meyer, Kalli, Tratalos, Jamie & Aldam, Chloe (1999). Effects of Recreational Scuba Diving on Caribbean Coral and Fish Communities. *Conservation Biology* 13, 888–897.

Holden, Andrew (2015). Evolving Perspectives on Tourism's Interaction with Nature During the Past 40 Years. *Tourism Recreation Research* 40(2), 133–143.

Jamal, Tazim & Stronza, Amanda (2009). Collaboration Theory and Tourism Practice in Protected Areas: Stakeholders, Structuring and Sustainability. *Journal of Sustainable Tourism* 17(2), 169–189.

Jessop, Bob (2010). Government and Governance. In Andy Pike, Andres Rodriguez-Pose & John Tomaney (Eds.) *Handbook of Local and Regional Development* (pp. 239–248). Routledge.

Lacher, Geoffrey R. & Nepal, Sanjay K. (2010). Dependency and Development in Northern Thailand. *Annals of Tourism Research* 37, 947–968.

Lehtonen, Heikki (1990). *Community*. Vastapaino (in Finnish).

Lenao, Monkgogi (2014). Rural Tourism Development and Economic Diversification for Local Communities in Botswana: The case of Lekhubu Island. *Nordia Geographical Publications* 4, 1–53.

Lenao, Monkgogi, Mbaiwa, Joseph & Saarinen, Jarkko (2014). Community Expectations from Rural Tourism Development at Lekhubu Island, Botswana. *Tourism Review International* 17, 223–236.

Liburd, Janne J. & Becken, Susanne (2017). Values in Nature Conservation, Tourism and UNESCO World Heritage Site Stewardship. *Journal of Sustainable Tourism* 25, 1719–1735.

Lu, Jiaying & Nepal, Sanjay K. (2009). Sustainable Tourism Research: An Analysis of Papers Published in the Journal of Sustainable Tourism. *Journal of Sustainable Tourism* 17, 5–16.

Mathieson, Alister & Wall, Geoffrey (1982). *Tourism: Economic, Physical and Social Impacts*. Longman.

Mbaiwa, Joseph (2005). Enclave Tourism and its Socio-economic Impacts in the Okavango Delta, Botswana. *Tourism Management* 26, 157–172.

McClung, Maureen, Seddon, Philip J., Massaro, M. & Setiawan Alvin, N. (2004). Nature-based Tourism Impacts on Yellow-Eyed Penguins Megadyptes Antipodes: Does Unregulated Visitor Access Affect Fledging Weight and Juvenile Survival? *Biological Conservation* 119, 279–285.

McMurry, Kenneth C. (1930). The Use of Land for Recreation. *Annals of Association of the American Geographers* 20, 7–20.

Mosedale, Jan (2011, Ed.). *Political Economy of Tourism*. Routledge.

Mowforth, Martin & Munt, Ian. (1998). *Tourism and Sustainability: A New Tourism in the Third World*. Routledge.

Murphy, Alexander B. (2013). Territory's Continuing Allure. *Annals of the Association of American Geographers* 103(5), 1212–1226.

Murphy, Peter. (1985). *Tourism – A Community Approach*. Methuen.

Nash, Roderick F. (1967). *Wilderness and the American Mind*. Yale University Press.

Nelson, Fred (2010, Ed.). *Community Rights, Conservation and Contested Land*. Earthscan.

Nepal, Sanjay, Saarinen, Jarkko and McLean-Purdon Eric (2016). Tourism and Political ecology: Introduction. In Sanjay Nepal & Jarkko Saarinen (Eds.) *Political Ecology and Tourism* (pp. 1–15). Routledge.

Neumann, Roderick P. (1998). *Imposing Wilderness: Struggles over Livelihood and Nature Preservation in Africa*. University of California Press.

Neumann, Roderick P. (2005). *Making Political Ecology*. Hodder Education.

Oates, John F. (1999). *Myth and Reality in The Rainforest: How Conservation Strategies are Failing West Africa*. University of California Press.

Ostrom, Elinor (1990). *Governing the Commons*. Cambridge University Press.

Peet, Richard, Robbins, Paul & Watts, Michael J. (2011, Eds.). *Global Political Ecology*. Routledge.

Poteete, Amy R. (2009). Defining Political Community and Rights to Natural Resources in Botswana. *Development and Change* 40, 281–305.

Public Law. Public Law 88–577. 88th Congress, September 3, 1964.

Ramutsindela, Maano (2007). *Transfrontier Conservation in Africa: At the Confluence of Capital, Politics and Nature*. CABI.

Ramutsindela, Maano (2014, Ed.). *Cartographies of Nature: How Nature Conservation Animates Borders*. Cambridge Scholars Publishing.

Robbins, Paul (2012 [2019]). *Political Ecology: A Critical Introduction*. Wiley-Blackwell (3rd Edition).

Rutherford, Stephanie (2011). *Governing the Wild: Ecotours of Power*. University of Minnesota Press.

Saarinen, Jarkko (1998). Wilderness, Tourism Development and Sustainability: Wilderness Attitudes and Place Ethics. In Alan E. Watson, Greg H. Aplet and John C. Hendee (Eds.) *Personal, Societal, and Ecological Values of Wilderness: Sixth World Wilderness Congress Proceedings on Research, Management, and Allocation* (Vol. I) (pp. 29–34). General Technical Report, USDA Forest Service, Rocky Mountain Research Station.

Saarinen, Jarkko. (2014). Critical Sustainability: Setting the Limits to Growth and Responsibility in Tourism. *Sustainability* 6, 1–17.

Saarinen, Jarkko (2019). What are Wilderness Areas for? Tourism and Political Ecologies of Wilderness Uses and Management in the Anthropocene. *Journal of Sustainable Tourism* 27, 472–487.

Saarinen, Jarkko (2021). Is Being Responsible Sustainable in Tourism? Connections and Critical Differences. *Sustainability* 13, 6599.

Saarinen, Jarkko & Nepal, Sanjay (2016). Conclusions: Towards a Political Ecology of Tourism – Key Issues and Research Prospects. In Sanjay Nepal & Jarkko Saarinen (Eds.) *Political Ecology and Tourism* (pp. 253–264). Routledge.

Saarinen, Jarkko, Rogerson, Christian M. & Manwa, Haretsebe (2013., Eds.). *Tourism, Development and Millennium Development Goals*. Routledge.

Saarinen, Jarkko & Wall-Reinius, Sandra (2019). Enclaves in Tourism: Producing and Governing Exclusive Spaces for Tourism. *Tourism Geographies* 21, 739–748.

Sack, Robert D. (1986). *Human Territoriality: Its Theory and History.* Cambridge University Press.

Scheyvens, Regina (2011). *Tourism and Poverty.* Routledge.

Scheyvens, Regina & Hughes, Emma (2019). Can Tourism Help to End Poverty in all its Forms Everywhere? The Challenge of Tourism Addressing SDG1. *Journal of Sustainable Tourism* 27, 1061–1079.

Scheyvens, Regina & Laeis, Gabriel (2021). Linkages between Tourist Resorts, Local Food Production and the Sustainable Development Goals. *Tourism Geographies* 23(4), 787–809,

Scheyvens, Regina & Russell, Matt (2012). Tourism, Land Tenure and Poverty Alleviation in Fiji. *Tourism Geographies* 14, 1–25.

Seyfi, Siamal & Hall, Michael C. (2019). *Tourism, Sanctions and Boy*cotts. Routledge.

Sharpley, Richard (2009). *Tourism Development and the Environment: Beyond Sustainability?* Earthscan.

Sharpley, Richard (2013). Responsible Tourism: Whose Responsibility? In Andrew Holden & David Fennell (Eds.) *The Routledge Handbook of Tourism and Environment* (pp. 382–391). Routledge.

Siakwah, Pius, Musavengane, Regis & Llewellyn, Leonard (2020). Tourism Governance and Attainment of the Sustainable Development Goals in Africa. *Tourism Planning & Development* 17, 355–383.

Spenceley, Anne & Meyer, Dorothea (2012). Tourism and Poverty Reduction: Theory and Practice in Less Economically Developed Countries. *Journal of Sustainable Tourism* 20, 297–317.

Spiteri, Arian & Nepal, Sanjay (2006). Incentive-based Conservation Programs in Developing Countries: A Review of Some Key Issues and Suggestions for Improvements. *Environmental Management* 37, 1–14.

Spiteri, Arian & Nepal, Sanjay (2011). Linking Livelihoods and Conservation: An Examination of Local Residents' Perceived Linkages Between Conservation and Livelihood Benefits around Nepal's Chitwan National Park. *Environmental Management* 48, 727–738.

Stonich Susan C. (1998). Political Ecology of Tourism. *Annals of Tourism Research* 25, 25–54.

Stone, Moren T. & Nyaupane, Gyan P. (2014). Rethinking Community in Community-based Natural Resource Management. *Community Development* 45, 17–31.

Swatuk, Larry (2005). From Project to Context: Community Based Natural Resource Management in Botswana. *Global Environmental Politics* 5, 95–124.

Sæþórsdóttir, Anna D., Hall, Michael C. & Saarinen, Jarkko (2011). Making Wilderness: Tourism and the History of the Wilderness Idea in Iceland. *Polar Geography* 34, 249–273.

UNWTO (United Nations World Tourism Organization) (2006). *UNWTO Declaration on Tourism and the Millennium Goals: Harnessing Tourism for the Millennium Development Goals.* UNWTO.

Urry, John (1990). *The Tourist Gaze: Leisure and Travel in Contemporary Societies.* Sage Publications.

Walton, John K. (2005). *Histories of Tourism.* Channelview Publications.

Watts, Michael J. (2000). Political Ecology. In Trevor J. Barnes & Eric Sheppard (Eds.) *A Companion to Economic Geography* (pp. 257–275). Blackwell.

WCED (World Commission on Environment and Development) (1987). *Our Common Future.* Oxford University Press.

Weaver, Adam (2019). Selling bubbles at Sea: Pleasurable Enclosure or Unwanted Confinement? *Tourism Geographies* 21, 785–800,

Ying, Tianyu, Norman, William & Zhou, Lingqiang (2016). Is Social Class Still Working? Revisiting the Social Class Division in Tourist Consumption. *Current Issues in Tourism* 19, 1405–1424.

Zapata, María J., Hall, Michael C., Lindo, Patricia & Vanderschaeghe, Mieke (2011). Can Community-based Tourism Contribute to Development and Poverty Alleviation? *Current Issues in Tourism* 14, 725–749.

PART V

Digital Transformation, Platform Economy and Tourism Geographies

25
DIGITAL TOURISM GEOGRAPHIES

Maartje Roelofsen

Introduction

Researching and Conceptualising the Digital in (Tourism) Geographies

A glance at the emerging disciplinary subfields, journal titles, academic events, funding schemes, research groups and educational programmes that contain the word 'digital' may suggest that a 'digital turn' is well underway in the production and circulation of geographical knowledge (Ash et al., 2018). Like other geography sub-disciplines, the 'digital' has become integral to how 'tourism geography' is understood and shaped as a field of knowledge and as scholarly practice. Digital technologies, software, and digital data have become almost imperative to how tourism spaces and spatialities take shape today and to how knowledge on these spaces and spatialities is generated and shared.

Many tourism scholars, educators, and students have become crucially dependent upon computerised- and networked devices and software to collaborate, communicate, teach and learn, to present and publish research, and to collect, store and analyse data (Munar & Bødker, 2014). At the same time, epistemologies and methodologies in tourism research have become progressively digitally mediated. Commonly used methods such as surveys, interviews, and participant observation are increasingly carried out with the support of computerised devices and cloud-based software, throwing up new questions regarding ethics, privacy, data storage and researcher reflexivity (Rakić & Chambers, 2012; Jeffrey et al., 2021).

'Digital methods', on the other hand, are used to study (new) digitally mediated tourism spaces of cultural production and social interaction, making use of available digital objects such as hyperlinks, tags, likes and other forms of web data (Rogers, 2019: 3; Wilson et al., 2021). Methodological approaches pertaining to the digital are often driven by ontological discussions about what can be known as "natively digital" and what has become "digitised" (Rogers 2013: 19), the latter referring to "the conversion of analogue data and processes into a machine-readable format" (OECD, 2020: 60). This also accounts for studying travel spaces enabled by digital applications such as *Tinder, Pokémon Go,* and Virtual Reality technology (Kaelber, 2007; Bauder & Hackenbroch, 2018), which are both a source and an object of study.

DOI: 10.4324/9781003286301-30

Similarly, digital transformations of tourism have inspired the development of entirely new fields of enquiry that extend beyond the mature fields of 'e-tourism' and 'tourism and ICT' (Gretzel et al., 2020). Most notably "critical digital tourism studies", which aim to examine the societal circumstances that give rise to digital technologies in tourism and consider the specific constraints and possibilities that these technologies afford (Munar & Gyimóthy, 2013: 256).

A few conceptual orientations might be in place to help frame 'the digital' as an evolving and complex field of study and subject of research in tourism geography. Geographers James Ash, Rob Kitchin and Agnieszka Leszczynski (2019) suggest four dimensions through which the digital can variously be approached and understood: ontics, logics, aesthetics and discourse. Throughout this chapter, each dimension will be discussed and illustrated through examples in tourism geographies research.

Computerised Infrastructures and Spaces of Tourism Commerce

As *ontics* – that which is conceived of as having a real being – 'the digital' are material technologies characterised by binary computing architectures. These computing architectures, or, digital systems, have the capacity to "translate all inputs and outputs into binary structures of 0s and 1s, which can be stored, transferred, or manipulated at the level of numbers, or 'digits'" (Lunenfeld, 1999: xv, cited in Ash et al., 2018: 3). Hardware (physical computer devices) and software (computer coding programs that provide instructions) function together using such digital, binary code (Lupton, 2014). Numerous *analogue* systems and technologies – which record and transmit information through continuously variable quantities – have undergone processes of digitisation over the past decades, such as phones, compasses, cameras and many other tools used in tourism and travel. Progressively these digital technologies have become interwoven into the fabric of tourism and travel, greatly affecting the operations, structures, and strategies of tourism enterprises, and shaping how places are represented, understood, produced and consumed.

Air travel is often alluded to as exemplary of the complex assemblages of computerised infrastructures and digitally coded processes that enable global (tourism) mobility and, effectively, the production of tourism space (Kitchin and Dodge 2014). This is because aviation was one of the first of many sectors to digitise its business processes in the early 1960s, most notably through the development and implementation of the Computer Reservation System (CRS); the fruit of a joint project between American Airlines and IBM (International Business Machines Corporation) in the United States. The CRS became a primary means of providing air travel information to airline-sales distribution systems, and replaced analogue data processing practices which, until then, were carried out by operators who used record books and index cards to process flights manually (Truitt et al. 1991). The CRS was designed to cope with the vastly increasing number of global air traffic passengers, a development that was brought about by another major invention: the first trans-oceanic passenger jet in 1958. Combined, the CRS and jet-propelled aeroplanes opened an era of "first-world expansionism" led by innovation and "techno-utopic fantasies", also known as the Jet Age (Yano, 2011: 47).

Historical analyses of the CRS contend that the development of multinational and regional CRSs outside the United States have been driven by nationalistic desires to secure control of the distribution of travel and related tourism products in their own markets (Truitt et al., 1991), reflecting the geopolitical importance of the CRS as a means to acquire dominance over (digitised) spaces of tourism commerce. The CRS later became an indispensable system for travel agents, tour operators and other intermediaries globally to offer a full range of travel services to their customers. Now more commonly referred to as the 'global distribution system' it has been used to store and distribute information about myriad other tourism products and services to the public, including hotel rooms, rental cars, train rides, package tours, and excursions.

Disintermediation and Algorithmic Decision-making

Beyond use in travel agencies and other commercial entities, networked computerised devices and software have become available to the broader public with the arrival of the personal computer, the modern Internet (or, commercial Internet), and the World Wide Web through the 1980s and 1990s (Buhalis & Law, 2008). The first online booking applications emerged in the late 1990s and early 2000s, allowing prospective travellers with access to a computer and the Internet to search for and book their trips independently rather than through traditional intermediaries. Brick-and-mortar travel agencies, which have historically performed a crucial role in the supply and exchange of analogue travel information to local communities, have struggled to survive these processes of digital 'disintermediation' (Novak & Schwabe, 2009; Law et al., 2015). Their position in the competitive chain of tourism production and distribution fundamentally altered with the rise of online travel agencies (OTA) such as Expedia and Booking.com. These operate through websites and device applications (software) that are underpinned by algorithms – "sets of defined steps structured to process instructions/data to produce an output" (Kitchin, 2017: 14). They execute travellers' search queries (data inputs) without human oversight and provide recommendations (output) regarding flights, hotels, homes, and travel experiences.

OTA algorithms have thus become endowed with significant power since they affect how, when and where people travel, and effectively reconfigure tourism space and global mobility. In yet other ways, algorithms – which underpin all software – shape tourist and tourism workers' behaviour towards specific ends and target their vulnerabilities, often merely to increase the profit ratios of enterprises. In the case of short-term rental platform Airbnb, for example, the algorithms that structure review and ratings systems work as a form of social regulation. Through notifications in 'data dashboards' they nudge hosts and guests on the platform to act upon and improve their individual digital reputation according to a set of constantly changing standards that correspond with ideas of cleanliness and hospitable behaviour (Minca & Roelofsen, 2019).

Algorithms thus also determine how touristic spaces are valued and shared and, concurrently, influence the visibility of these spaces in online search results and the chances that these spaces are booked/visited. Whilst many software developers have claimed that forms of algorithmic decision-making minimise human error and bias, studies have shown that algorithms in fact deepen social inequality by perpetuating harmful racial and gender stereotypes (Gebru, 2020; Noble, 2018).

Mediation of Tourism Spaces and Spatialities

When viewed as *logics*, "the digital" routinely configures and structures "orderings of everyday life that result from our spatial engagement with digital mediums" (Ash et al., 2018: 26). Through interactions with their users, these digital devices mediate travellers' interactions *with* and perceptions *of* the world through code. Tourist bodies, luggage and companion-animals continuously interact with digital technologies in order to travel between one place and the other, producing (embodied) data in the process. The coding of travellers' bodies and their objects not only aim for a controlled and structured flow of travellers, they also allow for pervasive and invasive forms of surveillance, monitoring and influencing behaviour (Amoore, 2011).

Many aspects of travel are crucially dependent on and configured through software and similarly structure the practices and interactions of those working to provide tourists. Aeroplanes today are flown by computers and guided by software-dependent air traffic control (ATC). Similarly, aeroplane cabins have increasingly become digitised in the provision of entertainment and services on board. Cabin crew operate mobile computing devices such as tablets to store and receive relevant passenger details in real-time, such as itineraries and meal preference. These devices are also important means for all staff on board to communicate with each other and to receive information and communicate with staff on the ground. Arguably, computerised systems and tourism spaces have become so crucially integral to another that they can be defined as 'code/spaces', for without one another, they would not exist (Kitchin & Dodge, 2014).

Over the years, computerised devices have transformed in size and capacity, allowing them to be carried around in pockets or even worn on the body when travelling, as is the case for smartphones, smart watches and 'wearable' technology (Tussyadiah et al., 2017). Equipped with social media and telecommunication technology, many of these devices have afforded travellers to sustain sociable arrangements with other travellers, distant friends and family members and coworkers, wherever they are. They have given rise to 'new forms of mobile and mediated co-presence', also termed 'absent presence', 'virtual proximity' or 'digital elasticity' (Germann Molz & Paris, 2015). Processes of digitisation have also become ever so intimately entangled with the mundane materialities and logistics of tourism space. Door locks, thermostats, housekeeping trolleys, bins and fridges in hotels are few of many 'things' that now operate through software, also referred to as the 'Internet of Things' (Gretzel et al., 2015). They can be tracked and controlled through the Internet allowing for the mediation and regulation of everyday tourism spaces from a distance, supposedly enhancing tourist experiences and energy-efficiency. The collection of data through these digitised devices and applications has turned into a commercial quest for many tourism businesses, and they have been used for consumer profiling and manipulation. For these reasons the incorporation of digital technologies and data collection has been subject to intense scrutiny. Data are not always collected with consent, raising issues regarding the invasion of privacy, safety and data security (Gössling, 2021).

The digital as logics similarly prevails when tourists bring their daily digital habits on holidays. Maintaining social relations with people at home and at work through communication software has become commonplace, further blurring divides between home and away, work and leisure, and the material and digital. Whilst for some, staying in touch whilst on the road can offer mobile sociality and a sense of security and community (Germann Molz & Paris, 2015), for others, being present in multiple occupational 'spheres' can generate anxiety, addiction, worries about being out of contact, sense of having wasted time, and/or mental exertion, and generally keep tourists from experiencing the destination (Tribe & Mkono, 2017). Such development has engendered new tourism spaces of 'digital disconnection' evidenced in tourism niches such as 'Digital Free' tourism or 'digital or wifi-free' zones in certain holiday destinations (Cai & McKenna, 2021).

Augmented Tourism Spaces

As *aesthetics*, digital technologies have the capacity to shape how spaces and spatialities are experienced, understood and valued (Ash et al., 2018). When software and code become embedded in (tourist) spaces and places, they fundamentally change the nature of that space (Leszczynski, 2014). Many popular tourist sites have now become fully 'augmented' through layers of digital content, which affect how these places are accessed, navigated and perceived by travellers and residents alike (Yung & Khoo-Lattimore, 2017; Rodríguez-Díaz & Pulido-Fernández, 2018). Think, for example, about museum visits guided and narrated by interactive location-aware smartphones, Quick Response (QR) codes that facilitate payment, access and information to attractions and parks, or interactive bus shelters which provide tourists with timetables and tourism information (Gretzel et al., 2015).

Many of these digital media have a specific spatial orientation, allowing for geographic information content forms to be (co)produced by its users. FourSquare, for instance, is an app which operates on smartphones and other digital devices and allows users to 'check-in' and share their geolocation in real-time with connected users so they can be alerted of other users in close vicinity. FourSquare and other digital spatial technologies mediate tourists' navigation of spaces and places, suggesting alternative routes and providing information about spaces that might have otherwise been inaccessible (Leszczynski, 2014). Yet, how spaces are sensed and made (in)visible through digital interfaces and code is also political, evidenced perhaps most clearly in search results produced through algorithms underpinning applications such as Tripadvisor, Google Maps or Booking.com. Whether searching for accommodation, restaurants, or points of interest, the underlying code of related geo-referenced software represents "a set of socially constructed values embodying a range of political, economic, and cultural imperatives" (Zook & Graham, 2007: 480).

Those places and businesses that are made visible through algorithms are often already entrenched with certain advantages, reinstating mobility and consumption patterns by directing people to the establishments that appear nearer to the top of the search results chain (ibid.). As aesthetics, Virtual Reality (VR) technologies have, instead, allowed for the rendering of 'real' and imaginary geographies in digital code, giving rise to three-dimensional representations of tourist space that people can engage with. In many instances, these 'virtual' spaces can be accessed from almost anywhere, including from home, affecting users' internal senses of space and spatiality, movement, and position. Rather than replacing travel to the 'material space' of a tourist destination, these embodied virtual reality experiences mostly offer VR users to 'glimpse in' or prepare for travel to a destination (Roelofsen & Carter-White, 2022). As such VR has often been used by tourism enterprises to generate interest among potential customers (Cheong, 1995).

In yet other ways, tourists have been afforded a more central role in 'augmenting and circulating enacted versions of destinations' through their interactions with digital technologies during and after travel (Munar & Gyimóthy, 2013). Digital (action) cameras and smartphones, and their related software have enabled people – beyond creative professionals – to engage in the production and dissemination of creative content, such as digital representations of tourist destinations. Digitally produced maps, brochures, and holiday pictures can be conceived of as digital cultural objects, which 'travel'; they come to have different meaning to different audiences and within the different contexts in which they circulate (Rose, 2016). Unlike their analogue predecessors, digital objects 'do' different things depending on the social spaces within which they have their 'afterlives', including at home (Urry & Larsen, 2011).

Social media, for example, afford new sociability for producing and consuming photographs collaboratively (Larsen, 2008: 148). What tourists share on social media such as Instagram and Facebook is carefully selected and potentially edited to elicit desired reactions in their audiences. Yet, tourists' motivations to create and share digital content in these spaces can upset or even harm those whose stories, places and histories are represented. Volunteer tourists' social media accounts, for example, have often reproduced colonial stereotypes: by sharing digital images of children from the Global South as in need of care by volunteer tourists from the Global North, these digital spaces help circulate metaphors for entire geographical areas as helpless and in need of saving (Mostafanezhad, 2014; Bandyopadhyay & Patil, 2017). At the same time, such harmful digital practices have given rise to new sites of contestation and digital activism. Satirical Instagram account 'barbiesaviour', for example, aims to create awareness among young women of the neo-colonial dynamics of volunteer tourism (Wearing et al., 2018).

Twitter, on the other hand, has been used by residents and community groups to contest the extractive practices of digital short-term rental platform Airbnb in their neighbourhoods (Wilson et al., 2021). In other instances, creation and use of digital images has been disputed directly by making cases against their producers. This was evidenced in Parks Australia's case against Google Maps which allowed for the unauthorised digital rendering of Uluru – a sacred indigenous site in Australia. Whilst visitors were already banned from physically climbing the material site in 2019 in accordance with Uluru's traditional owners (the Anangu people), Google Maps' street view function continued to allow visitors to take a virtual walking tour (Roberts, 2020).

Discourses of Digitalisation

Finally, the digital can be understood as *discourse*, as a taken-for-granted term that refers to many different things and as such obscures "more than it reveals about what are highly heterogeneous sets of objects, practices and processes" (Ash et al., 2018: 3). Discursive labelling of the digital in tourism is ubiquitous and the term is all too often (a-critically) associated with economic progress, innovation, problem-solving, and efficiency (Hannam, 2019). Processes of 'digitalisation', which refer to "the use of digital technologies and data as well as interconnection that results in new activities or changes to existing activities" have arguably resulted in a large-scale digital transformation" of economies and societies (OECD, 2020: 60). Notions of the transformative capacities of the digital have been instrumentalised over the past few decades to advance government and industry agendas of continued competitiveness and capitalist growth of the tourism sector (ibid.).

'Smart tourism destinations', for instance, connote places in which traditional infrastructures have been augmented by networked ICT and digital devices, allowing for data-driven forms of tourism governance and development (ibid.). A recent example is the *Public Eye* initiative in Amsterdam, the Netherlands, which relies on digital cameras, networked sensors, and algorithms to monitor and regulate the number of visitors in certain areas of the city, nudging tourists and residents, through an app, to take alternative routes to reduce overcrowding (Amsterdam, 2021). Similarly, the *Tourist Data Hub* in Barcelona (Spain) aims to analyse visitor demand and to manage the influx of tourists in certain areas relying on digital data (AgentTravel, 2021). These and other initiatives that 'smart tourism destinations' implement (Gretzel et al., 2015) supposedly lead to high-quality tourist experiences and "successful business results for destination stakeholders" (Jovicic, 2019: 279).

Within such technocratic and neo liberal accounts of digital technologies in tourism, individuals are often defined and studied as consumers, users, or data sources rather than affective human beings embedded and embodied in physical and virtual communities and places (Gretzel et al., 2020). If digital technologies are cast as nothing more than "tools of creative tourist exploitation and control" (Munar & Gyimóthy, 2013: 254), designed for value extraction and capital accumulation, the risk is that broader social, cultural, political, and environmental effects of the digital on society are obfuscated (Gretzel et al., 2020). Whilst the supposedly positive effects of digital transformations are normally celebrated by the tourist authorities and the industry, the 'digital turn' in tourism has also raised important questions in terms of equity and the production of socio-spatial (in)equalities.

Examples are already existing colonial and heteropatriarchal practices in sex tourism, which are reproduced through apps like Tinder and which position women "as commodities to be consumed by male tourists from powerful and wealthy countries". (James et al., 2019: 52). In a similar vein, those who benefit from 'sharing economy' platforms in tourism (see Van der Zee, Chapter 28 of this volume) are usually those who are already endowed with certain privileges. In the Airbnb platform economy, for example, homes that are most frequently booked are situated in the more affluent urban areas, and consequently, those who profit the most are usually white, middle-class, and asset-owning members of society (Roelofsen, 2022).

Uneven Digital Tourism Geographies and Future Directions

To understand who benefits from the 'digital turn' in tourism geographies and who does not, a narrower analysis is required of the uneven socio-spatial spread, uptake, and impacts of digital technologies across countries and within societies. The 'digital divide' tends to follow existing axes of social in/exclusion such as class, gender, education and age, and geographical delineations such as urban and rural areas and income-rich and income-poor countries (Kleine, 2019). Differences in access to digital technologies pertain to computing hardware and software, as well as to Internet connections (Graham, 2011).

With the widespread availability and increased affordability of smartphones and other hand-held computerised devices, desktop computers are no longer the main gateway to the Internet. Yet, computers tend to have different and more complex functions than handheld devices and allow their users to carry digitised work in a safe and comfortable way (Lutz, 2019). In 2019, only 63 per cent of urban households had computer access at home, versus 25 per cent of households in rural areas (ITU, 2020). And despite the widespread availability of mobile Internet, researchers of digital (in)equalities tend to focus on access to broadband Internet, since "mobile Internet access offers lower functionality in terms of speed, memory, and storage capacity" (Lutz, 2019: 142).

Globally, about 72 per cent of households in urban areas had access to the Internet at home, almost twice as much as in rural areas (ITU, 2020). The 'cost of connection' is also highly uneven across regions, with broadband Internet costing up to 18.6 per cent of monthly gross national income (GNI) in Africa, versus 1.3 per cent in Europe (ITU, 2021). A more nuanced picture of digital divides emerges when zooming in on the regional and country level. The Digital Economy and Society Index (DESI) shows that Nordic countries in the European Union (EU) perform better in relation to access to/use of the Internet, digital skills, digital public services, and 'e-business' than most Southern and Eastern countries in the EU (European Commission, 2021). But the availability of digital infrastructure, mobile broadband coverage and the Internet are not the only dimensions of access that make up the 'digital divide'. Generational and gender inequalities also remain key hurdles in the strive for digital inclusion.

Globally, women still use the Internet less than men, and again, this gender divide is geographically more uneven and pronounced when zooming in on a regional and country level. Moreover, whilst 71 per cent of the world's population between 15–24 years of age uses the Internet, only 57 per cent of all other age groups use the Internet (ITU, 2021). Digital skills and competences prove to be other important constraints, and so are literacy, language skills, and content availability (Minghetti & Buhalis, 2009). When looking for information about tourism destinations in search engines, for example, users of the Internet will likely be presented with online content that is written in a dominant language such as English.

This phenomenon has given rise to digital hegemony whereby producers in a few countries, usually wealthy and well-connected countries, "get to define what is read by others. The United States, in particular, is a dominant content-producing force" (Ballatore et al., 2017: 141). Being digitally literate and being able to use and maintain digital devices thus often requires users to have knowledge of a dominant language. In regard to digital competence, Small and Medium

Enterprises (SME) in tourism are faced with a number of additional challenges. Operating in a highly competitive digitised economy dominated by tech giants such as Expedia and Booking.com, these enterprises often struggle to find the economic means to keep pace with rapid changes pertaining to the constant innovation of digital technologies in business. SME also often face difficulties finding digitally-skilled employees who know how to operate the related hardware and software (Dredge et al., 2019).

Other crucial aspects that are still often overlooked in tourism scholarship are the labour and resources that constitute and sustain the 'digital turn' in tourism geographies. Many of the earlier mentioned digital technologies and platforms are primarily designed to enable efficiency-focused tourism economies that maximise company profits. Yet, digital devices are manufactured from materials such as metals, glass and plastics, which have to be mined, made or recycled *somewhere*, often in places far away from where these devices are bought. Jessica McLean (2020) notes the importance of being in touch with the material conditions that support or enable 'the digital', for "the people and places that are at the beginning of the commodity chains that make and remake digital technologies are often marginalized from the benefits that their labour accrues for corporate entities" (2020: 2).

Digital devices are powered by electricity, which is, still, predominantly produced in coal-fired power plants. Similarly, information, or data, is stored in data centres or 'clouds' which rely on water for cooling and, again, energy, and as such contribute significantly to carbon emissions and the related climate catastrophe (ibid.). In a similar vein, Dorothea Kleine reminds us how digital developments can be the cause or stimulant of social and political conflict and inequalities. Minerals for mobile phones may be sourced in countries such as the Democratic Republic of Congo where "trade fuels internal armed conflict" (Kleine, 2019: 231). Moreover, the (deliberate) short lifespan of many digital devices leads to vast amounts of e-waste, which are likely to be shipped off, disposed of and recycled in countries "under conditions which often endanger worker health, water, and soil" (ibid.).

Thus, the idea that digital technologies provide 'solutions' and symbolise 'progress' represents only a partial and instrumental view of technology as a tool that serves tourism businesses but does not always consider other underlying structural social, environmental, economic and political issues. The 'digital turn' has benefited tourism providers and consumers in highly uneven ways, and in its current exploitative nature, is based on and productive of geographies of inequality.

Future studies in tourism geographies should therefore cautiously approach notions of 'digital solutionism' by critically examining how digital developments are always already grounded in certain social, cultural, political and economic circumstances. Various examples elicited in this chapter have demonstrated how digital spaces have become just as 'real' as other spaces to achieve political, social, cultural, and justice-based concerns (McLean, 2020). As such, tourism scholars and geographers alike may continue to query and map tourism spaces and practices enabled by digital infrastructures, whilst staying attentive to the generative forces that challenge and resist the power inequalities involved in the production of these spaces.

References

AgentTravel (2021). *Nace Amazing Barcelona, un centro de inteligencia turística que optará a los fondos europeos.* www.agenttravel.es/noticia-041934_Nace-Amazing-Barcelona-un-centro-de-inteligencia-turistica-que-optara-a-los....html (accessed February 2024)

Amoore, L. (2011). Data derivatives: On the emergence of a security risk calculus for our times. *Theory, Culture & Society* 28(6), 24–43. https://doi.org/10.1177/0263276411417430

Amsterdam (2021). *Public Eye: An Open-Source Crowd Monitoring Solution.* www.amsterdam.nl/innovatie/mobiliteit/public-eye/ (accessed February 2024)

Ash, J., Kitchin, R., & Leszczynski, A. (2018). Digital turn, digital geographies? *Progress in Human Geography* 42(1), 25–43. https://doi.org/10.1177/0309132516664800

Ash, J., Kitchin, R., & Leszczynski, A. (2019). *Digital Geographies.* Sage.

Ballatore, A., Graham, M., & Sen, S. (2017). Digital hegemonies: The localness of search engine results. *Annals of the American Association of Geographers* 107(5), 1194–1215. https://doi.org/10.1080/24694452.2017.1308240

Bandyopadhyay, R., & Patil, V. (2017). 'The white woman's burden' – the racialized, gendered politics of volunteer tourism. *Tourism Geographies* 19(4), 644–657. https://doi.org/10.1080/14616688.2017.1298150

Bauder, M., & Hackenbroch, K. (2018). Public space as Code/Space: The geographies of Pokémon Go. *FreiDok Plus*, 1–14. https://doi.org/10.6094/UNIFR/14750

Buhalis, D., & Law, R. (2008). Progress in information technology and tourism management: 20 years on and 10 years after the Internet—The state of eTourism research. *Tourism Management* 29(4), 609–623. https://doi.org/https://doi.org/10.1016/j.tourman.2008.01.005

Cai, W., & McKenna, B. (2021). Power and resistance: Digital-free tourism in a connected world. *Journal of Travel Research*, https://doi.org/10.1177/00472875211061208

Cheong, R. (1995). The virtual threat to travel and tourism. *Tourism Management* 16(6), 417–422. https://doi.org/https://doi.org/10.1016/0261-5177(95)00049-T

Dredge, D., Phi, G. T. L., Mahadevan, R., Meehan, E., & Popescu, E. (2019). Digitalisation in Tourism: In-Depth Analysis of Challenges and Opportunities. Executive Agency for Small and Medium-sized Enterprises (EASME), European Commission.

European Commission (2021). *Digital Economy and Society Index (DESI) 2021.*

Gebru, T. (2020). Race and gender. In M. D. Dubber, F. Pasquale, & S. Das (Eds.), *The Oxford Handbook of Ethics of AI* (pp. 252–269). Oxford University Press. https://doi.org/10.1093/oxfordhb/9780190067397.001.0001

Germann Molz, J., & Paris, C. M. (2015). The social affordances of flashpacking: Exploring the mobility nexus of travel and communication. *Mobilities* 10(2), 173–192. https://doi.org/10.1080/17450101.2013.848605

Gössling, S. (2021). Tourism, technology and ICT: A critical review of affordances and concessions. *Journal of Sustainable Tourism* 29(5), 733–750. https://doi.org/10.1080/09669582.2021.1873353

Graham, M. (2011). Time machines and virtual portals: The spatialities of the digital divide. *Progress in Development Studies* 11(3), 211–227.

Gretzel, U., Fuchs, M., Baggio, R., Hoepken, W., Law, R., Neidhardt, J., Pesonen, J., Zanker, M., & Xiang, Z. (2020). e-Tourism beyond COVID-19: A call for transformative research. *Information Technology & Tourism* 22(2), 187–203. https://doi.org/10.1007/s40558-020-00181-3

Gretzel, U., Sigala, M., Xiang, Z., & Koo, C. (2015). Smart tourism: Foundations and developments. *Electronic Markets* 25(3), 179–188. https://doi.org/10.1007/s12525-015-0196-8

Hannam, K. (2019). Smart cities, smart tourism, and smart mobilities. In D. T. Timothy (Ed.), *Handbook of Globalisation and Tourism* (pp. 260–268). Edward Elgar Publishing. https://doi.org/10.4337/9781786431295.00033

ITU (2020). *Measuring Digital Development. Facts and Figures 2019.* International Telecommunication Union.

ITU (2021). *Measuring Digital Development. Facts and Figures 2021.* International Telecommunication Union.

James, D., Condie, J., & Lean, G. (2019). Travel, tinder and gender in digitally mediated tourism encounters. In C. J. Nash, & A. Gorman-Murray (Eds.), *The Geographies of Digital Sexuality* (pp. 49–68). Springer Singapore. https://doi.org/10.1007/978-981-13-6876-9_4

Jeffrey, H. L., Ashraf, H., & Paris, C. M. (2021). Hanging out on Snapchat: Disrupting passive covert netnography in tourism research. *Tourism Geographies* 23(1–2), 144–161. https://doi.org/10.1080/14616688.2019.1666159

Jovicic, D. Z. (2019). From the traditional understanding of tourism destination to the smart tourism destination. *Current Issues in Tourism* 22(3), 276–282. https://doi.org/10.1080/13683500.2017.1313203

Kaelber, L. (2007). A memorial as virtual traumascape: Darkest tourism in 3D and cyber-space to the gas chambers of Auschwitz. *E-Review of Tourism Research* 5(2), 24–33.

Kitchin, R. (2017). Thinking critically about and researching algorithms. *Information Communication and Society* 20(1), 14–29. https://doi.org/10.1080/1369118X.2016.1154087

Kitchin, R., & Dodge, M. (2014). *Code/space: Software and Everyday Life*. MIT Press.

Kleine, D. (2019). Development. In J. Ash, R. Kitchin, & A. Leszczynski (Eds.), *Digital Geographies* (pp. 225–237). Sage Publications.

Larsen, J. (2008). Practices and flows of digital photography: An ethnographic framework. *Mobilities* 3(1), 141–160. https://doi.org/10.1080/17450100701797398

Law, R., Leung, R., Lo, A., Leung, D., & Fong, L. H. N. (2015). Distribution channel in hospitality and tourism. Revisiting disintermediation from the perspectives of hotels and travel agencies. *International Journal of Contemporary Hospitality Management* 27(3), 431–452. https://doi.org/10.1108/IJCHM-11-2013-0498

Leszczynski, A. (2014). Spatial media/tion. *Progress in Human Geography* 39(6), 729–751. https://doi.org/10.1177/0309132514558443

Lupton, D. (2014). *Digital Sociology*. Routledge.

Lutz, C. (2019). Digital inequalities in the age of artificial intelligence and big data. *Human Behavior and Emerging Technologies* 1(2), 141–148. https://doi.org/https://doi.org/10.1002/hbe2.140

McLean, J. (2020). *Changing Digital Geographies*. Palgrave Macmillan.

Minca, C., & Roelofsen, M. (2019). Becoming Airbnbeings: On datafication and the quantified Self in tourism. *Tourism Geographies* 23(4), 743–764. https://doi.org/10.1080/14616688.2019.1686767

Minghetti, V., & Buhalis, D. (2009). Digital divide in tourism. *Journal of Travel Research* 49(3), 267–281. https://doi.org/10.1177/0047287509346843

Mostafanezhad, M. (2014). Volunteer tourism and the popular humanitarian gaze. *Geoforum* 54, 111–118. https://doi.org/10.1016/j.geoforum.2014.04.004

Munar, A. M., & Bødker, M. (2014). Information Technologies and Tourism: The Critical Turn in Curriculum Development. In D. Dredge, D. Airey, & M. J. Gross (Eds.), *The Routledge Handbook of Tourism and Hospitality Education* (pp. 105–117). Routledge.

Munar, A. M., & Gyimóthy, S. (2013). Critical digital tourism studies. In *Tourism Social Media: Transformations in Identity, Community and Culture* (Vol. 18, pp. 245–262). Emerald Group Publishing Limited. https://doi.org/10.1108/S1571-5043(2013)0000018016

Noble, S. U. (2018). *Algorithms of Oppression*. New York University Press.

Novak, J., & Gerhard Schwabe. (2009). Designing for reintermediation in the brick-and-mortar world: Towards the travel agency of the future. *Electronic Markets* 19, 15–29. https://doi.org/10.1007/s12525-009-0003-5

OECD (2020). Chapter 9: Digital Innovation. In OECD Digital Economy Outlook 2020. OECD (accessed June 2024). www.oecd-ilibrary.org/sites/bb167041-en/1/3/9/index.html?itemId=/content/publication/bb167041-en&_csp_=509e10cb8ea8559b6f9cc53015e8814d&itemIGO=oecd&itemContentType=book

Rakić, T., & Chambers, D. (2012). *An Introduction to Visual Research Methods in Tourism*. Routledge.

Roberts, G. (2020, September 23). Google maps asked to stop users "walking" on Uluru through street view function. *ABC*. www.abc.net.au/news/2020-09-23/google-asked-to-stop-walking-on-uluru-through-street-view/12693670 (accessed February 2024).

Rodríguez-Díaz, B., & Pulido-Fernández, J. I. (2018). Selecting the best route in a theme park through multi-objective programming. *Tourism Geographies* 20(5), 791–809. https://doi.org/10.1080/14616688.2018.1485728

Roelofsen, M. (2022). *Hospitality, Home and Life in the Platform Economies of Tourism*. Palgrave Macmillan.

Roelofsen, M., & Carter-White, R. (2022). Virtual reality as a spatial prompt in geography learning and teaching. *Geographical Research*, 60(4), 625–636. https://doi.org/10.1111/1745-5871.12551

Rogers, R. (2013). *Digital Methods*. Massachusetts Institute of Technology.

Rogers, R. (2019). *Doing Digital Methods*. Sage Publications.

Rose, G. (2016). Rethinking the geographies of cultural 'objects' through digital technologies: Interface, network and friction. *Progress in Human Geography* 40(3), 334–351. https://doi.org/10.1177/0309132515580493

Tribe, J., & Mkono, M. (2017). Not such smart tourism? The concept of e-lienation. *Annals of Tourism Research* 66, 105–115. https://doi.org/https://doi.org/10.1016/j.annals.2017.07.001

Truitt, L. J., Teye, V. B., & Farris, M. T. (1991). The role of computer reservations systems. International implications for the travel industry. *Tourism Management* (March), 21–36.

Tussyadiah, I. P., Jung, T. H., & tom Dieck, M. C. (2017). Embodiment of wearable augmented reality technology in tourism experiences. *Journal of Travel Research* 57(5), 597–611. https://doi.org/10.1177/0047287517709090

Urry, J., & Larsen, J. (2011). The Tourist Gaze 3.0. https://doi.org/10.4135/9781446251904

Wearing, S., Mostafanezhad, M., Nguyen, N., Nguyen, T. H. T., & McDonald, M. (2018). 'Poor children on Tinder' and their Barbie Saviours: Towards a feminist political economy of volunteer tourism. *Leisure Studies* 37(5), 500–514. https://doi.org/10.1080/02614367.2018.1504979

Wilson, J., Garay-Tamajon, L., & Morales-Perez, S. (2021). Politicising platform-mediated tourism rentals in the digital sphere: Airbnb in Madrid and Barcelona. *Journal of Sustainable Tourism* 30(5), 1080–1101. https://doi.org/10.1080/09669582.2020.1866585

Yano, C. R. (2011). *Airborne Dreams: "Nisei" Stewardesses and Pan American World Airways*. Duke University Press.

Yung, R., & Khoo-Lattimore, C. (2017). New realities: A systematic literature review on virtual reality and augmented reality in tourism research. *Current Issues in Tourism* 22(17), 2056–2081. https://doi.org/10.1080/13683500.2017.1417359

Zook, M. A., & Graham, M. (2007). Mapping DigiPlace: Geocoded internet data and the representation of place. *Environment and Planning B: Planning and Design* 34(3), 466–482. https://doi.org/10.1068/b3311

26

GEOGRAPHIES OF EXCLUSION IN 'SMART' TOURISM PLACES

Towards a Critical Research Agenda

Antonio Paolo Russo and Josep Antoni Ivars Baidal

Introduction

Now more than ever, contemporary cities are facing a key challenge for social inclusion, which is the transition from being sites of residence and work for stable, delimited communities, to dwelling places for mobile populations. The challenge resides mainly in traditional social structures – characterised as 'stickier' or less mobile resident populations – unable to cope with emerging forms of liquid, transient use of places, and losing control and access to their habitats. This is possibly a consequence of the sheer dimension and diversity of temporary populations, characterised by a material and economic power over urban resources that outcompetes less motile and adaptive populations, inhibiting their social reproduction. This chapter tackles the geographies of mobile empowerment stemming from the widespread adoption of digital technologies in tourist places: 'Smart destination' (SD) systems are thus connoted as enabling socio-technical regimes, widening the access gap between resident and visitor communities.

This conceptual approach subscribes to the 'mobilities turn'. This conjunct of questions, theories and methods (Sheller & Urry, 2006) emerged in the early 2000s to denote an epistemological shift in the social sciences from society as *sedentary* towards a paradigm in which it is conceived as inherently mobile, and towards an understanding of social phenomena that are inextricable from that of the movement of people, objects and information (see also Lapointe, Chapter 6). The mobility turn clarifies that no place can be analysed, let alone managed, as a discrete and objective entity ontologically separate from the mobility flows that define it, destabilise it and ultimately continuously transform it (Smith & Hetherington, 2013, Dredge and Jamal, 2013). It thus invites us to engage in the analysis of mobilities to get a grip on place genealogies: what place is – and means – at any given moment in time. What it looks like and how it is represented, as well as how it changes, depends critically on the mobilities that operate within it; the scale at which they operate; how they connect, shield or feed off one another and fixed elements; power relations that are played out in the negotiation for moorings; and how people, objects, technology, capital and knowledge are dynamically related in such processes.

This approach makes better sense of the local-global interdependencies and takes in the intrinsic tensions between democratic representation (based on constituencies that are 'sedentary', i.e. spatially fixed) – on which traditional approaches to urban management are based – and populations of

DOI: 10.4324/9781003286301-31

place users that are increasingly mobile, unstable and fluid. Mobilities are not only, per se, a highly politicised matter, but they subsume a geometry of power-forces which are clearly stratified along lines such as class, gender, ethnic and physical condition, as clarified by Cresswell (2010), among others. For instance, the bargaining power of white skilled migrants, Western (male) tourists or the corporate traveller elites for the use of urban resources (housing, services, public space) can be greatly superior to that of resident populations, who are often excluded from their use (Ryan & Mulholland, 2014; Conlon, 2011).

Concern for the material and economic obstacles faced by resident urban communities in tourist places has been heightened by the debate on overtourism (Milano et al., 2019) that took over the research agenda in the 2010s and, more recently, by the uneven immobilisations that characterised overtouristed places during the COVID-19 crisis (Adey et al., 2021). The epistemological lens of mobilities invites us to decipher the processes through which mobile populations become hegemonic in the (re)production of spaces of inequality and injustice. Indeed, overtourism hints at the result of intensification of people's global mobility during the last decades and its effects on localised social and environmental structures, which the ongoing transformations of 'ways to be mobile' and dwell – albeit temporarily – in places, define as a whole new problem field for social cohesion and urban sustainability.

The technological enhancements of mobility management systems and services widely pursued by the travel and tourism industry – frequently part and parcel of smart initiatives – may well be a key dimension in the transformations under scrutiny. Mobilities are closely intertwined with digital technology, which is not only the enabling infrastructure for large-scale human mobilities, but also the agency and structuring device for the connections (physical, mental, social, financial) which drive them (Williams et al., 2008). Hence, digital systems facilitate the anchoring of mobilities and provide enhanced access to place and its 'moorings', such as lessening the cognitive barrier between travellers and unfamiliar spaces (Line et al., 2011), or facilitating the establishment of social relations at destinations (Kwan, 2007). A main concern of this chapter is indeed that the widespread gains from the introduction of Information and Communication Technologies (ICTs) as enabling technologies for navigating the city, which evoke an ideal of emancipation and democratisation of travel, may prove unsettling for less mobile groups in the contest for urban resources, and in particular for vulnerable, dependent groups. Urban destinations may be becoming 'smarter', but it is largely still to be proved how this represents an overall improvement for resident communities; indeed, the concern that negative effects are prevailing is high.

These lines of research nuance a critical geography of the SD, bringing out conceptual complexities and policy entrenchments, which to some extent represent a counterpoint to the hegemonic characterisation of smart initiatives in public and policy discourse, and also the methodological challenges which should be taken up in smart cities/destinations that aim to remain inclusive.

Social Exclusion in Tourist Places and Mobilities as Epistemological Lens

Social exclusion entails marginalisation from the labour market, democratic and legal systems, and welfare provisions, and from family and community systems of individuals (Atkinson & Davoudi, 2000: 440). Such exclusions may stem from imbalances in representation and redistribution, and they are ingrained in processes of social and place change driven by endogenous or exogenous factors. The growth of tourism activity may be considered one such driver, reflecting both the exogenous dimension of people's rising international mobility and the endogenous factors which attract, order and fix such mobilities in the geographical spaces of destinations.

A rich stream of literature in various related disciplines has tackled the imbalances, contradictions, borders, ideological entrenchments and material obstacles through which the benefits of tourism largely accrue unevenly for local communities, reproducing or exacerbating social exclusion. These include place-based impacts (Agarwal & Brunt, 2006; Mordue, 2005), gentrification (Zukin, 2008; Wachsmuth & Weisler, 2018), gender representations (Aitchison, 2001), social innovation (Novy & Colomb, 2013) and the marginalisation of vulnerable sections of the population – such as people with disabilities (Small & Darcy, 2010) and low-income families (Minnaert, 2012).

As these various strands of research suggest, tourism and, more widely, mobilities have deep social and territorial consequences, manifested as relational processes through which privileged and underprivileged groups, characterised by differential material, cognitive and economic capacities, come together in (tourist) space and transform it through their agency, resulting in a constellation of hegemonies and exclusions. Cities – and especially city centres as urban tourism destinations – are possibly the socio-spatial arenas where social diversity is most complex and dynamic, the velocity of change and adaptation highest, the results in terms of social inclusion/exclusion most dramatic. Cities attract multiple flows of people (migrants, workers, commuters, tourists, local visitors, social relations), objects (vehicles, goods), information and capital, each one enmeshed to, driving or constraining the others. The continuous re-articulation of such relationships through agency, political action and technological innovation comes to determine and transform the physical layers of the city and its socioeconomic fabric.

On the one hand, tourism may change the social morphology of cities through gentrification and touristification processes, hindering local – and more prominently less well-off – communities from accessing services of basic interest in core urban areas. On the other hand, tourism's financial and infrastructural barriers prevent less advantaged social groups from enjoying the benefits of recreational mobility. While not all forms of mobilities are privileged and have the same power of agency – and even within tourism mobilities there are huge variations in terms of social status, consumption power and access – the literature embracing mobilities as a field of performance and practice invites one to look beyond class struggle, and at the way in which space adapts to mobile lifestyles and biographies, including some and fending off others. For instance, low-cost tourist enclaves and nightlife clusters may prove very attractive for visitors on a low budget and offer residential opportunities for low-salary and precarious mobile workers, at the same time repelling middle-class residents, while exclusive commercial areas or prestigious city centres may be enticing for middle-class residents and lifestyle migrants but unaffordable for low-budget visitors.

Thus, it is important to look at how the production of urban inequalities (rent gaps, social and spatial polarisation, exclusion from employment, etc.) may be reframed in terms of mobilities. Manderscheid (2009) suggests that the study of the (re)production and contestation of urban inequalities cannot escape the mobilities dialogue: "mobilities (...) constitute a significant stratifying force through which unequal life chances are being continuously reproduced" (p. 7); and "conceptualising inequalities as emerging from power relations shaped in multiple social spaces opens the appreciation of their contingent (but not arbitrary) as well as political character" (p. 11).

The notion of access to activities, values and goods is the main way in which mobilities have entered the debate on social stratification and inequality (Urry, 2007). However, Urry argues that the analysis should focus on the social consequences of such mobilities, namely, being able to engender and sustain social relations with those people (and to visit specific places), who are mostly not physically nearby. While there is an observable increase in socio-economic polarisations within most (Western) countries, most notably within the EU, traditional accounts of such trends are frequently framed by a national perspective, which understands societies as territorially bounded. Yet a focus on urban inequalities needs to consider both the spatial and social interconnectedness between the analytic entities. Furthermore, Manderscheid (2009), quoting Jiron (2007: 49 ff.), warns that

> the mere focus on the distributional side of inequalities – in urban studies mainly on residential segregation – is not sufficient to understand the whole complexity of (urban) inequalities, because it leaves aside their everyday living implications and the daily practices of movement allowing access to resources, markets, institutions, social networks and other options.

In other words, if social relations, capital or resources outline specific spatialities, access to – or appropriation of – such capital rests on the distribution of power between unevenly empowered mobilities, which, in turn, require the disposition of certain resources, such as economic capital, and the necessary knowledge, capacities and skills (Cass et al., 2005). Further empirical investigation is needed in order to understand the mechanism of reproducing social inequalities by means of mobilities, or, as Urry (2007) phrases it, "how these multiple mobilities do in fact make a difference to the contemporary nature of social stratification, to entering the gates of heaven or hell" (p. 187).

Manderscheid (2009) then suggests various research strands that fully accommodate the mobilities debate. For instance, she invites us to consider the frequently mentioned relationship between movements and fixities, analysing, as does Adey (2006), the material and social moorings to which mobilities become ingrained in place, and the impact of the infrastructural frame on the degree of polarisation within the space. In light of the previously illustrated concepts, ICTs are also conceived as mobile infrastructure and the foundations of a sociotechnical urban regime: Digital technologies transport (knowledge, capital), connect, enable and anchor other (physical) mobilities, they order social spaces (for instance, through age, skills or connectivity divides), and they themselves travel – as product, political project or system. For this reason, framing our understanding of 'smart city' through a mobilities approach goes some way towards problematising the agency of ICT in relation to social exclusion.

In the next section, we look at how 'smart' has taken over the agenda of urban planning and management, marking out a new playing field for the agency of tourism in urban change.

Tourism, Cities and 'Smartisation': A Confusing Relationship

The relationship between the smart approach and tourism is progressively being shaped from two perspectives: The impact of digitalisation on tourism activity and the smart city as a new paradigm for urban management. The wide influence of technological developments in tourism has opened a new paradigm of operation and value-generation for the tourism industry, defined as 'smart tourism' by Gretzel et al. (2015), through the following characteristics: Integration of the physical and digital world; use of ubiquitous technologies (mainly mobile devices) throughout the entire travel cycle (inspiration, booking, purchase, stay at destination and post-trip); harnessing new sources of data, including information from a variety of sensors; generating and exploiting of big data; replacement of the traditional value chain, exemplified in linear distribution (supplier-travel agency-consumer), by complex ecosystems that include the emergence of new players such as digital platforms; and increasing co-creation of technology-mediated experiences. The disruptive nature of smart tourism, the start of whose rise can be situated in 2015, brought with it a new scenario for tourism management that finds a useful reference in the emerging paradigm of smart cities (Ivars et al., 2019). In fact, it seems clear that the concept of SD is largely derived from that of smart city. However, the relationship between the two concepts is unclear, as is the role of tourism in smart city initiatives.

Because smart city models are diverse and evolving, it could be argued that the singularity of tourism is not well resolved in the main theoretical constructions of the smart city, probably due to its specificity and transversal nature, when it is an activity with an unquestionable territorial and socio-economic impact on urban environments, especially in the period of expansion of demand prior to the emergence of COVID-19. Giffinger et al. (2007), in one of the first theoretical formulations of the smart city based on a system of indicators, include the 'Touristic Attractivity' factor in the Smart Living dimension (Quality of Life), paradoxically one of the aspects affected in neighbourhoods with greater visitor pressure, and do not consider other tourism variables related to the dimension of mobility, economy or environment. This situation is replicated in international standards for smart and sustainable cities (International Organization for Standardization (ISO); International Telecommunication Union (ITU); European Telecommunications Standards Institute, (ETSI)), and indicator-based smart city assessment tools and ranking systems, where tourism is attributed a marginal role (Ivars et al., 2021). More importantly, in the practice of urban and tourism management, there is, in general, a dissociation between smart city initiatives and tourism governance.

In this context, the SD has emerged as a differentiated approach adapted to the needs of tourism management, especially in urban environments with a greater weight of tourism in their economic structure. Yet, just as the smart city is subject to controversy in its conception and implementation and to different critical interpretations, the SD can be considered a fuzzy concept, also used rhetorically and for advertising by corporate or institutional interests (Gretzel and Jamal, 2020), because it lacks empirical and theoretical clarity (Gelter et al., 2020).

A number of consolidated institutional SD development projects can shed more light on these inconsistencies. The Spanish smart destinations network, which emerged as a pilot project in 2012 and currently includes more than 400 destinations, has served as an international benchmark, mainly in Latin American countries such as Mexico, Argentina and Colombia. The destinations in the network develop an action plan based on five areas (governance, technology, innovation, accessibility and sustainability) and an indicator system used for an initial diagnosis and subsequent monitoring of the development of the smart strategy. At the European level, the Capital of Smart Tourism initiative integrates three similar categories (sustainability, accessibility and digitalisation) and includes cultural heritage and creativity. Other institutional initiatives can be seen in Asia, in countries such as China and South Korea, where ambitious government-led smart city development programmes are being implemented that also include SD projects, mainly focussing on the application of new technologies, such as big data, the Internet of Things and artificial intelligence to improve the tourist experience and provide more effective marketing that reinforces competitiveness.

Whatever the case, the integration of tourism into the smart city strategy is an unresolved issue: A problem that can be extended to the difficulty of integrating tourism into the urban model in contexts of observable negative social impacts of the pressure of the visitor economy and tourism-related mobilities, and which has led cities such as Amsterdam and Barcelona to re-evaluate their dominant tourism development and promotion models. In this sense, Gretzel & Koo (2021) advocate the use of technology to manage the mix of work, leisure and mobility activities that overlap in urban spaces through the smart tourism city concept, a new model of governance in which the postulates of the smart city and the SD converge. Compared with existing conceptions of the SD, it has a more holistic scope, where the primary focus is not on residents (as in the original epistemology of the smart city) or tourists (as in the hegemonic operational developments of the SD).

The "capacity to be mobile – motility – involves access to different forms and degrees of mobility; competence in recognising and making use of access; and appropriation of a particular choice" (Kaufmann et al., 2004: 750). Arguably, most SD domains are oriented towards augmenting the motility of visitors; they can access information and services from digital platforms, reassemble them according to their needs, situate themselves in 'unfamiliar' spaces as consumers just like any other 'permanent' citizen, and in turn spread this information in 3.0 channels, influencing the choices of other city users. In short, the presence and mobility of tourists in cities, once bound by prescription and regulation, is today not substantially discernible from that of citizens; however, their power of negotiation over the city's spaces and resources can be far superior. Empowered visitors resituate the 'dominated', slow sectors of local urban societies, which, per se, represent a stratification to which analyses of social exclusion need to bring to light.

The analytic toolbox of mobilities can be deployed to further excavate the role of ICT-enhanced or 'smart' tourism in the social ecology of contemporary cities, tackling in particular the relational nature of tourism with respect to the situation of all other economic agents operating in the destination. The literature depicts a more knowledgeable and connected visitor as an urban explorer, who is likely to use residential and everyday spaces as moorings for his/her urban experience. For instance, a clear trend of preferring accommodation in private homes as an integral part of such urban experience has been observed in the burgeoning literature on P2P hospitality, particularly highlighting how tourist activation of everyday spaces through digital platforms (and their algorithm-based recommendation systems), such as homes and commercial spaces, represents a key shift in the scale, reach and effects of traditional population change processes (López-Gay et al., 2021), producing notable socio-spatial stratifications (Celata et al., 2020; Goyette, 2021; Ferreri & Sanyal, 2022).

Hence, a critical approach towards the planning and consumption of technology linked directly and indirectly to mobilities draws new geographies that adapt to the dynamics of the growing support for such flows (tourists and what goes with them: workers, lifestyle migrants, investments, etc.) rather than the interests of a resident community, and more sedentary groups within them. This inevitably produces a clash between new and old livelihoods and ways of life.

The Nature, Scope and Ambiguous Social Impacts of Smart Initiatives

The roll-out of data-driven digital services, core components of smart tourism and thereby the basis for the operational development of many SD initiatives, has not only favoured the sheer growth of tourism mobilities but also, more remarkably, facilitated their mooring to destinations, lessening cognitive and material barriers in their 'contest for urban resources'. Thus, digitally empowered tourism can be seen to contribute to the social exclusion processes mentioned earlier. By contrast, the smart tourism city concept would seek to exploit the benefits of digitalisation and correct the negative impacts caused by the intensification of inter- and intra-urban mobility. However, public action is not neutral and its effects cannot be balanced between improving destination competitiveness and tourist experience, and preserving the quality of life of resident populations. For this reason, it is interesting to examine the fundamental purpose of smart initiatives 'in the real world' and their relationship with processes of social inclusion and exclusion.

The lack of conceptual accuracy regarding the smart city and the diverse ways in which it has been adopted in cities make it hard to measure the full range of impacts expected from smart initiatives and analyse how they balance out from a social perspective. For this reason, a diagnostic of place examining emerging geographies of social exclusion requires closer scrutiny of the type and effects of smart initiatives, which have the capacity to directly or indirectly influence practices and patterns in tourist consumption and mobility. We use here recent evidence from the SMARTDEST project,[1] with information on such programmes conducted over the last decade (2010–2020) from different databases and official information available online, for a relevant characterisation of eight European cities: Amsterdam, Barcelona, Edinburgh, Jerusalem, Lisbon, Ljubljana, Turin and Venice.

Figures 26.1 and 26.2 show the distribution of European smart city projects and smart tourism initiatives at the local level, according to their main focus. The 78 smart city projects selected are distributed unevenly by city, with a couple of cities with well-established smart strategies, such as Amsterdam and Barcelona, accounting for 61 per cent of the projects, followed by Lisbon and Turin (32% of the total number of projects) and the rest of the cities considered. These differences demonstrate the differing degrees of implementation of smart strategies and the inappropriateness of referring to smart city policies as a uniform phenomenon (Birenboim et al., 2022). In an even more detailed analysis, changes in the orientation of smart policies in the same city can be observed. Barcelona illustrates this reorientation by having moved from a stage based on a top-down model, with the emphasis on technology and economic development (2011–2015), to a smart city approach focused more on technological sovereignty, social innovation and digital democracy, promoted by a new left-wing local government from 2015 onwards.

This evolutionary perspective is also relevant in the focus of the projects examined, since technological, mobility and energy efficiency projects predominate at the beginning of the decade, in accordance with the programmes promoted by the European Union, while social innovation

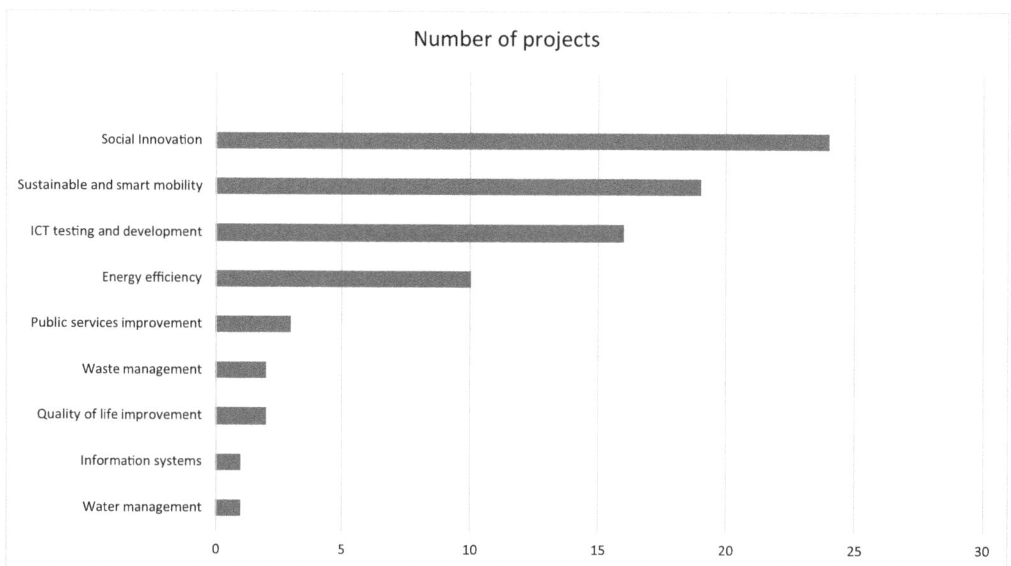

Figure 26.1   Distribution of smart city projects by main focus.

Source: Compiled by the author from data collected in the SMARTDEST project, 2022.

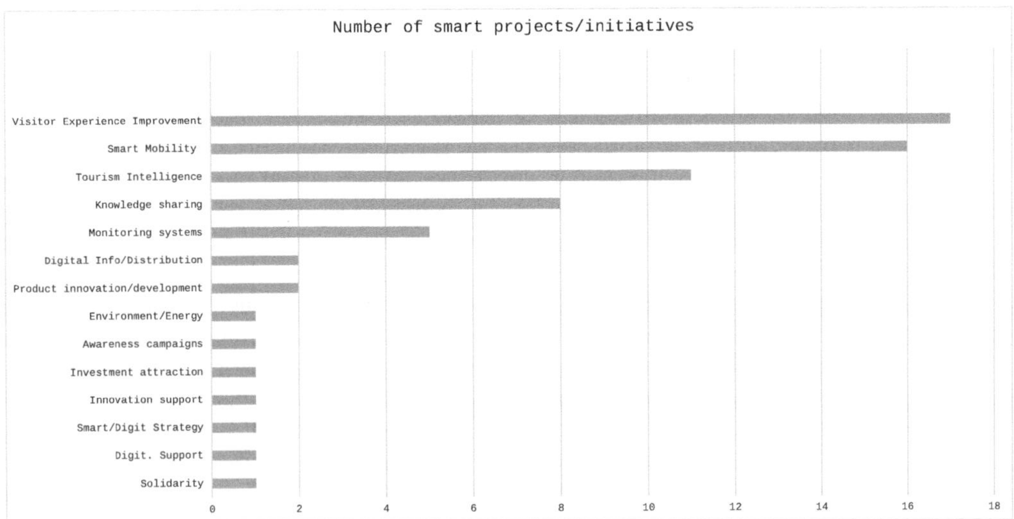

Figure 26.2   Distribution of smart tourism initiatives by main focus.
Source: Compiled by the author from data collected in the SMARTDEST project, 2022.

projects steadily grow to become the most numerous. Projects classified as social innovation include initiatives related to open innovation, entrepreneurship and social participation in the regeneration of urban areas. In many cases they are implemented in the format of Living Labs, a trend used to counter criticisms of the neoliberal, top-down approach to smart cities, repositioning smart initiatives as citizen- or community-centric (Cardullo, 2021).

Projects related to intra-urban mobility include a remarkable set of initiatives that combine the smart and sustainable perspectives, by applying a user-centric approach (use of data and mobile applications, participatory processes in pilot neighbourhoods, developments involving shared mobility or mobility as a service) and reducing the environmental impact of mobility (electric vehicles, alternative modes of transport, etc.). The most directly ICT-related projects stand out for their experimental nature and the establishment of infrastructure and connectivity networks that enable measurement of different types of urban parameters and open up new possibilities for providing services to citizens or support for open innovation initiatives. Finally, environmentally oriented projects (energy efficiency or waste and water management), generally applied at district level, tend to converge in the paradigm of the circular economy, in which the city of Amsterdam is particularly notable.

Smart initiatives tend to be more participatory and inclusive, including programmes such as Techconnect in Amsterdam, aimed at increasing equality in the technology labour market and making tech education and jobs accessible to everyone. However, views critical of the citizen-centric approach to the smart city as a neo-liberal idea highlight the limitations of citizen participation. The bias in favour of tech savvy citizens seems clear, as well as other effects worth investigating, such as the concentration of actions in central city neighbourhoods and alternative mobility services that could aggravate gentrification processes.

As far as smart tourism initiatives are concerned, it is worth noting that the degree of coordination between smart city strategy and tourism governance is relatively low. Furthermore, although five of the cities analysed can be considered to have an active smart city strategy (Amsterdam,

Barcelona, Edinburgh, Lisbon and Turin), none of the cities link their tourism governance to a specific smart strategy based on the SD concept. This does not exclude the development of initiatives that can be assimilated into smart projects, which are the ones considered in this section. Figure 26.2 shows the prominence of initiatives aimed at improving the tourist experience through various technologies (real-time information systems; tourist service booking platforms; smart cards; mobile applications; immersive experiences through virtual and/or augmented reality; information on routes or urban areas through sensors, generally beacons, etc.).

Together with improving the tourism experience, tourism intelligence and monitoring systems as destination management tools are a key area of action. These initiatives are aimed at greater efficiency, predictive capacity and personalisation of digital marketing, but they can also help prevent and manage the impact of tourism on social exclusion processes through crowd monitoring systems in public spaces, systems for controlling the supply of unregulated accommodation, and the development of online awareness campaigns to improve tourists' behaviour. Knowledge-sharing projects are usually connected to tourism intelligence and greater cooperation among destination stakeholders. This is an underdeveloped SD premise, with the exception of initiatives by destination marketing organisations (DMOs), a central actor in tourism governance whose role is subject to review in the cities most stressed by tourism, given the need for a tourism policy more closely integrated into urban management and oriented towards increasingly selective visitor attraction.

Yet the question of how social inclusion plays out in such systems is more complex, and in fact extends to any contexts in which the smart city concept is developed (not just tourism places), if the relational, multi-scalar approaches, advocated by Manderscheid (2009), mentioned above, and Ohnmacht et al. (2009), is taken seriously. In this regard, Graham and Marvin (2002) first pointed out how a narrow, technocratic focus of the smart city concept may lead to underestimating the possible negative effects of developing the technological and networked infrastructures needed for a city to become 'smart'.

The exploration of the potential entanglements of 'smart' with social inclusion at destinations needs to be contextualised in the geography and evolution of urban destinations. By way of example, we include here an identification and delimitation of projects, policy programmes and corporate initiatives implemented or currently in progress in the city of Barcelona, which together have justified the city's high rankings as a smart city, SD or attractive destination. Thus, Table 26.1 presents an outline of different domains of SD, specific projects and initiatives, the agents involved and the outcomes expected from their adoption, also highlighting the adverse effects they could have on social inclusion. It should be noted that some of these initiatives are not necessarily implemented as part of broader smart city strategy and fall outside local government initiatives; nevertheless, city or regional governments may provide an 'enabling environment' by, for instance, adapting regulatory frameworks and legal barriers, or providing the necessary digital infrastructure.

The design and impact of these systems and their evaluation in terms of social inclusion and inclusiveness should provide a means of analysing and balancing the type of effects registered (or at least reported) in the last three columns of the table. Effects on user satisfaction and resource or system efficiency should generally be positive; differing levels of digital inclusion may prevent at least some of the 'local' users from competing for access to the standards generally attainable by the visiting and mobility-empowered population (Brandajs, 2021). In particular, a study currently under review (SMARTDEST, 2022) analyses how the neighbourhoods in Barcelona with a more abundant supply of smart mobility services are those characterised by the most intense population change dynamics in years following implementation of these plans, with low-rent households generally displaced to other neighbourhoods. Such effects, which may unfold in distinct socio-spatial

Table 26.1 'Smart' Projects and System Initiatives in Barcelona, Involved Agents, and Expected Effects

DOMAINS OF 'SMART DESTINATION'	AGENTS	SYSTEMS (reference to projects)	EFFECTS ON VISITORS' EXPERIENCE AND SATISFACTION	POTENTIAL EFFECTS FOR CITIZENS INCLUSION / DESTINATION COHESION	
				Positive	Negative
SMART MOBILITY	Public and PP transit providers, transport companies, parking managers	Flexible routing of public transport based on user demand (LIVE, MOBIlus Driverless metro)	Facilitated use of public transport for visitors	Lower use of public cars by visitors, shorter queues, better internal accessibility	Greater visitor pressure on public transport system in core tourist areas
		Road and access pricing schemes for non-resident vehicles (UGD Area, Connected Urban Car)	De-crowding of access roads, lighter traffic	Controlled traffic, revenue reinvested in infrastructure and service improvement	Non-tourist commuters affected, especially workers to central tourist facilities
		Public bike rental systems (eBicing)	Availability of cheap bikes to visitors, better mobility experience	Increased use of non-contaminating transport	Heavy tourist occupation of bike lanes and infrastructure used by workers
		Parking space locator services (Parking B:SM, ApparkB)	Easier and faster parking	Better ordered traffic, revenue from parking use	More replacement of private resident parking with visitor parking
TOURIST INFORMATION AND MANAGEMENT SYSTEMS	DMOs, attraction managers, planning authorities	User-activated personalised recommendations (Twentytú e-walk smart tour)	Increased accessibility of visitor attractions, more time-efficient and tailored visits	Visitor pressure more spread out, promotion of a wider set of attractions	Greater visitor pressure in residential neighbourhoods, everyday spaces
		Immersive experiences at various heritage sites	Increased comprehensibility of heritage, events, etc.	Better capacity to promote intangibles, more time/ money spent at sites	Capacity of representation out of citizens' hands
		Crowd control and redirection systems (IoT monitoring tourism in Barcelona: the Sagrada Familia)	Fewer risks from overcrowding, increased safety from robbery or terrorist attacks	Reduced risks, fewer incidents and lower related costs, more efficient surveillance	Infringements to privacy of vulnerable groups

(Continued)

Table 26.1 (Continued)

DOMAINS OF 'SMART DESTINATION'	AGENTS	SYSTEMS *(reference to projects)*	EFFECTS ON VISITORS' EXPERIENCE AND SATISFACTION	POTENTIAL EFFECTS FOR CITIZENS INCLUSION / DESTINATION COHESION	
				Positive	*Negative*
SHARING PLATFORMS	Corporate P2P platforms, private providers	P2P transport and vehicle sharing (Über, Galileo-Egnos) Hospitality platforms (Airbnb, etc.)	Greater accessibility to and security of private transport, cheaper taxis Larger stock of accommodation, better adaptation to demand, better services to families	Increased capacity of taxi system Flexible expansion of accommodation stock, promotion of 'community identity', revenue to citizens at risk of exclusion	More and unregulated cabs on the streets, aggravating precarious labour Airbnb's effects on labour and real estate market
		Free tours and visitor experiences (various platforms)	Opportunities for personalised visits	Promotion and valorisation of personal knowledge	Casualisation of labour, de-professionalisation of guides

Source: Compiled by the Authors.

contexts, according to a geography of exclusions also mediated by labour conditions, nationality and age cohorts, should be fully considered and possibly tackled in other areas of urban policy (such as urban planning or housing policy), which can still benefit from a certain degree of 'smartification' and distributed governance.

Final Reflections for a New Research Agenda on 'Smart' Tourist Places

The aim of this chapter is to provide a critical outlook on the roll-out of smart strategies (imagined or operational) in the geographical context of tourism cities. After framing this concern in the mobilities literature, we approached it through an examination of smart initiatives in a relevant sample of European cities. Finally, we focused on the city of Barcelona, which has undergone substantial (though not homogeneous) processes of socio-demographic transformation in the last few years, with generally adverse effects on incumbent resident populations or its more vulnerable communities.

Although the smart city imaginary evokes empowered and connected citizens, producing, consuming and moving around more efficiently with smaller environmental footprints, the epistemological toolbox of the mobilities turn invites us to examine more closely which citizenship we are talking about – that of long-term residents or that of global mobilities, including tourists, for which a larger supply of smart services inevitably bolsters their capacity to dwell in and navigate the city. Smart city projects have inevitably made the city more 'porous' to the intervention of global mobile populations and signs, and enhanced the negotiation capacity of mobile dwellers over 'sedentary' residents. In the emerging context of overtourism, the increasing penetration of enabled tourism in the daily life of cities could be expected to lead to faster, more intense social exclusion.

In terms of agency, as suggested by the critical smart city literature (e.g., Hollands, 2008; Söderström et al., 2014), it is matter for further scrutiny whether and to what degree smart destinations are deaf to social imperatives and skewed towards corporate interests – ranging from those of technology developers looking for big municipal contracts to real estate and financial conglomerates increasingly tuning into the mobilisation of urban assets for global consumers (Aalbers, 2019). It is however evident that local pro-commons agendas are underrepresented in smart strategies, and have little chance of being scaled up to a wider polity of urban resilience, as suggested by Martin (2016): for instance, in relation to the boom of the 'corporate sharing economy' in the hospitality sector.

A critical geography of 'smart places' and the related paradoxes and hindrances, taking in the ambivalence of technology as an enabler of mobility and an ally of smart and sustainable mobility, needs to tackle the difficulty of examining experimental projects in the domain of tourism and technology, which may present several problems of limited efficiency due to high cost and ambiguities related to technological dependence. In this sense, SDs operate in the same problematic domain and biases inherent to the smart city, and generate doubts about the sociospatial scope of their application. This deficit makes it advisable to delve deeper into other scales and territorial environments. The regional scale is thus fundamental for obtaining synergies and complementarities between cities (Gretzel, 2018); a key dimension for metropolitan areas, as evidenced by many smart city projects being implemented at this scale. However, adapting the smart approach to rural and natural spaces with lower population densities poses a challenge for overcoming the digital divide between rural and urban spaces, which act as a brake on local development and tourism competitiveness, with a clear concern for processes of population mobility and displacement triggered by the revaluation of successful core tourist areas (Valente & Russo, 2022).

A research agenda extending the inquiry to such dimensions should therefore be open to the following lines of research:

- Genealogy of the smart city/SD projects in terms of power relations, developing a practical approach to the scrutiny of material agencies beyond the lack of conceptual accuracy.
- A focus on the fine urban scale: Regeneration of historic districts, contribution to gentrification, entanglement and conflicts between visitor mobility and the everyday mobility of citizens and workers.
- The promotion of citizen engagement vis-à-vis digital exclusion of vulnerable communities which could hamper the effectiveness of 'smart' as a mechanism of empowerment and inclusion.
- Citizen privacy concerns, extending to questions of safety, surveillance, gender inclusion and other intersectional obstacles for the motility of the resident community.

In this sense, we find that the concept and proposed operationalisation(s) of the smart tourism city proposed by Gretzel and Koo (2021), in each one of these lines of inquiry offers a good starting point. This may hint at the fact that the smart city itself is an outdated concept, which should be replaced as an object of academic analysis by 'smarter' cities, or the development of wise cities (Young & Lieberknecht, 2019) as urban environments built on user intelligence conducive to extended and inclusive welfare.

Note

1 Cities as mobility hubs: Tackling social exclusion through 'smart' citizen engagement (H2020, ref. 870753).

References

Aalbers, M. B. (2019). Introduction to the forum: From third to fifth-wave gentrification. *Tijdschrift voor economische en sociale geografie*, 110(1), 1–11.

Adey, P. (2006). If Mobility is everything then it is nothing: Towards a relational politics of (Im)mobilities. *Mobilities*, 1(1), 75–94.

Adey, P., Hannam, K., Sheller, M., & Tyfield, D. (2021). Pandemic (im) mobilities. *Mobilities*, 16(1), 1–19.

Agarwal, S., & Brunt, P. (2006). Social exclusion and English seaside resorts. *Tourism Management*, 27(4), 654–670.

Aitchison, C. (2001). Theorizing Other discourses of tourism, gender and culture: Can the subaltern speak (in tourism)? *Tourist Studies*, 1(2), 133–147.

Atkinson, R., & Davoudi, S. (2000). The concept of social exclusion in the European Union: context, development and possibilities. *Journal of Common Market Studies*, 38(3), 427–448.

Birenboim, A., Anton Clavé, S., Ganzaroli, A., Bornioli, A., Vermeulen, S., Zuckerman Farkash, M., ... & Ivars Baidal, J. A. (2022). Touristification, smartization, and social sustainability in European regions. *Current Issues in Tourism*, 26(3), 353–357. DOI: 10.1080/13683500.2022.2051449.

Brandajs, F. (2021). The digital dimension of mobilities: Mapping spatial relationships between corporeal and digital displacements in Barcelona. *Information*, 12(10), 421.

Cardullo, P. (2021). *Citizens in the 'Smart City': Participation, Co-production, Governance*. Routledge.

Cass, N., Shove, E., & Urry, J. (2005). Social exclusion, mobility and access. *The Sociological Review*, 53(3), 539–555. https://doi.org/10.1111/j.1467-954X.2005.00565.x

Celata, F., Capineri, C., & Romano, A. (2020). A room with a (re) view. Short-term rentals, digital reputation and the uneven spatiality of platform-mediated tourism. *Geoforum*, 112, 129–138.

Conlon, D. (2011). Waiting: Feminist perspectives on the spacings/timings of migrant (im) mobility. *Gender, Place & Culture*, 18(3), 353–360.

Cresswell, T. (2010). Towards a politics of mobility. *Environment and Planning D: Society and Space*, 28(1), 17–31.

Dredge, D., & Jamal, T. (2013). Mobilities on the Gold Coast, Australia: implications for destination governance and sustainable tourism. *Journal of Sustainable Tourism* 21(4), 557–579.

Ferreri, M., & Sanyal, R. (2022). Digital informalisation: Rental housing, platforms, and the management of risk. *Housing Studies*, 37(6), 1035–1053.

Gelter, J., Lexhagen, M., & Fuchs, M. (2020). A meta-narrative analysis of smart tourism destinations: Implications for tourism destination management. *Current Issues in Tourism*, 24(20), 1–15.

Giffinger, R., Fertner, C., Kramar, H., Kalasek, R. et al. (2007). Smart cities: ranking of European medium-sized cities. www.smart-cities.eu/download/smart_cities_final_report.pdf (accessed February 2024)

Goyette, K. (2021). 'Making ends meet' by renting homes to strangers: Historicizing Airbnb through women's supplemental income. *City*, 25(3–4), 332–354.

Graham, S., & Marvin, S. (2002). *Splintering urbanism: Networked infrastructures, technological mobilities and the urban condition*. Routledge.

Gretzel, U. (2018). From smart destinations to smart tourism regions. *Investigaciones Regionales-Journal of Regional Research*, 42, 171–184.

Gretzel, U., & Jamal, T. B. (2020). Guiding principles for good governance of the smart destination. *Travel and Tourism Research Association: Advancing Tourism Research Globally*, 42. Available at: https://schol arworks.umass.edu/ttra/2020/research_papers/42 (accessed February 2024)

Gretzel, U., & Koo, C. (2021). Smart tourism cities: A duality of place where technology supports the convergence of touristic and residential experiences. *Asia Pacific Journal of Tourism Research*, 26(4), 1–13.

Gretzel, U., Werthner, H., Koo, C., & Lamsfus, C. (2015). Conceptual foundations for understanding smart tourism ecosystems. *Computers in Human Behavior*, 50, 558–563.

Hollands, R. G. (2008). Will the real smart city please stand up? *City*, 12(3), 303–320.

Ivars-Baidal, J. A., Celdrán-Bernabeu, M. A., Mazón, J. N., & Perles-Ivars, Á. F. (2019). Smart destinations and the evolution of ICTs: A new scenario for destination management? *Current Issues in Tourism*, 22(13), 1581–1600.

Ivars-Baidal, J. A., Vera-Rebollo, J. F., Perles-Ribes, J., Femenia-Serra, F., & Celdrán-Bernabeu, M. A. (2021). Sustainable tourism indicators: What's new within the smart city/destination approach? *Journal of Sustainable Tourism*, 31(7), 1556–1582. DOI: 10.1080/09669582.2021.1876075.

Jiron, P. M. (2007). Unravelling invisible inequalities in the city through urban daily mobility. The case of Santiago de Chile. *Schweiwrische Zeifschj3firr Soziologie*, 33(1), 45–68.

Kaufmann, V., Bergman, M. M., & Joye, D. (2004). Motility: Mobility as capital. *International Journal of Urban and Regional Research*, 28(4), 745–756.

Kwan, M. P. (2007). Mobile communications, social networks, and urban travel: Hypertext as a new metaphor for conceptualizing spatial interaction. *The Professional Geographer*, 59(4), 434–446.

Line, T., Jain, J., & Lyons, G. (2011). The role of ICTs in everyday mobile lives. *Journal of Transport Geography*, 19(6), 1490–1499.

López-Gay, A., Cocola-Gant, A., & Russo, A. P. (2021). Urban tourism and population change: Gentrification in the age of mobilities. *Population, Space and Place*, 27(1), e2380.

Manderscheid, K. (2009). Integrating space and mobilities into the analysis of social inequality. *Distinktion: Scandinavian Journal of Social Theory*, 10(1), 7–27.

Martin, C. J. (2016). The sharing economy: A pathway to sustainability or a nightmarish form of neoliberal capitalism? *Ecological Economics*, 121, 149–159.

Milano, C., Cheer, J. M. & Novelli, M. (2019). Introduction: Overtourism: An evolving phenomenon. In Milano, C., Cheer, J. M. & Novelli, M. (Eds.), *Overtourism: Excesses, discontents and measures in travel and tourism* (pp. 1–17). CABI.

Minnaert, L. (2012). The value of social tourism for disadvantaged families. In H. Schanzel, H. Schänzel, I. Yeoman E. Backer (Eds.), *Family tourism: Multidisciplinary perspectives* (Vol. 56) (pp. 93–105). Channel View Publications.

Mordue, T. (2005). Tourism, performance and social exclusion in "Olde York". *Annals of Tourism Research*, 32(1), 179–198.

Novy, J., & Colomb, C. (2013). Struggling for the right to the (creative) city in Berlin and Hamburg: new urban social movements, new 'spaces of hope'? *International Journal of Urban and Regional Research*, 37(5), 1816–1838.

Ohnmacht, T., Maksim, H., & Bergman, M. M. (Eds.). (2009). *Mobilities and inequality*. Ashgate.

Ryan, L., & Mulholland, J. (2014). Trading places: French highly skilled migrants negotiating mobility and emplacement in London. *Journal of Ethnic and Migration Studies*, 40(4), 584–600.

Sheller, M., & Urry, J. (2006). The new mobilities paradigm. *Environment and Planning A* 38(2), 207–226.

Small, J., & Darcy, S. (2010). Tourism, disability and mobility. In S. Cole & N. Morgan (Eds.), *Tourism and inequality: Problems and prospects* (pp. 1–21). CABI,

SMARTDEST (2022). *Scientific Report on "Local Effects of Tourism and Associated Mobilities as Drivers of Social Exclusion" at 8 case study cities.* Report under review, to be published in: https://smartdest.eu/.

Smith, R. J., & Hetherington, K. (2013). Urban rhythms: Mobilities, space and interaction in the contemporary city. *The Sociological Review*, 61(1_suppl), 4–16.

Söderström, O., Paasche, T., & Klauser, F. (2014). Smart cities as corporate storytelling. *City*, 18(3), 307–320.

Urry, J. (2007). *Mobilities.* Polity.

Valente, R., & Russo, A. P. (2022). Tourism pressure as a driver of social inequalities: A BSEM estimation of housing instability in European urban areas. *European Urban and Regional Studies*, 29(3), 332–349.

Wachsmuth, D., & Weisler, A. (2018). Airbnb and the rent gap: Gentrification through the sharing economy. *Environment and Planning A*, 50(6), 1147–1170.

Williams, A., Anderson, K., & Dourish, P. (2008). Anchored mobilities: Mobile technology and transnational migration. In J. van der Schijff & G. Marsden (Eds.) *Proceedings of the 7th ACM conference on designing interactive systems* (pp. 323–332). ACM.

Young, R. F., & Lieberknecht, K. (2019). From smart cities to wise cities: Ecological wisdom as a basis for sustainable urban development. *Journal of Environmental Planning and Management*, 62(10), 1675–1692.

Zukin, S. (2008). Consuming Authenticity: From outposts of difference to means of exclusion. *Cultural Studies*, 22(5), 724–748. https://doi.org/10.1080/09502380802245985

27

ANXIETY, FEAR AND VIOLENCE

Capitalist Configurations and the Surveillance of Tourism Spaces

Katie D. Dudley

Introduction

It is argued that tourism is, in its core, a capitalist practice (see Büscher & Fletcher, 2017; Fletcher, 2011; Higgins-Desbiolles, 2009; Mowforth & Munt, 2016; Robinson, 2008; Blazquez et al., Chapter 5 of this volume). While the argument was formally introduced some forty years ago by Britton (1982), who described it as "a product of metropolitan capitalist enterprise" (p. 331), and "a major internationalised component of Western capitalist economies" (Britton, 1991: 451), the global pandemic beginning in 2020 helped to illuminate the deep-seeded, and complex, relationship even further. Tourism scholars and practitioners alike took the forced pause to re-evaluate tourism and tourism development in pursuit of a more equitable landscape.

This included, among many other reflections and declarations, introducing concepts such as regenerative tourism (see Cave & Dredge, 2020), or suggesting a push to mainstream strategies rooted in sustainable tourism frameworks (see for example UNWTO, 2020, as cited in Fletcher et al., 2021). While commendable, notably the work of Cave and Dredge in introducing a diverse economies framework in pursuit of the co-existence of capitalist, alternative capitalist and non-capitalist practices, the question remains: How effective are such efforts in support for the common good when implemented within the dominant global economic system of capitalism?

Can such frameworks move us toward a more just system? Or will we find ourselves asking once more,

> why our supports for an equitable common good have become competition for services, further labelling of individuals, the practice of rejecting certain people from public spaces, and increased inequities between those who consume tourism experiences and those who produce them.
>
> (Dudley & Duffy, 2022: 15)

There are of course examples of tourism development thriving outside of the capitalist system (see for example Rowen, 2020) – and will be discussed in this chapter – but as a whole remain conceptual and "a demonstration more of wishful thinking than empirical evidence or theoretical rigour" (Fletcher et al., 2021: 3).

DOI: 10.4324/9781003286301-32

Rather than explore the nuance and evolution of the historical entanglement of tourism and capitalism, or serve simply as a list of critiques, this chapter discusses the ways in which capitalist logic permeates, shifts and evolves to support the emergence of broader patterns and societal norms within tourism. When referring to capitalist logic, this chapter adopts the description put forth by Dudley and Duffy (2022), wherein

> [t]he processes that structure the ever-evolving value relations are referred to as the *logic* underlying capitalist accumulation. Two definitive characteristics of such logic are *production for profit*, whereby the production of commodities (e.g. tourism products/leisure experiences) is primarily done by private companies that aim to generate profit, and *wage labor*, whereby most of the work done within the economy is completed by individuals who do not own the company and are instead given a wage in return for their labor.
>
> (p. 3)

The objective is to arm readers with the tools to recognise and critically analyse the ways in which even the noblest intentions risk perpetuating the violence inherent in tourism development as a modern expression of capitalist growth (see Andrews, 2014; Buscher & Fletcher, 2017; Devine, 2017; Farmer, 2004; Frenzel et al., 2012). The primary focus is what some argue is the dominant, modern configuration of capitalism: Surveillance capitalism. Zuboff (2019) defines surveillance capitalism in part as "a new economic order that claims human experience as free raw material for hidden commercial practices of extraction, prediction, and sales" (p. 2). In other words, while surveillance capitalism follows established capitalist logic such as competitive production, profit maximisation and growth, it introduces a distinct logic of data extraction for the use of behavioural prediction and modification. In tourism, we see examples of this through the use of Big Data to better predict and modify tourists' travel patterns, shopping behaviours, and much more by implementing gentle nudges such as inserting key words or phrases into social media newsfeeds, sending suggested places to eat or shop based on mobile technology location data, or timing the appearance of a BUY button. This same logic is also used to surveille, predict and modify behaviour of local residents, whether working in tourism-dependent jobs or moving about tourist centres, through the language of efficiency and security.

Laying the Foundation for Surveillance

Surveillance capitalism is fueled by a logic that provides infrastructure for companies harvesting data through modern surveillance technologies to predict, and in some cases modify, manipulate or shape, human behaviour (Zuboff, 2019). The progression of such logic, combined with surveillance technologies, and the implications for those navigating tourism spaces are the primary focus of this chapter. This expands the conversation of tourism and capitalism by providing an outline – and subsequently pathways for future explorations – to better understand the ways in which those navigating tourism spaces are impacted through the collection, buying and selling of surveillance-based data. Rather than championing subsets of tourism (e.g., mass versus niche) such as visiting friends and relatives, close-to-home/proximity tourism (see Li et al., 2021) or second-home tourism (see Seraphin & Dosquet, 2020), which remain "situated in the wider context of tourism development being fostered by the capitalist economy system for profit accumulation of multinational corporations and the global elite" (Higgins-Desbiolles, et al., 2019: 6), this chapter outlines the ways in which capitalist configurations have both evolved and shaped qualitative changes in the microphysics of power, which Foucault (1975/1995) describes as the mechanisms beyond penal institutions (e.g., law and police) that subjugate people at the individual level through discursive strategies rooted in fear, anxiety, and violence.

Contrary to keeping the prisoners on their best behaviour, Dudley and Duffy (2022) discuss the ways in which surveillance capitalism operates without walls, windows, towers or guards, but rather as a power that categorises certain behaviours as delinquent, and thus normalises their suppression; a concept Foucault (1991) terms governmentality. Governmentality posits the systematic construction and exercise of power and knowledge is internalised by those over whom power is being exerted. The idea is that people will mostly *do as they should*, not necessarily because of the physical force the government may bring to bear on them, but because what they *ought* to do is constantly being informed and reinforced through the workings of discursive power. Dudley and Duffy (2022) examine an exercise of this through the use of a tourism promotional campaign targeted toward residents of New Orleans rather than potential tourists. The campaign suggests if you are not fully *for* tourism development, you are not just against tourism, you are against New Orleans, and thus are not one of us. They go on to point out that "language is not neutral and objective, but rather constructs descriptions of the *real-world* using discourse as a form of regulation and domination" (p. 7; see also Kincheloe & McLaren, 2008; Gotham, 2007).

Capitalism has been defined as both an economic and sociopolitical system that has come to shape all aspects of life (see Bowles & Gintis, 2012), while globally, tourism continues to privilege those of the world's more affluent communities, living within modern capitalist societies and possessing the ability to "travel, unchallenged and unfettered by bureaucratic inconvenience or economic restrictions. ... Such privilege does not, of course, extend to the economically and socially excluded in developed economies or to most citizens of the Global South" (Baum & Hai, 2020: 2399). Yet, as Bianchi (2018) points out, the study of tourism has been "plagued by an ahistorical and static conceptualisation of capitalism rather than as a continuously evolving set of forces shaped by the dialectics of capital and labour, technological advances, changing market structures and the territorial arrangements of state power" (p. 90). Though Britton's (1991) seminal work argued the need to recognise the capitalist nature of tourism, exploring its role in capital accumulation, there remains a reluctance to recognise the inherent intersection of tourism and the life it breathes into the logic underpinning ever-evolving capitalist configurations; a logic many argue is fueled by the pursuit of both social and economic capital for some, and fear, anxiety, and violence for others (see Bianchi, 2018; Buscher & Fletcher, 2017; Devine & Ojeda, 2017; Fletcher, 2011; Fletcher & Neves, 2012; Higgins-Desbiolles, 2006; 2009; Ioannides & Petridou, 2016).

Anxiety, Fear and Violence

The rise and spread of capitalism is grounded in multiple forms of violence, both direct/physical and structural (Harvey, 2006). The threat of direct or physical violence is often overt and discursively wielded to instil the fear and anxiety upon which capitalist logic thrives, while discursive strategies to enact structural violence are often more covert. Yet the two remain in pursuit of a similar outcome: the growth of tourism. According to Salazar (2017), when asked about the links between tourism and violence, most conjure thoughts of physical altercations centred around tourist experiences or safety such as assault, robberies, or terrorist attacks on tourist areas (see also Andrews, 2014; Fletcher, 2011). Devine and Ojeda (2017) agree violence in tourism settings does come in the form of physical, horrific, and extraordinary events, however, it also "manifests in the spaces and silences of everyday life, in the loss of land, community and language" (Devine & Ojeda, 2017: 605). It is more often the physical, however, that captures the attention of government officials, practitioners, and academics alike. This is due in large part to the idea that if a place is deemed unsafe, people will not go there. According to Devine (2017) this profit-predicated-on-security equation justifies the enormous amount of resources utilised to *secure* tourism spaces. The securitisation of tourism is rarely only about the implementation of policies or technologies. It is also about less tangible notions such as geopolitical imaginaries and the courting of international travellers among strategic mobilisations of fear, anxiety and danger (Lisle, 2013).

In her work exploring the security of everyday spaces in Colombia, Ojeda (2013) found that tourism often lies at the heart of the securitisation process, resulting in a socio-spatial order determining whose fears the security is there to ease. Beyond state force, security and governance also operate in the unexpected and mundane worlds of everyday life through "the feelings, experiences, practices and actions of people outside the realm of formal politics" (Pain & Smith, 2008: 2). As Foucault describes in the concept of governmentality, and Dudley and Duffy describe in the New Orleans promotional campaign to residents, *deviant* behaviour that is believed to threaten the perception of clean and sanitised tourist zones is often reported and acted upon by local residents, business owners or tourists themselves resulting in either a physical or structural discipline of such behaviour (see also Ioannides & Petridou, 2016; Mosedale, 2016; Duggan, 2012). This occurs in tourism spaces such as parks and city streets, activities such as attending an event, or routines such as taking public transportation (Lisle, 2013; see also Amoore, 2007, 2009; Anderson, 2010; Nadesan, 2008). In other words, sources of enjoyment for some can result in both direct/physical and structural violence stemming from various forms of surveillance fueled by capitalist violence for others (Büscher & Fletcher, 2017).

Surveillance Capitalism

According to Zuboff (2019), we are in a transformation of capitalist logic wherein new technologies are being developed at an unprecedented pace and utilised by industries to transition from pursuing capital through serving users/consumers/visitors to surveilling them. It is not however, the development of new technologies alone that has driven this shift. In a modern capitalist society, "technology was, is, and always will be an expression of the economic objectives that direct it into action" (Zuboff, 2019: 16). As tourism becomes increasingly digitally mediated (Roelofsen, Chapter 25 of this volume), the logic driving the use of new technologies within it has the potential to generate new tourism realities by altering the ways in which we navigate social and material worlds (Minca & Roelofsen, 2019). In one example, Hannam et al. (2006) describe the ways in which airport terminals are increasingly becoming like cities while cities are simultaneously becoming like airports (see also Gottdiener, 2001; Pascoe, 2001). Surveillance technologies that become progressively normalised functions of our daily mundane movements such as closed-circuit television (CCTV) cameras in public spaces, GPS tracking systems, and WiFi hotspots are first trialled in airports. Although these technologies may now be part of our daily lives, Hannam et al. (2006) remind us that contrary to airports, cities are places of "fear and highly contingent ordering within the new world disorder" (p. 6).

Navigating Surveilled Spaces

Using the logic of disaster capitalists, the attacks on the World Trade Center on September 11, 2001, in the United States ushered in an acceptance of surveillance capabilities in the name of certainty during an uncertain time (Zuboff, 2019). Disaster capitalists acted swiftly, realising they could do as they pleased in the name of upholding liberty so long as it happened quickly while trauma and anxieties remained high (see Klein, 2007). The result was an ongoing prioritisation of security over privacy, which permanently transformed many societal systems, including the complex system of tourism (Hall & Page, 2014). Tourists are now accustomed to and versed in navigating both visible and invisible surveillance measures in the name of security. Using hotels as an example, Lisle (2013) describes how tourists are prepared in advance to become securitised subjects as well as collections of data. This includes "security guards, CCTV cameras … data surveillance linked to check-in, automated keys and key-activated elevators, biometric door locks, increased background checks on employees, detailed evacuation plans, emergency lighting, bomb-proof Kevlar wallpaper and shatter-proof glass" (Lisle, 2013: 136; see also Lyon, 2003; Morgan & Pritchard, 2005).

While analogue surveillance systems are not new, the quantitative advances in the technologies of digital surveillance, which can store enormous data volumes and much faster than in times of analogue systems, today, both analogue and digital surveillance work in tandem through automated processes to monitor mobilities, collecting enormous amounts of data, which is often used for predictive purposes (see Graham & Wood, 2003; Hugl, 2013). This presents a complex network of surveillance for those living and working in tourism spaces. While much of the policy discussion is rightly focused on data privacy concerns, such a lens remains consumer/tourist-centric and ignores the impact of omnipresent surveillance on others navigating tourism spaces such as labourers. The dangers posed by this fall heavily on the most vulnerable workers, which represents a majority of the tourism workforce (see Dudley et al., 2021; Robinson et al., 2019). If gone unchecked, this presents opportunities for those in positions of power to prey upon an array of vulnerabilities, while at the same time preventing workers from challenging what some describe as increasingly invasive practices (Zickuhr, 2021).

Labouring Under Surveillance

In recent years, the Internet and development of new technologies are largely responsible for an increase in employee surveillance (Bernhardt et al., 2021; Ball, 2010). According to Bernhardt et al (2021), with the advent of "flawed systems based on faulty data and pseudoscience and powerful technologies such as facial recognition, there is growing understanding that the potential harms of new technologies extend far beyond privacy" (p. 15). This is not to say workplace surveillance is new; it has been around since the early era of industrialisation (see Galič et al., 2017). Similar to the objectives of industrial-age workplace surveillance, a key application of data-driven technologies under surveillance capitalism is to monitor and manage worker productivity. This is not harmful in and of itself, but when an employer uses technology to track and relentlessly push workers using the capitalist language of efficiency, the negative effects compound, directly impacting workers' physical and mental wellbeing (see Bernhardt et al., 2021). For an example, again, look no further than one of the largest components of the tourism system: hotels.

The hotel industry has increasingly adopted a suite of technologies to monitor and manage workers, especially housekeepers. One such technology is the introduction of *panic buttons* with the aim of protecting housekeepers from sexual assault and harassment (for a more robust overview of sexual harassment in the tourism workplace see Vizcaino et al., 2020). Panic buttons are devices that housekeepers and other isolated workers carry with them while working, which when activated will notify security or emergency personnel of the worker's precise location. The technologies that allow for this to happen, such as Wi-Fi and GPS, are not inherently harmful to the workers – it is primarily the workers asking for the safety provided by panic buttons – but rather the use of their surveillance capabilities. For instance, features such as continuous real-time monitoring can be, and is, used by (some) employers for purposes other than worker safety, such as collecting data on workers' location on and off the clock, which can then be used to evaluate job performance (Bernhardt et al., 2021). This, coupled with the current capitalist growth paradigm within which most tourism resides, can result in a constant state of fear and anxiety for employees rather than the safe workplaces panic buttons are discursively described to create.

In addition to panic buttons, hotels are increasingly adopting service optimisation systems that automate task prioritisation and delegation (see Bernhardt et al., 2021; Zickuhr, 2021). These systems are designed to achieve a specified management objective such as room cleaning order, or personalised VIP services. When you or I are travelling and check out of our room or request services such as an early check-in, the system delegates the task based on such things as their proximity or current workload. Employees receive notifications regarding the updated task assignment via smartphone or tablet in real-time throughout the day. While implemented using capitalist discourse rooted in efficiency and product improvement, these systems can also lead to incoherent task prioritisation, unrealistic productivity expectations, and work intensification for jobs that are already physically demanding and prone to injuries.

Where to from Here?

While not the sole driver of capitalist logic, a major component is the necessary presence of anxieties and fears in *unprecedented times* (e.g., the modern swiftness of new technological developments) meaning transformation is often unrecognisable and therefore difficult to pinpoint (Zuboff, 2019). Rather than question transformations introduced during these times, we are drawn to interpreting things through the eyes of the familiar. Yet exploring such concepts as surveillance through dated and static lenses, outside the dominant economic and sociopolitical system within which the majority of global tourism resides, runs the risk of overlooking – or perpetuating – the very inequities many academics and practitioners are working to minimise. Fresh observation, analysis, and new naming are required if we are to grasp the unprecedented as a way to challenge the status quo. For instance, prior to the COVID-19 induced shutdown beginning in 2020, the relentless drive for growth, profits and expansion were evident in most parts of the world, with overtourism being a well noted symptom (see Higgins-Desbiolles, 2018; Jover & Díaz Parra, 2020). Now, in 2023, with over two years' time to reflect, and the pandemic still not officially over, overtourism is becoming a topic of concern once again. To better understand this phenomenon, and other negative impacts of tourism, Higgins-Desbiolles et al. (2019) echo the prior calls of Britton (1982; 1991) and Bianchi (2018) for exploration "situated in the wider context of tourism development being fostered by the capitalist economy system for profit accumulation of multinational corporations and the global elite" (p. 6; see also Fletcher, 2011; Higgins-Desbiolles, 2020).

According to Buscher and Fletcher (2017), in order to mitigate the fear and anxiety that accompanies both the physical and structural violence present in modern capitalist tourism development, the field must become part of a broader degrowth movement. Degrowth is a "project of radical socio ecological transformation focused especially on a sustainable downsizing of global consumption and production patterns" (Buscher & Fletcher, 2017, p. 664). Moving forward, tourism must not only adopt sustainable principles, but make radical shifts from a private and privatising activity under capitalism – in all its configurations – to one that more evenly contributes to the common and ceases to perpetuate the violence inherent in capitalist expansion (see for examples, Buscher & Fletcher, 2017; Devine & Ojeda, 2017; Hall et al., 2020; Salazar, 2017).

Although the connection between tourism and capitalist growth has been laid bare, there remains much to be done regarding the recognition of capitalist configurations and transformations as they are introduced into tourism spaces. Regarding surveillance capitalism, tourism research remains primarily through a managerial lens in the pursuit of harnessing the capabilities of rapidly developing technologies for surveillance capitalist outcomes (e.g., predictive behavioural data). Yet there are scholars exploring ways for tourism to exist outside of the capitalist system such as Rowen's (2020) exploration of the annual 'Burning Man' festival in Nevada. Originally developed as an anti-capitalist festival, it continues to serve as an alternative space for non- or not-fully capitalist forms of exchange. While acknowledging the critiques against the festival's ongoing corporatisation both in popular and academic literature, Mostafanezhad points out it remains an example of thinking differently about economic exchange, which will be an important way forward in a post COVID-19 tourism landscape (Brouder et al., 2020; see also Higgins-Desbiolles, 2020). Examples such as this can be used as pathways for exploration into how transformations of capitalist logic can be covertly embedded into anti-capitalist pursuits (see Fletcher, 2011; 2019 for a discussion of ecotourism as an example).

To this end, reflexive processes should be employed by researchers when making methodological choices in the age of surveillance capitalism; even when using technologies for disruptive purposes. Rapid advances of tracking technologies have proved enticing to tourism researchers as they provide insight and information that was unattainable less than a decade ago (see Jeffrey et al., 2021). Knowing the capitalist logic through which many of these technologies emerge however, creates a division between the excitement of access to uncharted data and the ethical dilemmas they present. One such dilemma is ensuring participants are not exposed to the forms of violence that are often embedded in digital technologies through surveillance capitalist logic. For instance, Duignan & McGillivray (2019) suggest employing activist research methodologies such as their use of digital platforms and #RioZones to create new lines of disruptive inquiry. In their research, the authors explore real-time impacts of the 2016 Rio Olympics on the host community using participatory digital methods. According to Shoval and Ahas (2016), however, using digital tools, even for disruptive purposes, creates its own challenges. The use of such technologies, which can obtain the exact locations of study participants at any given moment, may be infringing on the privacy of participants by adding a geographical dimension to the surveillance society (Lyon, 2001) and the ability to better track the digital individual (Curry, 1997). Before embracing new digital technologies into research, Shoval and Ahas suggest more research is needed regarding participant privacy and, more generally, the ethical aspects of such research.

When looking backward at the longitudinal reciprocal relationship between tourism and capitalist growth, the fear, anxiety and violence it is reliant upon, and forward into the transforming landscape of surveillance capitalism, Zuboff (2019) challenges us to reflect upon "what fresh legacy of damage and regret will be mourned by future generations?" (p. 11). In the time of continued calls for growth and *bringing back* tourism, paralleled with unprecedented technological advances, it is certainly important to reflect on the positive aspects of tourism, but the challenge is to explore and name the evolving ways in which tourism has and continues to perpetuate both physical and structural violence without dismissing the logic powering the tools used to do so under the modern configuration of surveillance capitalism.

References

Amoore, L. (2007). Vigilant visualities: The watchful politics of the War on Terror. *Security Dialogue* 38(2), 215–232.

Amoore, L. (2009). Algorithmic war: Everyday geographies of the War on Terror. *Antipode* 41(1), 49–69.

Anderson, B. (2010). Morale and the affective geographies of the 'war on terror'. *Cultural Geographies* 17(2), 219–236.

Andrews, H. (2014). *Tourism and violence*. Ashgate.

Ball, K. (2010). Workplace surveillance: An overview. *Labor History* 51(1), 87–106.

Baum, T., & Hai, N.T.T. (2020). Hospitality, tourism, human rights and the impact of COVID-19. *International Journal of Contemporary Hospitality Management* 32(7), 2397–2407.

Bernhardt, A., Kresge, L. & Suleiman, R. (2021). The case for worker technology rights. *UC Berkeley Labor Center* [accessed February 2024]. https://laborcenter.berkeley.edu/wp-content/uploads/2021/11/Data-and-Algorithms-at-Work.pdf

Bianchi, R. (2018). The political economy of tourism development: A critical review. *Annals of Tourism Research* 70, 88–102. https://doi.org/10.1016/j.annals.2017.08.005

Bowles, S., & Gintis, H. (2012). *Democracy and capitalism: Property, community, and the contradictions of modern social thought*. Routledge.

Britton, S. (1982). The political economy of tourism in the third world. *Annals of Tourism Research* 9, 331–358.

Britton, S. (1991). Tourism, capital, and place: Towards a critical geography of tourism. *Environment and Planning D 9*, 451–478.

Brouder, P., Teoh, S., Salazar, N.B., Mostafanezhad, M., Pung, J.M., Lapointe, D., … & Clausen, H.B. (2020). Reflections and discussions: Tourism matters in the new normal post COVID-19. *Tourism Geographies* 22(3), 735–746.

Büscher, B., & Fletcher, R. (2017). Destructive creation: Capital accumulation and the structural violence of tourism. *Journal of Sustainable Tourism* 25(5), 651–667. https://doi.org/10.1080/09669582.2016.1159214

Cave, J., & Dredge, D. (2020). *Reworking Tourism: Diverse Economies in a Changing World*. Routledge.

Curry, M.R. (1997). The digital individual and the private realm. *Annals of the Association of American Geographers* 87(4), 681–699.

Devine, J.A. (2017). Colonizing space and commodifying place: Tourism's violent geographies. *Journal of Sustainable Tourism* 25(5), 634–650.

Devine, J., & Ojeda, D. (2017). Violence and dispossession in tourism development: A critical geographical approach. *Journal of Sustainable Tourism* 25(5), 605–617.

Dudley, K.D., & Duffy, L.N. (2022). Tourism discourse and surveillance: Situational analysis of Post-Katrina New Orleans. *Leisure Sciences*, Advanced online publication. 1–19.

Dudley, K.D., Duffy, L.N., Terry, W.C., & Norman, W.C. (2021). The historical structuring of the US tourism workforce: A critical review. *Journal of Sustainable Tourism* 30(12), 2823–2838. https://doi.org/10.1080/09669582.2021.1952417

Duggan, L. (2012). *The twilight of equality?: Neoliberalism, cultural politics, and the attack on democracy*. Beacon Press.

Duignan, M.B., & McGillivray, D. (2019). Walking methodologies, digital platforms and the interrogation of Olympic spaces: The '# RioZones-Approach'. *Tourism Geographies* 23(1–2), 275–295. https://doi.org/10.1080/14616688.2019.1586988

Farmer, P. (2004). An anthropology of structural violence. *Current anthropology* 45(3), 305–325.

Fletcher, R. (2011). Sustaining tourism, sustaining capitalism? The tourism industry's role in global capitalist expansion. *Tourism Geographies* 13, 443–461.

Fletcher, R. (2019). Ecotourism after nature: Anthropocene tourism as a new capitalist 'fix'. *Journal of Sustainable Tourism* 27(4), 522–535. https://doi.org/10.1080/09669582.2018.1471084

Fletcher, R., & Neves, K. (2012). Contradictions in tourism: The promise and pitfalls of ecotourism as a manifold capitalist fix. *Environment and Society* 3(1), 60–77.

Fletcher, R., Blanco-Romero, A., Blázquez-Salom, M., Cañada, E., Murray Mas, I., & Sekulova, F. (2021). Pathways to post-capitalist tourism. *Tourism Geographies* 25(2–3), 707–728.

Foucault, M. (1991). Politics and the study of discourse. In G. Burchell, C. Gordon, & P. Miller (Eds.), *The Foucault Effect, Studies in Governmentality, with Two Lectures by and an Interview with Michel Foucault* (pp. 53–72). The University of Chicago Press.

Foucault, M. (1995). *Discipline and punish: The birth of the prison* (A. Sheridan, Trans.). Vintage. (Original work published 1975).

Frenzel, F., Koens, K., & Steinbrink, M. (2012, Eds.). *Slum tourism*. Taylor & Francis.

Galič, M., Timan, T., & Koops, B.J. (2017). Bentham, Deleuze and beyond: An overview of surveillance theories from the Panopticon to participation. *Philosophy & Technology* 30(1), 9–37. https://doi.org/10.1007/s13347-016-0219-1

Gotham, K.F. (2007). (Re) branding the big easy: Tourism rebuilding in post-Katrina New Orleans. *Urban Affairs Review* 42(6), 823–850.

Gottdiener, M. (2001). *Life in the Air*. Rowman and Littlefield.

Graham, S., & Wood, D. (2003). Digitizing surveillance: Categorization, space, inequality. *Critical Social Policy* 23, 227–248.

Hall, C.M., & Page, S.J. (2014). *The geography of tourism and recreation: Environment, place and space*. Routledge.

Hall, C.M., Lundmark, L., & Zhang, J. (Eds.). (2020). *Degrowth and tourism: New perspectives on tourism entrepreneurship, destinations and policy*. Routledge.

Hannam, K., Sheller, M., & Urry, J. (2006). Mobilities, immobilities and moorings. *Mobilities* 1(1), 1–22.

Harvey, D. (2006). *Spaces of global capitalism: A theory of uneven geographical development*. Verso.

Higgins-Desbiolles, F. (2006). More than an 'Industry': The forgotten power of tourism as a social force. *Tourism Management* 27, 1192–1208.

Higgins-Desbiolles, F. (2009). *Capitalist globalisation, corporated tourism and their alternatives*. Nova Science Publishers.

Higgins-Desbiolles, F. (2018). Sustainable tourism: Sustaining tourism or something more? *Tourism Management Perspectives* 25, 157–160.

Higgins-Desbiolles, F. (2020). Socialising tourism for social and ecological justice after COVID-19. *Tourism Geographies* 22(3), 610–623. https://doi.org/10.1080/14616688.2020.1757748

Higgins-Desbiolles, F., Carnicelli, S., Krolikowski, C., Wijesinghe, G., & Boluk, K. (2019). Degrowing tourism: Rethinking tourism. *Journal of Sustainable Tourism* 27(12), 1926–1944. https://doi.org/10.1080/09669582.2019.1601732

Hugl, U. (2013). Workplace surveillance: Examining current instruments, limitations and legal background issues. *Tourism & Management Studies* 9(1), 58–63.

Ioannides, D., & Petridou, E. (2016). Contingent neoliberalism and urban tourism in the United States. In J. Mosedale (Ed.), *Neoliberalism and the political economy of tourism* (pp. 21–35). Routledge.

Jeffrey, H.L., Ashraf, H., & Paris, C.M. (2021). Hanging out on Snapchat: Disrupting passive covert netnography in tourism research. *Tourism Geographies* 23(1–2), 144–161.

Jover, J., & Díaz-Parra, I. (2020). Who is the city for? Overtourism, lifestyle migration and social sustainability. *Tourism Geographies* 24(1), 9–32. https://doi.org/10.1080/14616688.2020.1713878

Kincheloe, J. L., & McLaren, P. (2008). Rethinking critical theory and qualitative research. In N. K. Denzin & Y. S. Lincoln (Eds.), *The Sage handbook of qualitative research* (pp. 303–342). Sage Publications.

Klein, N. (2007). *The shock doctrine: The rise of disaster capitalism*. Macmillan.

Li, X., Gong, J., Gao, B., & Yuan, P. (2021). Impacts of COVID-19 on tourists' destination preferences: Evidence from China. *Annals of Tourism Research* 90, 103258. https://doi.org/10.1016/j.annals.2021.103258

Lisle, D. (2013). Frontline leisure: Securitizing tourism in the War on Terror. *Security Dialogue* 44(2), 127–146.

Lyon, D. (2001). *Surveillance society: Monitoring everyday life*. McGraw-Hill Education.

Lyon, D. (2003). Airports as data filters: Converging surveillance systems after September 11th. *Journal of Information, Communication and Ethics in Society* 1(1), 13–20.

Minca, C., & Roelofsen, M. (2019). Becoming airbnbeings: On datafication and the quantified self in tourism. *Tourism Geographies* 23(4), 743–764.

Morgan, N., & Pritchard, A. (2005). Security and social 'sorting': Traversing the surveillance–tourism dialectic. *Tourist Studies* 5(2), 115–132.

Mosedale, J. (2016, Ed.). *Neoliberalism and the political economy of tourism*. Routledge.

Mowforth, M. & Munt, I. (2016). *Tourism and sustainability: Development, globalisation and new tourism in the third world* (4th ed.).Routledge.

Nadesan, M.H. (2008). *Governmentality, biopower and everyday life*. Routledge.

Ojeda, D. (2013). War and tourism: The banal geographies of security in Colombia's "retaking". *Geopolitics* 18(4), 759–778.

Pain, R., & Smith, S.J. (2008). *Fear: critical geopolitics and everyday life*. Ashgate.

Pascoe, D. (2001). *Airspaces*. Reaktion.

Robinson, R.N., Martins, A., Solnet, D., & Baum, T. (2019). Sustaining precarity: Critically examining tourism and employment. *Journal of Sustainable Tourism* 27(7), 1008–1025.

Robinson, W. (2008). *Latin America and global capitalism: A critical globalization perspective*. John Hopkins University Press.

Rowen, I. (2020). The transformational festival as a subversive toolbox for a transformed tourism: Lessons from Burning Man for a Covid-19 world. *Tourism Geographies* 22(3), 695–702.

Salazar, N.B. (2017). The unbearable lightness of tourism… as violence: An afterword. *Journal of Sustainable Tourism* 25(5), 703–709.

Seraphin, H., & Dosquet, F. (2020), "Mountain tourism and second home tourism as post COVID-19 lockdown placebo?", *Worldwide Hospitality and Tourism Themes* 12(4), 485–500. https://doi.org/10.1108/WHATT-05-2020-0027

Shoval, N., & Ahas, R. (2016). The use of tracking technologies in tourism research: The first decade. *Tourism Geographies*, *18*(5), 587–606.

UNWTO (2020). Tbilisi Declaration: Actions for a Sustainable Recovery of Tourism (accessed June 2024). www.unwto.org/actions-for-a-sustainable-recovery-of-tourism

Vizcaino, P., Jeffrey, H., Eger, C., & Turkoglu, H. (2020). The silenced phenomenon: sexual harassment in the hospitality and tourism workplace. In P. Vizcaino, H. Jeffrey & C. Eger (Eds.), *Tourism and gender-based violence: challenging inequalities* (pp. 48–64). CABI.

Zickuhr, K. (2021). Workplace surveillance is becoming the new normal for US workers. *Washington Center for Equitable Growth.* [accessed February 2024]. https://equitablegrowth.org/research-paper/workplace-surveillance-is-becoming-the-new-normal-for-u-s-workers/

Zuboff, S. (2019). *The age of surveillance capitalism*. PublicAffairs.

28

THE PLATFORM ECONOMY AS A GAME CHANGER FOR TOURISM GEOGRAPHIES

Egbert van der Zee

Introduction

Sharing in many different forms has always been a key element of human interactions (Belk, 2007; Frenken & Schor, 2017). In tourism, sharing rides, meals or providing temporary access to one's home to accommodate paying or non-paying guests have been common practice since well before the 'sharing economy' became a fashionable term to describe these activities. Frenken and Schor (2017, 4–5) define the sharing economy as "consumers granting each other temporary access to under-utilised physical assets ('idle capacity'), possibly for money". The rise of social media, acceptance of online platforms and a wider shift in society from valuing ownership to having access to goods caused the sharing economy to flourish (Dredge & Gyimóthy, 2015; Guttentag, 2015; Andreu et al., 2020). Specifically, online platforms played a crucial role as facilitators connecting supply and demand. Platforms not only function as online marketplaces where supply and demand meet, they also actively enhance connecting supply and demand through mechanisms such as review systems that generate trust between strangers (Tussyadiah, 2016).

During the last decade, the use of platforms expanded rapidly, raising two issues. First, because of its rapid growth, the supply of goods and services offered via platforms has become more diverse. This supply now ranges from sharing idle capacity to providing access to commercially exploited goods and services (Frenken & Schor, 2017). Second, due to a combination of its rapid expansion, sheer size and increasingly commercialising offer, platforms are increasingly seen as a disruptive innovation which can have strong but also adverse impacts on both incumbent industries and local communities (Guttentag, 2015). These two issues set the stage for this chapter, where I aim to provide a conceptual overview of the development of the sharing economy and the role of platforms in the context of tourism and hospitality, as well as discuss how this development affects tourism geographies at different scale levels. I provide a brief overview of the expanding body of research on this topic and identify a number of caveats in current research which I translate into a research agenda.

DOI: 10.4324/9781003286301-33

To Share or not to Share? Defining Peer-to-peer Accommodation Provision via Platforms

While the term 'sharing economy' is the most widely used term to describe activities that involve the provision of goods and services between peers facilitated by platforms, it might not be the term that best fits these activities. In conceptual discussions, a distinction is made between the sharing of temporary idle capacity and on-demand services or permanently available goods, whereby the latter is considered an 'on-demand' or 'gig' economy (Frenken & Schor, 2017). What both activities have in common is the important role online platforms play in facilitating peer-to-peer sharing of goods or service provision between strangers (Tussyadiah, 2016). Granting access to or sharing goods or services between peers coordinated by online platforms without transferring ownership is a broad and frequently applied definition of this 'platform economy' (Belk, 2007; Klarin & Suseno, 2021).

In this chapter, I use the term 'platform economy' instead of 'sharing economy' for a number of reasons. First, platforms play a crucial role in the development and rapid growth of peer-to-peer provision of goods and services in tourism and hospitality (Guttentag, 2015). It is argued that the role of these platforms goes beyond providing an online place where peers can meet and match offer and supply. The trust mechanism provided by platforms, through review systems, support and insurance, facilitates these interactions (Tussyadiah, 2016). Apart from this, platforms themselves are powerful actors, as they can mediate how these interactions take place by applying algorithmic management affecting search results or advising pricing strategies (Cheng & Foley, 2019; Gibbs et al., 2018). Lastly, there is much debate as to whether providers of these types of services, such as Airbnb and VRBO, are actually facilitating sharing of 'idle capacity' or are rather providing a more commercialised 'micro-entrepreneurialism' via their platforms (Belk, 2007; Frenken & Schor, 2017). Dann et al. (2019, 459) note, for example, that platforms are much more than just booking portals or network facilitators, as they are "a blueprint for an entire category of novel business models". The platform economy is therefore used as an overlapping concept which "is any type of digital platform that uses the Internet to connect dispersed networks of individuals to facilitate digital interactions between people" (Zysman & Kenney in Klarin & Suseno, 2021: 257).

Platforms as True Disruptors of Tourism Geographies

Platforms in hospitality and tourism have become a central area of academic inquiry. In their extensive review of the literature and bibliometric analysis of the sharing economy in general, Klarin and Suseno (2021) identified hospitality and tourism as one of the main research themes in this field.

The growing attention on this topic can be explained by the fact that hospitality and tourism are regarded as one of the industries most affected by the development of – specifically – the platform economy (Sigala, 2017; Kuhzady et al., 2021). Within the platform economy in tourism and hospitality, peer-to-peer accommodation sharing is the type of service which has received the most attention in the media (Huertas et al., 2021), in policy and regulation debates (Nieuwland & van Melik, 2020) and in academia (Klarin & Suseno, 2021; Kuhzady et al., 2021). In their comprehensive literature review, Kuhzady et al. (2021) found that more than 90 per cent of the included papers solely discussed accommodation sharing and an overwhelming share of 83 per cent focused on Airbnb. Because of its early presence, innovative business model, platform design and sheer size, Airbnb is referred to as a 'poster child' in this field (Dann et al., 2019). The term Airbnb is used both by academics as well as practitioners as an overarching concept when referring to offering peer-to-peer accommodation, as it "offers structure to asynchronous, disparate research efforts around one common construct, and identifies areas of investigation that require attention in future" (Dolnicar, 2019: 249).

A little more than a decade since Airbnb started operating its business, it can be concluded that this phenomenon has gained a central position in the geography of tourism. The platform economy, and specifically peer-to-peer accommodation provision, can without doubt be defined as a game changer in contemporary tourism geographies (Guttentag, 2019). The metaphor of a perfect storm may be used to describe how several circumstances came together at the right time to cause unprecedented growth of the platform economy in tourism and hospitality. Fuelled by an ever-growing number of tourists looking for 'authentic' experiences, or simply affordable accommodation options (Guttentag, 2019), in just a few years peer-to-peer accommodation platforms took over a significant market share from the incumbent industry (Kuhzady et al., 2021). Peer-to-peer accommodation provision even grew to be the largest segment of the wider platform economy in Europe (Andreu et al., 2020). The current market leader, Airbnb, offered more than 6 million listings in more than 100,000 destinations in March 2022 (Airbnb, 2022), which is less than the 7 million listings offered at its peak in 2019, probably due to the COVID-19 pandemic (Adamiak, 2022). VRBO, part of the Expedia group and one of Airbnb's main competitors, offered more than 2 million listings in 2022 (VRBO, 2022). Other companies, such as Booking.com and TripAdvisor, have also recently started operating as peer-to-peer accommodation platforms. It should be noted that part of the offer is active on multiple platforms at the same time (Rodríguez-Pérez de Arenaza et al., 2022). However, Airbnb is still the main focus of academic inquiry, and information on these other platforms is still largely lacking.

Peer-to-peer accommodation platforms quickly grew from being the 'new kid on the block' to an integral part of tourism and hospitality, outperforming the largest hotel chains, such as Marriott and Hilton, in size and market share (Dogru et al., 2019). It is still debatable whether the platform economy is truly a disruptive innovation (Guttentag, 2015; Muller, 2020). However, there are numerous accounts stating that the impact of platform economies on tourism geographies should not be underestimated: The platform economy is believed to have an impact on the traditional accommodation sector (Zervas et al., 2017; Dogru et al., 2019), on destinations and their communities (Nofre et al., 2018), on the housing market (Gurran & Phibbs, 2017) and on the spatial extent (Ioannides et al., 2020) and performance of tourism in general (Farmaki et al., 2020).

A Decade of Platform Economy Research in Tourism and Hospitality Studies

Since 2015, the body of literature discussing peer-to-peer accommodation has grown at a rapid pace. In 2015, the Scopus database lists a total of 19 newly published papers using the term 'Airbnb' in their title, keywords or abstract. The number of papers published in 2021 was 355 (Figure 28.1). Between 2015 and 2020, the number of publications on Airbnb and more general peer-to-peer accommodation followed a linear growth pattern. In relative terms, the number of papers on the topic increased faster than papers on tourism in general. The growing academic attention given to Airbnb does, however, seem to have stagnated since 2020, while the number of papers on tourism has continued its upward trend. A possible explanation could be the effects of the COVID pandemic, which shifted attention to other topics within tourism studies, or the fading novelty of the topic in general.

This rapidly expanding body of literature has recently been synthesised in a number of literature reviews (see Table 28.1). By systematically analysing the literature, these reviews found similar recurring overarching research themes. For this chapter, themes were synthesised into 13

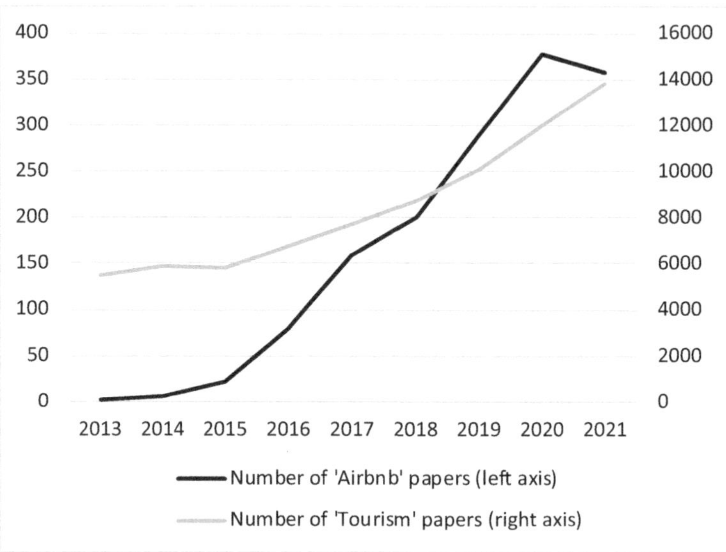

Figure 28.1 Number of papers on Scopus having 'Airbnb' or 'tourism' in the title, key words or abstract (published between 2014 and 2021).

Source: Compiled by the author.

labels and placed within a conceptual representation of the platform economy in tourism geographies (Figure 28.2 and Table 28.1). From this, I can firstly distinguish between three different scale levels which simultaneously affect and are affected by platform economies. Even though peer-to-peer accommodation provision facilitated by platforms can be seen as a global phenomenon, many of the actual activities caused by these platforms take place at the micro-scale level. This level consists of individuals, their interactions and transactions, and the space in which these take place, such as the home where hosts welcome guests (Farmaki et al., 2020).

On the meso level, places and their local contexts simultaneously shape and are shaped by platform economies. These places can be different types of already-popular tourist destinations, such as historic city centres, but also gentrifying suburbs or rural communities where tourism had had only a marginal impact until the arrival of peer-to-peer accommodation (Ioannides et al., 2019). In these places, stakeholders such as local policymakers and planners, the traditional accommodation sector and local communities have different ways of responding to the development of platform economies. While protests by local communities and/or lobbying activities by the incumbent industry have led to strict policy measures and regulations in some places, other places are characterised by a laissez-faire attitude or even embracing the activities of platforms (Nieuwland & van Melik, 2020).

At the macro level, the development of information and communication technologies, the widespread adoption of social media, the development of data analytics and algorithms, rapidly growing international tourism numbers and image formation on a global scale via international media and other supra-national institutions have facilitated the global expansion of platforms. Following this conceptualisation, the platform economy can be seen as a complex system in which stakeholders interact on and between different scale levels.

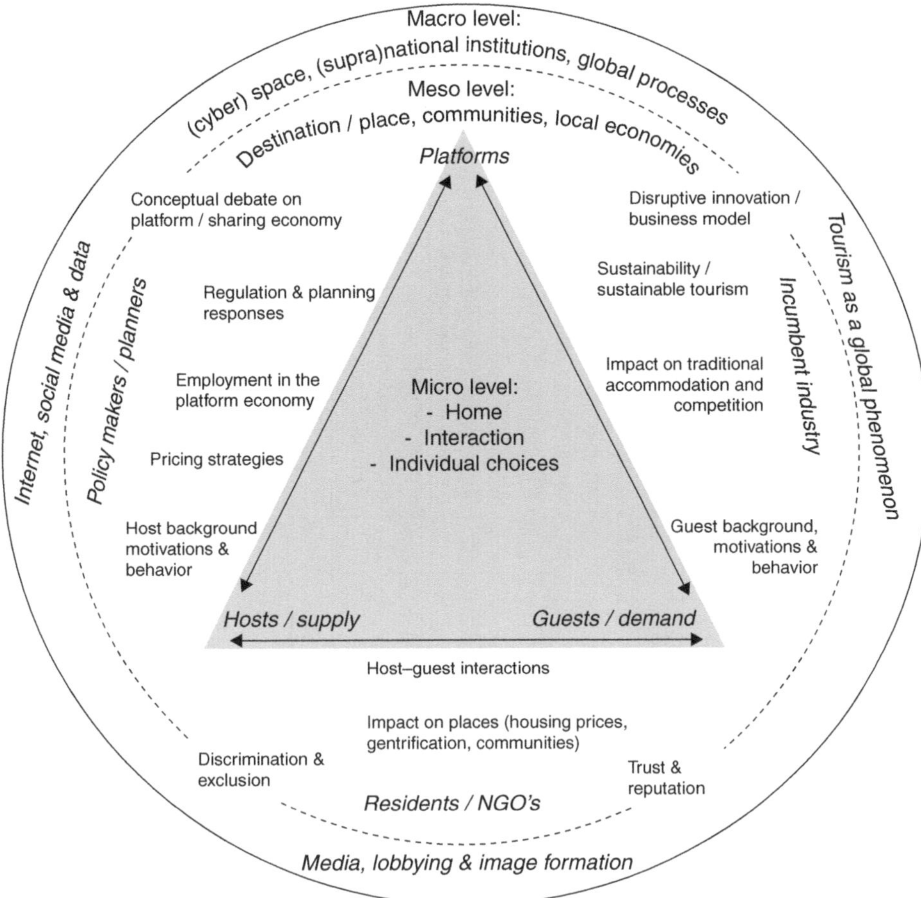

Figure 28.2 Schematic overview of the current body of research into platform economies in tourism.
Source: Compiled by the author.

The overarching research themes found in the different literature reviews have been placed in the schematic conceptualisation of the platform economy in tourism. Research initially focused on peer-to-peer accommodation sharing as a new business model (Oskam & Boswijk, 2016) and discussed how this business model evolved by looking at what motivates individuals from the supply and demand side to participate in the sharing economy (Ikkala & Lampinen, 2015). Guest motivation and behaviour is currently the most discussed topic in this field of research (Dann et al., 2019; Guttentag, 2019; Belarmino & Koh, 2020). These studies often find that value, authenticity, uniqueness and social benefits are important drivers for using peer-to-peer accommodations (Guttentag et al., 2018; Tussyadiah, 2016). Trust (Ert et al., 2016) and host-guest relations are also a widely discussed topic in this stream of research, underlining the new conceptual approach in this business model where platforms are more than just online meeting places. Around the same time, studies emerged to support and understand the practical development of platforms as a new business model. The topics of accommodation pricing strategies (Chen & Xie, 2017; Gibbs et al.,

Table 28.1 Overview of Literature Reviews on the Platform Economy within Tourism Studies.

Authors (year)	Title	Field / focus	Timespan	Number of papers included	Themes
Prayag & Ozanne (2018)	A systematic review of peer-to-peer (P2P) accommodation sharing research from 2010 to 2016: progress and prospects from the multi-level perspective	Peer-to-peer accommodation	2010–2016	71	Conceptual development; Regulation; Regime responses; Macro level impacts; Host behaviour; Guest/host experiences; Marketing issues;
Altinay & Tahery (2019)	Emerging themes and theories in the sharing economy: a critical note for hospitality and tourism	Not specified	Not specified	Not specified	Socially excluded customers; Pricing strategies; Trust and reputation; Disruptive behaviour; Choice and segmentation; Personality; Satisfaction;
Dann et al., (2019)	Poster child and guinea pig—insights from a structured literature review on Airbnb	Airbnb	2013–2018	118	Legal and regulatory aspects; Economic and media impact; Price and pricing; Reputation systems; User motives and types; Text reviews; Profile images;
Dolnicar (2019)	A review of research into paid online peer-to-peer accommodation: Launching the Annals of Tourism Research Curated Collection on peer-to-peer accommodation.	Airbnb	2013–2019	122	P2P & tourism industry; Environment; Policy; Society;
Guttentag (2019)	Progress on Airbnb: a literature review	Airbnb	2013–2018	132	Nature of the sharing economy; Airbnb regulation; Airbnb's impact on the tourism sector; Airbnb supply and its impacts on destinations; Airbnb guests; Airbnb hosts;

Ozdemir & Turker (2019)	Institutionalisation of the sharing in the context of Airbnb: a systematic literature review and content analysis	Airbnb & institutionalisation	2013–2017	56	The Airbnb company; Legal impacts; Economic impacts; Motives;
Andreu et al. (2020)	Airbnb research: an analysis in tourism and hospitality journals	Peer-to-peer accommodation	2015–2019	129	Ethics and sustainability; Legal and regulatory aspects; Impacts on destinations; Pricing and revenue management; Reputation/trust/mistrust; Consumer behaviour; Guest/host experience; Airbnb supply; Quantitative and qualitative methods;
Belarmino & Koh (2020)	A critical review of research regarding peer-to-peer accommodations	Peer-to-peer accommodation	2010–2017	107	Conceptualising P2P accommodations; Sharing economy; Legal issues; Revenue management; Trust and mistrust; Affordable housing concerns; P2P accommodations and hotels; Consumer behaviour; Owner motivations; emerging fields;
Sainaghi (2020)	The current state of academic research into peer-to-peer accommodation platforms	Peer-to-peer accommodation	2010–2019	189	P2P platforms; Regulation; Economic impact studies; Social impact; Guest-host relationships; Demand side studies; Hosts; Literature reviews
Sainaghi & Baggio (2020)	Clusters of topics and research designs in peer-to-peer accommodation platforms	Peer-to-peer accommodation (focus on applied methods)	2010–2019	189	P2P platforms; Regulation; Economic impact studies; Social impact; Guest-host relationships; Demand side studies; Hosts; Literature reviews

(*Continued*)

Table 28.1 (Continued)

Authors (year)	Title	Field / focus	Timespan	Number of papers included	Themes
Hati et al. (2021)	A decade of systematic literature review on Airbnb: the sharing economy from a multiple stakeholder perspective	Airbnb	2009–2020	282	Government/regulators; Communities; Competitors; Guest; Host; Employees;
Kuhzady et al. (2021)	Sharing economy in hospitality and tourism: a review and the future pathways	Sharing economy in tourism and hospitality	2010–2019	486	The sharing economy as a disruptive business model; Sharing economy impacts and corresponding regulatory issues; Drivers to participate in the sharing economy;
Mody et al. (2021a)	Sharing economy research in hospitality and tourism: A critical review using bibliometric analysis, content analysis and a quantitative systematic literature review	Sharing economy in tourism and hospitality	2010–2020	308	Pricing strategies; Rent impacts; Impacts of the sharing economy on hotels; Sharing – consumption practices; Psychological aspects. Methods.

Source: Compiled by the author.

2018) and case studies explaining spatial patterns in the use of platforms for accommodation provision (Quattrone et al., 2016) are discussed.

The disruptive nature of the platform economy (Guttentag, 2015) is a major topic that started to dominate the literature in the second part of this decade, with a number of studies on the impact of the platform economy on the traditional accommodation sector (Zervas et al., 2017) and on destinations and their communities, including residents' attitudes and perspectives (Wachsmuth & Weisler, 2018; Cocola-Gant & Gago, 2021; Suess et al., 2020). Zervas et al. (2017), for example, showed that Airbnb had a negative effect on hotel revenues in the state of Texas. However, other case studies produced contrasting results. For instance, Dogru et al. (2019) found in their study that the presence of Airbnb had a positive effect on employment in the hospitality, tourism and leisure industries.

Contrasting results have also been found regarding the impact on neighbourhoods and communities. In various case studies, the presence of Airbnb listings is related to rising housing prices and gentrification (see, e.g., Gurran & Phibbs, 2017) while others find no such effects in their case studies (see Dann et al., 2019 for an overview). Suess and colleagues (2021) found that individuals who have previously stayed in peer-to-peer accommodations tend to be more positive towards the presence of such accommodation in their own neighbourhood. This indicates that the impact of platform economies can differ between places, but also at micro scale. A growing number of scholars and practitioners have highlighted that the development and impact of peer-to-peer platforms can differ strongly between geographical contexts (Gyódi, 2019; Wegman & Jiao, 2017; Dolnicar, 2019). Findings such as these fuel the debate on how to regulate the platform economy and how this phenomenon should be regarded from an urban planner's perspective (see, e.g., Gurran & Phibbs, 2017 or Ferreri & Sanyal, 2018).

Most recent review studies focus on contemporary social issues such as (social) sustainability and questions regarding employment in the platform economy. A popular new topic is how the media portrays peer-to-peer accommodation provision (Mody et al., 2021b; Huertas et al., 2021), while the impact of COVID-19 on peer-to-peer accommodation is also discussed in recent contributions (Lee & Deale, 2021).

Although the development of the platform economy and its impact on tourism geographies depends on processes occurring on and between different scale levels, most studies focus on the meso level (Prayag & Ozanne, 2018). These studies often apply a case study perspective to a specific part of this complex system, focusing on a specific geographic context, most often in North America or Europe (Guttentag, 2019; Mody et al., 2021a). This recent boom in academic studies on peer-to-peer accommodation has produced a wide range of topics, conceptualisations and methodological approaches, which mainly provide 'first-order knowledge' and lack theoretical rigour and cause-and-effect relationships (Dolnicar, 2019). Half of the studies included in the literature review by Belarmino & Koh (2020) have no clear theoretical framework, while the other half apply 37 different theories, an impressive amount considering the review included a total of 107 studies. Similar results were found by Mody et al. (2021a: 1730), who conclude that "theoretical application and development is overshadowed by a 'number-crunching' mentality that limits deeper understanding of the sharing economy phenomenon". Consequently, there is still a lack of understanding of 'the DNA of this phenomenon' (Sainaghi, 2020).

Discussion of Applied Methods and Research Design

The dominant position of number crunching can also be identified when reviewing the types of research designs and methodologies applied in studying the platform economy. Prayag and Ozanne

(2018) found that most of the 71 studies included in their literature review applied a quantitative methodology, while a quarter of the studies applied qualitative methods. Sainaghi and Baggio (2020) and Belarmino and Koh (2020) found similar results, noting that the most common form of quantitative research are studies applying quantitative modelling, such as regression analysis and structural equation modelling (see Suess et al., 2021). These studies apply survey-based research designs to the motivations and behaviour of guests and, to a lesser extent, hosts. Data for these studies are most often obtained from online panel services such as Amazon's Mechanical Turk (Mody et al., 2021a).

Analysis of spatial patterns in peer-to-peer accommodation provision is another frequently applied method. It ranges from descriptive analysis to spatial autocorrelation (Quattrone et al., 2016) and spatial modelling, such as geographically weighted regressions (Lagonigro et al., 2020). In these studies, the data analysed are most often collected through web-scraping. This data are either collected by the authors or obtained from openly available platforms such as Inside Airbnb or from commercial providers such as AirDNA (Mody et al., 2021a). Although analysis of this type of data can provide detailed and meaningful insights into spatial patterns, determinants of accommodation pricing, and can explain which places are impacted more than others, a number of gaps in the research still exist. First, there is a strong geographical bias in the present body of literature, as most studies focus on large cities and metropolitan areas in North America and Europe. This geographic bias does not reflect the geography of Airbnb, as in 2019 only a third of accommodation active on Airbnb was to be found in big cities (Adamiak, 2022). While roughly two-thirds of accommodation active on Airbnb in 2019 were located in Europe or the United States, "the geographical centre of Airbnb is moving from Europe towards other continents" (Adamiak, 2022: 3142). Due to the dominance of meso-level case studies and their geographical bias, it is important to keep in mind that generalising existing knowledge to other cities and regions, or to the entire platform, is problematic and in need of more geographically diverse input (Dann et al., 2019: 461). Second, Guttentag (2019) calls for more ambitious ways to tackle present research questions, for example by applying more qualitative methods or multi-method and mixed-method research designs. Last, Mody and colleagues (2021a) add that, in this sense, the 'what' of the sharing economy has been widely discussed, while the 'why' and 'how' remain underexposed.

To overcome these issues, a geographic perspective on the development and impact of the platform economy that takes into account the multiple scale levels that affect and are affected by the platform economy can provide novel insights (Prayag & Ozanne, 2018; Mody et al., 2021a). The platform economy and, specifically, peer-to-peer accommodation provision have grown to become an important part of tourism geographies, as it is simultaneously shaped by and shaping the interrelation between tourism and space on different scale levels (see Figure 28.2). This ranges from the micro scales, such as the home, the guest and the host (Farmaki et al., 2020), to meso scales, such as destinations (Cocola-Gant & Gago, 2021), their communities and businesses (Zervas et al., 2017), and macro scales, as platforms such as Airbnb operate globally and are shaped by cultures, societies and political and economic systems (Mody et al., 2021a). However, the micro and macro scales, in particular, are currently understudied (Kuhzady et al., 2021). Although this geographic perspective seems evident, Farmaki et al. (2020) note that "despite the proliferation of studies on peer-to-peer accommodation, there is a paucity of research on the spatial dimensions of the phenomenon" (Farmaki et al., 2020: 1).

**A Research Agenda towards a Geographical Perspective on the
Impact and Implications of Airbnb**

Approximately half a decade of academic inquiry into the platform economy has taught us many things. We know more about spatial development patterns, development over time, pricing mechanisms, user motivations, and its impact on incumbent industries. However, to better understand how the platform economy is transforming tourism geographies, in the coming decade or so we should start to pose new research questions, develop matching multiple or mixed-method research and integrate this topic into the wider academic debate on how tourism is transforming places and affecting communities (Cave & Dredge, 2020). Regarding the field of tourism geographies, the impact of platforms and their different types of users on the places where they are active are important topics for further investigation. While some authors have explored how accommodation-sharing platforms contribute to the financialisation of housing and act as drivers of gentrification (see, e.g., Grisdale, 2021), more work is needed on the interrelation between platform economies and the right to the city. In addition to broadening the conceptual and methodological scope (as discussed above), I suggest a number of routes that merit further exploration:

- The future of quantitative (spatial) data analysis in the platform economy requires the development of a critical perspective. Single-city case studies need to be placed in a broader perspective and stories behind the data need to be uncovered by including more qualitative methods in this stream of research.
- More insights are needed into relationships and encounters between hosts and guests. This includes research on (embodied) experiences and how interactions lead to mutual understanding and/or more mindful tourism.
- The role platform economies play in processes such as gentrification, financialisation of the housing market and changing geographies of leisure and work and how policy-makers or planners should respond to this.
- The role platform economies could or should play in moving towards a more sustainable type of tourism, for example by actively enhancing resilient communities and stimulating the development of regenerative tourism (Cave & Dredge, 2020).
- The development of platform economies outside peer-to-peer accommodation, and specifically Airbnb.
- Hybridisation of the platform economy, uncovering the many shades between sharing idle capacity and professional accommodation and service provision.
- The role of lobbying by different stakeholders and of framing and image formation by the media, and the impact these activities have on policy formation.
- Getting an idea of how planners, policy makers, destination management organisations (DMOs) and other stakeholders (should) react to disruptive innovations such as the platform economy.
- The role platforms themselves play as a 'new intermediary': As an agent of image formation and by influencing choice behaviour due to the working of the platforms and their algorithms.

References

Adamiak, C. (2022). Current state and development of Airbnb in 167 countries. *Current Issues in Tourism*, 25(19), 3131–3149. https://doi.org/10.1080/13683500.2019.1696758

Airbnb (2022). Newsroom – about us. Online source: https://news.airbnb.com/about-us/ (accessed February 2024)

Altinay, L. & Tahery, B. (2019). Emerging themes and theories in the sharing economy: a critical note for hospitality and tourism. *International Journal of Contemporary Hospitality Management*, 31(1), 180–193.

Andreu, L., Bigne, E., Amaro, S., & Palomo, J. (2020). Airbnb research: An analysis in tourism and hospitality journals. *International Journal of Culture, Tourism and Hospitality Research*, 14(1), 2–20. https://doi.org/10.1108/IJCTHR-06-2019-0113

Belarmino, A., & Koh, Y. (2020). A critical review of research regarding peer-to-peer accommodations. *International Journal of Hospitality Management*, 84, 102315. https://doi.org/10.1016/j.ijhm.2019.05.011

Belk, R. (2007). Why not share rather than own? *The Annals of the American Academy of Political and Social Science*, 611(1), 126–140. https://doi.org/10.1177/0002716206298483

Cave, J., & Dredge, D. (2020). Regenerative tourism needs diverse economic practices. *Tourism Geographies*, 22(3), 503–513. https://doi.org/10.1080/14616688.2020.1768434

Chen, Y., & Xie, K. (2017). Consumer valuation of Airbnb listings: A hedonic pricing approach. *International Journal of Contemporary Hospitality Management*, 29(9), 2405–2424. https://doi.org/10.1108/IJCHM-10-2016-0606

Cheng, M., & Foley, C. (2019). Algorithmic management: The case of Airbnb. *International Journal of Hospitality Management*, 83, 33–36. https://doi.org/10.1016/j.ijhm.2019.04.009

Cocola-Gant, A., & Gago, A. (2021). Airbnb, buy-to-let investment and tourism-driven displacement: A case study in Lisbon. *Environment and Planning A: Economy and Space*, 53(7), 1671–1688. https://doi.org/10.1177/0308518X19869012

Dann, D., Teubner, T., & Weinhardt, C. (2019). Poster child and guinea pig–insights from a structured literature review on Airbnb. *International Journal of Contemporary Hospitality Management*, 30(1), 427–473. https://doi.org/10.1108/IJCHM-03-2018-0186

Dogru, T., Makarand, M., & Suess, C. (2019). Adding evidence to the debate: Quantifying Airbnb's disruptive impact on ten key hotel markets. *Tourism Management*, 72, 27–38. https://doi.org/10.1016/j.tourman.2018.11.008

Dolnicar, S. (2019). A review of research into paid online peer-to-peer accommodation: Launching the annals of tourism research curated collection on peer-to-peer accommodation. *Annals of Tourism Research*, 75, 248–264. https://doi.org/10.1016/j.annals.2019.02.003

Dredge, D., & Gyimóthy, S. (2015). The collaborative economy and tourism: Critical perspectives, questionable claims and silenced voices. *Tourism Recreation Research*, 40(3), 286–302. https://doi.org/10.1080/02508281.2015.1086076

Ert, E., Fleischer, A., & Magen, N. (2016). Trust and reputation in the sharing economy: The role of personal photos in Airbnb. *Tourism Management*, 55, 62–73. https://doi.org/10.1016/j.tourman.2016.01.013

Farmaki, A., Christou, P., & Saveriades, A. (2020). A Lefebvrian analysis of Airbnb space. *Annals of Tourism Research*, 80, 102806. https://doi.org/10.1016/j.annals.2019.102806

Ferreri, M., & Sanyal, R. (2018). Platform economies and urban planning: Airbnb and regulated deregulation in London. *Urban Studies*, 55(15), 3353–3368. https://doi.org/10.1177/0042098017751982

Frenken, K., & Schor, J. (2017). Putting the sharing economy into perspective. *Environmental Innovation and Societal Transitions*, 23, 3–10. https://doi.org/10.1016/j.eist.2017.01.003

Gibbs, C., Guttentag, D., Gretzel, U., Yao, L., & Morton, J. (2018). Use of dynamic pricing strategies by Airbnb hosts. *International Journal of Contemporary Hospitality Management*, 30(1), 2–20. https://doi.org/10.1108/IJCHM-09-2016-0540

Grisdale, S. (2021). Displacement by disruption: Short-term rentals and the political economy of "belonging anywhere" in Toronto. *Urban Geography*, 42(5), 654–680. https://doi.org/10.1080/02723638.2019.1642714

Gurran, N., & Phibbs, P. (2017). When tourists move in: How should urban planners respond to Airbnb? *Journal of the American Planning Association*, 83(1), 80–92. https://doi.org/10.1080/01944363.2016.1249011

Guttentag, D. (2015). Airbnb: Disruptive innovation and the rise of an informal tourism accommodation sector. *Current Issues in Tourism*, 18(12), 1192–1217. https://doi.org/10.1080/13683500.2013.827159

Guttentag, D. (2019). Progress on Airbnb: A literature review. *Journal of Hospitality and Tourism Technology*, 10(4), 814–844. https://doi.org/10.1108/JHTT-08-2018-0075

Guttentag, D., Smith, S., Potwarka, L., & Havitz, M. (2018). Why tourists choose Airbnb: A motivation-based segmentation study. *Journal of Travel Research*, 57(3), 342–359. https://doi.org/10.1177/0047287517696980

Gyódi, K. (2019). Airbnb in European cities: Business as usual or true sharing economy? *Journal of Cleaner Production*, 221, 536–551.

Hati, S. R. H., Balqiah, T. E., Hananto, A. & Yuliati, E. (2021) A decade of systematic literature review on Airbnb: the sharing economy from a multiple stakeholder perspective. *Heliyon*, 7(10), e08222. https://doi.org/10.1016/j.heliyon.2021.e08222.

Huertas, A., Ferrer-Rosell, B., Marine-Roig, E., & Cristobal-Fransi, E. (2021). Treatment of the Airbnb controversy by the press. *International Journal of Hospitality Management*, 95, 102762. https://doi.org/10.1016/j.ijhm.2020.102762

Ikkala, T., & Lampinen, A. (2015, February). Monetizing network hospitality: Hospitality and sociability in the context of Airbnb. *Proceedings of the 18th ACM conference on computer supported cooperative work & social computing, 2019*, 1033–1044. https://doi.org/10.1145/2675133.2675274

Ioannides, D., Röslmaier, M., & Van Der Zee, E. (2019). Airbnb as an instigator of 'tourism bubble' expansion in Utrecht's Lombok neighbourhood. *Tourism Geographies*, 21(5), 822–840. https://doi.org/10.1080/14616688.2018.1454505

Klarin, A., & Suseno, Y. (2021). A state-of-the-art review of the sharing economy: Scientometric mapping of the scholarship. *Journal of Business Research*, 126, 250–262. https://doi.org/10.1016/j.jbusres.2020.12.063

Kuhzady, S., Olya, H., Farmaki, A., & Ertaş, Ç. (2021). Sharing economy in hospitality and tourism: A review and the future pathways. *Journal of Hospitality Marketing & Management*, 30(5), 549–570. https://doi.org/10.1080/19368623.2021.1867281

Lagonigro, R., Martori, J. C., & Apparicio, P. (2020). Understanding Airbnb spatial distribution in a southern European city: The case of Barcelona. *Applied Geography*, 115, 102136. https://doi.org/10.1016/j.apgeog.2019.102136

Lee, S. H., & Deale, C. (2021). Consumers' perceptions of risks associated with the use of Airbnb before and during the COVID-19 pandemic. *International Hospitality Review*, 35(2), 225–239. https://doi.org/10.1108/IHR-09-2020-0056

Mody, M. A., Hanks, L., & Cheng, M. (2021a). Sharing economy research in hospitality and tourism: A critical review using bibliometric analysis, content analysis and a quantitative systematic literature review. *International Journal of Contemporary Hospitality Management*, 33(5), 1711–1745. https://doi.org/10.1108/IJCHM-12-2020-1457

Mody, M., Suess, C., & Dogru, T. (2021b). Does Airbnb impact non-hosting Residents' quality of life? Comparing media discourse with empirical evidence. *Tourism Management Perspectives*, 39, 100853. https://doi.org/10.1016/j.tmp.2021.100853

Muller, E. (2020). Delimiting disruption: Why Uber is disruptive, but Airbnb is not. *International Journal of Research in Marketing*, 37(1), 43–55. https://doi.org/10.1016/j.ijresmar.2019.10.004

Nieuwland, S., & Van Melik, R. (2020). Regulating Airbnb: How cities deal with perceived negative externalities of short-term rentals. *Current Issues in Tourism*, 23(7), 811–825. https://doi.org/10.1080/13683500.2018.1504899

Nofre, J., Giordano, E., Eldridge, A., Martins, J. C., & Sequera, J. (2018). Tourism, nightlife and planning: Challenges and opportunities for community liveability in La Barceloneta. *Tourism Geographies*, 20(3), 377–396. https://doi.org/10.1080/14616688.2017.1375972

Oskam, J., & Boswijk, A. (2016). Airbnb: The future of networked hospitality businesses. Journal of *Tourism Futures*, 2(1), 22–42. https://doi.org/10.1108/JTF-11-2015-0048

Ozdemir, G., & Turker, D. (2019). Institutionalization of the sharing in the context of Airbnb: a systematic literature review and content analysis. *Anatolia*, 30(4), 601–613. https://doi.org/10.1080/13032917.2019.1669686

Prayag, G., & Ozanne, L. K. (2018). A systematic review of peer-to-peer (P2P) accommodation sharing research from 2010 to 2016: Progress and prospects from the multi-level perspective. *Journal of Hospitality Marketing & Management*, 27(6), 649–678. https://doi.org/10.1080/19368623.2018.1429977

Quattrone, G., Proserpio, D., Quercia, D., Capra, L., & Musolesi, M. (2016). Who benefits from the 'sharing' economy of Airbnb? *Proceedings of the 25th international conference on world wide web, 2016*, 1385–1394. https://doi.org/10.1145/2872427.2874815

Rodríguez-Pérez de Arenaza, D., Hierro, L. Á., & Patiño, D. (2022). Airbnb, sun-and-beach tourism and residential rental prices. The case of the coast of Andalusia (Spain). *Current Issues in Tourism*, 25(20), 3261–3278. https://doi.org/10.1080/13683500.2019.1705768

Sainaghi, R. (2020). The current state of academic research into peer-to-peer accommodation platforms. *International Journal of Hospitality Management*, 89, 102555. https://doi.org/10.1016/j.ijhm.2020.102555

Sainaghi, R., & Baggio, R. (2020). Clusters of topics and research designs in peer-to-peer accommodation platforms. International Journal of Hospitality Management, 88, 102393. https://doi.org/10.1016/j.ijhm.2019.102393

Sigala, M. (2017). Collaborative commerce in tourism: Implications for research and industry. *Current Issues in Tourism*, 20(4), 346–355. https://doi.org/10.1080/13683500.2014.982522

Suess, C., Woosnam, K. M., & Erul, E. (2020). Stranger-danger? Understanding the moderating effects of children in the household on non-hosting residents' emotional solidarity with Airbnb visitors, feeling safe, and support for Airbnb. *Tourism Management*, 77, 103952. https://doi.org/10.1016/j.tourman.2019.103952

Suess, C., Woosnam, K., Mody, M., Dogru, T., & Sirakaya Turk, E. (2021). Understanding how residents' emotional solidarity with Airbnb visitors influences perceptions of their impact on a community: The moderating role of prior experience staying at an Airbnb. *Journal of Travel Research*, 60(5), 1039–1060. https://doi.org/10.1177/0047287520921234

Tussyadiah, I. P. (2016). Factors of satisfaction and intention to use peer-to-peer accommodation. *International Journal of Hospitality Management*, 55, 70–80. https://doi.org/10.1016/j.ijhm.2016.03.005

VRBO (2022). Get to know VRBO. Online source: www.vrbo.com/about/ (consulted February 2024).

Wachsmuth, D., & Weisler, A. (2018). Airbnb and the rent gap: Gentrification through the sharing economy. *Environment and Planning A: Economy and Space*, 50(6), 1147–1170. https://doi.org/10.1177/0308518X18778038

Wegmann, J., & Jiao, J. (2017). Taming Airbnb: Toward guiding principles for local regulation of urban vacation rentals based on empirical results from five US cities. *Land Use Policy*, 69, 494–501. https://doi.org/10.1016/j.landusepol.2017.09.025

Zervas, G., Proserpio, D., & Byers, J. W. (2017). The rise of the sharing economy: Estimating the impact of Airbnb on the hotel industry. *Journal of Marketing Research*, 54(5), 687–705. https://doi.org/10.1509/jmr.15.0204

PART VI

Geographies of Tourism Mobilities

29

TOURISM AND (IM)MOBILITIES

During and Post-Pandemic

Maria Thulemark and Tara Duncan

Introduction

'Mobility' is – in essence – the ability to move freely, while 'tourism' is about exploring places and experiencing atmospheres other than 'home', which often includes some kind of movement of the corporeal body (see Everingham et al., Chapter 11). As such, tourism geography is very much about mobility in time and space, closely interlinked with time geography (Hall, 2008; Shoval, Chapter 33). As can be seen in Hall's (2005) description of the evolution of tourism geographies, tourism is closely linked to corporeal movement in time-and-space/distance and as such is interlinked with many academic disciplines (consciously or unconsciously). 'Mobilities' have moved from being a necessity of corporeal movement characterised by 'dead time', where the need for travelling between places was 'lost' in life. Pre-COVID-19, ideas of the new mobilities turn included, but were not limited to, an idea of travelling time – journey time – as filled with psychological and social meaning (cf. Urry, 1999). During times when the Coronavirus swept around the world, it was highly noticeable for many just how much we value our ability to be mobile. We were "all made aware of the fundamental premise on which modern societies are built: constant but unequal movement" (Adey et al., 2021: 2). When linking this movement and ability to be mobile to pre-COVID-19 tourism development, it becomes evident that tourism is a "form of extractive economy founded upon uneven mobilities and the consumption of labour, natural resources, and produced spaces as places for some to play" (Sheller & Urry, 2004, cited in Adey et al., 2021: 13).

In times of crises, such as during a pandemic, the severe interruption of our ability to move freely implies that current time geography models need adjustment and additional dimensions to explain the (im)mobilities prevalent in our understanding of tourism. The previous conceptualisation of tourism as comprising overnight stays or 24 hours away from home is no longer sufficient. In pandemic times, tourism turns into more of an 'experiencescape' whereby, for many, our home is the only place we are allowed to 'experience'. That said, the eagerness to see and experience still exists even though the ability to move is limited. Major tourism attractions and tourism destinations were empty, yet other parts of the tourism industry were running invariably on, at times, a lower speed and, at other times, with much higher volumes than normal. This was often at the cost of, for example, precarious workers or ecologically sensitive nature.

DOI: 10.4324/9781003286301-35

This chapter moves beyond explaining the relationships between tourism and mobility discussed in the first volume of this handbook (Duncan, 2012) by giving examples of how changing patterns of (im)mobility, caused by the COVID-19 pandemic, relate to changes in the tourism system (in which we adopt a Western viewpoint). In a special issue, edited by Adey et al. (2021) on *Pandemic (Im)mobilities,* leading researchers from the field of mobilities studies addressed issues of how the COVID-19 pandemic suddenly changed mobile social practices. Tourism as an activity highly dependent on corporeal mobility becomes an obvious area of interest. Tourism was transformed during the pandemic into being constrained by the encounters that are allowed in an (im)mobile setting. Furthermore:

> The pandemic immobilization of travel has not only severely interrupted tourism, but may in fact be leading to a shift in the entire apparatus of tourism, meaning the practices, places, and genres of existing styles of tourism, which may be experiencing the kind of generational and historical shift that only happens when many interconnected technological, social, cultural, economic, and governmental practices are forced to change in an interlocking manner.
>
> (Adey et al., 2021:13)

By using two stand-alone cases, the chapter emphasises the impact of COVID-19 on the contemporary globalised tourism industry and sheds light upon some of the pros and cons of (im) mobilities. The chapter thus examines the links between (im)mobility, collaborative economy and precarious work, as well as discusses whether the pandemic has forced the tourism industry to become more domestic-centred, initiating a shift in its path-dependency towards a more resilient tourism system. We do not make any assumptions about the future; rather we take notice of what happened during the COVID-19 pandemic through a mobilities lens, reflecting upon how this might cause opportunities for action to turn the tourism industry towards a more sustainable praxis. The chapter will conclude by discussing a new role for the 'mobilities turn' (Cresswell, 2006; Hannam et al., 2006; see also Gale, 2008), evolving around a new way of looking at the 'mobilities paradigm' (Sheller & Urry, 2006) with a future-centred focus of what post-pandemic mobilities might look like and how global tourism – previously highly dependent on corporeal mobility – will change. We conclude by considering how a new mobilities turn might be formed as our ability to be mobile became disrupted by the pandemic, and the everyday mobilities argued to be at the core of 'the mobilities paradigm' (Sheller & Urry, 2006) could no longer be taken for granted.

(Im)mobility, Collaborative Economy and Precarious Work

Lockdowns and quarantines during the COVID-19 pandemic forced many of us to stay at home or remain fixed in certain locations, with limited freedom. As our homes or outdoor neighbourhoods turned into the only places to stay safe and secure, many searched for new, albeit highly restricted, ways of 'experiencing'. One opportunity to experience was through food. Restaurants in Sweden, and in many other countries, were still open, albeit with restrictions in opening hours, numbers of guests and seating options. The eagerness to experience in concession to lockdown and social distancing led to a growing market for (food) delivery services and, as such, a rise in the mobilities of precarious groups of workers (including migrants and low-skilled workers); an uneven development that made it possible for 'us' to still explore and experience the world while remaining 'safely' at home.

This exemplifies how mobility through food perpetuates the collaborative economy[1] and provides the opportunity, often at the expense of workers in precarious conditions, to experience different places/cultures gastronomically. Ever since the global financial crisis in 2008/2009, the collaborative economy has emerged as an economic system and model within the business landscape (Gyimóthy et al., 2020; Van der Zee, Chapter 28), profoundly changing the tourism and hospitality industry, having an even greater impact and prominence during the pandemic.

While some stayed safely at home, navigating the opportunities generated by the collaborative economy to be able to create experiences (through food), others were the ones forced to deliver these experiences as mobile food-delivery staff. On foot or by bicycle, scooters, electric kick-bikes, motorbikes and cars, these workers delivered food and goods to the safe home environment of the customer. As more food businesses innovated, diversified and offered food delivery services, the experiential value of such food increased, even as 'we' still moderated our purchases due to continuing worries about exposure to COVID-19 (Gavilan et al, 2021). Through the delivery staff's high mobility levels, the immobile customer could maintain their desire for new experiences, highly connected to the basics of the tourism industry.

Corporeal mobility remained a constant and important part of the tourism (or experience) industry during these times, in part down to the fooddelivery staff. Yet, this very mobility was also a health risk – whereby it was part of a workforce's precarious task, precisely to be mobile. The commercialised manifestation of the collaborative economy (mediating commercial exchanges, involving platforms owned by global corporations) is very far from a truly sharing-based economy (based on non-commercial, even cooperative initiatives), with far from promising outcomes when it comes to the UN's Sustainable Development Goals (SDGs) (Gössling & Hall, 2019). This type of collaborative economy business creates new challenges for achieving decent work (Bianchi & de Man, 2021; Dredge & Gyimóthy, 2015) and a more sustainable tourism industry (Gössling & Hall, 2019). While the work of food delivery drivers and riders (amongst others) helped ensure business survival and generally lowered the risk of exposure to COVID-19, it also highlighted their precarity and accentuated the health risks to the workers themselves (see for example Tran et al., 2022). Thus, through our immobility of being 'tourists at home', others –food delivery staff working under precarious conditions – had to perform our mobilities, being at the beck-and-call of technology to answer our demands for new (food) experiences delivered to our door.

Mobility, Domestic Travelling and Sustainability

Stemming from a question raised by Ioannides and Gyimóthy (2020: 627) – "Could we envision a scenario where our endless neophilia and unquenching thirst for (often irresponsible) adventure in far-flung places are substituted by travel and leisure activities much closer to home?" – this section discusses how restricted mobility might cause positive and negative outcomes for domestic tourism. The restricted ability to move freely was described and conceptualised by researchers as an 'anthropause' (Rutz et al, 2020) in which we could see a slowing of modern human activities, such as travel. This anthropause led to opportunities in which, for example, human–wildlife interactions could be studied (Rutz et al., 2020), and nature-based tourism could be critically examined in relation to the anthropocene.

When facing less governmental restrictions in the wake of the pandemic, country borders were (in most cases) still closed, and domestic tourism was, for many, the only option when planning for leisure travel and activities. Staycations, holistays and domestic tourism more generally were promoted and grew exponentially (Jacobsen et al., 2021), not only due to geographical restrictions – but also to economic restrictions – for families, in part due to financial constraints that far outlive governmental restrictions on geographical mobility.

(Domestic) tourism with an outdoor focus was growing and both anthropause and 'post-anthropause' tourism (referring to the time after lockdowns) utilised nature for tourism purposes (Bustad et al., 2022) and for psychologically therapeutic outcomes (Buckley & Westaway, 2020), whereby visitors continued to consume nature through visiting, for example, national parks, as they were perceived as ´safe´ destinations with low probabilities of infection during the pandemic (and beyond) (Templeton et al., 2021; Rogowski, 2022). However, this increase in nature-based leisure activities led to ever-growing problems of overtourism in sensitive natural areas, even though such issues were often already seen prior to the pandemic (Cheer et al., 2019). As McGinlay et al. (2020) point out, some European national parks saw visitation increases of almost 100 percent on certain days, as compared to previous years. While national parks were seen as 'safe' spaces in which to spend leisure time, overcrowding became a not uncommon problem, and many park authorities raced to enact health and safety measures and recommendations to allow parks users to enjoy these spaces more safely (McGinlay et al, 2020).

In other places, however, issues of overcrowding were not the main concern during the pandemic. For example, in Indonesia, visiting the country's geoparks had been a growing domestic tourism trend before the pandemic, and overcrowding again became an issue as the geoparks reopened, despite visitation essentially having been halted completely due to the pandemic. While the lack of visitors might imply some degree of recovery of nature and increased wildlife sightings, both factors are not predicted to be long-term trends; it is the activities with a more negative impact, including illegal logging and forest loss and the poaching of wildlife – increased during the pandemic – that are also worth noting (Cahyadi & Newsome, 2021). The reduction in tourism mobilities (and due to such knock-on economic impacts as a decrease in park ranger patrols) empowered a different type of mobility that highlights other negative aspects of the pandemic.

One of the most documented negative behaviours in natural areas during the pandemic was the mistreatment of these often ecologically sensitive regions. Tourism, with its traditionally built-in human activity, has been a driving force for the conservation of fragile natural environments due to the use of nature as a commodity (Saarinen, 2019; Saarinen & Nepal, Chapter 24). During the last few years, numerous news articles documented a mistreatment of nature reserves, national parks and protected areas (McGinlay et al., 2020). The presence of new profiles of visitor "unaware of the main regulations that are in place and of widely accepted norms of behaviour in conservation areas" (McGinlay et al., 2020: 5), generated new impacts in natural areas, including unprecedented increases in littering, new paths being tramped, unsanctioned camping, the lighting of fires, and biking in prohibited areas. We can add to this mix the tension between local populations and domestic visitors, many of whom chose to drive rather than use public transport (often due to the inaccessibility of natural spaces open for visitation). It became obvious that the supposed immobility of the pandemic in fact often left overcrowded natural areas in a state of (anthropogenic) shock.

Another 'Mobilities Turn'?

Our immobility during the pandemic has in many ways allowed us more leisure time in our homes and local communities, allowing (re)exploration of different forms of mobility, such as walking or cycling or discovering new forms of social (leisure) capital through gardening, discovering local architecture or undertaking hobbies such as photography. Becoming domestic tourists in our own cities, regions and in our own homes became the 'norm'. Yet, despite the many positives to be found in staying closer to home, experiencing the local and becoming a tourist in our own backyards, the bounce-back narrative may not be 'better', as many would perhaps hope or claim.

Going back to the question raised by Ioannides and Gyimóthy (2020: 627), before the anthropause, asking if tourists' endless neophilia and thirst for adventure will take place closer to home, is also somehow asking whether corporeal mobility is actually a necessity for a tourist experience? Or, indeed, it asks whether we still can use our homes, backyards and local communities as experiencescape arenas? What is obvious, as Gössling and Schweiggart (2022) say, is that tourists have been the most resilient "component in the tourism system" (p. 924), for instance, who have most easily adapted to domestic travel. Hence, while the consumption of tourism elements such as restaurant food contributed to experiences (through food) while being safe at home as tourism turned into more of an experiencescape, the home of the tourist became the only place in which the eagerness to experience could be performed. What seems to be still missing from many of the discussions, though, is the way(s) in which the corporeal mobility of platform workers in precarious labour situations, hinted at earlier in the chapter, has contributed to the changing tourism system.

Alongside this, we can recall the pre-pandemic research, which had shifted the focus onto the climate footprint of leisure travel (Cohen et al 2011, Gössling et al 2012, Mkono & Huges, 2020, Jacobson et al 2020). With an almost two-year break in long distance travelling for leisure purposes, we now face a real (tourism) mobility shift from long-haul to domestic tourism post-pandemic, at least for a while. Yet, one of Adey et al.'s (2021: 6), (many) thought-provoking questions revolved around the use of fossil fuel and whether, as countries move towards recovery, they will "seek to return to the high-mobility, high-energy, high-carbon economy of the past?" If we take this down to the level of the individual tourist, we can perhaps ask similar questions. Did the state of immobility during lockdown cause an increased demand for movement and mobility of the corporeal body at more local levels at the cost of, for instance, nature and preservation? And for how long will we see a lower number of air transport journeys for leisure purposes? Will the current preferences for domestic tourism alternatives remain?

Gyimóthy et al (2022) suggest that tourists' health and security concerns have garnered stronger support for domestic tourism, though at the same time, they worry, as 'locals' themselves, about the other domestic tourists around them. This, however, does not reflect what the UNWTO (2022) calls the 'strong pent-up demand for international travel', which saw tourism numbers globally increase back to 60 percent of pre-pandemic numbers in the first half of 2022.

These discussions arguably highlight a shift in tourism mobilities thinking. Adey et al. (2021: 13) discuss the "pandemic related uneven mobilities", reiterating the historically differential power dynamics within tourism. What they hint at (and we also argue here) is that these power dynamics, on some scales, have been reversed. For those with the appropriate 'mobility capital', and perhaps, social capital, becoming immobile became the 'safe' way to be a tourist. (Re)discovering the local through staycations and the like became the focus of the industry and media who extolled the positives of such experiences.

Yet while being immobile or travelling locally suddenly became the popular way to be a tourist in the constrained environments brought about by COVID-19, there were many people who became more mobile, such as those engaged in food delivery platform work. While re-envisaged as important front-line workers, their precarious working conditions remained central, and their mobility, or the rate of their mobility, increased. Here lie the "emergent conflicts over (im) mobilities, [where] we can detect the uneven and differential capabilities" (Adey, et al., 2021: 13) and where the reversal of being mobile means the reversal of some of the very power dynamics inherent in mobilities.

This potential shift in power dynamics, and the shift back as the world continues to open up to tourism, increases the necessity to argue for mobility justice and liberty, social justice, and the right to move – all imperative arguments within and outside of the tourism academy that have also grown during the pandemic (Adey et al., 2021; Higgins-Desbiolles, 2020). With increasing calls for degrowth (Higgins-Desbiolles et al., 2019; Blazquez et al., Chapter 5), regenerative tourism (Cave & Dredge, 2020; Bellato et al., Chapter 23), and community-centred tourism (Higgins-Desbiolles, 2020), what is becoming increasingly obvious is that a new mobilities turn is needed to further our understanding and equip ourselves better as we move into a post-COVID-19 future.

Note

1   As there is no definite conceptualisation, the collaborative economy is sometimes conceptualised as the sharing economy which is widely used interchangeably with 'participative economy,' "peer economy,' 'gig economy' or 'platform economy' (see Gyimóthy & Dredge, 2017).

References

Adey, P., Hannam, K., Sheller, M., & Tyfield, D. (2021). Pandemic (Im)mobilities. *Mobilities*, 16(1), 1–19.

Bianchi, R.V., & de Man, F. (2021). Tourism, inclusive growth and decent work: A political economy critique. *Journal of Sustainable Tourism*, 29(2–3), 353–371.

Buckley, R., & Westaway, D. (2020). Mental health rescue effects of women's outdoor tourism: A role in COVID-19 recovery. *Annals of Tourism Research*, 85, 103041.

Bustad, J.J., Clevenger, S.M., & Rick, O.J. (2022). COVID-19 and outdoor recreation in the post-anthropause. *Leisure Studies*, 42(1), 85–99.

Cahyadi, H.S., & Newsome, D. (2021). The post COVID-19 tourism dilemma for geoparks in Indonesia. *International Journal of Geoheritage and Parks*, 9(2), 199–211.

Cave, J., & Dredge, D. (2020). Regenerative tourism needs diverse economic practices. *Tourism Geographies*, 22(3), 503–513.

Cheer, J.M., Milano, C., & Novelli, M. (2019). Tourism and community resilience in the Anthropocene: Accentuating temporal overtourism. *Journal of Sustainable Tourism*, 27(4), 554–572.

Cohen, S.A., Higham, J.E., & Cavaliere, C.T. (2011). Binge flying: Behavioural addiction and climate change. *Annals of Tourism Research*, 38(3), 1070–1089.

Cresswell, T. (2006). *On the Move: Mobility in the Modern Western World*. Routledge.

Dredge, D., & Gyimóthy, S. (2015). The collaborative economy and tourism: Critical perspectives, questionable claims and silenced voices. *Tourism Recreation Research*, 40(3), 286–302.

Duncan, T. (2012). The mobilities turn and the geography of tourism. In J. Wilson (Ed.), *The Routledge Handbook of Tourism Geographies* (pp 113–120). Routledge.

Gale, T. (2008). 'The end of tourism, or endings in tourism?'. In P.M. Burns & M. Novelli (Eds.), *Tourism and Mobilities* (pp. 1–14). CABI.

Gavilan, D., Balderas-Cejudo, A., Fernández-Lores, S., & Martinez-Navarro, G. (2021). Innovation in online food delivery: Learnings from COVID-19. *International Journal of Gastronomy and Food Science*, 24, 100330.

Gyimóthy, S., Braun, E., & Zenker, S. (2022). Travel-at-home: Paradoxical effects of a pandemic threat on domestic tourism. *Tourism Management*, 93, 104613.

Gyimóthy, S. & Dredge, D. (2017). *Collaborative Economy and Tourism: Perspectives, Politics, Policies and Prospects*. Springer.

Gyimóthy, S., Pérez, S.M., Meged, J.W., & Wilson, J. (2020). Contested spaces in the sharing economy. *Scandinavian Journal of Hospitality and Tourism*, 20(3), 205–211.

Gössling, S., & Hall, C. M. (2019). Sharing versus collaborative economy: how to align ICT developments and the SDGs in tourism? *Journal of Sustainable Tourism*, 27(1), 74–96. https://doi.org/10.1080/09669 582.2018.1560455

Gössling, S., & Schweiggart, N. (2022). Two years of COVID-19 and tourism: What we learned, and what we should have learned. *Journal of Sustainable Tourism*, 30(4), 915–931.

Gössling, S., Scott, D., Hall, C.M., Ceron, J.P., & Dubois, G. (2012). Consumer behaviour and demand response of tourists to climate change. *Annals of Tourism Research*, 39(1), 36–58.

Hall, C.M. (2005). Reconsidering the geography of tourism and contemporary mobility, *Geographical Research*, 43(2), 125–139.

Hall, C.M. (2008). 'Of time and space and other things: Laws of tourism and the geographies of contemporary mobilities', in P.M. Burns & M. Novelli (Eds.), *Tourism and Mobilities* (pp. 15–32). CABI.

Hannam, K., Sheller, M. & Urry, J. (2006). Mobilities, immobilities and moorings. *Mobilities,* 1(1), 1–22.

Higgins-Desbiolles, F. (2020). Socialising tourism for social and ecological justice after COVID-19. *Tourism Geographies*, 22(3), 610–623.

Higgins-Desbiolles, F., Carnicelli, S., Krolikowski, C., Wijesinghe, G., & Boluk, K. (2019). Degrowing tourism: Rethinking tourism. *Journal of Sustainable Tourism*, 27(12), 1926–1944. https://doi.org/10.1080/09669582.2019.1601732

Ioannides, D., & Gyimóthy, S. (2020). The COVID-19 crisis as an opportunity for escaping the unsustainable global tourism path. *Tourism Geographies*, 22(3), 624–632.

Jacobsen, J.K.S., Farstad, E., Higham, J., Hopkins, D., & Landa-Mata, I. (2021). Travel discontinuities, enforced holidaying-at-home and alternative leisure travel futures after COVID-19. *Tourism Geographies*, 25(2–3), 615–633.

Jacobson, L., Åkerman, J., Giusti, M., & Bhowmik, A.K. (2020). Tipping to staying on the ground: Internalized knowledge of climate change crucial for transformed air travel behavior. *Sustainability*, 12(5), 1994.

McGinlay, J., Gkoumas, V., Holtvoeth, J., Fuertes, R.F.A., Bazhenova, E., Benzoni, A., Botsch, K., et al. (2020). The impact of COVID-19 on the management of European protected areas and policy implications. *Forests* 11(11), 1214. doi: https://doi.org/10.3390/f11111214

Mkono, M., & Hughes, K. (2020). Eco-guilt and eco-shame in tourism consumption contexts: Understanding the triggers and responses. *Journal of Sustainable Tourism*, 28(8), 1223–1244.

Rogowski, M. (2022). The impact of COVID-19 pandemic on nature-based tourism in National Parks. Case studies for Poland. *Journal of Environmental Management and Tourism,* (Volume XIII, Spring), 2(58), 572–585. DOI:10.14505/jemt.v13.2(58).25

Rutz, C., Loretto, M.C., Bates, A.E., Davidson, S.C., Durate, C.M., Jetz, W., Johnson, M., Kato, A., Kays, R., Mueller, T., & Rb, P. (2020). COVID-19 lockdown allows researchers to quantify the effects of human activity on wildlife. *Nature Ecology and Evolution*, 4, 1156–1159. https://doi.org/10.1038/s41559-020-1237-z

Saarinen, J. (2019). What are wilderness areas for? Tourism and political ecologies of wilderness uses and management in the Anthropocene. *Journal of Sustainable Tourism*, 27(4), 472–487. https://doi.org/10.1080/09669582.2018.1456543

Sheller, M., & Urry, J. (2004). *Tourism Mobilities: Places to Play, Places in Play*. Routledge.

Sheller, M., & Urry, J. (2006). 'The new mobilities paradigm', *Environment and Planning A*, 38(2), 207–226.

Templeton, A.J., Goonan, K., & Fyall, A. (2021). COVID-19 and its impact on visitation and management at US national parks. *International Hospitality Review*, 35(2), 240–259.

Tran, N.A.T., Nguyen, H.L.A., Nguyen, T.B.H., Nguyen, Q.H., Huynh, T.N.L., Pojani, D., … & Nguyen, M.H. (2022). Health and safety risks faced by delivery riders during the Covid-19 pandemic. *Journal of Transport & Health*, 25, 101343.

UNWTO (2022, September 22nd). International Tourism Back To 60% Of Pre-pandemic Levels in January-July 2022. www.unwto.org/news/international-tourism-back-to-60-of-pre-pandemic-levels-in-january-july-2022#:~:text=An%20estimated%20474%20million%20tourists,same%20two%20months%20last%20year.

Urry, J. (1999). Sociology beyond Societies: Mobilities for the Twenty-First Century. Routledge. https://doi.org/10.4324/9780203021613

30

DIGITAL NOMADISM AND TOURISM MOBILITIES

Olga Hannonen

Introduction

The contemporary processes of construction of global markets, and the development of transportation and communication technologies have led to new social formations, patterns and opportunities. The digitalisation and incorporation of mobility into everyday life (both as physical relocation and technological connectivity) have resulted in the expansion of leisure and mobile lifestyles both nationally and internationally (Duncan, 2012; Hannonen, 2020; Hannonen et al., 2023; Urry, 2007). The emergence of digital nomadism as a new form of neo-nomadic and lifestyle mobility is tied to technological advancement, globalisation, individualisation, increased international experiences and mobility, and flexibility of working lives (Hannonen, 2020; 2022; Makimoto & Manners, 1997; Woldoff & Litchfield, 2021). This phenomenon is also a result of financial pressures and the higher costs of living in their home countries that have forced some to seek cheaper destinations and alternative, more affordable ways of life (Hannonen, 2020; Thompson, 2018). This chapter focuses on the emerging body of literature on digital nomads, addressing the main perspectives on the phenomenon and offering future research directions. The chapter starts by situating digital nomadism within conceptual and empirical developments in tourism geographies, including the mobilities and experience turn, (re)construction of the tourist gaze as well as place attachments and belonging. Next, it outlines current conceptual and methodological approaches to digital nomadism, and the chapter concludes with suggestions for future research perspectives.

The term digital nomad was introduced by Makimoto and Manners in 1997 to describe the outcomes of technological advancement on people's lives (Makimoto & Manners, 1997). They described how mobile and portable technologies would augment work and leisure and produce a new lifestyle, in which "people are freed from constraints of time and location" (Makimoto 2013, p. 40). The term digital nomad refers to a rapidly emerging class of highly mobile professionals, who perform their work remotely from anywhere in the world, utilising digital technologies. Thus, they work while travelling on a (semi)permanent basis and vice versa. As such, the mobile lifestyle that is developed by these highly mobile location independent professionals is referred to as 'digital nomadism' (Hannonen, 2020).

DOI: 10.4324/9781003286301-36

The outbreak of COVID-19 intensified discussions on digital nomadism. With the mandate to work from home, many office workers were turned into location-independent (i.e., footloose) professionals. What has transpired to be the world's largest experiment in remote working has demonstrated that it is now possible for working remotely to become normalised. Surveys show that many individuals enjoy being remote and some would like to remain remote after the pandemic (Eurofound, 2022), on the basis that being remote creates fruitful grounds for location independence and digital nomadism. It is anticipated that with the ease of travel restrictions, some location-independent professionals will contribute to the phenomenon of digital nomadism. Indeed, the 'State of Independence in America' survey suggests that in the United States alone 16.9 million American workers describe themselves as digital nomads, with a 131 per cent increase from 2019, the pre-pandemic year (MBO Partners Report, 2022). Moreover, during the pandemic, a shift to remote work has developed new travel patterns and lifestyles, such as 'remote work travel' or 'workations' (Hannonen et al., 2023) and long-term relocation to different destinations both domestically and internationally. Thus, these short-term remote-work-travel relocations will prevail, diversifying the modes and formats of digital nomadism.

Digital Nomadism, Mobilities and Experiences

Digital nomads are both a product and an example of the ubiquity of mobilities in everyday lives (Hannonen 2020; Thulemark & Duncan, Chapter 29). They continue to embody international mobility trends that are driven by personal desires for a change in lifestyle, freedom of choice and self-fulfilment. These lifestyle-led mobilities have become a worldwide trend since the 1980s. They have taken a number of forms and include, although not limited to, second home/residential tourism, seasonal and lifestyle migration, global/neo-nomadism, worldschooling and presently – digital nomadism (Cohen, 2010; Cohen, et al., 2015; D'Andrea, 2016; Hannonen, 2016; 2018; 2020; Kannisto, 2014; Green, 2015; Germann Molz, 2020; Thompson, 2018). These lifestyle-led mobilities have become an important research subject for different disciplines, such as geography (Hannonen, 2016; 2018), tourism and travel research (Cohen et al., 2015; Kannisto, 2014), sociology (Germann Molz, 2020; Thompson 2018) and anthropology (D'Andrea, 2016; Cook, 2020; 2022; Green, 2015; Korpela, 2019), among others.

Makimoto and Manners (1997), original authors of the concept, described digital nomadism as an evolution of mobility – from trains to automobility, to aeromobility, and to technological development. A decade later, John Urry (2007) published *Mobilities,* in which he thoroughly described societal changes and development of mobilities from walking to flying. He defined connectivity and proximity at a distance as peculiar features of contemporary mobilities, and further conceptualised the new mobilities paradigm – a novel perspective by which to examine societies and movement. The mobilities theory and the new mobilities paradigm describe the social world as being defined and transformed by constant movement and flow reflected in mass geographical mobilities and cosmopolitanisation (Sheller, 2020; Sheller & Urry, 2006; Urry, 2007). Therefore, digital nomadism fosters the discussion on adopting the 'mobilities turn' approach by tourism geographers (Duncan, 2012; Hannonen, 2022a).

Digital nomads have become a clear example, and an empirical application of the 'mobilities turn', in which mobile lifestyles are a part of everyday life. In such a manner, they have overstepped the facets of tourism in its traditional understanding in which leisure activity is opposite to organised work (Urry 1990). Following the United Nations World Tourism Organisation (UNWTO)'s definition, a tourist is defined as someone who is taking a trip with on overnight stay to a "destination outside his/her usual environment, for less than a year, for any main purpose (business, leisure or other personal purpose) other than to be employed by a resident entity in the country or place visited" (UNWTO, 2010: 10). Scholars argue that "to define when someone is outside of their 'usual environment' is conceptually vague and inherently problematic" (Hoogendoorn and Hammett 2021).

Similar to other lifestyle-led mobilities, in digital nomadism the boundary between 'home' and 'away' is blurred (Cohen et al., 2015; Hannonen, 2020). The travel destinations of digital nomads become their usual environment; their home. As de Loryn (2022: 103) states, "people feel at home when travelling with a loved one or by surrounding themselves with objects of emotional value". For many digital nomads, 'home' is a social and emotional construct, a network of social relations rather than some geographical location (de Loryn, 2022). Moreover, they are distinct through their "length of travel and decision not to have a home base" (Nash et al., 2018: 212).

While the nature of travel for digital nomads falls outside the definition of a tourist, their travel motivations are often touristic. They "select their location choice based on leisure and lifestyle expectations, not work" (Thompson, 2018: 3). Studies show that they long for exotic locations and are eager to visit new places that have attractive leisure characteristics (Nash et al., 2018; Reichenberger, 2018). It is argued that digital nomads bring leisure into their working environments and even perceive work as leisure (Reichenberger, 2018), thus distancing digital nomads from tourists in this respect. Indeed, several researchers suggest that, like other types of lifestyle travellers, digital nomads should not be fully regarded as tourists per se (Cook, 2020; Hannonen, 2020; Mancinelli, 2020). Empirical studies show that digital nomads want to distance themselves from tourists both terminologically and spatially, emphasising the different reasons and purposes for their travel (Woldoff & Litchfield, 2021).

Digital nomads also exemplify the 'experience turn' in tourism, in which consumers value experiences more than goods and services (Pine & Gilmore, 2013); a concept that has become a major research topic in tourism research during the last two decades. Experiences have become an economic offering and an output in tourism and travel – in the experience economy, not only data and information, but stories assist individuals, communities and businesses thrive (Pine & Gilmore, 2013). Following the pathways of the experience economy, digital nomads represent both consumers, marketers, and experience sellers. As Mancinelli (2020: 433) accurately encapsulates the situation: "the need to multiply streams of income turned digital nomads' personal biographies into potential commodities. Circulated through social media, the personal biographies of (allegedly) successful nomads inspired other people to try this lifestyle, influencing their movement". Personal blogs, social media channels and books have become a marketplace for the commodification and monetisation of personal experiences as well as a further popularisation of the lifestyle. Some destinations are rapidly becoming hotspots for digital nomads through hefty online presence, electronic word of mouth and nomads' ratings (such as nomadlist.com and other platforms).

Visual representations of destinations that are created and reproduced by digital nomads contribute to the construction of a tourist gaze. The tourist gaze involves daydreaming and anticipation of new experiences that are constructed through encounters with various media content (Urry, 1990). Thus, the development of the gaze has shifted from professional experts to digital nomads; both visitors and marketers of destinations through personal social media channels. Local service providers at destinations feel and adapt to this change, as an owner of a coworking and coliving space notes, nowadays everything should be instagrammable (Hannonen et al., 2023). Indeed, many 'tourist spaces' are organised around 'staged authenticity', something that is portrayed to and performed for visitors as local and authentic (Urry, 1990; see also Willment, 2020).

With regard to digital nomads and the reproduction of the gaze, it is claimed that they capture their experiences with "other nomads, tourists, and expats, not the local community in which they are embedded" (Thompson, 2019a: 74). In such a manner they reproduce the 'staged' images, "an insulated bubble-like existence", often excluding the local population and cultural contexts (Thompson, 2019a: 28). The staged authenticity approach has been noted in studies on digital nomads' online representations (Miguel et al., 2023; Willment, 2020).

Digital nomadism accumulates societal changes in relation to consumption in tourism. As Meethan (2012) notes in the first edition of this Handbook, shifting identities and growing individualisation have become a result of the de-industrialisation process and the increase in physical mobility. Indeed, in post-traditional societies individual lives are no longer determined by a person's position in society, such as status and class (Mancinelli, 2020). This creates possibilities for personal identity-based choices rather than navigation within prescribed social roles. For Mancinelli (2020: 427), digital nomads believe that routine life "cuts down on the choices that everybody is expected to make, trapping the individual in a series of expectations for professional and economic success".

Opposing these societal arrangements, the flourishing individualisation has opened a way for the development of different lifestyles and the growth of niche consumption and individually-tailored travel (Meethan, 2012). The digital nomadic lifestyle is regarded as a manifestation of freedom of choice and a disruption of conventional societal structures (Hannonen, 2020; Cook, 2022); digital nomads often portray their lifestyle as a happier and more fulfilling life of location independent living and working (Miguel et al., 2023; Willment, 2020). Digital nomadism as a lifestyle is associated with travel that improves the life circumstances and work conditions of digital nomads (Jacobs & Gussekloo, 2016; Woldoff & Litchfield, 2021). Thus, from an experiential marketing perspective, they are making the transition to the next stage – the customisation of experiences that leads to transformative travel (Pine & Gilmore, 2013). It is however yet to be seen whether this travel trend will become mainstream and bring transformative travel to mass levels.

In relation to digital nomadism, we can see the development of bespoke services and products that are specifically crafted for their needs. Among these are (co-)living and (co-)working spaces, digital nomad house rentals, leisure programmes, conferences, banking services, healthcare insurance, magazines and even mobile applications (Hannonen, 2020; Hannonen et al., 2023). These developments have been defined as the commodification of digital nomadism (Aroles et al., 2020). Destinations around the world have strategically responded to digital nomadism and started to market their locations as ideal settings for this lifestyle segment to live and work. Indeed, the number of countries that have introduced digital nomad visas and residence schemes to attract this growing customer segment is rapidly increasing (Chevtaeva & Denizci-Guillet, 2021; Hannonen, 2021; 2022). This can facilitate a change in destinations and an alteration in popularity among digital nomads.

Digital nomads challenge many geographical concepts that have sedentarist connotations, such as place attachment and belonging, citizenship, and notions of home. The outbreak of COVID-19 has shown how deeply modern societies are rooted in the logic of sedentarism and territorial belonging. With the outbreak of the pandemic, national governments effectively called every citizen back 'home'. Thus, in the state of global lockdown, many digital nomads and other lifestyle travellers were compelled to return or remain in their 'home' nation state. Others found themselves in a foreign destination, and having to decide on a safe location, evaluate the duration of visa validity and travel insurance capacity against the (in)efficiency of their homeland's medical care, and a possible (non-)return 'home'. At the same time many destinations could not accommodate foreign visitors due to limitations on their length of stay. This has reaffirmed that nation states are often unable to reconcile modern mobile societies, long-term travel and lifestyle arrangements that are not fixed in one place.

Simultaneously, suffering from the impact of economic decline and the absence of economic impact of tourism, a number of destinations introduced digital nomad visa schemes that would secure year-round flow of travellers (Chevtaeva & Denizci-Guillet, 2021; Hannonen, 2021; 2022). The latter shows that in addition to diversifying the landscape of the private sector services, digital nomads have also augmented public sector infrastructures to meet the needs of long-term travellers.

The growing communities of digital nomads have created their own hotspot destinations in different parts of the world. Digital nomads usually relocate to more affordable destinations, in which they can be more affluent on their salaries, escaping the high cost of living in Western countries (Holleran, 2022; Mancinelli, 2020; Woldoff & Litchfield, 2021). In this way, they continue the practice of geoarbitrage that has been performed and pursued in lifestyle migration and second home ownership, implying the utilisation of opportunities for affordable properties and cheaper costs of living in the destination countries (Hayes, 2019; Janoschka, 2009; Hannonen, 2016). In the context of digital nomadic mobilities, this is referred to as 'lifestyle hacking' or 'geo-hacking' (Cook, 2022).

The agglomeration of digital nomads in specific locations has brought both positive and negative impacts on local communities. On one hand, digital nomads have facilitated an increase in businesses that cater specifically for their needs while on the other the negative impacts associated with mobile lifestyles and tourism have become widely discussed. Among these impacts are the displacement of locals through an increase in rental prices and the proliferation of short-term rentals in cities, the displacement of traditional businesses as well as gentrification (Cook et al., 2022; Hayes & Zaban, 2020; Holleran, 2022; Woldoff & Litchfield, 2021). These impacts are collectively defined as place-based transformations that are induced by increased and intensified mobile lifestyles (Hayes & Zaban, 2020).

Digital Nomadism: An Emerging Research Inquiry

As a research category, digital nomads have appeared in academic publications only during the last decade (see, e.g., first publications by Makimoto, 2013; Müller, 2016). The number of studies and academic discussions on the phenomenon in different parts of the world have been steadily growing. Despite the rapid growth of digital nomadism, scholars argue that it is still, as yet, an under-theorised phenomenon in tourism and travel research (Chevtaeva & Denizci-Guillet, 2021; Hannonen, 2020).

Initial studies on digital nomads have focused mainly on definitions and the motivations behind digital nomadism (see Prabawa & Pertiwi, 2020; Reichenberger, 2018; Schlagwein, 2018). Thematically, such studies have approached digital nomadism from such perspectives as: the work-life balance (Cook, 2020; Reichenberger, 2018; Thompson, 2019a), employment and relationship angles (Cook, 2023; Hall et al., 2019; Thompson, 2018; 2019b), working conditions (Aroles et al., 2020; Cook, 2020; Orel, 2019) mobility and travel (Green, 2020; Hannonen, 2022; Mancinelli, 2020; Matos & Ardévol, 2021; Schlagwein & Jarrahi, 2020), and most recently from the perspective of COVID-19's impact on the phenomenon (De Almeida et al., 2021; Ehn et al., 2022; Holleran, 2022; Holleran & Notting, 2023) and the development of digital nomad visa schemes (Bednorz, 2023; Mancinelli & Molz, 2024; Sánchez-Vergara et al., 2023).

Emerging approaches on digital nomads can be broadly divided into two camps. One being an understanding of them as a work-related phenomenon, and the other as a tourism and travel-related phenomenon. Though most of the studies define both work and travel as essential components of digital nomadism, the emphasis on either one of these two camps is often conditioned by a research goal or the particular research discipline of a given researcher. Müller (2016: 344) defines digital nomads as a "social figure of current work life", while Woldoff and Litchfield (2021) introduce the concept of work tourism. They regard digital nomads as a work phenomenon – individuals, who engage in international travel to work. In a similar manner, Nash et al. (2018) see digital nomads as primarily working professionals, whose work includes four intertwining elements: Digital, gig and nomadic work as well as global adventure travel. They stress that to define someone as a digital nomad, all four elements should be present.

Cook (2023) suggests categorising digital nomads based on their relationship to work and the work context, including freelance, business owners, salaried, experimental and armchair digital nomads. Bonneau, Aroles and Estagnasié (2023) propose four archetypes for the online identities of digital nomads. As lifestyle promoters they are categorised as inspirers, teachers, community managers and influencers. Digital nomads are also defined as a type of location-independent workforce; as a new form of work and employment in the digital economy (Orel, 2019; Wang et al., 2018), and as a career path choice (Aroles et al., 2020).

From the tourism and travel perspective, digital nomads have been categorised invariably as a multifaceted travel phenomenon and a new customer segment (Hannonen et al., 2023), a leisure activity (Reichenberger, 2018), and a form of slow tourism (Putra & Agirachman, 2016). These approaches collectively suggest that digital nomads engage in long-term travel that includes both touristic and work-related activities that are distinct in relation to other modes of tourism. Other categorisations conceptualise digital nomads as an emerging economic activity and a cultural phenomenon that blurs the boundaries of consumption and production, establishing new flexible arrangements of working and travelling that are predominantly based on individual freedom (Wang et al., 2018).

Digital nomadism has also been conceptually framed within mobilities and lifestyle-led mobilities research (Green, 2020; Hannonen, 2020; 2022; Mancinell, 2020; Schlagwein & Jarrahi, 2020). Schlagwein and Jarrahi (2020) applied a mobilities perspective to digital nomadism to explain the diverse mobilities that constitute this lifestyle: administrative, content-related, temporal and spatial mobilities. Hannonen (2020; 2022) and Mancinelli (2020) indicate that lifestyle mobilities is a useful conceptual lens for examining various forms of contemporary nomadic mobilities, including digital nomadism. Lifestyle-led mobility is defined as voluntary relocations of varying durations in which individuals engage in long-term travel and relocate to places that provide alternative or different lifestyles (Cohen et al., 2015; Korpela, 2019).

Despite the conceptual fit of digital nomadism into this approach, Hannonen (2020; 2022) argues that lifestyle mobility acknowledges the existence of several 'homes' that are visited and returned to in a preferred manner. Thus, in relation to digital nomadism, she proposes to look at a possible return home from the perspective of the life course rather than through seasonality or circulation between lifestyle and 'home' locations. Mancinelli (2020) also points out a limited application of the lifestyle mobilities concept to digital nomadism, as it fails to fully account for the role of remote work and its possible consequences on the future of employment (see also Hannonen, 2020). However, based on the number of studies and conceptual approaches to digital nomadism, the mobilities perspective certainly prevails at present. Nevertheless, to position digital nomads better as a mobile category, there is a need for more empirical insights on the different facets of their particular mobilities in different socio-spatial contexts. Some of these research directions are discussed at the end of this chapter.

Methodological Approaches

The accelerating growth of mobile populations and the changing pace, scale and modes of mobilities present a challenge for traditional social scientific methods. The mobilities turn has caused the social and emotional dimensions of mobilities to surface (de Neergard & Jensen, 2020) and scholars now argue that studying mobilities requires going beyond a methodological exercise to capture the issues of, among other things, class and gender, scale and ethics, geographic boundaries and social imagination (Salazar et al., 2016). Furthermore, the proliferation of mobile technologies in everyday life has resulted in the interconnectedness of offline and online domains without clear-cut boundaries (Kedzior, 2014). As a result, to capture mobile lives and what occurs on the move, it is important to capture the myriad aspects that are at play in mobilities, such as social structures, meanings, emotions, communities and many more (de Neergard & Jensen 2020). Thus, to capture these diverse mobilities of digital nomads, they should ideally be accessed both on the ground and online.

Online communities and blogs are an important source of information about social interactions, mobilities and extended social networks in virtual space (Germann Molz, 2012; Thompson, 2018). As Mancinelli (2020: 423) highlights: "While digital nomads' physical travel spans nearly the entire globe, the majority of their social and business interactions occur through communication technologies and social media". Omitting the online presence of digital nomads can result in a partial image of the phenomenon without capturing the experience communicated via the Internet, which is an important context for digital nomadic mobility. Thompson (2018), who studied digital nomads in three nomadic events, proposes that future research should be conducted both online and in the geographical hotspots of digital nomads.

The combination of offline and online domains when studying digital nomads would bring a multi-sited-ness to the field and allow us to go beyond the description of a particular group in a given geographical context at a given time. Thus, digital nomads, as contemporary mobile subjects, would extend the concept of multi-sited ethnography to the online domain, converting it into an additional research site. As Germann Molz (2020: 196) puts it: "The multi-sited-ness of the field includes multiple physical destinations as well as online forums, social media feeds, websites and digital platforms where social relations are enacted". However, the inclusion of both domains has been implemented by a very limited number of researchers (see Mancinelli, 2020; Mancinelli & Molz, 2024; Schlagwein, 2018; Willment, 2020). Thus, there is room to apply a multi-sited research lens in future studies on digital nomads.

To date, research on digital nomads includes a variety of qualitative and quantitative methods and different combinations of the two via mixed-method and multi-method approaches. Due to the high mobility of digital nomads, one of the most popular research methods is to conduct online surveys or interviews that provide high flexibility for the researcher and the respondents, irrespective of their location. Reaching out for potential participants for surveys and interviews has usually taken place in the online domain through social media platforms (see Bonneau et al., 2023; Chevtaeva & Denizci-Guillet, 2021; Holleran & Notting, 2023; Miguel et al., 2023; Nash et al., 2018; Reichenberger, 2018; Willment, 2020). Surveys and online interviews provide rich but yet fragmented and one-sided information about digital nomads' everyday life, which currently lacks the social and emotional components of their mobilities and related contexts.

Another form of data collection includes participation in joint activities with digital nomads, such as retreats, conferences or staying in coworking spaces (see Orel, 2019; Thompson, 2018; Toivanen, 2023). This indeed provides a deeper immersion into the life of digital nomads, though only in a particular context and for a short period of time. As a digital and online phenomenon, digital nomads have been studied through forums and social media (Aroles et al., 2020; De Almeida et al., 2021; Ehn et al., 2022; Holleran, 2022; Holleran & Notting, 2023; Parreño-Castellano et al., 2022). Qualitative analysis of online interactions, shared opinions and comments provides access to a rich pool of data that allows the uncovering of experiences, perspectives, desires, emotions and problems through digital nomads' self-expression.

An increasing number of researchers have been immersing themselves in the digital nomads' way of life while at their travel destinations. Several ethnographic studies have been conducted in some of the world's major digital nomad hotspots, such as Chiang Mai in Thailand (see Cook, 2020; de Loryn, 2022; Green, 2020; Mancinelli, 2020; Mancinelli & Molz, 2024) and Bali, Indonesia (Prabawa & Pertiwi, 2020; Woldoff & Litchfield, 2021). Ethnographic observations and interviews have uncovered important aspects of the everyday life of digital nomads. In addition to classic ethnographic observations, digital nomads have been studied through the means of autoethnography (see Hall et al., 2019), longitudinal ethnographic research (Cook, 2020), netnography (see Mancinelli, 2020, Schlagwein, 2018), and photo elicitation interviews (Miguel et al., 2023; Willment, 2020).

The extensive travels and strong online presence of digital nomads have resulted in diverse methodological approaches that address various aspects of digital nomadism in their online or offline domains, or through a combination of the two. These diverse approaches combine classic ethnographic observations with interviewing in the field, online surveys and interviews, and increasingly include analysis of online communication and mobilities through forums and social media as well as netnography.

Future Research Directions

Existing approaches to digital nomadism go beyond the theoretical scholarship of a particular discipline. This suggests that future examination of the phenomenon should remain multidisciplinary. In this concluding part of the chapter, I emphasise the importance of looking at mobility regimes and travel destinations in order to understand the production and viability of digital nomadic lifestyles in more depth.

Studies show that the choice of location-independent lifestyles and engagement with a life on the move is based on both class and citizenship (as digital nomads are predominantly Western and holders of 'powerful' passports) (Hannonen, 2020). On the other hand, societal changes, the precariousness of employment in Western societies, and pandemic-driven unemployment have all contributed to the production of alternative lifestyles (Hannonen, 2020; Reichenberger, 2018; Thompson, 2018). The issue of mobility justice (Sheller, 2018; 2020) has been a dominant discourse in mobilities research in relation to the uneven privilege of access to mobility and mobility resources. The pandemic outbreak and the urban exodus to the countryside rekindled these discussions, and researchers pondered on who has the right to mobility, to the countryside and to alternative lifestyles (Pitkänen et al., 2020). As Sheller (2018: 14) puts it: "mobility justice is an overarching concept for thinking about how power and inequality inform governments and control of movement, shaping the patterns of unequal mobility and immobility in the circulation of people, resources and information". Reflecting colonial legacies, mobility justice is manifested through mobility regimes (Kesselring 2015; Sheller 2018).

Discussions on mobility regimes have appeared in some studies on contemporary nomadic and lifestyle mobilities (Green, 2015; Hannonen, 2020; Kannisto, 2014; Korpela, 2019; Mancinelli, 2020; Mancinelli & Molz, 2024). However, these issues still require further examination through a wider empirical base, in order to address the dependence of digital nomadism on structural constraints in more depth (Hannonen, 2021; Thompson, 2018). Mobility regimes address the mobility strategies of digital nomads and their negotiation of mobility structures, as both citizens of a particular state and employees of a certain organisation. The pandemic outbreak has reinforced mobility regimes and the physical presence of borders through international travel-related non-pharmaceutical interventions[1]; these interventions have created their own routes and travel corridors, channelling contemporary travellers towards certain destinations while restricting access to others (Hannonen, 2021; Hannonen & Prokkola, 2022). As already noted, the pandemic has also facilitated the rapid launch of nomad residence schemes for longer stays or digital nomad visas in a number of states worldwide. While the new visa stream creates considerable possibilities for many non-Western nationalities, it still runs a filtering process though an income threshold that defines who can be regarded as a digital nomad and who may not (Hannonen, 2021). Thus, the issue of regimes in digital nomadism requires further research attention.

The upsurge of remote stay programs, (co-)living and (co-)working spaces that accommodate the needs of digital nomads and their lifestyles, had been rapidly transforming destinations before the pandemic (Hannonen, 2020; Thompson, 2018). Woldoff and Litchfield (2021: 8) state that for many it is hard to understand "why someone would travel 8,000 miles just to work in a coworking space or café. After all, they can plug their computer into a Starbucks at home." This emphasises the importance of looking at destinations in which digital nomads stay. I argue that destinations are at the core of understanding the nomadic lifestyles. Nevertheless, research on the destinations of digital nomads is still very limited (see Almeida & Belezas, 2022; Hannonen et al., 2023; Parreño-Castellano et al., 2022; Thompson, 2018).

In relation to digital nomadic mobilities, it is important to understand how destinations attract and support these lifestyles as well as how destinations perceive the opportunities and aim to benefit from these lifestyles. A complex set of interactions among various stakeholders, travellers and local inhabitants has the potential to transform the character, scale, and structure of destinations. Thus, another important aspect is to reveal how digital nomads affect local communities, their economies and social structures.

Digital nomadism, as a growing travel phenomenon has generated considerable empirical and theoretical development in tourism geographies. It has challenged and disrupted sedentaristic perspectives to place, home, citizenship and belonging, exemplifying the mobilities turn in social sciences. The phenomenon has diversified methodological perspectives in mobilities research by means of combining online and offline domains as research sites, and uncovering differing ways to examine the diverse sides of digital nomadism. The uneven geographical distribution and societal impacts of digital nomadism require additional academic scrutiny in relation to mobility justice and mobility regimes as well as the travel destinations of digital nomads. Such studies would further contribute to the geographical approaches to mobilities, and provide information on spatial discontinuities, destination development and the use of social and economic resources.

Note

1 Non-pharmaceutical interventions are a set of measures that are aimed to slow the spread of infectious diseases. They include traveller screenings and testing at borders for symptoms of illness, quarantine requirements, travel restrictions between particular states or groups of states, and border closures (Hannonen & Prokkola 2022).

References

Almeida, J. & Belezas, F. (2022). The rise of half-tourists and their impact on the tourism strategies of peripheral territories. In J. Leitao, J., Ratten, V., & Braga, V. (Eds.), *Tourism Entrepreneurship in Portugal and Spain* (pp. 181–191). Springer Nature.

Aroles, J., Granter, E., & de Vaujany, F.-X. (2020). 'Becoming mainstream': The professionalisation and corporatisation of digital nomadism. *New Technology, Work and Employment, 35*(1), 114–129.

Bednorz, J. (2023). Digital Nomad Visas with Jan Bednorz. *Academia Experience Podcast.* Available at: https://www.youtube.com/watch?v=oL-LJ4Rdkts

Bonneau, C., Aroles, J., & Estagnasié, C. (2023). Romanticisation and monetisation of the digital nomad lifestyle: The role played by online narratives in shaping professional identity work. *Organization, 30*(1), 65–88.

Chevtaeva, E., & Denizci-Guillet, B. (2021). Digital nomands lifestyle's and coworkation. *Journal of Destination Marketing and Management, 21*, 100633

Cohen, S. (2010). Re-conceptualising lifestyle travellers: Contemporary 'drifters'. In K. Hannam & A. Diekmann (Eds.), *Beyond Backpacker Tourism: Mobilities & Experiences* (pp. 64–84). Channel View.

Cohen, S. A., Duncan, T., & Thulemark, M. (2015). Lifestyle mobilities: The crossroads of travel, leisure and migration. *Mobilities, 10* (1), 155–172.

Cook, D. (2020). The freedom trap: Digital nomads and the use of disciplining practices to manage work/leisure boundaries. *Information Technology & Tourism, 22*, 355–390.

Cook, D. (2022). Breaking the contract: Digital nomads and the state. *Critique of Anthropology, 42*(3), 304–323.

Cook, D. (2023). What is a digital nomad? Definition and taxonomy in the era of mainstream remote work. *World Leisure Journal, 65*(2), 256–275.

Cook, D., Silva, R., & Muellbauer, J. (2022). Cost of living crisis: Are digital nomads pushing up rental prices? Interviewed by David Foster, *TRT World*, 3 November. Available at: www.trtworld.com/video/roundtable/cost-of-living-crisis-are-digital-nomads-pushing-up-rental-prices/6363cd0a903ac000116999b5

D'Andrea, A. (2016). Neo-nomadism: A theory of post-identitarian mobility in the global age. *Mobilities. 1*(1), 95–119.

De Almeida, M. A., Correia, A., & Schneider, D. (2021). COVID-19 as opportunity to test digital nomad lifestyle. *Proceedings of the 2021 IEEE 24th International Conference on Computer Supported Cooperative Work in Design.* Available at: https://ieeexplore.ieee.org/document/9437685

De Neergaard, M., & Jensen, H. L. (2020). Embodied ethnography in mobilities research. In M. Büscher, et al. (Eds.), *Handbook of Research Methods and Applications for Mobilities* (pp. 374–381). Edward Elgar.

De Loryn, B. (2022). Not necessarily a place: How mobile transnational online workers (digital nomads) construct and experience 'home'. *Global Networks, 22*(1), 103–118.

Duncan, T. (2012). The 'mobilities turn' and the geography of tourism. In J. Wilson (Ed.), *The Routledge Handbook of Tourism Geographies* (pp. 113–119). Routledge.

Ehn, K., Jorge, A., & Marques-Pita, M. (2022). Digital nomads and the COVID-19 pandemic: Narratives about relocation in a time of lockdowns and reduced mobility. *Social Media + Society, 8*(1), 20563051221084958.

Eurofound. (2022). *Fifth Round of the Living, Working and COVID-19 E-survey: Living in a New Era of Uncertainty.* Publications Office of the EU.

German Molz, J. (2012). *Travel Connections: Tourism, Technology, and Togetherness in a Mobile World.* Routledge.

German Molz, J., (2020). MoVE: Mobile virtual ethnography. In M. Büscher et al. (Eds.), *Handbook of Research Methods and Applications for Mobilities* (pp. 193–201). Edward Elgar.

Green, P. (2015). Mobility regimes in practise: Later-life westerners and visa runs in South-East Asia. *Mobilities, 10*(5), 748–763.

Green, P. (2020). Disruptions of self, place and mobility: Digital nomads in Chiang Mai, Thailand. *Mobilities, 15*(3), 431–445.

Hall, G., Sigala, M., Rentschler, R., & Boyle, S. (2019). Motivations, mobility and work practices: The conceptual realities of digital nomads. In J. Pesonen & J. Neidhardt (Eds.), *Information and Communication Technologies in Tourism* (pp. 437–449). Springer.

Hannonen, O. (2016). *Peace and Quiet Beyond the Border: The Trans-border Mobility of Russian Second Home Owners in Finland.* Doctoral Dissertation. UEF.

Hannonen, O. (2018). Second home owners as tourism trend-setters: A case of residential tourism in Gran Canaria. *Journal of Spatial and Organizational Dynamics, 6*(4), 345–359.

Hannonen, O. (2020). In search of a digital nomad: Defining the phenomenon. *Information Technology and Tourism, 22*, 335–353.

Hannonen, O. (2021). *Mobility Regime and Digital Nomadism*. Conference paper presented at the Mobility Humanities Conference, October 29, 2021 (Seoul, Korea and online).

Hannonen, O. (2022). Towards an understanding of digital nomadic mobilities. *Transfers, 12*(3), 115–126.

Hannonen, O., Aguiar, T., & Lehto, X. (2023). A supplier side view of digital nomadism: The case of destination Gran Canaria. *Tourism Management, 97*, 104744.

Hannonen, O., & Prokkola E. K. (2022). Physical access and perceived constraints: Borders as barriers to travel mobilities and tourism development. In D. J. Timothy & A. Gelbman (Eds.), *The Routledge Handbook of Tourism and Borders* (pp 112–125). Routledge.

Hayes, M. (2019). *Gringolandia: Lifestyle Migration Under Late Capitalism*. University of Minnesota Press.

Hayes, M., & Zaban, H. (2020). Transnational gentrification: The crossroads of transnational mobility and urban research. *Urban Studies, 57*(15), 3009–3024.

Holleran, M. (2022). Pandemics and geoarbitrage: Digital nomadism before and after COVID-19, *City, 26*(5–6), 831–847.

Holleran, M., & Notting, M. (2023). Mobility guilt: Digital nomads and COVID-19. *Tourism Geographies, 25*(5), 1341–1358.

Hoogendoorn, G., & Hammett, D. (2021). Resident tourists and the local 'other'. *Tourism Geographies, 23*(5–6), 1021–1039.

Jacobs, E., & Gussekloo, A. (2016). *Digital Nomads: How to Live, Work and Play Around the World*. Self-Published.

Janoschka, M. (2009). The contested spaces of lifestyle mobilities: Regime analysis as a tool to study political claims in Latin American retirement destinations. *Die Erde, 140*(3), 1–20.

Kannisto, P. (2014). *Global Nomads: Challenges of Mobility in the Sedentary World*. Tilburg University.

Kedzior, R. (2014). *How Digital Worlds Become Material: An Ethnographic and Netnographic Investigation in Second Life*. Publications of the Hanken School of Economics, no 281.

Kesselring, S. (2015). Corporate mobilities regimes: Mobility, power and the socio-geographical structurations of mobile work. *Mobilities, 10*(4), 571–591.

Korpela, M. (2019). Searching for a countercultural life abroad: Neo-nomadism, lifestyle mobility or bohemian lifestyle migration? *Journal of Ethnic & Migration Studies, 46*(15), 3352–3369.

Makimoto, T. (2013). The age of digital nomad – Impacts of CMOS innovation. *IEEE Solid-State Circuits Magazine, 5*(1), 40–47.

Makimoto, T., & Manners, D. (1997). *Digital Nomad*. John Wiley & Sons.

Mancinelli, F. (2020). Digital nomads: Freedom, responsibility and the neoliberal order. *Information Technology & Tourism, 22*, 417–437.

Mancinelli, F., & Germann Molz, J. (2024). Moving with and against the state: Digital nomads and frictional mobility regimes. *Mobilities, 19*(2), 189–207.

Matos, P., & Ardévol, E. (2021). The potentiality to move: Mobility and future in digital nomads' practices. *Transfers, 11*(3), 62–79.

MBO Partners Report. (2022). *Working from the Road: The Aspiration and Reality for Digital Nomads*. Available at: www.mbopartners.com/state-of-independence/digital-nomads/

Meethan, K. (2012). Tourism, individuation and space. In J. Wilson (ed.) *The Routledge Handbook of Tourism Geographies* (pp. 61–66). Routledge.

Miguel, C., Lutz, C., Majetić, F., & Perez-Vega, R. (2023). Working from paradise? An analysis of the representation of digital nomads' values and lifestyle on Instagram. *New Media & Society*. https://doi.org/10.1177/14614448231205892

Müller, A. (2016). The digital nomad: Buzzword or research category? *Transnational Social Review, 6*(3), 344–348.

Nash, C., Jarrahi M. H., Sutherland, W., & Phillips, G. (2018). Digital Nomads beyond the buzzword: Defining digital nomadic work and use of digital technologies. In G. Chowdhury, J. McLeod, V. Gillet & P. Willet (Eds.), *Transforming Digital Worlds* (pp. 207–217). Springer.

Orel, M. (2019). Coworking environments and digital nomadism: Balancing work and leisure whilst on the move. *World Leisure Journal, 61*(3), 215–227.

Parreño-Castellano, J., Domínguez-Mujica, J., & Moreno-Medina, C. (2022). Reflections on digital nomadism in Spain during the COVID-19 pandemic – Effect of policy and place. *Sustainability, 14,* 16253.

Pine, B. J., & Gilmore, J. H. (2013). The experience economy: Past, present and future. In J. Sundbo & F. Sørensen (Eds.), *Handbook on the Experience Economy* (pp. 21–44). Edward Elgar Publishing.

Pitkänen, K., Hannonen, O., Toso, S., Gallent, N., Hamiduddin, I., Halseth, G., Hall, C. M., Müller, D. K., Treivish, A., & Nefedova, T. (2020). Second homes during corona – safe or unsafe haven and for whom? Reflections from researchers around the world. *Finnish Journal of Tourism Studies, 16*(2), 20–39.

Prabawa, I. W. S. W., & Pertiwi, P. R. (2020). The digital nomad tourist motivation in Bali: Exploratory research based on push and pull theory. *Athens Journal of Tourism, 7*(3), 161–174.

Putra, G. B., & Agirachman, F.A. (2016). *Urban Coworking Space: Creative Tourism in Digital Nomads Perspective.* Conference Paper: Arte-Polis 6 Int Conference 4–6.8.2016, Bandung.

Reichenberger, I. (2018). Digital nomads – A quest for holistic freedom in work and leisure. *Annals of Leisure Research, 21*(3), 364–380.

Salazar, N. B., Elliot, A., & Norum, R. (2016). Studying mobilities: Theoretical notes & methodological queries. In A. Elliot, R. Norum & N. Salazar (Eds.), *Methodologies of Mobility: Ethnography and Experiment* (pp. 1–24). Berghahn Books.

Sánchez-Vergara, J. I., Orel, M., & Capdevila, I. (2023). "Home office is the here and now." Digital nomad visa systems and remote work-focused leisure policies. *World Leisure Journal, 65*(2), 236–255.

Schlagwein, D. (2018). "Escaping the rat race": Justifications in digital nomadism. *Research-in-Progress Papers.* 31. Available at: https://aisel.aisnet.org/ecis2018_rip/31

Schlagwein, D., & Jarrahi, M. H. (2020). The mobilities of digital work: The case of digital nomadism. *ECIS 2020 Research-in-Progress Papers.* 89. Available at: https://aisel.aisnet.org/ecis2020_rip/89

Sheller, M. (2018). *Mobility Justice: The Politics of Movement in the Age of Extremes.* Verso.

Sheller, M. (2020). Mobility justice. In M. Büscher et al. (Eds.), *Handbook of Research Methods and Applications for Mobilities* (pp. 11–20). Edward Elgar.

Sheller, M., & Urry, J. (2006). The new mobilities paradigm. *Environment & Planning A, 38*(2), 207–226.

Thompson, B. Y. (2018). Digital nomads: Employment in the online gig economy. *Glocalism: Journal of Culture, Politics and Innovation,* DOI:https://doi.org/10.12893/gjcpi.2018.1.11

Thompson, B. Y. (2019a). The digital nomad lifestyle: (Remote) work/leisure balance, privilege, and constructed community. *International Journal of the Sociology of Leisure, 2,* 27–42.

Thompson, B. Y. (2019b). 'I get my lovin' on the run': Digital nomads, constant travel, and nurturing romantic relationships". In C. Nash & A. Gorman-Murray (Eds.), *The Geographies of Digital Sexuality* (pp. 69–90). Palgrave Macmillan.

Toivanen, M. (2023). Countercultural lifestyle no more? Digital nomadism and the commodification of neo-nomadic mobilities. *Mobility Humanities, 2*(2), 70–89.

Wang, B., Schlagwein, D., Cecez-Kecmanovic, D., & Cahalane, M. C., (2018). Digital work and high-tech wanderers: Three theoretical framings and a research Agenda for digital nomadism. *ACIS 2018 Proceedings.* 55. Available at: https://aisel.aisnet.org/acis2018/55

Willment, N. (2020). The travel blogger as digital nomad: (Re-)imagining workplace performances of digital nomadism within travel blogging work. *Information Technology and Tourism, 22,* 391–416.

Woldoff, R. A., & Litchfield, R. C. (2021). *Digital Nomads: In Search of Freedom, Community, and Meaningful Work in the New Economy.* Oxford University Press.

Urry, J. (1990). *The Tourist Gaze: Leisure and Travel in Contemporary Societies.* SAGE.

Urry, J. (2007). *Mobilities.* SAGE.

UNWTO. (2010). *International Recommendations for Tourism Statistics 2008.* UNWTO.

31

TOURISM MOBILITIES AND URBAN CHANGE

Geographies of Transnational Gentrification

Franz Buhr and Agustín Cocola-Gant

Introduction

The relationship between tourism and urban change has recently attracted unprecedented levels of scholarly and public attention. Since the publication of the *Routledge Handbook of Tourism Geographies* in 2012, much debate has been held around the relevance of tourism and its impact on processes of urban transformation, which are frequently motivated by its perceived exclusionary effects (Gravari-Barbas & Guinand, 2017; Yrigoy, 2019; González-Pérez, 2020; Pinkster & Boterman, 2017; Cocola-Gant, 2018). The last ten years have reshaped tourism industries, tourist practices, and also the very cities where tourism takes place. Likewise, the concept of gentrification has advanced beyond academic circles and it is now a key notion around which civil society and activists mobilise to denounce the inequalities produced in contemporary touristic cities. As this chapter shows, debates around the impact of tourism in urban landscapes have converged increasingly with gentrifying processes, signalling their interdependencies and cross-fertilisation.

Airbnb's trajectory in recent years is probably the most remarkable example of how tourism became so deeply entangled with gentrification processes. While, a decade ago, Airbnb was still a sharing economy business, where most hosts would offer an extra room to visitors, now the company has become the world's largest accommodation service, with over 6 million active listings in more than 100,000 cities (Airbnb, 2022). A large portion of the Airbnb supply caused a withdrawal of millions of apartments from traditional housing markets. This had major impacts across cities and has become the subject of local political campaigns, social mobilisations, and anti-tourism protests (Novy & Colomb, 2019; Garay, Morales & Wilson, 2020). Tourism rapidly penetrated remote urban spaces, whether they were working-class neighbourhoods in the Global North (Quaglieri-Domínguez & Scarnato, 2017, Cocola-Gant & Gago, 2021), or even favelas in the Global South (Freire-Medeiros, 2012; Cummings, 2015).

Yet, tourism has only recently begun to be seen as a conduit for gentrification. Although the concept of gentrification has been used in very different contexts – perhaps stretching well beyond its reach (Maloutas, 2012; 2018) – its common concern is the inflow of middle-class residents into working-class areas, resulting in the displacement and exclusion of original populations. In the case of neighbourhoods dominated by short-term rentals, in which 'new residents' are often

     DOI: 10.4324/9781003286301-37

tourists or temporary dwellers, gentrification assumes new forms, yet can produce similar displacement processes. Terms such as tourism gentrification (Gotham, 2005; Cocola-Gant, 2018), touristification (Jover & Díaz-Parra, 2020; Salerno, 2022; Tulumello & Allegretti, 2021), or even *Airbnbfication* (Peters, 2016; Rozena & Lees, 2021) have been coined to shed light on the particular ways transient populations, and the real estate investment opportunities they bring, impact urban neighbourhoods.

In this chapter, we turn to the recent scholarship addressing the notion of transnational gentrification. First used in Sigler and Wachsmuth (2016) to describe the neighbourhood change generated by the inflow of relatively affluent foreign retirees to Panama's Casco Antíguo, the term transnational gentrification has since encompassed an array of transnational mobilities which constitute gentrifying processes. Besides retired migrants and tourists, certain cities have also been attracting other kinds of lifestyle migrants: 'expats', digital nomads, international students, and second-home owners, among other transient populations whose place consumption patterns may often overlap (Novy, 2018).

On the one hand, it is precisely their consumption patterns – their lifestyle preferences, the kinds of restaurants they go to, the local brands they are interested in, the coffee shops they work from – that illustrate how urban space is increasingly adapting to cater for the needs of a globally mobile gentrifier class (Bridge 2007). On the other hand, the arrival of transnational mobile populations opens up investment opportunities in the real estate market. For instance, developers and property owners have been creating accommodations such as short-term rentals, co-living spaces, or luxury student housing. This highlights a convergence of the hospitality and property industries, which significantly exclude local populations.

This chapter focuses mainly on scholarly work published over the last decade. The common thread is the nexus between increasingly mobile lifestyles and urban transformation. First, the chapter looks into the relationship between short-term rentals and gentrification processes. Second, it explores who transnational gentrifiers (usually) are and how their place consumption practices are reshaping urban territories. The chapter's final section briefly presents two promising topics for future urban research.

Short-term Rentals, Housing, and Tourism-led Gentrification

The initial steps of the vacation rental industry in the twentieth century were linked to the leasing of second homes in coastal and rural sites (Nicod, Mungall & Henwood, 2007). In Europe, the leading company, Interhome, was founded in 1965 and in the United States VRBO (Vacation Rentals by Owner) appeared in 1995. However, the short-term rental (STR) industry has been popularised worldwide in the last decade thanks to the creation of digital platforms such as Airbnb. In a context of low-cost travel and state-led tourism promotion as a strategy for economic growth after the 2008 financial crisis, it was the use of the digital platform business model which became the turning point for the industry spreading ubiquitously.

Digital platforms offer, fundamentally, a technological infrastructure through which property owners can easily list their flats on a global scale and reach international guests. By listing their properties on platforms, owners have free access to services such as marketing, reservation and payment systems, which enable them to effortlessly compete with traditional accommodation enterprises (Guttentag, 2015; Shaw, 2020). The result has been an exponential rise in the supply of short-term rentals, which has reduced the cost of travelling. The period of overtourism registered in the years before COVID-19 (Milano et al., 2019) has much to do with this phenomenon (see also Cruz, Chapter 2 of this volume).

The unprecedented growth in the number of visitors who consume urban landscapes and who are willing to pay high prices for short stays have been paralleled by a growth of research concerning the socio-spatial impacts of the process. In this regard, it is not surprising that gentrification and housing scholars have recently paid attention to the effects of tourism (Gotham, 2005; Cocola-Gant, 2018, 2016; Jover & Díaz-Parra, 2020; Yrigoy, 2019; Wachsmuth & Weisler, 2018). Gentrification research has long demonstrated how the arrival of new users and capital to certain places drives displacement processes of local communities (Lees et al., 2008), and scholars have applied such a conceptual framework to explain the effects of STRs.

In short, by renting on digital platforms such as Airbnb, property owners obtain higher profits than in the long-term rental market, which encourages a process of tenant replacement by visitors (Wachsmuth & Weisler, 2018; Cocola-Gant, 2016). Although Airbnb claims that hosts are home-sharers earning an additional income, in central urban areas of major tourist destinations, the tendency of the market is toward increased commercialisation of entire apartments available all year round. Indeed, most Airbnb revenues are generated through entire homes supplied by multi-listing hosts (Deboosere et al., 2019; Dogru et al., 2020).

The proliferation of STRs has affected places in different geographies, from global cities to rural places, in both core accumulation regions and peripheral economies. In the last decades, many rural and coastal areas have been restructured into having a primarily touristic economic base (Phillips, 2002; Freeman and Cheyne, 2008). Here both recreational facilities and the expansion of second homes played a crucial role (Müller, 1999). These places have become valued as leisure facilities serving both resident and visiting middle-class people, thus driving processes of gentrification. However, this leisure-led gentrification of post-productive rural and coastal landscapes has been widely accelerated by the expansion of short-term rentals predominantly because a part of Airbnb's supply consists of second homes that are offered on the platform while they are not in use (Morales, Garay & Troyano, 2022).

In terms of gentrification dynamics in cities, it is interesting to see how the phenomenon evolved in peripheral places that were not on the global tourism map and that indeed did not experience classical gentrification in previous decades, meaning that central areas were inhabited by working-class populations living in degraded built environments. This is the case for instance in Mediterranean cities such as Naples (Esposito 2020), Porto (Carvalho et al. 2019), Lisbon (Barata-Salgueiro et. al., 2017; Cocola-Gant & Gago, 2021), Seville (Jover & Diaz-Parra, 2020) and Thessaloniki (Katsinas, 2021) among others.

On the one hand, STRs offer the sufficient infrastructure required to host visitors. On the other hand, STRs provide an opportunity for investors and property owners to extract profits from urban spaces and, consequently, real estate capital started to flow in (Montezuma & McGarrigle, 2019). In a very short period, these places moved from a state of 'decline' to a new phase of physical upgrading, escalating property prices and social change. Therefore, in many places, the progression of gentrification has been less related to the consumption power of local middle classes and more to the effects of the increased local demand that tourism provides.

A central topic of research has been to decipher how STRs impact housing markets. Authors show that, in central locations of tourist destinations, STRs have severely reduced the supply of apartments available for long-term occupation, potentially driving the increase of rental prices (Barron et al., 2020; Garcia-López et al., 2020). Other authors have estimated how much housing stock has been lost to Airbnb. For instance, Wachsmuth and Weisler (2018) found that there are several neighbourhoods where Airbnb removed more than 3 percent of all housing off the primary residential market. Other studies have measured the issue at a finer spatial scale in areas under high tourism pressure and found a tacit conversion of the rental market into STRs (Cocola-Gant

& Gago, 2021). This is, furthermore, the case in some areas of Berlin, where Schäfer and Braun (2016) found that all of the available flats are let out to tourists. On the one hand, the process has led to the displacement of tenants.

On the other hand, the reduction of long-term rentals means that finding affordable rental housing has become extremely difficult in certain places. In conclusion, the growth of STRs substantially reduces housing alternatives for residents due to both price increase and a lack of stock available in the long-term rental market. Although these consequences may be particularly intense in central areas, the process puts pressures on the rest of the city as well. For instance, as it has been noted in Los Angeles (Lee, 2016), it creates a gentrifying domino effect because middle-income residents displaced from or unable to find accommodation in central areas tend to move to more affordable neighbourhoods, adding market pressures in other geographies of the city as well.

So far, we have explored literature that focuses on how STRs affect the right to housing and can cause the displacement of residents. Additionally, qualitative research on the impacts of STRs found that the sharing of apartment buildings with tourists causes pronounced disruption for residents (Törnberg, 2022; Rozena & Lees, 2021; Gurran & Phibbs, 2017). The most frequent issue is noise and difficulties in terms of resting and sleeping at night. In this regard, closer attention must be paid to changes at a neighbourhood level and how the mutation of residential areas into spaces for visitors and mobile populations affects the daily lives of residents. In cities such as Amsterdam, authors found that residents are moving from tourist areas not only because of the lack of housing alternatives, but because of the daily disruptions that tourism and STRs create (Pinkster & Boterman, 2017).

In other words, the excessive growth of tourism causes daily disruptions that make everyday life increasingly unpleasant, particularly due to noise, a lack of stores that residents use, the overcrowding of public spaces as well as the loss of community life and social bonds. These changes in the daily lives of residents have implied that some tourist areas experienced a progressive population decline (Celata & Romano, 2020; López-Gay, Cocola-Gant, & Russo, 2021). Interestingly, this was already suggested in the Doxey's (1975) 'irritation index' model, by which communities living in tourist destinations experience four stages, the final one being an antagonism towards visitors.

Finally, it is worth noting that during the COVID-19 pandemic, the STR market survived by adapting itself to mid-term rentals for floating young professionals (see Hannonen, Chapter 30 of this volume). In terms of supply dynamics, Airbnb advised hosts to offer discounts for mid-term stays, and, during the pandemic, Airbnb adopted several measures to allow listings to be available for longer rental periods (Roelofsen & Minca, 2021). In addition, professional STR property managers use digital technologies, such as channel managers, to list their portfolio on several platforms (Cocola-Gant et al. 2021). Beyond Airbnb, the pandemic has consolidated an ecosystem of platforms for mid-term rentals, such as AltoVita, SpotaHome, NomadX, and Uniplaces among others. These platforms connect landlords with professionals searching for accommodation, usually for business purpose travels. As such, the process is significantly linked to transnational gentrification. We develop this point in the next section.

Gentrifiers on the Move

Picture this: tourists gathering in front of a celebrity architect-designed building. Digital nomads sharing coworking tables at specialty coffee shops. Hipsters browsing for vintage clothes at the newest pop-up store. 'Expats' buying organic produce at the corner bodega. International students celebrate the end of exam period at the new hip artisanal brewery, while well-off couples shop for ceramics in a minimalist boutique – the list could go on. This scene is not unknown to anyone having recently visited Berlin, London, New York, Lisbon, or Athens. But the list *should* go on. In all these places, long-term residents face the risk of eviction, elderly residents see their go-to grocery shops disappear, rents skyrocket, and everyday items such as coffee or bread now cost many times what they used to.

The socio-economic change of cities is a matter of complex causality, spanning wider the present chapter's ambition. Yet, the description above conveys the rather familiar image of a gentrified landscape crosscut by various kinds of mobilities. As Gravari-Barbas and Guinand (2017) have pointed out, tourism often follows urban gentrifiers to their 'trendy' urban areas, their farmers' markets, and gourmet shops (see also Opillard, 2016). Tourists' never-ending quest for 'authentic' and 'off-the-beaten-path' experiences (Maitland & Newman, 2009) arguably fuels gentrification processes, as well as takes them further into more remote or peripheral urban neighbourhoods, which, in turn, makes those neighbourhoods more desirable and expensive. The dynamics producing the next 'soon-to-be-hip place' colonises urban space with amenities and real estate investment, inducing not only residential gentrification, but also commercial gentrification (Atkinson & Bridge, 2005).

Yet, scholars have argued that such wide-ranging processes of urban transformation may hardly be attributed to the impact of tourism alone. Writing in 2014 about Berlin, Fuller and Michel (2014: 1306) anticipated that 'the effects of tourism on urban neighbourhoods will be quite difficult to distinguish from general processes of urban change and commodification'. Indeed, four years later Novy (2018) argued precisely that in order to understand recent socio-spatial change in Berlin, we would need to account for the many mobilities intersecting with tourism which also result in place transformation. He evokes the growing presence of 'temporary city users' (Martinotti, 1993) whose middle-class lifestyles and consumption practices cannot be easily distinguished from those of local gentrifiers.

Novy (2018) suggests that the relationship between (tourism) mobilities and place consumption should account for the following five interrelated dimensions: (1) (urban) tourism; (2) (temporary) 'lifestyle' migration; (3) (temporary) migration for work/education; (4) residents exploring places in their own cities 'as if tourists'; and (5), leisure and place consumption as a practice of everyday life.

In practice, scholars have noticed that several kinds of mobile populations also consume and live (albeit temporarily) in gentrified neighbourhoods. International students (Malet-Calvo, 2018), digital nomads (Polson, 2020), residential tourists (Domínguez-Mujica, McGarrigle & Parreño-Castellano, 2021), expats (Kunz, 2016), young migrants (King, 2018), and lifestyle migrants (Hayes, 2020) increasingly participate in the visitor economy, while simultaneously contributing to the expansion of the leisure-led restructuring of urban space. In their study about the Gothic quarter in Barcelona, López-Gay and colleagues (2021) found that tourist areas are also deemed to be attractive to young transnational gentrifiers, who prefer to live in more 'cosmopolitan' environments. The authors call for urban research

> to move from the implicitly assumed distinction between residents and visitors to consider instead how the population restructuring of central areas in contemporary cities could be the result of an assemblage of emerging forms of temporary dwelling, among which tourism is a powerful driver.
>
> (López-Gay, Cocola-Gant & Russo, 2021: 2)

The recognition of the various kinds of mobilities and mobile life practices intersecting with tourism largely stems from urban scholarship's increasing engagement with the new mobilities paradigm (Sheller & Urry, 2006). The 'mobilities turn' (Sheller, 2017) in social sciences drew attention to the ways in which everyday life is inherently mobile, to varying extents and at different scales (Urry, 2000) (see also Thulemark & Duncan, Chapter 29). As Duncan (2012) points out, through the lens of mobilities, scholars have been able to address specific practices that obscure conventional definitions of travel, migration, and dwelling. Studies about lifestyle migration are exemplary of that. Usually depicted as retirees from richer countries moving somewhere sunny in lower latitudes, lifestyle migrants often carry out a 'poly-topical' existence (Stock, 2006), as (temporary) residents of multiple places, and infusing everyday life with leisure and tourism-related activities.

Beyond migrant retirees, lifestyle migration literature has looked at an array of transnational mobilities performed by the relatively privileged, from business 'expatriates' to bohemian migration (Korpela, 2020). When revisiting their original contribution to the field (Benson & O'Reilly, 2009), Benson and O'Reilly concluded that lifestyle as a concept offers a way of introducing both choice and consumption into discussions about migration (Benson & O'Reilly, 2015; 14). This is crucial as lifestyle migrants are often able to maximise their purchase power by resettling in lower-cost destinations – what Hayes (2014) calls 'geoarbitrage' – thus impacting local destinations with their higher spendable income and consumption practices. In this sense, the convergence of various kinds of place-consumption practices by tourists, lifestyle migrants, and other transient populations, has helped reshape the notion of transnational gentrification (Cocola-Gant & López-Gay, 2020). Scholars now argue that transnational gentrification currently encompasses more than only the socio-spatial transformation caused by foreign home-buyers, but also by those resulting from new place consumption patterns by a wider, more global, mobile class (Sigler & Wachsmuth, 2020).

The concept of transnational gentrification thus elucidates the ways urban landscapes are transformed in order to cater for the consumption habits of a mobile clientele (Alexandri & Janoschka, 2020) and, as a consequence, lead to the social and spatial dispossession and displacement of long-term residents. In particular, it shows that what is usually called gentrification in touristic cities seems to be produced by a global collection of privileged mobile dwellers, tourists, short-term residents of different profiles, and both transnational and local gentrifiers. The worldwide proliferation of consumptionscapes, such as the 'global Brooklyn' (Halawa & Parasecoli & Halaw, 2021), including the pop-up of specialty coffee shops (Bantman-Masum, 2020), brunch eateries, and design boutiques (Zukin, Kasinitz & Chen, 2016) are testament to how the old 'ABC' of gentrification (Art galleries, Boutiques, and Cafés) (Zukin, 1998) are now crosscut by, and embedded in, global mobilities.

Pathways for Future Research

The last decade witnessed both the most international tourist arrivals in the world's history (1.46 billion in 2019) and its most drastic halt, due to the COVID-19 pandemic outbreak and its ensuing travel restrictions. The pandemic exposed some of the urban fragilities of cities, which are (overly) dependent on tourism, and renewed the urgency for a constant examination of its impacts over local livelihoods, housing markets, and sustainability (Sheller, 2021). In the following paragraphs, we highlight two emerging phenomena that seem to have been accelerated by the pandemic and which have not yet received enough scholarly attention.

The first phenomenon is the growing professionalisation of the STR market and how it can further extend the reach of transnational gentrification. As Cocola-Gant et al. (2021) found, amateur hosts have progressively outsourced the management of their properties to 'corporate hosts', that is, property managers who carry out all the aspects of running a STR, from interior design and responding to guests' enquiries, to changing linen and doing check-ins. By making the STR business more efficient, these commercial operators have become key players and have initiated a phenomenon of market concentration as they have the ability to secure high occupancy rates and, thus, profitability. In the United States, companies such as Vacasa, manage more than 30,000 properties.

The success of 'corporate hosts' testify to the definitive end of a so-called 'sharing economy' model, and to its replacement by what authors have called 'platform real estate' (Field & Rogers, 2019; Shaw, 2018); characterised by the use of digital innovation and platform business models to ensure even more profitable real estate investment. The proliferation of listings managed by corporate hosts has implications which remain to be explored. In particular, by using digital technology such as channel managers to commercialise the listings on different platforms, corporate hosts have made the STR market more resilient during the pandemic. While many hotels closed down, STRs have been able to adapt to other forms of rentals, particularly to mid-term rentals for professionals who were able to work remotely. In terms of demand, the pandemic accelerated an emerging phenomenon, that is, the increased mobility of well-paid childless professionals working from home and able to change location as they wish. The result is the blurring of work and leisure activities which has the potential to further drive a process of gentrification in which some destinations experience an increase in transnational mobile populations.

In relation to this, the second phenomenon is the increasing importance of digital nomads among mobile populations (see Hannonen, Chapter 30). Digital nomads are professionals who work while travelling. Motivated by the possibility to combine travel, leisure, and work, this new class of transnational workers is composed of relatively privileged individuals who are able to decide their next location choice based on lifestyle criteria (Thompson, 2018). Given that the pandemic has generalised and intensified practices of remote work, it is likely that larger cohorts of remote freelance workers will turn to digital nomadism as a possible lifestyle. Indeed, countries such as Greece, Costa Rica, Croatia and Portugal among others, have established visa or fiscal regimes aiming to attract international remote workers.

In contrast with many of the populations analysed by lifestyle migration scholarship, work/production is as central to digital nomads as leisure/consumption. This entails a very specific need to secure work and business infrastructures, making remote/collaborative work viable, such as coworking spaces (Grazian, 2020; Merkel, 2019) or specialty coffee shops (Jung & Buhr, 2021). In this sense, digital nomads' impact over urban landscapes may partially overlap with that of tourists, other lifestyle migrants, and global gentrifiers in terms of consumption, but it may also reach other infrastructures related to the provision of work-related support and professional sociability. In addition, understanding how cities accommodate digital nomads' work practices is also relevant, as some of the spaces catering for privileged remote workers may increasingly impact less privileged, non-mobile, and precarious local remote workers, thus channelling urban transformation even further.

Whether in terms of housing or commercial landscape transformation, urban infrastructures seem to increasingly respond to a broad spectrum of mobile life practices and (more or less) temporary forms of dwelling, which cannot be neatly disaggregated. A challenge ahead is to take stock of those short/mid-term mobilities in terms of urban planning, which entails making them visible in formal records and official statistics.

References

Airbnb. (2022). *About us.* [Online]. https://news.airbnb.com/about-us/ [accessed February 2024].

Alexandri, G. & Janoschka, M. (2020). 'Post-pandemic' transnational gentrifications: A critical outlook. *Urban Studies* 57(15), 3202–3214.

Atkinson, R. & Bridge, G. (Eds.) (2005). *Gentrification in a global context: The new urban colonialism.* Routledge.

Bantman-Masum, E. (2020). Unpacking commercial gentrification in central Paris. *Urban Studies* 57(15), 3135–3150

Barata-Salgueiro, T., Mendes, L. & Guimarães, P. (2017). Tourism and urban changes: Lessons from Lisbon. In Gravari-Barbas, M., & Guinand, S. (Eds.), *Tourism and Gentrification in Contemporary Metropolises* (pp. 255–275). Routledge.

Barron, K., Kung, E., & Proserpio, D. (2020). *The Sharing Economy and Housing Affordability: Evidence from Airbnb.* SSRN: https://ssrn.com/abstract=3006832.

Benson, M. & O'Reilly, K. (2009). Migration and the search for a better way of life: A critical exploration of lifestyle migration. *The Sociological Review* 57(4), 608–625.

Benson, M. & O'Reilly, K. (2015). From lifestyle migration to lifestyle in migration: Categories, concepts and ways of thinking. *Migration Studies* 4(1), 20–37.

Bridge, G. (2007). A global gentrifier class? *Environment & Planning A* 39(1), 32–46.

Carvalho, L., Chamusca, P., Fernandes, J., et al. (2019). Gentrification in Porto: Floating city users and internationally-driven urban change. *Urban Geography* 40(4), 565–572.

Celata, F. & Romano, A. (2020). Overtourism and online short-term rental platforms in Italian cities. *Journal of Sustainable Tourism* 30(5), 1020–1039.

Cocola-Gant, A. (2016). Holiday rentals: The new gentrification battlefront. *Sociological Research Online* 21(3), 10.

Cocola-Gant, A. (2018). Tourism gentrification. In Lees, L. & Phillips, M. (Eds.), *Handbook of Gentrification Studies* (pp. 281–293). Edward Elgar Publishing.

Cocola-Gant, A. & Gago, A. (2021). Airbnb, buy-to-let investment and tourism-driven displacement. A case study in Lisbon. *Environment and Planning A* 53(7), 1671–1688.

Cocola-Gant, A., Jover, J., Carvalho, L. & Chamusca, P. (2021). Corporate hosts: The rise of professional management in the short-term rental industry. *Tourism Management Perspectives* 40, 100879.

Cocola-Gant, A. & López-Gay, A. (2020). Transnational gentrification, tourism and the formation of 'foreign only' enclaves in Barcelona. *Urban Studies* 57(15), 3025–3043.

Cummings, J. (2015). Confronting favela chic: The gentrification of informal settlements in Rio de Janeiro, Brazil. In Lees, L., Shin, H. & López-Morales, E.(Eds.), *Global Gentrifications: Uneven Development and Displacement* (pp. 81–100). Policy Press. https://doi.org/10.1332/policypress/9781447313472.003.0005

Deboosere, R, Kerrigan, D.J., Wachsmuth, D., et al. (2019). Location, location and professionalization: A multilevel hedonic analysis of Airbnb listing prices and revenue. *Regional Studies, Regional Science* 6(1), 143–156.

Dogru, T., Mody, M., Suess, C., et al. (2020). Airbnb 2.0: Is it a sharing economy platform or a lodging corporation? *Tourism Management* 78, 104049.

Domínguez-Mujica, J., McGarrigle, J. & Parreño-Castellano, J. (Eds.) (2021). *International Residential Mobilities. From Lifestyle Migrations to Tourism Gentrification.* Springer.

Doxey, G. (1975). A causation theory of visitor-resident irritants: Methodology and research inferences. *Conference Proceedings of the Travel Research Association* (pp. 195–198). San Diego.

Duncan, T, (2012). The 'mobilities turn' and the geography of tourism. In Wilson, J. (Ed.) *The Routledge Handbook of Tourism Geographies* (1st Edition) (pp. 113–119). Routledge.

Esposito, A. (2020). La città turistica e la ristrutturazione digitale della rendita urbana. *Archivio di Studi Urbani e Regionali* 129, 183–208.

Fields, D. & Rogers, D. (2019). Towards a critical housing studies research agenda on platform real estate. *Housing, Theory and Society* 38(1), 72–94. https://doi.org/10.1080/14036096.2019.1670724

Freeman, C. & Cheyne, C. (2008). Coasts for sale: Gentrification in New Zealand. *Planning Theory & Practice* 9(1), 33–56.

Freire-Medeiros, B, (2012). *Touring Poverty.* Routledge.

Fuller, H., & Michel, B. (2014). Stop being a tourist!' New dynamics of urban tourism in Berlin-Kreuzberg. *International Journal of Urban and Regional Research* 38, 1304–1318.

Garay, L., Morales, S., & Wilson, J. (2020). Tweeting the right to the city: Digital protest and resistance surrounding the Airbnb effect. *Scandinavian Journal of Hospitality and Tourism* 20(3), 246–267.

Garcia-López, M.À., Jofre-Monseny, J., Martínez-Mazza, R. & Segú, M. (2020). Do short-term rental platforms affect housing markets? Evidence from Airbnb in Barcelona. *Journal of Urban Economics* 119, 103278.

González-Pérez, J.M. (2020). The dispute over tourist cities. Tourism gentrification in the historic Centre of Palma (Majorca, Spain). *Tourism Geographies* 22(1), 171–191.

Gotham, K.F. (2005). Tourism gentrification: The case of New Orleans' Vieux Carre (French Quarter). *Urban Studies* 42(7), 1099–1121.

Gravari-Barbas, M. & Guinand, S. (2017). Introduction: Addressing tourism-gentrification processes in contemporary metropolises. In Gravari-Barbas, M. & Guinand, S. (Eds.) *Tourism and Gentrification in Contemporary Metropolises. International Perspectives* (pp. 1–22). Routledge.

Grazian, D. (2020). Thank God it's Monday: Manhattan coworking spaces in the new economy, *Theory and Society* 49(5–6), 991–1019.

Gurran, N. & Phibbs, P. (2017). When tourists move in: How should urban planners respond to Airbnb? *Journal of the American Planning Association* 83(1), 80–92.

Guttentag, D. (2015). Airbnb: Disruptive innovation and the rise of an informal tourism accommodation sector. *Current Issues in Tourism* 18(12), 1192–1217.

Hayes, M. (2014). "We gained a lot over what we would have had": The geographic arbitrage of North American lifestyle migrants to Cuenca, Ecuador. *Journal of Ethnic and Migration Studies* 40(12), 1953–1971.

Hayes, M. (2020). Mobilités et urbanisme patrimonial. Les modes de vie mobiles et leurs implications territoriales. *Anthropologie et Sociétés* 44(2), 147–166.

Jover, J. & Díaz-Parra, I. (2020). Gentrification, transnational gentrification and touristification in Seville, Spain. *Urban Studies* 57(15), 3044–3059.

Jung, P. & Buhr, F. (2021). Channelling mobilities: Migrant-owned businesses as mobility infrastructures, *Mobilities* 17(1), 119–135. DOI: 10.1080/17450101.2021.1958250

Katsinas, P. (2021). Professionalisation of short-term rentals and emergent tourism gentrification in post-crisis Thessaloniki. *Environment and Planning A* 53(7), 1652–1670.

King, R. (2018). Theorising new European youth mobilities. *Population, Space and Place* 24(1), e2117.

Korpela, M. (2020). Searching for a countercultural life abroad: Neo-nomadism, lifestyle mobility or bohemian lifestyle migration? *Journal of Ethnic and Migration Studies* 46(15), 3352–3369.

Kunz, S. (2016). Privileged mobilities: Locating the expatriate in migration scholarship. *Geography Compass*, 10, 89–101.

Lee, D. (2016). How Airbnb short-term rentals exacerbate Los Angeles's affordable housing crisis: Analysis and policy recommendations. *Harvard Law & Policy Review* 10, 229–254.

Lees, L., Slater, T. & Wyly, E. (2008). *Gentrification*. Routledge.

López-Gay, A.; Cocola-Gant, A. & Russo, AP. (2021). Urban tourism and population change: Gentrification in the age of mobilities. *Population, Space and Place*, 27 (1), e2380.

Maitland, R. & Newman, P. (Eds.) (2009). *World Tourism Cities. Developing Tourism Off the Beaten Track.* Routledge

Malet-Calvo, D. (2018). Understanding international students beyond studentification: A new class of transnational urban consumers. The example of Erasmus students in Lisbon (Portugal). *Urban Studies* 55(10), 2142–2158

Maloutas, T. (2012). Contextual diversity in gentrification research. *Critical Sociology* 38(1), 33–48.

Maloutas, T. (2018). Travelling concepts and universal particularisms: A reappraisal of gentrification's global reach. *European Urban and Regional Studies* 25(3), 250–265

Martinotti, G. (1993). Metropoli: *La nuova morfologia sociale della città.* Il Mulino.

Merkel, J. (2019). 'Freelance isn't free.' Co-working as a critical urban practice to cope with informality in creative labour markets. *Urban Studies* 56(3), 526–547

Milano, C., Cheer, J.M. & Novelli, M. (2019). *Overtourism: Excesses, Discontents and Measures in Travel and Tourism.* CABI.

Montezuma, J., & McGarrigle, J. (2019). What motivates international homebuyers? Investor to lifestyle 'migrants' in a tourist city. *Tourism Geographies* 21(2), 214–234.

Morales, S., Garay, L. & Troyano, X. (2022). Beyond the big touristic city: Nature and distribution of Airbnb in regional destinations in Catalonia (Spain). *Current Issues in Tourism* 25(20), 3381–3394.

Müller, D.K. (1999). *German second home owners in the Swedish countryside.* (PhD Thesis), Umeå University.

Nicod, P., Mungall, A. & Henwood, J. (2007). Self-catering accommodation in Switzerland. *International Journal of Hospitality Management* 26(2), 244–262.

Novy, J. (2018). 'Destination' Berlin revisited: From (new) tourism towards a pentagon of mobility and place consumption. *Tourism Geographies* 20(3), 418–442.

Novy, J. & Colomb, C. (2019). Urban tourism as a source of contention and social mobilisations: A critical review. *Tourism Planning & Development* 16(4), 358–375.

Opillard, F. (2016). From San Francisco's 'Tech Boom 2.0' to Valparaíso's UNESCO World Heritage Site: Resistance to tourism gentrification from a comparative political perspective. In Colomb, C. & Novy, J. (Eds.) *Protest and Resistance in the Tourist City* (pp. 129–151). Routledge

Parasecoli, F. & Halawa, M, (2021). *Global Brooklyn: Designing Food Experiences in World Cities.* Bloomsbury.

Peters, D. (2016). Density wars in Silicon Beach: The struggle to mix new spaces for toil, stay and play in Santa Monica, California. In C. Colomb & J. Novy (Eds.), *Protest and Resistance in the Tourist City* (pp. 104–120). Routledge.

Phillips, M. (2002). The production, symbolization and socialization of gentrification: Impressions from two Berkshire villages. *Transactions of the Institute of British Geographers* 27(3): 282–308.

Pinkster, F.M. & Boterman, W.R. (2017). When the spell is broken: Gentrification, urban tourism and privileged discontent in the Amsterdam canal district. *Cultural Geographies* 24(3), 457–472.

Polson, E. (2020). The aspirational class 'mobility' of digital nomads. In Polson, E. Clark, L. & Gajjala, R. (Eds.) *Routledge Companion to Media and Class* (pp. 168–179). Routledge.

Quaglieri-Domínguez, A. & Scarnato, A. (2017). The penetration of everyday tourism in the dodgy heart of the Raval. In Gravari-Barbas, M & Guinand, S. (Eds.) *Tourism and Gentrification in Contemporary Metropolises. International Perspectives* (pp. 107–133). Routledge.

Roelofsen, M. & Minca, C. (2021). Sanitised homes and healthy bodies: Reflections on Airbnb's response to the pandemic. *Oikonomics*, 15, 1–10.

Rozena, S. & Lees, L. (2021). The everyday lived experiences of Airbnbification in London. *Social & Cultural Geography* 24(2), 253–273. DOI: 10.1080/14649365.2021.1939124

Salerno, G.M. (2022). Touristification and displacement. The long-standing production of Venice as a tourist attraction. *City* 26(2–3), 519–541.

Schäfer, P. & Braun, N. (2016). Misuse through short-term rentals on the Berlin housing market. *International Journal of Housing Markets and Analysis* 9(2), 287–311.

Shaw, J. (2020). Platform Real Estate: Theory and practice of new urban real estate markets. *Urban Geography* 41(8), 1037–1064.

Sheller, M. (2021). Mobility justice and the return of tourism after the Pandemic. *Mondes du Tourisme* 19. DOI: https://doi.org/10.4000/tourisme.3463

Sheller, M. (2017). From spatial turn to mobilities turn. *Current Sociology* 65(4), 623–639.

Sheller, M. & Urry, J. (2006). The new mobilities paradigm. *Environment and Planning A* 38 (2), 207–226.

Sigler, T. & Wachsmuth, D. (2016). Transnational gentrification: Globalisation and neighbourhood change in Panama's Casco Antiguo. *Urban Studies* 53(4), 705–722.

Sigler, T. & Wachsmuth, D. (2020). Tourism-led, state-led, and lifestyle-led urban transformations: New directions in transnational gentrification. *Urban Studies* 57(15), 3190–3201.

Stock, M (2016). *L'hypothèse de l'habiter poly-topique: pratiquer les lieux géographiques dans les sociétés à individus mobiles*, EspacesTemps.net [Online]. www.espacestemps.net/en/articles/hypothese-habiter-poly topique/ [accessed February 2024].

Thompson, B.Y. (2018). Digital nomads: Employment in the online gig economy. Glocalism: *Journal of Culture Politics & Innovation* 1, 1–27. DOI:10.12893/gjcpi.2018.1.11

Törnberg, P. (2022). Platform placemaking and the digital urban culture of Airbnbification. *Urban Transformations* 4(1), 1–26.

Tulumello, S. & Allegretti, G. (2021). Articulating urban change in Southern Europe: Gentrification, touristification and financialisation in Mouraria, Lisbon. *European Urban and Regional Studies* 28(2), 111–132.

Urry, J. (2000). Mobile sociology. *The British Journal of Sociology* 51(1), 185–203.

Wachsmuth, D. & Weisler, A. (2018). Airbnb and the rent gap: Gentrification through the sharing economy. *Environment and Planning A* 50(6), 1147–1170.

Yrigoy, I. (2019). Rent gap reloaded: Airbnb and the shift from residential to touristic rental housing in the Palma Old Quarter in Mallorca, Spain. *Urban Studies* 56(13), 2709–2726.

Zukin, S. (1998). Urban life styles: Diversity and standardization in spaces of consumption. *Urban Studies* 35, 825–839.

Zukin, S., Kasinitz, P. & Chen, X. (Eds.) (2016). *Global Cities, Local Streets: Everyday Diversity from New York to Shanghai*. Routledge.

32

THE MESSINESS OF TOURISM AND TRANSPORTATION GEOGRAPHIES

David Timothy Duval and John Macilree

Introduction

This chapter evaluates the current state of transportation networks and flows, and their implications for global tourism geographies. Our focus is air transport, and we offer a partial assessment of the implications of the COVID-19 pandemic in 2020/2021 for air transport geographies. A shift began in 2020, where concepts and phrases such as 'vaccine passports', 'quarantine' and 'aerosolisation' became part of common vernacular when considering the current state, and indeed future, of both tourism and transport. The first 15 months of COVID-19 forced acknowledgement of the close connections between tourism and transport for those who had previously been unaware of their strength. It thrusts into the spotlight several realities that shifted and disrupted the symbiotic (Duval, 2013) relationship between tourism and transport. Indeed, Hall et al (2020: 590) conclude cogently that "[c]hanges to tourism as a result of COVID-19 will be uneven in space and time".

Following Ackoff (1979), we describe the multiple interfaces between tourism and transport as *messy*. Ackoff's concept of 'messes', which he (1979: 99) frames as "dynamic situations that consist of complex systems of changing problems that interact with each other", helps disentangle tourism and transport interfaces during periods of uncertainty and disruption. Acknowledging a mess should lead to better policy, meaningful implementation and clearer outcomes.

Existing scholarship sufficiently addresses competition (e.g., Wang et al., 2020, Su et al., 2019, Zhang et al., 2019) and substitution (Koo et al. 2010) between transport modes, the complex range of behavioural variables associated with modal integration (e.g., Yuan et al., 2021) and how integration can contribute to ecological sustainability (see Potter and Skinner, 2000). For this chapter, however, the focus is on (1) government support/subsidies that seek to maintain mobility networks as well as (2) the existential crisis that climate change brings to both tourism and transport. Geographic research touching these two themes has certainly expanded in recent years to incorporate concepts relating to, for instance, social equity (e.g., Guzman et al., 2017, Bocarejo & Ovideo, 2012), affordability (e.g., Guzman & Ovideo, 2018), active transportation (e.g. Millward et al., 2013) and sustainable mobilities (e.g., Gössling & Cohen, 2014, Schwanen, 2019).

DOI: 10.4324/9781003286301-38

Messy: Post-pandemic Government Responses
Affecting Tourism Transport Networks

The COVID-19 pandemic hammered the tourism and transport sectors internationally. The International Air Transport Association's (IATA) (2021) review of 2020 showed a 70 per cent reduction in global RPKs (revenue passenger kilometres, a common measure of revenue-related output). Despite some evidence of strong demand in certain regions, bookings also decreased roughly 70 per cent (IATA 2021). Airports and support industries suffered. The Airports Council International (ACI) is not hopeful of a return to pre-pandemic levels before 2023 (ACI, 2021) in areas where domestic traffic is normally robust.

The presence, importance and role of government regulation in tourism and air transportation is well established (Duval, 2013). Airlines, in particular, are heavily regulated. Apart from the obvious safety parameters around which operations are undertaken, the more 'silent' or unseen regulatory oversights address market access and traffic freedoms (e.g., van Brandenburg-Kulkarni, 2015, Lee & Horton, 2018), ownership (Linzell, 2010) and designation (van Fenema, 2006). Both individually and together, these play a stronger role in the spatial scale and scope of tourism than perhaps any other set of variables. On the ground, even airport regulation, ranging from light-handed approach (e.g., Littlechild, 2012, see also Gillen, 2011) to more heavy-handed oversight (e.g., Phang, 2016), helps determine how national/international accessibility and connectivity are arranged.

It is within this regulatory backdrop that the post-pandemic response of governments worldwide can be situated and examined, especially in terms of the impact on international transportation geographies as well as the direction(s) of scholarly inquiry necessary to unpack associated complexities. Geographical research aimed at tourism finds itself at an inflection point in this regard. It will be called upon to synthesise critical (and fluctuating) values in the spatial (and socio-spatial) and time (e.g., Neutens et al., 2011) relevance of mobility, but also to incorporate constantly revising, government-led policies of access and protection. It will need to be cognisant of increasing calls for 'new normals' and regenerative efforts, but even these will require sensitivities to economic realities as governments at all levels seek to (re)build economies during a period of, at the time of writing at least, increased inflation and pronounced economic uncertainty. There exists, as ever, an inherent tension, still unresolved in some respects, between economic development boosted by inward tourism flows (and carried through international transport) and human-made ecological and medical disruptions. Such disruptions have proven in the past to be temporary. This was identified by Brouder (2020: 486), who writes: "Notwithstanding the changes to travel security (after 2001) and airline restructuring (after 2008), the long-term development of international tourism has continued without any real evidence of systemic change (i.e., more people continue to travel and spend more money than ever before)".

Ensuring Connectivity Through Capital Assistance

International transport providers suffer when governments, out of necessity, close borders. Mendes de Leon (2020) remarks that the COVID-19 crisis has allowed state governments to invoke airspace closures as a 'last resort', which is a feature of the 1944 Chicago Convention and through which most bilateral (even multilateral/plurilateral) access internationally is governed and overseen (see Macilree & Duval, 2020). Other modes are also impacted. For example, in recognising the importance of connectivity there have been calls for both the UK and French governments to ensure the Eurostar rail link between the two countries is maintained, despite significant losses and the added complexity of Brexit (Bloomberg, 2021).

Many major airlines worldwide required (or still require) some form of direct capital assistance and wage subsidies from their respective governments to ensure survival as a result of the pandemic. Some airlines faced difficulties paying lessors for leased or financed aircraft (Hanley, 2020). In a collapsed market, it falls to national governments to recapitalise (Abate et al., 2020) through capital aid in order to ensure mobility options are retained for their populations and vital tourist markets. Such public bailouts (direct or otherwise) introduce complexities with respect to capital markets (and the perceptions of the airline industry subsequently held by investors) and existing liquidity. As well, many international carriers do not have unrestricted access to global capital markets due to restrictions relating to air service arrangements, thus in almost all cases they are (and have been) dependent almost entirely on their own national governments.

Despite the near universal acknowledgement of the value of air transport, the question was raised yet again during the pandemic: Should public money be used to capitalise private transport firms? Claims of economic discrimination are not new (e.g., Piermartini and Rousova, 2013), particularly in relation to state aid. It cannot be ignored that financial assistance introduces complexities relating to industrial economics, notably the structure of competition. Ryanair argued in April 2021, for instance, that it would file an appeal to a recent ruling that aid received by SAS and Finnair by the EU (in the amount of US\$3 billion) was 'unfair' as it distorted competition (Simple Flying, 2021).

Public Service Obligations: A New Normal?

In some instances, government injections of liquidity can come with specific service requirements. For example, Air Canada was offered a series of loans amounting to C\$5.9 billion dollars in the first quarter of 2021 on the condition that certain routes to secondary and tertiary points in the country were reinstated (Government of Canada, n.d.). It is somewhat unusual (in recent years, anyway) for a Canadian government to impose service requirements on a flag carrier. Similar conditions have been seen elsewhere. A 12 February 2021 letter to the Chair of the Board of Air New Zealand from the New Zealand deputy prime minister and minister of finance argued that the Crown holds a majority shareholding in the airline because of "the importance of a strong domestic air travel network for economic and social development purposes" (NZX, 2021). This follows a capital raise that was signalled previously and also follows the New Zealand Government's established 'Maintaining International Air Connectivity' (NZ Government, 2021) policy in the face of the pandemic. With the amounts and frequency of state aid and/or subsidies flowing toward airlines, consideration must be given to what impact, if any, this will have on competitive structures and, ultimately, mobilities. These types of questions have been posed before (e.g. Wassenbergh, 1997, Bergamasco, 2015), but the rate and level of aid that has flowed (so far) in 2020/2021 worldwide is unprecedented. Truxal (2020) examines the framework of, and provision for, state aid in the EU with respect to airlines, noting that questions pertaining to a 'level playing field' arise:

With a view to maintaining a level playing field within the internal market, European state aid rules have evolved and modernized in 'better' times before COVID-19. Now, the crisis exposes an imbalance in public support offered by the different States. Take for instance the incredible volume of state aid provided by Germany and France. This is mirrored specifically in public support schemes for air transport: state recapitalization of privatized carrier Lufthansa; and state-backed loans for already state-owned Air France.

(Truxal, 2020: 80)

Network Mobilities in a Post-pandemic Era

Beyond financial requests for capital, what will the scale, scope and shape of networks and mobilities be like post-pandemic and, as discussed later in the chapter, how might these influence future research directions? The economic and operational value of hubbing has been explored in the last few decades (e.g., Dresner & Windle, 1995; and Lenaerts et al., 2021). Lanaerts et al (2021: 8) argue that

> connections to hubs bring more onward connectivity than connections to non-hubs. This effect is likely to be larger (i) for smaller airports that have fewer connections, so the capacity of each direct flight to generate onward connectivity matters more, and (ii) in long-haul markets where intercontinental destinations are often reached by indirect connections.

Will hubbing remain critical in terms of network design and revenue management? Wong et al (2019) found, for instance, that there was even evidence at the global level of "hub bypassing" during pre-pandemic times. Related to this is the precariousness of business models such as low-cost long-haul (LCLH) (Albers et al., 2020). LCLH models are inherently designed to bypass hubs, at least in the spirit of Wong et al's (2019) analysis, and by doing so they offer lengthy point-to-point flights. Point-to-point models are often associated with low-cost carriers (Alderighi et al., 2007). Bauer et al. (2020) posit that, post-COVID, ultra long-haul services (but not necessarily low cost) will accelerate, given that optimisation rather than scale will become an operational efficiency standard. Bauer et al. (2020: 7) suggest that "the key selling points of Ultra Long-Haul aviation projects appear to not only maintain the characteristics necessary to thrive after the COVID-19 era, but to become increasingly attractive as the industry attempts to return to some new form of normalcy". However, the format of the LCLH model in the post-pandemic short-term may change, especially when competition increases as airlines operating multiple business models offer capacity into markets in order to secure early first-mover advantages. The LCLH model may also be affected by the phasing out of extra large aircraft such as the A380 (following the A340 and B747), which was designed primarily as a hub-to-hub aircraft (e.g., Lohmann et al., 2009) and was instrumental in the recovery following the events of 11 September 2001 in the United States (King, 2007). Changes in LCLH delivery could ultimately reflect strategic operational decisions made by carriers due to inconsistent and variable demand, stronger economic rationale for point-to-point services with a small gauge of aircraft (to start), health policies at destinations that preclude 'mass' levels of tourism or multiple combinations and iterations of the above. These questions do not just affect LCLH models. The president of United Airlines indicated in April 2021 that profitability hinges on the return of business travel and long-haul travel (Wolfsteller, 2021).

Health, Liability and New Passports

For passengers, 'travel fear' was always a phenomenon. It is perhaps more amplified by the COVID-19 pandemic (see Zheng et al., 2021, Piccinelli et al., 2021) despite proactive measures taken by some public health organisations (Sousa Uva & Ratajczyk, 2020). Tuchen et al (2020) argue that the user experience, from a holistic perspective, will be at the centre of strategic planning going forward. That experience, however, undoubtedly contains health considerations.

Proof of vaccination has previously been necessary for travel to certain areas/countries of the world, but with COVID-19 the discussion of vaccine passports has been amplified and may serve to govern tourism mobilities more than other factors (e.g., economic conditions which influence demand, fares, competition, etc.). In 2022, the International Air Transport Association (IATA) has announced a Travel Pass designed to allow travellers to share information about vaccination and testing as well as receive information about health requirements before travelling. The International Civil Aviation Organization (ICAO) and various governments have also been active in considering vaccine passports

Vaccine passports are part of a wider framework of legal issues relating to transport and tourism. This has required a degree of international cooperation perhaps not yet seen. Naboush and Alnimer (2020), for example, review instances in which a carrier may be liable for transmission of COVID-19 and call for uniform application of established international standardised operation procedures. In acknowledging that aviation is largely responsible for the global distribution of COVID-19, Cassar (2020) reviews legal parameters around which air transport is regulated. He (Cassar, 2020: 14) notes that Standard 8.16 of the International Health Regulations of the World Health Organization requires that states introduce a national aviation plan, but suggests that,

> the fact that the aviation industry contracted as COVID-19 proliferated demonstrates that States did not have any national aviation plans in place, as required by this Standard. This can also be elicited from the hodgepodge of actions undertaken by various States across various continents.

Messier: The Legal Geography of Tourism Transport Emissions

While the COVID-19 pandemic is a serious concern, a comparatively longer-term issue is the negative externalities of tourism-related transport on climate. Academic treatments of the topic are robust, and the science is clear. Lee et al. (2021) highlight this starkly, with concern raised over non-CO_2 radiative forcings from aviation – such as contrails (which can have an impact three times greater than CO_2 alone) – not being included in either the Paris Agreement or in the EU Emissions Trading Scheme. How policy communities should respond is still under debate. As aviation was left out of the Paris negotiations in favour of allowing ICAO to continue to address the issue, and amid calls for coordinated collective action (Higham et al., 2019), precisely how the problem of aviation emissions can be addressed globally rests on complex matrices of regulation (international or national), technological (e.g., alternative fuels) and political communities (e.g., regional vs international approaches; comparative contribution of transport and infrastructure investment).

Early efforts by the European Union to address emissions from aviation resulted in the '2008 Directive' as a means of folding aviation into the existing EU Emissions Trading Scheme. Paragraph 16 of the Directive provides the rationale: "In order to avoid distortions of competition and improve environmental effectiveness, emissions from all flights arriving at and departing from Community aerodromes should be included from 2012" (European Commission, 2009). It was contentious. Chalifour and Besco (2016) point in particular to Erling's Introductory Note on the 2011 ruling by the European Court of Justice on Air Transport Association (US) et al. v. Secretary of State for Energy and Climate Change with respect to the compatibility of the Directive with international law. Erling (2012) notes that, while the Court found that the inclusion of aviation emissions in the EU ETS does not violate the three principles of customary international law, various states (China, Russia, the United States) still criticised the applicability of the Directive. The European Commission decided (European Commission, 2013) to defer further progression of enforcing the 2008 Directive and, instead, allow ICAO to address the issue internationally.

Proposed at the 39th ICAO Assembly in 2016, the result is ICAO's Carbon Offset and Reduction Scheme for International Aviation (CORSIA) programme as a market-based measure, government-led, international programme designed to curtail the environmental impacts of air transport. Using the base years of 2019 and 2020 (with flexibility to change as more states join the scheme), emissions that exceed this require offset costs (in the form of carbon credits) to be incurred by a participating airline operating a route between two participating States. Winchester (2019: 1) raises concerns about CORSIA's efficacy on pure economic grounds:

> From an industry point of view, the level of abatement is efficient when the cost of the last unit abated is equal to the price of offsets. However, as the costs of each airline's emissions are partially borne by other airlines under the CORSIA, the agreement will not incentivize airlines to implement all abatement options that cost less than the price of offsets until 2030 at the earliest and perhaps not at all.

Chalifour and Besco (2016: 596) offer similar arguments when they suggest that

> the system's initially voluntary nature and slow start suggests that additional action is needed to ensure the international aviation sector plays its part in contributing to the global goal of keeping average temperatures from rising beyond one and a half or two degrees Celsius.

Of course, CORSIA does not stand alone. Alternative 'green taxes' attached to transport have been utilised at the state level (e.g., the UK Air Passenger Duty). The Dutch Government introduced legislation in 2007 that required departing passengers to pay an €11.25 tax per passenger. Legal proceedings were initiated from plaintiff groups which included airlines, airports and travel agents (Havel & Van Antwerpen, 2009a). The substance of the legal complaint was that the tax violated Article 15 of the Chicago Convention, which speaks to non-discriminatory charges specified as fees, dues or other charges. The Dutch Government's perspective was that the tax itself is non-discriminatory in that it applied to all flights. The lower Court ultimately determined that this was indeed the case, and this was affirmed by the Dutch Supreme Court upon appeal (Havel and Van Antwerpen 2009b). Nevertheless, Havel and Van Antwerpen (2009b: 451) signaled a cautionary tale:

> The fact that, following the introduction of the Ticket Tax, nearly 1 million passengers rerouted their flights through airports outside The Netherlands is strong evidence that States should tread carefully when considering similar measures. This is particularly true for States whose neighbours have nearby airports.

The question thus remains: How do governments reconcile the parallel and equally prescient need to preserve economic structures and foundations with the obvious and increasing need to curtail negative environmental externalities? This is a legal, economic and philosophical question, with an added technological dimension.

Impressive incremental technological innovations have driven progress toward more sustainable air transport operations. Fuel use (and costs) have been reduced thanks to improvements in jet engine technology (Zheng and Rutherford, 2020, see also Peeters et al., 2005). This has had the added advantage of reducing noise and, based on recent evidence, a reduction in contrails (NASA, 2021). Investment and deployment of these new technologies, however, is contingent on airline financial health and/or the availability of direct government subsidies. In some instances, new technologies have been applied to entire aircraft, as is the case of Boeing's B737MAX. Safety-related issues have delayed its uptake more widely and have introduced questions regarding certification authority (see Nunn, 2020). Navigational performance has also been utilised to reduce emissions, as more direct tracks between points can be utilised (e.g., CANSO, 2020). More efficient air traffic management can reduce contrails, which have been shown to be particularly damaging (e.g., Lee et al., 2021). Broader utilisation requires investment in ground-based and satellite-based air transport infrastructure such as ATM/CNS. The use of sustainable aviation fuels (SAFs) is promising, but these require global adoption and coordination and must address cost and land-use concerns (e.g., Zhao et al., 2021). As the ICAO Secretary General noted in speech to the World Economic Forum in May 2021, "the move to SAFs must be a global transition, taking place in both developed and developing countries, effective knowledge sharing and capacity building will be just as essential to successful outcomes in some cases as the most advanced technological breakthroughs" (ICAO, 2021).

Alternative means of transportation continue to be championed, and sometimes on the basis that they bring ecological sustainable outcomes. For example, when Air France was provided with capital in the early phases of the COVID-19 pandemic in 2020, conditions were attached that called for a reduction/ban on domestic flights and the promotion of rail for journeys under 2.5 hours (Euronews 2021). Research by Reiter et al. (2022) in the German market indicates that emissions savings are indeed possible, although there is some risk of increases in travel times and thus a social cost. Indeed, high speed rail (HSR) is known to redistribute traffic in relation to airports (where they can be winners and losers depending on the level of connectivity) (Zhang et al., 2019).

The 2022 decision at the 41st Assembly of the ICAO to adopt a long-term aspiration goal of net-zero emissions for the sector builds on the previously discussed CORSIA scheme, and is a reflection that aviation and climate change has mainstream awareness. Indeed, it is difficult politically to ignore increasing attention directed to movements such as 'flight shaming'. The idea re-entered public space in some countries beginning 2019–2020 following initial appearances (primarily in the EU) in the mid-2000s. If successful, this drive to raise awareness of the ecological damage caused by aircraft to sensitive atmospheric layers may drive down demand (which is arguably the intent) and, consequently, transform the economic contribution of international tourism in some destinations. Calls for more regenerative tourism (Cave & Dredge, 2020; Bellato et al., Chapter 23 of this volume), however, may ameliorate that, with attention directed toward more responsible and sustainable (long term) tourism sectors. Another issue that requires untangling is the hesitancy in many developed countries over airport expansion. While such expansions would permit more efficient operations and positive net contributions to economies, they are opposed on the basis of direct ecological impacts and the optics of encouraging inefficient and unsustainable forms of transport. To put this in perspective, however, one can look at China – where there are construction plans for over 200 additional airports (*South China Morning Post*, 2020) – and India (e.g., *Economic Times*, 2019), where the construction of new airports could likely offset (real or optically) the gains made from holding back on expansion elsewhere.

Conclusions

As this chapter is being written in the immediate post-COVID times, the shape and trajectory of air transport's recovery is far from certain. If anything, COVID-19 represented yet another in a long line of significant exogenous shocks to modern airline operations. None has been more deep, nor longer lasting, than COVID-19, however. If anything can be concluded, it is that global transportation, and perhaps transportation at all levels, is perhaps far more fragile than originally thought. A subsequent consideration could perhaps be framed around the implications of this for long-term tourism and mobilities? What, for example, might the patterns of mobility look like three, five and ten years from today? In particular, how might/could geography attempt to rationalise and explain these shifting patterns? Multidisciplinary approaches (see also Pucci & Vecchio, 2019) will continue to be useful, for they stand the best chance of capturing inherent complexities with accuracy and vigour.

We see two directions for future research. First, shifting understandings of globalisation and the global legal framework(s) around which air transport mobility is shaped will increasingly deserve attention. This is not to suggest that spatial and temporal treatments of mobilities and their relationship to tourism and transport ought to be decommissioned, for they still hold their rightful place of knowledge relevance (see, for instance, Wang et al., 2018). Rather, recent movements toward de-globalisation give rise to re-shaping of transport networks; the re-shoring of some production (Delis et al., 2019) can result in less of an emphasis on transport of finished products and a somewhat diminished need for global connections. Further still, prevailing economic conditions will influence demand patterns for activities where transport networks are critical. COVID-19 rudely presented itself to the world with unprecedented consequences. It certainly impacted hyper-local transportation demand (e.g., Gramsch et al., 2022, Shortall et al., 2022), although the longer-term impact on global mobilities has still not been fully established.

A second direction for research would do well to incorporate a legal lens. While linkages to competitive pressures exacted by new forms of mobilities (e.g., high-speed rail, see Mizutani & Sakai, 2021, for instance) will assist with an understanding of the commercial side of mobility and transport, their uptake will be governed in large part by legal geographies of policy, implementation and sustainability. Despite the existence of a relatively healthy literature covering modal substitutions (e.g., Dobruszkes, 2011, Givoni & Dobruszkes, 2013), geographies of public policy will need to assess whether rivalrous consumption of transport one form or another ought to be privileged for environment, social and economic reasons, or some combination thereof, and in varying amounts, which respond to hyperlocal realities and situations.

Questions of sustainability (social and ecological) have been raised for years, but given the reset brought on by the COVID pandemic, calls for 'regenerative tourism' are coming a time when we also have 'vaccine tourism' initiatives from destinations such as Alaska and New York City (Newton 2021) and Russia (Moscow Times, 2021). We are seeing patterns of motivations for mobility and tourism never before experienced. It remains to be seen whether transportation can ever be functionally sustainable from an ecological perspective. Indeed, Hall et al (2020: 590) question whether anything will be learned from this crisis with respect to ongoing sustainability efforts: "While the COVID-19 crisis resulted in, quite literally, parked carbon assets in the form of cruise ships and aircraft, and a substantial drop in emissions, the possibilities of that being sustained, even at much more moderate levels, are doubtful." Forward-thinking guidance may cast doubt on changes to the status quo, but it is likely the relationship between transport and tourism will long be 'messy'.

Acknowledgement

This chapter was written in the memory of the late Professor Paul Wilkinson (York University, Canada): Geographer, mentor, friend.

References

Abate, M., Christidis, P. & Purwanto, A.J. (2020). Government support to airlines in the aftermath of the COVID-19 pandemic. *Journal of Air Transport Management* 89, 101931.

Ackoff, R.L. (1979). The future of operational research is past. *The Journal of the Operational Research Society* 30(2), 93–104.

Airports Council International. (2021). ACI World updates guidance on restart and recovery for airports. https://aci.aero/news/2021/03/16/aci-world-updates-guidance-on-restart-and-recovery-for-airports/ [accessed February 2024].

Albers, S., Daft, J., Stabenow, S. & Rundshagen, V. (2020). The long-haul low-cost airline business model: A disruptive innovation perspective. *Journal of Air Transport Management* 89, 101878.

Alderighi, M., Cento, A., Nijkamp, P. & Rietveld, P. (2007). Assessment of new hub-and-spoke and point-to-point airline network configurations. *Transport Reviews* 27(5), 529–549.

Bauer, L.B., Bloch, D., & Merkert, R. (2020). Ultra Long-Haul: an emerging business model accelerated by COVID-19. *Journal of Air Transport Management* 89, 101901.

Bergamasco, F. (2015). State subsidies and fair competition in international air services: The European perspective. *Issues in Aviation Law and Policy* 15(1), 29–75.

Bloomberg. (2021). *Keep the Eurostar running after Covid and Brexit.* Available at www.bloomberg.com/opinion/articles/2021-04-08/eurostar-is-worth-saving-after-covid-and-brexit-the-u-k-and-eu-should-agree [accessed February 2024].

Bocarejo, J.P. & Ovideo, D.R. (2012). Transport accessibility and social inequities: A tool for identification of mobility needs and evaluation of transport investments. *Journal of Transport Geography* 24, 142–154.

Brouder, P. (2020). Reset redux: Possible evolutionary pathways towards the transformation of tourism in a COVID-19 world. *Tourism Geographies* 22(3), 484–490.

CANSO. (2020). *Joint effort to deliver additional environmental benefits during the traffic downt*urn. Available at: https://canso.org/joint-effort-to-deliver-additional-environmental-benefits-during-the-traffic-downturn/ [accessed February 2024].

Cassar, R. (2020). Evolution or devolution: Aviation law and practice after COVID-19. *Air and Space Law* 45, 3–16.

Cave, J. & Dredge, D. (2020). Regenerative tourism needs diverse economic practices. *Tourism Geographies* 22(3), 503–513.

Chalifour, N. & Besco, L. (2016). Taking flight: Federal action to mitigate Canada's GHG emissions from aviation. *Ottawa Law Review* 48, 575–623.

Delis, A., Driffield, N. & Temouri, Y. (2019). The global recession and the shift to re-shoring: Myth or reality? *Journal of Business Research* 103, 632—643.

Dobruszkes, F. (2011). High-speed rail and air transport competition in Western Europe: A supply-oriented perspective. *Transport Policy* 18(6), 870–879.

Dresner, M. & Windle, R. (1995). Are US air carriers to be feared? Implication of hubbing to North Atlantic competition. *Transport Policy* 2(3), 195–202.

Duval, D.T. (2013). Critical issues in air transport and tourism. *Tourism Geographies* 15(3), 494–510.

Economic Times. (2019). *India plans to open 100 airports in five years.* Available at: https://economictimes.indiatimes.com/industry/transportation/airlines-/-aviation/india-plans-to-open-100-airports-in-five-years/articleshow/71821181.cms [accessed February 2024].

Erling, U.M. (2012). Introductory note to the court of justice of The European Union: Air Transport Association Of America (Ata) et al. *v.* secretary of state for energy & climate change. *International Legal Materials* 51(3), 535–562.

Euronews. (2021). *French lawmakers approve ban on short domestic flights to cut carbon emissions.* Available at: www.euronews.com/2021/04/12/french-lawmakers-approve-ban-on-short-domestic-flights-to-cut-carbon-emissions?utm_source=Twitter&utm_medium=Social [accessed February 2024].

European Commission. (2009). *Directive 2008/101/EC of the European Parliament and of the Council of 19 November 2008 amending Directive 2003/87/EC so as to include aviation activities in the scheme for greenhouse gas emission allowance trading within the Community,* OJ L 8, 13.1.2009, 3–21.

European Commission. (2013). *Decision No 377/2013/EU of the European Parliament and of the Council of 24 April 2013 derogating temporarily from Directive 2003/87/EC establishing a scheme for greenhouse gas emission allowance trading within the Community,* OJ L 113, 25.4.2013, 1–4.

Gillen, D. (2011). The evolution of airport ownership and governance. *Journal of Air Transport Management* 17, 3–13.

Givoni, M. & Dobruszkes, F. (2013). A review of ex-post evidence for mode substitution and induced demand following the introduction of high-speed rail. *Transport Reviews* 33(6), 720–742.

Gramsch, B., Angelo Guevara, C., Munizaga, M., Schwartz, D. & Tirachini, A. (2022). The effect of dynamic lockdowns on public transport demand in times of COVID-19: Evidence from smartcard data. *Transport Policy* 126, 136–150.

Gössling, S. & Cohen, S. (2014). Why sustainable transport policies will fail: EU climate policy in the light of transport taboos. *Journal of Transport Geography* 39, 197–207.

Guzman, L.A. & Oviedo, D. (2018). Accessibility, affordability and equity: Assessing 'pro-poor' public transport subsidies in Bogotá. *Transport Policy* 68, 37–51.

Guzman, L.A., Oviedo, D. & Rivera, C. (2017). Assessing equity in transport accessibility to work and study: The Bogotá region. *Journal of Transport Geography* 58, 236–246.

Hall, C.M., Scott, D. & Gössling, S. (2020). Pandemics, transformations and tourism: Be careful what you wish for. *Tourism Geographies* 22(3), 577–598.

Hanley, D. (2020). COVID-19 and International aircraft financing law. *Air and Space Law* 45, 155–172.

Havel, B. & van Antwerpen, N. (2009a). The Dutch ticket tax and article 15 of the Chicago convention, *Air and Space Law* 34(2), 141–146.

Havel, B. & van Antwerpen, N. (2009b). Dutch ticket tax and article 15 of the Chicago convention (Continued). *Air and Space Law* 34(6), 447–451.

Higham, J., Ellis, E. & MacLaurin, J. (2019). Tourist aviation emissions: A problem of collective action. *Journal of Travel Research* 58(4), 535–548.

International Air Transport Association. (2021). *COVID-19 Weak year-end for air travel and outlook is deteriorating.* Available at: www.iata.org/en/iata-repository/publications/economic-reports/weak-year-end-for-air-travel-and-deteriorating-outlook/ [accessed February 2024].

International Civil Aviation Organisation (ICAO). (2021). *ICAO addresses WEF Ministerial meeting on sustainable aviation fuels.* Available at: www.icao.int/Newsroom/Pages/ICAO-addresses-WEF-ministers-meeting-on-sustainable-aviation-fuels.aspx [accessed February 2024].

King, J.M.C. (2007). The Airbus 380 and Boeing 787: A role in the recovery of the airline transport market. *Journal of Air Transport Management* 13, 16–22.

Koo, T.T.R., Wu, C.-L., & Dwyer, L. (2010). Transport and regional dispersal of tourists: Is travel modal substitution a source of conflict between low-fare air services and regional dispersal? *Journal of Travel Research* 49(1), 106–120.

Lee, D.S., Fahey, D.W., Skowron, A., et al. (2021). The contribution of global aviation to anthropogenic climate forcing for 2000 to 2018. *Atmosphere Environment* 244, 117834.

Lee, J.W. & Horton, W. (2018). Fifth freedom traffic rights: Formidable threat or theoretical concern? *Air and Space Law* 43(3), 303–318.

Lenaerts, B., Allroggen, F. & Malina, R. (2021). The economic impact of aviation: A review on the role of market access. *Journal of Air Transport Management* 91, 102000.

Linzell, S. (2010). Ownership and control restrictions in US aviation law. *Air and Space Law* 35(6), 379–407.

Littlechild, S.C. (2012). Australian airport regulation: Exploring the frontier. *Journal of Air Transport Management* 21, 50–62.

Lohmann, G., Albers, S., Koch, B. & Pavlovic, K. (2009). From hub to tourist destination – An explorative study of Singapore and Dubai's aviation-based transformation. *Journal of Air Transport Management* 15, 205–211.

Macilree, J. & Duval, D.T. (2020). Aeropolitics in a post-COVID-19 world. *Journal of Air Transport Management* 88, 101864.

Mendes de Leon, P .(2020). National reflexes following the COVID-19 outbreak: Is sovereignty back in the air? *Air and Space Law* 45, 17–38.

Millward, H., Spinney, J. & Scott, D. (2013). Active-transport walking behavior: Destinations, durations, distances. *Journal of Transport Geography* 28, 101–110.

Mizutani, J. & Sakai, H. (2021). Which is a stronger competitor, high speed rail, or low cost carrier, to full service carrier? – Effects of HSR network extension and LCC entry on FSC's airfare in Japan. *Journal of Air Transport Management* 90, 101965.

Moscow Times. (2021). *Putin wants 'Vaccine tourism' plan by End-June.* Available at: www.themoscowtimes.com/2021/06/07/putin-wants-vaccine-tourism-plan-by-end-june-a74117 [accessed February 2024].

Naboush, E. & Alnimer, R. (2020). Air carrier's liability for the safety of passengers during COVID-19 pandemic. *Journal of Air Transport Management* 89, 101896.

NASA. (2021). *NASA-DLR study finds sustainable aviation fuel can reduce contrails*. Available at: www.nasa.gov/press-release/nasa-dlr-study-finds-sustainable-aviation-fuel-can-reduce-contrails [accessed February 2024].

Neutens, T., Schwanen, T., & Witlox, F (2011). The prism of everyday life: Towards a new research agenda for time geography, *Transport Reviews* 31(1), 25–47.

Newton, J. (2021). A Global Search for Vaccines Fuels a Different Kind of Tourism. *Afar*. Available at: www.afar.com/magazine/vaccine-tourism-is-on-the-rise [accessed February 2024].

New Zealand Government. (2021). Maintaining International Air Connectivity scheme announced. Available at: www.transport.govt.nz/area-of-interest/air-transport/government-support-for-the-transport-sector/ [accessed February 2024].

Nunn, D.H. (2020). Grounded: How the 737 MAX crashes highlight issues with FAA delegation and a potential remedy in the federal tort claims act. *Journal of Air Law and Commerce* 85, 703–731.

NZX. (2021). *Letter from Crown on Air NZ capital raise (UPDATE)*. Available at: www.nzx.com/announcements/367462 [accessed February 2024].

Peeters, P.M., Middel, J & Hoolhoost, A. (2005). *Fuel efficiency of commercial aircraft: An overview of historical and future trends*. National Aerospace Laboratory NLR-CR-2005-669. Available at: www.transportenvironment.org/uploads/files/2005-12_nlr_aviation_fuel_efficiency.pdf [accessed February 2024].

Phang, S.Y. (2016). A general framework for price regulation of airports. *Journal of Air Transport Management* 51, 39–45.

Piccinelli, S., Moro, S., & Rita, P. (2021). Air-travelers' concerns emerging from online comments during the COVID-19 outbreak. *Tourism Management* 85, 104313.

Piermartini, R. & Rousova, L. (2013). The sky is not flat: How discriminatory is the access to international air services? *American Economic Journal: Economic Policy* 5(3), 287–319.

Potter, S. & Skinner, M.J. (2000). On transport integration: A contribution to better understanding. *Futures* 32, 275–287.

Pucci, P. & Vecchio, G. (2019). Trespassing for mobilities. Operational directions for addressing mobile lives. *Journal of Transport Geography* 81, 102536.

Reiter, V., Voltes-Dorta, A. & Suau-Sanchez, P. (2022). The substitution of short-haul flights with rail services in German air travel markets: A quantitative analysis. *Case Studies in Transport Policy* 10, 2025–2043.

Schwanen, T. (2019). Transport geography, climate change and space: Opportunity for new thinking, *Journal of Transport Geography* 81, 102530.

Shortall, R., Mouter, N. & Van Wee, B. (2022). COVID-19 passenger transport measures and their impacts, *Transport Reviews* 42(4), 441–466.

Simple Flying. (2021). *Ryanair set to appeal against the EU's decision on state aid*. Available at: https://simpleflying.com/ryanair-eu-state-aid-appeal/ [accessed February 2024].

Sousa Uva, R.S. & Ratajczyk, M. (2020). COVID-19 pandemic and the measures taken by the European Union aviation safety agency. *Air and Space Law* 45, 95–106.

South China Morning Post. (2020). *China forges ahead with airport construction binge, despite signs of slowing air traffic growth*. Available at: www.scmp.com/economy/china-economy/article/3074291/china-forges-ahead-airport-construction-binge-despite-signs [accessed February 2024].

Su, M., Luan, W. & Sun, T. (2019). Effect of high-speed rail competition on airlines' intertemporal price strategies. *Journal of Air Transport Management* 80, 101694.

Truxal, S. (2020). State Aid and Air Transport in the Shadow of COVID-19. *Air and Space Law* 45, 61–82.

Tuchen, S., Arora, M. & Blessing, L. (2020). Airport user experience unpacked: Conceptualizing its potential in the face of COVID-19. *Journal of Air Transport Management* 89, 101919.

van Brandenburg-Kulkarni, P. (2015). 'Sixth freedom' revisited in the Twenty-first century. *Air and Space Law* 40(1), 55–64.

Van Fenema, P. (2006). EU horizontal agreements: Community designation and the 'free rider' clause. *Air and Space Law* 31(3), 172–195.

Wang, D., Niu, Y. & Qian, J. (2018). Evolution and optimization of China's urban tourism spatial structure: A high speed rail perspective. *Tourism Management* 64, 218–232.

Wang, J., Huang, J. & Jing, Y. (2020). Competition between high-speed trains and air travel in China: From a spatial to spatiotemporal perspective. *Transportation Research Part A* 133, 62–78.

Wassenberg, H. (1997). The regulation of state-aid in international air transport. *Air and Space Law* 22(3), 158–165.

Winchester, N. (2019). A win-win solution to abate aviation CO2 emissions. *Journal of Air Transport Management* 80, 101692.

Wolfsteller, P. (2021). *Profitability will return with business and long-haul travel: United's Kirby, Flight Global.* www.flightglobal.com/strategy/profitability-will-return-with-business-and-long-haul-travel-unit eds-kirby/143535.article [accessed February 2024].

Wong, W.G., Cheung, T., Zhang, A. & Wang, Y. (2019). Is spatial dispersal the dominant trend in air transport development? A global analysis for 2006–2015. *Journal of Air Transport Management* 74, 1–12.

Yuan, Y., Yang, M., Feng, T., et al. (2021). Heterogeneity in passenger satisfaction with air-rail integration services: Results of a finite mixture partial least squares model. *Transportation Research Part A* 147, 133–158.

Zhang, A., Wan, Y., & Yang, H. (2019). Impacts of high-speed rail on airlines, airports and regional economies: A survey of recent research. *Transport Policy* 81, A1–A19.

Zhao, X., Taheripour, F., Malina, R., et al. (2021). Estimating induced land use change emissions for sustainable aviation biofuel pathways. *Science of the Total Environment* 779, 146238.

Zheng, D., Luo, Q. & Ritchie, B.R. (2021). Afraid to travel after COVID-19? Self-protection, coping and resilience against pandemic 'travel fear'. *Tourism Management* 83, 104261.

Zheng, X.S. & Rutherford, D. (2020). *Fuel burn of new commercial Jet aircraft: 1960 to 2019, White Paper, International Council on Clean Transportation.* Available at: https://theicct.org/sites/default/files/publi cations/Aircraft-fuel-burn-trends-sept2020.pdf [accessed February 2024].

33

TIME GEOGRAPHY AND TOURISM

Noam Shoval

Introduction

Sightseeing, walking, shopping, and sitting in restaurants and cafés are widely recognised as the major activities undertaken by tourists. Although these activities appear to be clearly set in time and space, so far relatively little attention has been paid to visitor mobility within the fields of human geography and tourism research (Shaw et al., 2000).

The dearth of research that exists on this subject can be attributed primarily to the methodological complexity involved in studies of this kind. First, it is difficult to locate the tourist when he or she enters and leaves a city or region, due to the absence of defined entry and exit points. Second, the term 'tourist' includes a wide variety of different types of tourists that are distinguishable from one another by their interests, the purpose of their visits, and their time budgets, among other factors, so that in order to illustrate the tourist's spatial behaviour, different types of tourists in the region must first be identified. Third, the funding requirements for such surveys have "restricted the wide implementation of empirical research on tourists' time-space activities" (Forer, 2002: 24).

In this chapter, I argue that a better understanding of the logic of visitor activities in time and space could not only serve a number of practical purposes in tourism industries, planning, and management, but also develop the existing concept of time geography and considerably enlarge the theoretical foundations of tourism research.

Time Geography

Time geography, which focuses on the constraints and trade-offs that occur when people find themselves having to divide a limited amount of time between various activities in space, is one of the earliest analytical perspectives used to analyse patterns of human activity (Miller, 2005: 17). Time geography was the brainchild of the Swedish geographer Torsten Hägerstrand, who in the 1960s, 1970s, and early 1980s, together with his associates at the University of Lund (known collectively as the Lund School) developed time geography's basic tenets (Gregory, 2000: 830).

Beyond Sweden's borders, researchers such as Allan Pred, Nigel Thrift, and Anthony Giddens helped with the international diffusion of time-geographic thought. In particular, Giddens, with his structuration theory and thoughts on space-time, made time geography known to a wider circle of researchers (Lenntorp, 1999: 57). As a result, analysis of human activities in space-time

DOI: 10.4324/9781003286301-39

burgeoned, not only amongst geographers but amid transport researchers as well (Kwan, 2004). During the 1990s, however, interest in the field (at least among geographers) gradually faded, while in transport studies there were a number of 'activity based analysis' projects that explicitly drew upon time-geographic notions (e.g. Kitamura et al., 1990).

As time has passed, attitudes towards time geography have been somewhat ambivalent. It was praised for its representational capabilities and for offering a conceptual and methodological basis for empirical research. It has also, however, been strongly criticised for being too 'physicalistic', 'reductionistic', 'objectivistic' and 'masculinistic' (Gren, 2001: 209). Giddens, having to a certain extent tempered his initial enthusiasm, questioned time geography's 'naïve and defective conception of the human agent', which meant that it would often consider individuals 'independently of the social settings, which they confront in their day-to-day activities' (Giddens 1984). Time geography, he concluded, is no more than 'a weakly developed theory of power' (for various critiques of time geography, see also in Harvey, 1989: 211–213; and a summary of the critiques in Gregory, 2000: 832–833. For some responses to those critiques, see Gregory 1994: 245–254; Kwan, 2002: 653–654).

The past two decades have seen resurgence in time-geographic studies. The spread of increasingly sophisticated geographical information systems (GIS) and AI-based machine learning tools, capable of providing detailed computational representations of very large datasets and much more precise measurements of basic time-geographic entities, including space-time paths and prisms, have persuaded a growing number of researchers to return to the time geography fold (Mieli et al., 2024; Miller, 2005). The development of new digital information technologies and platforms, such as mobile smart phones, wireless Personal Digital Assistants (PDAs), Location Based Services (LBS), social media, Global Positioning System (GPS) receivers, and radiolocation methods, multiplied the volume and improved the spatio-temporal resolutions of the empirical data obtained to a degree previously unimagined and, thus, aided the recent revival of the field (Miller, 2005). For more on these technologies' contributions to geographic research, see also Kwan (2004), Shoval and Isaacson (2006) and Roelofsen, Chapter 25 of this volume).

Hägerstrand's concept of time geography captured the way in which space and time constrains individuals in their day-to-day lives (Hägerstrand, 1970). As time geography has it, human activities occur in specific locations and for limited time periods only. Thus, transport systems, by enabling people to travel from one place (activity) to another, allow them to make more efficient use of their time-space restrictions by trading time for space. According to Hägerstrand, an individual's ability to move from one location (activity) to another is dependent upon the combined effect of three types of constraints.

(1) Capability constraints, which include such things as: the need for a minimum amount of sleep, which, of course, limits the amount of time available for travelling; the kind of transport used – bicycle, car, or train – which marks the boundaries of the territory within which people pursue their daily life paths.
(2) Coupling constraints, which is the need to meet and team up with other individuals or groups at particular locations and for set time periods. This, naturally, limits an individual's ability to participate in activities taking place in other locations.
(3) Authority constraints, defined as the right and ability of sundry public or private authorities to limit or regulate access to various locations at different points in time.

The specific constraints that are relevant in the case of tourists will be discussed later in this chapter.

Visualisation of Space-time Paths

One of Hägerstrand's most profound professional and disciplinary achievements was the ability to represent space and time in a single diagram (Gren, 2001) unlike an ordinary map, but rather like a snapshot, that reproduces a moment frozen in time. The result was his now famous time-geographic diagrams: notational (representational) systems, which formed the basis of much of the subsequent work in the field of time geography, particularly in the realm of analysis and interpretation. The diagrams (Figure 33.1, adapted from Gregory, 1989) consist, as a rule, of two axes: a time-axis and a space-axis, thus making it possible to trace in graphic terms individual time budgets. The effective range of each person is described by a prism, or a series of prisms, whose

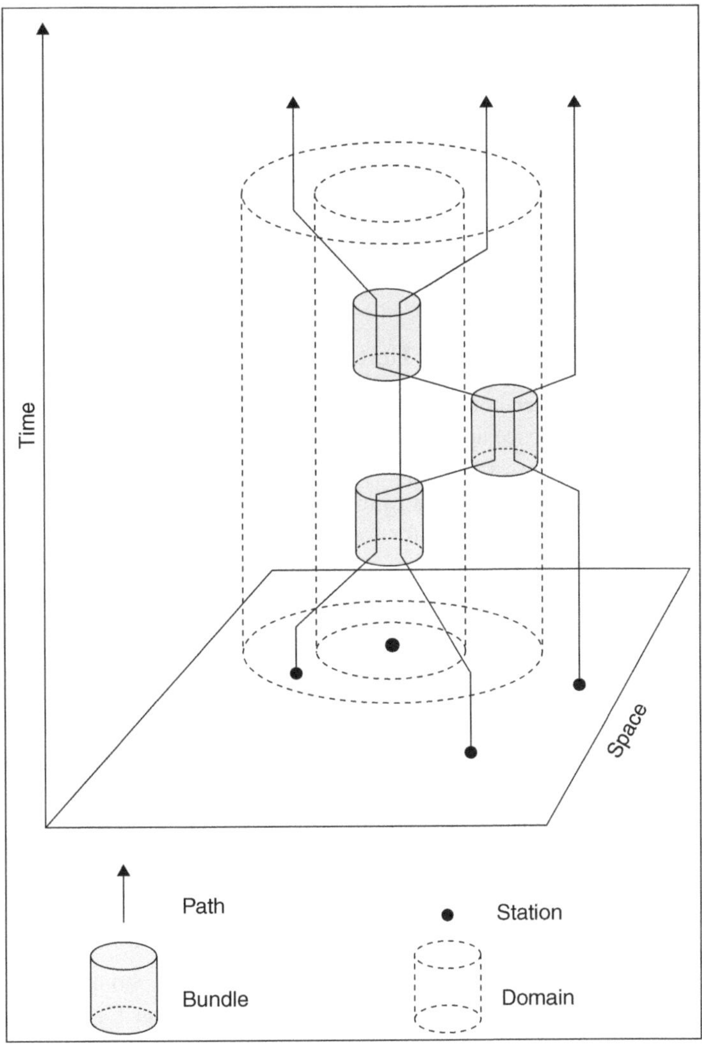

Figure 33.1   Hägerstrand's time-geographical diagram.
Source: Compiled by authors.

shape is dependent upon the aforementioned capability constraints. Hence, every pause, regardless of the activity involved, will cause the prism's (or sub-prism's) range to shrink in direct proportion to the time spent at said stop. But there are also other wider structural features specific to the social systems within which individuals operate which, as has long been recognised, help shape people's time budgets and activity patterns.

As long as most of the work in time geography was primarily theoretical, there was little problem in presenting time-space activity in such a manner. Indeed, the majority of studies published in the 1970s and 1980s presented one, two or, at the most, four sequences (see, for example, Parkes & Thrift, 1980). The method was until recently less suited for empirical research purposes, where there is often a need to incorporate a large number of sequences on the dual axis. In such cases, several problems arise: Some visual, others analytical. To begin with, including a great many sequences within a single graph results in an end product, this resembles a particularly chaotic junction. This creates severe visual difficulties, as it is virtually impossible to distinguish one strand from the other. Then there is the fact that it is difficult if not impossible to construct, using this method, a typology of sequences based on the data amassed. In other words, one is left with a lot of data and information, but without the ability to generalise. One method to tackle this challenge by using sequence alignment methods was presented by Shoval and Isaacson (2007b). More recently, Mieli et al. (2024) incorporated the concept of 'phygitality' into time geography research on tourism contexts, in order to understand how digital technologies are used in physical space, and how the interaction between the physical and the digital reconfigures tourists' projects, paths, bundles and constraints.

Potential Contribution of Time Geography to Tourism Research

The second part of this chapter commences with presenting the adaptation of time geography constraints for tourists' time-space activity:

(1) Capability constraints, the need for a minimum amount of sleep and food, which of course limits the amount of time available for travelling and touring; the kind of transport used – bicycle, rented car, train or aeroplane – which marks the boundaries of the territory within tourists' pursuit of their trips. In the case of tourists, the length of stay in the destination varies and therefore the length of visit is a constraining factor that changes the spectrum of opportunities for the tourists.

(2) Coupling constraints, which constitute the need to meet and team up with other individuals in the visitor group or, in the case of an organised group, at particular locations and for set time periods. The spatial activity of individual tourists and the geographic range of their activities in a destination will be completely different from that of organised groups, as they are personally responsible for selecting the particular tourist sites to be visited. This is the most absolute distinction between types of tourists in terms of space-time activity. The differences between these two types stem primarily from the 'rigidity' of the organised groups' itineraries in contrast with the greater freedom the individual tourist experiences when choosing sites to visit.

(3) Authority constraints – for example the existence of opening hours for museums and visitor attractions, train timetables. However, the central parameter for segmenting tourists into types, employed by national and international agencies and in academic research, is the 'main purpose of the visit' (Page & Hall, 2003). *Purpose of visit* has a direct impact on the spectrum of possibilities available to the tourist: tourists who travel for business or to visit friends and

relatives will be less likely to visit tourist sites than tourists who travel for the specific purpose of touring and sightseeing.

The 'constraining' variables and the principles underlying the spatial activity of local residents and tourists are very different. This stems from the fact that the activity of the local resident tends to be the outcome of long-term decisions, whereas the activity of the tourist tends to originate from short-term decisions. The day-to-day activity of the local resident is determined by decisions regarding career choices, selection of workplace, place of residence, and various other factors. Such decisions are largely influenced by the individual's education, age, gender, and income level. For example, factors such as gender and income, which are important in explaining the spatial activity of individuals in everyday life (Hanson & Hanson, 1981), will be less meaningful in explaining the activity of these same individuals in a destination in which they visit.

In contrast to the local resident, the tourist is unencumbered by such long-term decisions because, as implied by his or her tourist identity, he or she will stay in the city only for a limited duration. The tourist does not need to take care of a household – this is taken care of by the staff of the hotel or, alternatively, by the family he or she is staying with. If children are taken along for the trip, they do not need to attend school or other educational frameworks and, therefore, the family is free to plan visits to tourist sites that will satisfy all members of the family or tour group.

Debbage (1991: 266) found that the major determinants of spatial behaviour were temporal constraints (length of stay, mobility levels) and the spatial structure of the resort environment. However, Debbage found that some of the more traditional explanations for spatial behaviour did not seem to be applicable in his case study of Paradise Island (Bahamas). For example, factors that were not significant included the socioeconomic characteristics of the individual (income, education, age, and occupation), the type of travel arrangement (packaged tour versus independent traveller), familiarity levels (first-time or repeat visitors), and party size. Part of the reasoning for this may be because research in other fields (intra-urban commuting patterns, consumer shopping behaviour, and residential location decisions) may not be directly transferable to tourist behaviour (Debbage, 1991: 266).

It is possible to use this approach in order to explain a tourist's spatial activity, though many of the factors existing for the tourist differ from those of the individual in his or her natural surroundings. In addition, in contrast with Hägerstrand's model, the tourist has much choice and a lot of free time, and therefore via the tourist activity expresses more of their cultural background and personality and less of the 'traditional' factors as understood by Hägerstrand.

In place of these factors there are other constraints that result from the length of the stay or the primary purpose of the visit. It would not be surprising if we found that an entire range of socioeconomic factors were of no importance whatsoever for the explanation of a tourist's spatial activity. It is indeed possible to relate to constraints such as income, gender, and age as factors that are influential on the person in his or her natural surroundings; however, upon leaving the natural surroundings for a short time, the person may be freed from the limitations these factors place on him or her.

In tourism research, the application of Hägerstrand's theoretical and pragmatic framework for visualisation and analysis of time-space activities (through, for example, space-time budgets of tourists) was only scarcely done despite its clear relevance to the field (see also Hall 2012 in Chapter 21 of this volume). Some exceptions are Dietvorst's (1994; 1995) works on tourism in historic cities and Knaap's (1999) study of tourists' activity patterns in two national parks in the Netherlands; Shoval's works in the historic city of Akko in Israel (2008); Forer's (2002) implementation regarding flows of visitors to New Zealand; Hardy's (2020) analysis of visitors flows in Tasmania and, especially, Hall's (2005) conceptual work on the application of Hägerstrand's thought to tourism research are all good examples of the potential of this framework for tourism studies.

Agenda for Advancing the Field in the Future

New methods for data collection and data analysis have enormous potential for increasing time-space studies in tourism research and, naturally, those studies will make use of the theoretical and visualisation advantages of time geography.

New Methods for Data Collection and Big Data Analysis

Relatively little attention has been paid to the spatial and temporal behaviour of tourists despite the proliferation of research in tourism over the past few decades. This is probably related to the various problems associated with the traditional methods employed in the gathering of information on tourists' spatial and temporal behaviour. The most common problems relate to the level of accuracy and/or the validity of the data collected. This dearth of research is especially troubling, given that it is widely recognised that the movement of tourists has profound implications for infrastructure and transport development, tourism product development, marketing strategies, the commercial viability of the tourism industry, and the management of the social, environmental, and cultural impacts of tourism. Past research has focused primarily on the movement of tourists between destinations or from source markets to destination areas, applying concepts of distance decay, market access, and the valuation of time.

Methodological problems have prevented most researchers from undertaking similar studies of smaller areas, such as urban destinations. In recent years, the rapid development and availability of small, cheap and reliable tracking devices has led to a growing volume of spatial research in general and in tourism studies in particular (see C.M. Hall's chapter in the first edition of this handbook). Global Positioning System (GPS) devices offer researchers the opportunity of continuous and intensive high-resolution data collection in time (seconds) and space (metres) for long periods of time; this was never before possible in spatial research (Shoval & Isaacson, 2010; Hardy, 2020). In recent years a growing body of work has demonstrated the potential and efficacy of using tracking technologies and big data analysis to explore leisure and tourist activities, see for example, Ahas et al., 2007; 2008; Arrowsmith & Chhetri, 2003; Chen et al., 2022; Harder et al., 2008; Modsching et al., 2008; Shoval, 2007; Shoval & Isaacson, 2007a; 2007b; 2010; Shoval & Ahas, 2016; Spek, 2008).

New Methods for Data Analysis that Enable Time Geography Quantitative Analysis

The introduction of geographic information systems (GIS) in tourism research about a decade ago created a spectrum of opportunities for handling time-space data and enabled advanced analysis of time-space data. However, one fundamental problem in space-time analysis is the aggregation of space-time paths to create generalised types composed of varied activities in order to identify patterns fashioned on a quantitative basis while taking into account the sequential element. This problem could not be dealt with using GIS software. Previous attempts with quantitative pattern aggregation methods, mainly by transport researchers (for example see Schlich & Axhausen, 2003) did not manage to tackle the issue of the sequential element. Understanding the sequence of activities in space and time allows one to understand an additional integral dimension of activity and to recognise patterns that exist within this dimension.

One example of analysis that has very promising potential for creating typologies of tourists based on their spatial behaviour while taking into account the sequence of locations can be seen in sequence alignment methods (SAM). These methods, which have been used since the 1980s, were introduced to the social sciences by Abbott (1995) and Wilson (1998) and to the spatial sciences by Shoval and Isaacson (2007b), Shoval et al. (2015), Choe et al. (2022) and Wilson (2008). These methods, which have developed with time and have been refined to more accurately compare sequences, have tremendous potential as tools for creating typologies of tourists by analysing their spatial activity. Figure 33.2 presents the outcome of such analysis that identified a typical group of visitors to the Old City of Akko in Israel. It demonstrates the average time-space path of a group of visitors and is calculated taking into account the time and the order in which they visited the different parts of the city.

The introduction of tracking devices, including GPS (Global Positioning Systems) and analytic software, such as GIS software (Geographical Information Systems) and sequence alignment methods (SAM) have created new opportunities to obtain and analyse accurate information on the time-space movements of tourists at a detailed level. It has much potential; not least in opening up

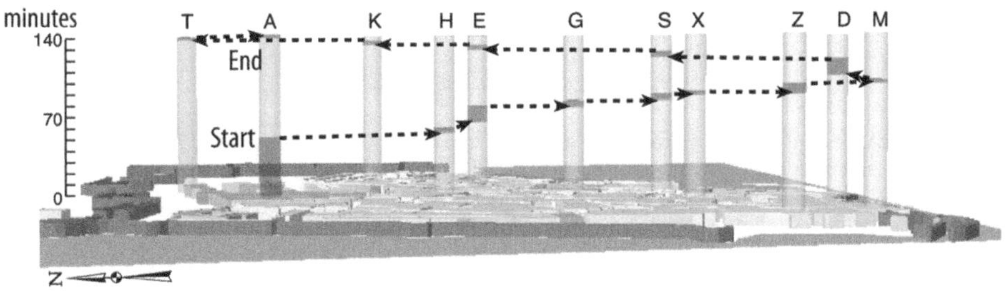

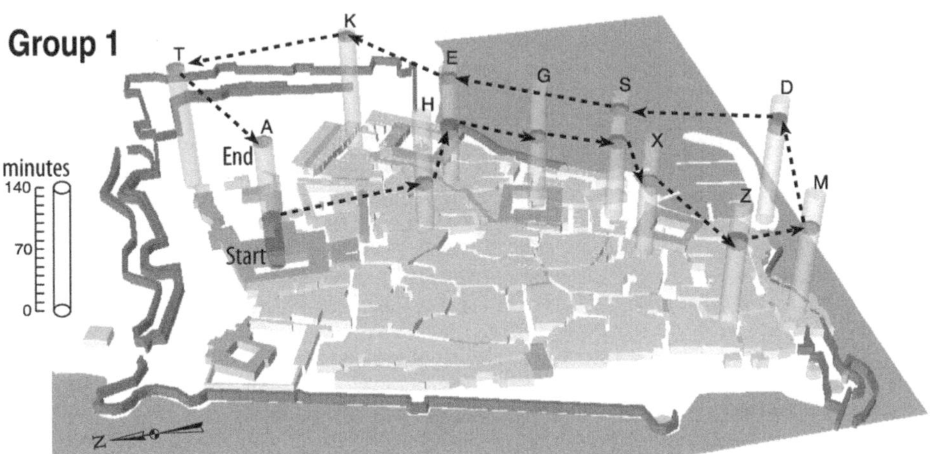

Figure 33.2　A group of visitors to the Old City of Akko in Israel as defined by SAM.

Source: Compiled by authors.

new and previously unfeasible lines of inquiry in the field of geography in general, but also that of time geographies of tourism in particular.

References

Abbott, A. (1995). Sequence analysis: New methods for old ideas. *Annual Review of Sociology* 21, 93–113.

Ahas, R., Aasa, A., Mark, U., Pae, T., & Kull, A. (2007). Seasonal tourism spaces in Estonia: Case study with mobile positioning data, *Tourism Management* 28(3), 898–910.

Ahas, R., Aasa, A., Roose, A., Mark, U., & Silm, S. (2008). Evaluating passive mobile positioning data for tourism surveys: An Estonian case study. *Tourism Management* 29(3), 469–86.

Arrowsmith, C. & P. Chhetri. (2003). *Port Campbell National Park: Patterns of Use. A report handed to parks Victoria visitor research*, Parks Victoria.

Chen, J., Becken, S., & Stantic, B. (2022). Harnessing social media to understand tourist mobility: The role of information technology and big data. *Tourism Review* 77(4), 1219–1233.

Choe, Y., Lee, C-K., Choi, J., Kim, M., & Sim, K.-W. (2022). Identifying tourist spatial and temporal patterns using GPS and sequence alignment method. *Journal of Travel Research* 62(6), 1181–1201. https://doi.org/10.1177/00472875221127685

Debbage, K. (1991). Spatial behavior in a Bahamian resort. *Annals of Tourism Research* 18, 251–68.

Dietvorst, A.G.J. (1994). Cultural tourism and time-space behavior. In G. Ashworth, & P. Larkham (Eds.), *Building a New Heritage: Tourism, Culture and Identity in the New Europe* (pp. 69–89). Routledge.

Dietvorst, A.G.J. (1995). Tourist behavior and the importance of time-space analysis. In G. Ashworth, & A.G.J. Dietvorst (Eds.), *Tourism and Spatial Transformations* (pp. 163–181). CAB International.

Forer, P. (2002). Tourist flows and dynamic geographies. In D.G. Simmons, & J. Fairweather (Eds.), *Understanding the Tourism Host-Guest Encounter in New Zealand: Foundations for Adaptive Planning and Management* (pp. 21–56). EOS Ecology.

Giddens, A. (1984). *The Constitution of Society: Outline of the Theory of Structuration.* University of California Press.

Gregory, D. (1989). Areal differentiation and post-modern human geography. In D. Gregory, & R. Walford (Eds.), *Horizons in Human Geography* (pp. 67–96). Barnes and Noble Books.

Gregory, D. (1994). *Geographical Imaginations.* Blackwell.

Gregory, D. (2000). Time-geography. In R.J. Johnston, D. Gregory, G. Pratt, & M. Watts (Eds.), *The Dictionary of Human Geography* (pp. 830–833). Blackwell.

Gren, M. (2001). Time-geography matters. In J. May, & N. Thrift (Eds.), *Timespace: Geographies of Temporality* (pp. 208–225). Routledge.

Hägerstrand, T. (1970). What about people in regional science? *Papers of the Regional Science Association* 24(1), 7–21.

Hall, C.M. (2005). Reconsidering the geography of tourism and contemporary mobility, *Geographical Research* 43(2), 125–139.

Hall, C.M. (2012). Spatial analysis: A critical tool for tourism geographies. In J. Wilson (Ed.), *The Routledge Handbook of Tourism Geographies* (pp. 163–173). Routledge.

Hanson, S. & Hanson, P. (1981). Travel-activity patterns of urban residents: Dimensions and relationships to socio-demographic characteristics. *Economic Geography* 54, 332–47.

Harder, H., P. Bro, N. Tradisauskas, & T. S. Nielsen, (2008). Tracking visitors in public parks experiences with GPS in Denmark. In J. van Schaick, & S. C. van der Spek (Eds.), *Urbanism on Track: Application of Tracking Technologies in Urbanism* (pp. 65–77). IOS Press BV.

Hardy, A. (2020). *Tracking Tourists: Movement and Mobility.* Goodfellow Publishers.

Harvey, D. (1989). *The Condition of Postmodernity: An Enquiry into the Origins of Cultural Change.* Blackwell.

Kitamura, R., Nishii, K., & Goulias, K. (1990). Trip chaining behavior by central city commuters: A causal analysis of time-space constraints. In P. Jones (Ed.), *Developments in Dynamic and Activity Based Approaches to Travel Analysis* (pp. 145–170). Avebury.

Knaap, W. G. M. van der (1999). Research report: GIS-oriented analysis of tourist time-space patterns to support sustainable tourism development. *Tourism Geographies* 1(1), 56–69.

Kwan, M.P. (2002). Time, information technologies, and the geographies of everyday life. *Urban Geography* 23(5), 471–82.

Kwan, M.P. (2004). GIS methods in time-geographic research: Geocomputation and geovisualization of human activity patterns. *Geografiska Annaler B* 86(4), 267–80.

Lenntorp, B. (1999). Time-geography – at the end of its beginning. *GeoJournal* 48(3), 155–8.

Mieli, M., Zillinger, M. & Nilsson, J.H. (2024). Phygital time geography, or: What about technology in tourists' space-time behaviour? *Tourism Geographies.* DOI: 10.1080/14616688.2024.2318722

Miller, H.J. (2005). A measurement theory for time geography. *Geographical Analysis* 37(1), 17–45.

Modsching, M., Kramer, R., Ten Hagen, K., & Gretzel, Y. (2008). Using location-based tracking data to analyze the movements of city tourists. *Information Technology & Tourism* 10(1), 31–42.

Page, S.J. & Hall, M.C. (2003). *Managing Urban Tourism,* Prentice Hall.

Parkes, D.N. & Thrift, N. (1980). *Times, Spaces and Places: A Chronogeographic Perspective,* Wiley.

Schlich, R. & Axhausen, K.W. (2003). Habitual travel behaviour: Evidence from a six-week travel diary. *Transportation* 30(1), 13–36.

Shaw, G., Agarwal, S., & Bull, P. (2000). Tourism consumption and tourism behavior: A British perspective. *Tourism Geographies* 2, 264–89.

Shoval, N. (2007). Sensing human society. *Environment and Planning B: Planning and Design* 34, 191–5.

Shoval, N. & Ahas, R. (2016). The use of tracking technologies in tourism research: The first decade. *Tourism Geographies* 18(5), 587–606.

Shoval, N. & Isaacson, M. (2006). The application of tracking technologies to the study of pedestrian spatial behaviour, *The Professional Geographer* 58(2), 172–83.

Shoval, N. & Isaacson, M. (2007a). Tracking tourists in the digital age. *Annals of Tourism Research* 34(2), 141–59.

Shoval, N. & Isaacson, M. (2007b). Sequence alignment as a method for human activity analysis. *Annals of the Association of American Geographers* 97(2), 282–97.

Shoval, N. & Isaacson, M. (2010). *Tourist Mobility and Advanced Tracking Technologies.* Routledge.

Shoval, N., McKercher, B., Birenboim, A., & Ng, E. (2015). The application of a sequence alignment method to the creation of typologies of tourist activity in time and space. *Environment and Planning B: Planning and Design* 42(1), 76–94.

Spek, S. van der (2008). Spatial metro: Tracking pedestrians in historic city centres. In J. van Schaick, & S. van der Spek (Eds.), *Urbanism on Track: Application of Tracking Technologies in Urbanism* (pp. 79–102). IOS Press.

Wilson, C. (1998). Activity pattern analysis by means of sequence-alignment methods. *Environment and Planning A* 30(6), 1017–38.

Wilson, C. (2008). Activity patterns in space and time: Calculating representative Hagerstrand trajectories. *Transportation* 35, 485–99.

PART VII

Economic, Entrepreneurship and Business Perspectives on Tourism Geographies

34

HIDDEN IN PLAIN SIGHT

Evolving Geographies of Business Innovation and Tourism

Tim Coles

Introduction

Tourism geography has always had an uneasy relationship with economic geography. Around a quarter of a century ago, Ioannides and Debbage (1997) encouraged greater reciprocity between the two. In addition to encouraging further consideration of a key sector in the global economy, they observed the potential of theory, concept and method developed in economic geography for enhancing our understanding of how tourism enterprises, destinations and visitor economies function (see also Ioannides & Brouder, Chapter 36 of this volume).

In retrospect, it is easy to read this solely as a call for rapprochement between two distinctive fields of study within a discipline, geography, that was struggling with its identity (Turner, 2002). Each had grown separately and with divergent intellectual trajectories notwithstanding they should have shared imperatives. After all, even back then, travel and tourism together constituted one of the major sectors of the global economy. Rhetoric describing travel and tourism as the 'world's largest industry' directly countered early critiques of the sector as lacking substance, a form of 'candyfloss economics' (Williams & Shaw, 1988).

Yet economic geography stubbornly refused to engage with tourism and leisure. With a proliferation of small-scale, under-capitalised enterprises associated with weak labour markets, the sector was routinely alleged to lack productivity, competitiveness and dynamism (Shaw and Williams 1994). Since then, the persistent typecasting of tourism as lacking innovation, innovativeness and the capacity for innovation, has been the routine rationale for many a study in a burgeoning inter-disciplinary body of knowledge on the subject, including the original publication and subsequent revision of a landmark text (Hall and Williams, 2008; 2019).

This chapter examines recent research on business innovation and tourism from a geographical perspective. It covers developments broadly since the previous edition of this collection in 2010. By virtue of the way in which the topic has been studied and innovation has come to be known, a major focus is on innovation by, and among, private sector organisations; that is, privately owned and operated enterprises, businesses and firms. As largely synonymous and context-dependent terms, these are routinely and uncritically used interchangeably in the discourse of innovation. Of course, innovation is not restricted solely to private sector organisations in tourism. Public sector

DOI: 10.4324/9781003286301-41

and voluntary sector organisations, including destination management organisations (Reinhold et al. 2019a), may also innovate their practices and processes, for instance through evolving forms of social enterprise involving tourism. Before we turn to consider recent progress, the chapter first outlines several major conceptual and theoretical foundations relating to the subject (and how they have recently emerged).

Business Innovation – Some Key Ideas

While it may sound like a straightforward term, 'innovation' is a fuzzy concept that defies simple definition and deserves careful delimitation (Coles et al., 2008; Hjalager, 2002; 2010). Often confused with invention, innovation is about the development of new ideas, often following the identification of opportunities, and putting these into widely used practice; as a result it is often conceptually coupled with entrepreneurship (Bessant & Tidd, 2009). Innovation can be radical or incremental depending on the nature and extent of change. Drawing on foundational work by Joseph Schumpeter (1934), innovation may be described as 'disruptive' in the sense that it rejects the previous, established (status quo) ways of working, processes and operations in a business or organisation (Christensen et al., 2015). There is a strong implication that it equates to fundamental, step changes in practice, and 'disruptors' are variously depicted as the 'game changers', 'thought leaders' and 'rain makers' among the many neologisms that characterise business vocabulary. Yet innovation can also be 'incremental' in the sense that change may be more gradual, subtle or modest in scope but no less effective or potent because of it.

Within early studies of business innovation and tourism, attempts were made to identify the nature and types of innovation encountered in the sector. Drawing on the original *Oslo Manual* for collecting, using and reporting data on innovation (OECD and Eurostat 2018) as well as foundational contributions in management studies, Hjalager (2002: 465–466) originally distinguished five broad types of innovation, namely: product, process, management, logistics and institutional innovations. In a later revision, the classifications were revised to clarify 'product or service innovations' as 'changes directly observed by the customer' (Hjalager 2010: 2), including an expanded definition of process innovations to incorporate logistics. Managerial innovations were distinguished from management innovations; the former covering 'new ways of organising internal collaboration, directing and empowering staff' while the latter denoted innovations associated with the collective management of tourism, for instance through loyalty cards (Hjalager, 2010: 3).

As diagnostically helpful as such typologies can be, work of this nature has been more valuable for two important, but often overlooked, reasons. First, it stimulated further detailed empirical research that has changed how we know and think about tourism innovation, as her own work perhaps best attests. Hjalager (2015) points to rich histories of innovation in and by those in the tourism sector as well as the uptake and further adaptation of innovations perhaps more usually associated with other forms, or sectors, of economic activity. Among the former may be the traveller's cheque or the guidebook while among the latter arguably the best example is the mobile (now smart) phone which is transforming the way that very many people travel (see Hjalager, 2015: 7–8).

Empirical documentation of numerous cases goes some way towards refuting the argument that the tourism sector is or has not been subject, nor the generator, of innovation in the past. In fact, more recent work stresses that tourism, as a form of human behaviour and a sector of economic activity, continues to be a crucible for innovation (Hall & Williams, 2019). Second, then, work on categorisation reminds us that not all tourism innovation is obvious or visible, either to customers or users (or researchers for that matter). Not only may innovation refer to obvious, tangible physical changes to a business or an organisation – for instance to the layout, design and branding. It may also refer to intangible changes in business systems and/or to back office functions that facilitate visitor experiences.

Business Model Innovation

Extending this idea further, within tourism studies a further distinct strand of research has started to emerge around the innovation of business models (Reinhold et al., 2017; 2019b; Coles, 2022a). At their most basic, business models are 'a framework or recipe for making money – for creating and capturing value' (Afuah, 2014: 4). In contrast to a business plan, a business model is a higher level concept regarding how organisations 'do business' and relate to their markets and customers. It is a template for how an 'enterprise delivers value to customers, entices customers to pay for value, and converts those payments to profit' (Teece, 2010: 172).

According to Velu et al. (2015), three connected ideas – value propositions, value capture, and value creation – define how organisations relate to their end users, the customers, through their business models. Value capture refers to the benefits that both parties will realise (and its measurement). In the strictest of commercial logic, it may be the ultimate priority for some managers, owners and investors. Value creation is an intermediate stage that 'focuses on how products and services will be created/provided' (Velu et al., 2015: 9) while value propositions refer to the initial identification and subsequent exploitation of opportunities to particular markets or segments within them.

Business models are not the same as revenue models; they also include costs in the consideration of profit and success (Coles et al., 2017); and value is not understood solely as financial. They embed ideas of tangible and intangible experience as well as both the producer's and user's expectations of a product or service. The latter can be negotiated in co-production practices between producer and consumer as a feature of the so-called Service Dominant Logic in tourism and hospitality (Shaw & Agarwal, 2018). While more may be known about customers from the plethora of data available to contemporary organisations, co-productive approaches seek to involve end users more in the design and operation of the product or service.

Thus, business model innovation (BMI) is the means of 'creating and capturing value by doing things differently' (Afuah, 2014: 4). This may include using new and/or existing assets to identify new opportunities and to deliver altogether new or modified value propositions (Velu et al., 2015). As a concept, BMI does not map easily onto the categories for innovation that were noted above. In each particular instance BMI requires one or more types of innovation to realise, often in complex combinations and intricately linked. For example, a revised offer (product) requires new processes and adapted managerial systems. BMI can stem from new ways of thinking about how the organisation creates value, as evident in the values, beliefs, identities and even interests of producers and consumers. In other words, BMI is a social practice (i.e., it requires interaction among individuals to be agreed and enacted) and ultimately, it has to be grounded (i.e., somehow physically manifest) in the business.

Two recent reviews have surveyed the themes pursued in a rapidly growing body of knowledge on business model innovation in tourism (cf. Reinhold et al., 2017; 2019b). Just under a half the work identified by Reinhold et al. (2019b) had been published in the previous five years. However, it is pertinent to note the absence of overtly geographical foci within the research they surveyed or the themes they identified, notwithstanding the inherent spatialities associated with business model innovation. For example, BMI may impact upon the nature and production of space. Subtle innovations in the business models of major hotel chains such as IHG and Marriott may involve changes to the brand architecture and, with it, adjustments in the pricing, positioning, amenities and services, and to the physical fabric to achieve closer customer orientation. In turn, as Niewiadomski's (2014) work on the internationalisation of hotel businesses makes clear, they are widely reproduced across the world.

Business model innovation may also result in new flows of knowledge, ideas and capital, as well as localised adaptation and impacts on visitor economies, communities and the environment. One of the most high-profile and enduring instances of this is the rise, growth and consolidation of low-fares airlines (LFAs), especially in Europe. Originally based on innovating the business model of short-haul aviation developed by national flag carriers in order to pare back costs, charge lower ticket prices and benefit from ancillary revenues, many low-fares airlines used secondary airports, often some distance from the destinations they 'serve' (cf. Groß & Schröder, 2007). Although this practice has continued to court controversy, LFAs continue to configure new flows of visitors in previously harder-to-reach, often peripheral locations and sometimes with notable redistributive effects. Recent research suggests that reorientation among LFAs towards business travellers questions 'the future importance of secondary airports for LCCs [low cost carriers, a synonym for LFAs]' (Dziedzic & Warnock-Smith, 2016: 19).

Recent Progress in Understanding Business Innovation and Tourism from a Geographical Perspective

Before proceeding further, it is worth reiterating that business model innovation has become an important topic in its own right, forming a subset of innovation studies. Understanding BMI is driven by advances from scholars professionally homed in management studies, especially in business school settings (e.g., Teece, 2010, Velu et al., 2015) whose ideas tourism scholars have borrowed and in some cases adapted. Clearly then, as a term 'innovation' is multifaceted, and it is multiply interpreted and understood (see Hall & Williams, 2019) – and the way that innovation has been studied has itself been innovated! Research on tourism businesses may deal with innovation but may not necessarily label itself as such. Conversely, some studies may use the vocabulary of innovation, but this does not always mean that they are fully engaged with the most recent theoretical, conceptual or methodological advances in the field (Christensen et al., 2015). All of which makes the identification of recent trends and developments in the body of knowledge more challenging. This is more so from a geographical perspective because, while the subject matter may be of interest to tourism geography and geographers, the geographical implications of innovation are often not explicitly raised and, instead, have to be teased out or inferred from many studies.

A search of Scopus reveals the increasing interest in 'business innovation [and] tourism' in the widest sense. At the time of writing, 875 publications of different forms were associated with these keywords in their titles, abstracts and/or among their keywords. Although a coarse indicator of progress and interest, the number of such publications had increased sixfold from 22 in 2012 to 133 in 2020. However, when the term 'geograph*' was added, just 28 publications were identified.

Clearly, this chapter is not intended to be a full, exhaustive review of the burgeoning body of knowledge on tourism and innovation, nor its development over time (see instead Hjalager, 2010; Gomezelj, 2016; Pikkemaat et al., 2019). Similarly, no attempt is made to assess in any systematic way the geographical 'credentials' or 'appeals' of particular studies, either where geography is named or in the considerably more numerous studies where it is inferred or implied. Nevertheless, in the past decade tourism research on business innovation from a more literal geographical perspective has pursued three major directions.

First, extending the tradition in tourism studies for identifying and recognising innovations (Hjalagar, 2002; 2010; 2015), a major strand of work has been involved with documenting where innovation takes place and explaining why and how it occurs. The most obvious example is the rise of peer-to-peer accommodation sites and Airbnb in particular (Guttentag, 2015, Coles, 2022b; see also van der Zee, Chapter 28 in this volume). Although platform providers such as Airbnb comprised radical innovations of themselves, they have subsequently inspired business model innovation among online travel agencies (OTAs) and among accommodation providers who also use third-party platforms to dispose of otherwise unused capacity. Perhaps somewhat paradoxically, the processes of innovation and evolution may have led to convergence, not divergence among a much wider array of organisations that distribute or broker property to visitors for impermanent occupation (Coles, 2022b).

Beyond documentation, elaborating on theorisations from the 2000s, business innovation is variously read as a function of the intrinsic characteristics of the firm as well as the external conditions framing business change. In the case of the former, a notable shift in emphasis has been to focus on innovation as a process, as something to be fostered, grown or developed, rather than solely on innovation as an outcome, of and by itself, resulting in entities to be documented and classified. In this respect, the absorptive capacity of the organisation, including the capability of the various actors to stimulate and manage (i.e., the employees, managers, directors, even investors), has come under the spotlight (Thomas & Wood, 2014; Wu, 2020).

Drawing on the principal tenets of the Resource-based View of the Firm, originating in Coasian economics and widely employed in management studies (Armstrong & Shimizu, 2007), readings of this nature view competitive advantage and commercial performance as a function of the organisation's attributes (e.g., the capacity to innovate or even the locational settings they find themselves in; see below) and ability to harness them effectively (Walsh et al., 2011). The dual and connected ideas of co-production and the Service Dominant Logic – again emerging from management studies (Shaw & Agarwal, 2018) – have added further compelling layers of understanding and interpretation. The ability to harness knowledge of, and the views from, end users is a major resource to deploy in shaping future innovation, and for configuring service experiences and encounters which take different forms than traditional producer-customer relationships.

Thus, the process of tourism business innovation and its relationship with place and space comprises the second direction for recent research. The innovative organisation is theorised as being capable of learning, of being agile and capable of responding effectively to fast-moving markets, changing stimuli and trends. Organisations do not, though, learn in isolation or act in exclusively siloed ways. Rather, in the tourism sector, like many other forms of economic activity, there is increased evidence of 'open innovation', which stresses innovation through leveraging external resources and relationships in a purposive manner (Iglesias-Sánchez et al., 2019). As knowledge exchange across the 'boundaries' of an organisation is required, as it were, there are inherent spatialities associated with open innovation.

Continuing to draw on ideas first evident in management studies on clustering and the networked nature of organisational development and local competitive advantage (Sørensen, 2007; Lowe et al., 2012; Hall & Williams, 2020), tourism research variously continues to examine: how constellations of businesses can variously learn from, or learn alongside, one another in linked relationships (Tang et al., 2019); how the environment creates conditions conducive to learning and development (cf. Lowe et al., 2012; Booyens & Rogerson, 2017); and the various stakeholders and their roles in networks (Sørensen & Balsby, 2021) driving change. The configuration, operation and ultimately competitiveness of destinations, including interdependencies among different types of tourism businesses, is – put simply – greater than the sum of the individual parts.

The final and connected direction has been to examine the impacts of tourism business innovations, and their moorings in time and space. Once more, perhaps the most conspicuous manifestation of this relates to peer-to-peer accommodation platforms and Airbnb in particular (Guttentag, 2015). Alongside the considerable attention shown to the novelty of the offer and the opportunities afforded by recent platform technologies, their (sometimes insidious) effects on destinations, their offers (i.e., pricing and availability), and their local labour and housing markets (Coles, 2022b) have come under the spotlight.

Much of the blame for 'hollowing out' communities, higher rents and the lower availability of accommodation for local residents is placed at the door of the platform providers (i.e., 'the innovators' who enable the commercial transaction) rather than the individual businesses and property owners who use the platforms to bring their accommodation to market to achieve optimal yield (Coles, 2022b). In contrast, other work has examined innovation very specifically for more virtuous purposes, including the engineering of more sustainable tourism. Several contributions have explored how the identity, values and belief systems evident in individual businesses – especially among small and micro enterprises – have driven the development and introduction of sustainable practices (Garay et al., 2019; Bressan and Pedrini, 2021).

Of course, these are traditional – and some may allege 'dated' – ways of reading business innovation in tourism from a geographical perspective. The studies invoked are selective but emblematic, and no attempt is made to suggest which is more or less important based on the quantum of attention (as this is an almost impossible task). Be this as it may, beyond identifying these broad strands and pointing to their respective contributions, it is important to stress two salient features of recent knowledge production. The first is that research on business innovation in tourism is rarely published in geographical journals, including the leading outlets in the sub-disciplines of both tourism geography and economic geography. Of the 28 studies noted above, just two had been published in *Tourism Geographies*, as the dedicated sub-disciplinary journal, and just one in the past decade (Sørensen, 2007; Booyens & Rogerson, 2017). Among the two subject-leading journals in economic geography, there appear to have been just two studies where tourism innovation has been the subject of detailed or overt attention (cf. Lowe et al., 2012). Despite the obvious analytical potential of their reconciliation (Debbage, 2018), the estrangement between the sub-disciplines appears to continue, at least regarding this particular subject.

Second, advances in understanding the geographies of business innovation in tourism are mostly produced through bricolage; that is, the collective body of knowledge is characterised by the gradual piecing together of knowledge from separate, self-contained cross-sectional studies conducted on cases at particular moments or periods in time. In other words, most research on business innovation in tourism is historical in nature. There has been very little if any longitudinal research, tracking change continuously over extended periods. This is a crucial point: by being agnostic about time, much of this work fails to recognise that innovation is not instantaneous, immediate and rapid in many cases. This might be the romantic view of invention or of disruptive innovation (Christensen et al., 2015). By contrast, much innovation is incremental in nature. As firms and organisations evolve, they innovate gradually, almost as a process of continuous improvement which will – or should – extend into the future.

Indeed, Fayos-Solà and Cooper (2018: 1) argued before the Coronavirus pandemic, that 'the future of tourism is really a misnomer' and trying to understand it 'would imply an almost impossible introspective task'. Rather, the present is effectively the future unfolding and, in their view, 'innovation is the bridge to the future' such that businesses and organisations have to be cognisant of, and able to respond to, the opportunities afforded by (and threats associated with) trends, shifts and key strategic issues (e.g., climate change, development, the rise of automation and artificial intelligence). Indeed, one of the critical issues raised by the pandemic was the extent to which it interrupted or accelerated existing plans for, and processes of, innovation under previous 'business-as-usual' conditions. For instance, in one recent study of hospitality firms during the early stages of the pandemic, the ability and capacity to innovate (i.e., respond quickly) was argued to enhance business resilience in a highly competitive and congested sector, and moreover one that is especially vulnerable to crises and catastrophes (Brier et al., 2021).

Future Geographies of Business Innovation in Tourism

So, where next for geographical studies of business innovation and tourism? An obvious focus for future work may be the extension of the existing themes and foci. There is a clear case for greater engagement with business model innovation as an emergent theme with clear ramifications for the unfolding production of tourism space and place.

However, if the Coronavirus pandemic has taught us anything, it is that there will probably not be a return to the 'business-as-usual' of the 'old normal' and moreover this is not desired by some (Ateljevic, 2020; Benjamin et al., 2020; Ioannides & Gyimóthy, 2020). Rather, the disruptive effects of closing down visitor economies and the conditions associated with reopening them, may mean that many organisations are unlikely to return to ways of operating that are *exactly* the same as before. The critical issue – and the challenge for scholars – then is to understand the manner and extent of the (possibly forced) disruption and (potentially induced) change, and how innovation is valorised in the post-COVID world (Nunes & Cooke, 2021). Put another way, where, when and why have innovations taken place in the tourism sector as a result of or during the pandemic, and to what effect or impact?

Within the UK, as in other countries (OECD, 2020), there have been institutional innovations in the form of new policy interventions and instruments to support businesses and the labour force through challenging times (DCMS, 2021). The pandemic has variously been a time when some businesses have had to innovate as a response; others have enacted changes delayed through normal operating periods and procedures; and some experimented because of temporary closures (VB-VE, 2021). New public health guidelines required tourism businesses to develop product innovations (i.e., to comply with social distancing, lower capacity), process (i.e., track-and-trace registration of guests, enhanced risk assessments, cleaning regimes) and managerial innovations (i.e., managing staff, presences, absences, vaccination) (HMG, Government, 2021).

The practice-based literature is replete with examples of innovation during the pandemic. As yet though, we know little about which innovations succeeded, which failed and the reasons why? Moreover, as a consequence, will there be new geographies of production and consumption? Little is known about which innovations have been retained or will 'stick'? Or, for that matter, whether some of the recent trends and preferences – for instance, for contactless hotels (BVA-BDRC, 2021) – will continue and warrant further innovation and investment? Of course, the persistence of innovations is vital to appraising the transformative potential of the pandemic and whether the exhortations that it should be an opportunity to 'build back better', to eschew old unsustainable, excessive and irresponsible ways of working (cf., Ateljevic, 2020; Ioannides & Gyimóthy, 2020), have translated into commensurate and enduring action. And finally, for critical scholars, there are unresolved questions of who have been and who will be the beneficiaries of innovation? Another way of framing this is, reading the situation from a perspective of political economy, for whom and for what reason(s) has business innovation in tourism taken place?

Conclusion

This chapter started by reading the schism between tourism geography and economic geography as a function of the existential crisis geography was suffering a quarter of a century ago. Economic geographers had benefited from productive knowledge exchange with other scholars (principally economists), to develop new and powerful ways of knowing, researching and understanding businesses and organisations. Tourism geographers had yet to do so, and to their apparent disadvantage.

For some, the worth of historical retrospectives is limited and 'the past is a foreign country: … .'. Attributed to the novelist, L.P. Hartley (1953: 5), this aphorism is actually just the first part of a longer lament, the forgotten part of which is especially germane here: 'they do things differently there'. Where once the tourism sector may have been typecast as lacking innovation and the sector's credentials for fostering greater economic development were questioned, empirical insights produced in the past decade demonstrate both the range of innovative activity and how far the concept has penetrated the consciousness of practitioners and academics alike. The nature of academic engagement and knowledge exchange is also different these days. Instead of gravitating towards economic geography, the greatest advances on tourism innovation over the past decade have been in studies informed by major contributions in management studies: for instance through the lenses of innovation theory and Service Dominant Logic.

It is easy to lose sight of how knowledge we take for granted and use so routinely, is produced and originates. The crisis of legitimation in the late 1990s may be all but forgotten now, but another looms large. Business innovation in tourism is a major topic and one that should occupy geographers. So much work is hidden in plain sight and not necessarily identified as such. This triggers a further critical question. As the body of knowledge on this topic – grounded as it is in management studies –grows further, what should a distinctively geographical reading of tourism business innovation look like? Put another way: What will geographical approaches contribute to future understandings of business innovation in tourism? The recent COVID-19 pandemic has emphasised just how innovative, agile and responsive tourism organisations can be, and these attributes will be necessary moving forward. There will be new geographies of innovation and, rather than focus on the patterns (as a traditional geographical focus) and processes (as a traditional management studies interest), perhaps one major space for geographers to make a difference is to examine the spatialities of the outcomes and inequalities that result from business change.

References

Afuah, A. (2014). *Business Model Innovation: Concepts, Analysis, and Cases*. Routledge.

Ateljevic, I. (2020). Transforming the (tourism) world for good and (re)generating the potential 'new normal'. *Tourism Geographies*, 22(3), 467–475.

Armstrong, C., & Shimizu, K. (2007). A review of approaches to empirical research on the resource-based view of the firm. *Journal of Management*, 33(6), 959–986.

Benjamin, S., Dillette, A., & Alderman, D. H. (2020). "We can't return to normal": Committing to tourism equity in the post-pandemic age. *Tourism Geographies*, 22(3), 476–483.

Bessant, J., & Tidd, J. (2009). *Managing Innovation: Integrating Technological, Market and Organizational Change*. 4th ed. Wiley.

Booyens, I., & Rogerson, C. (2017). Networking and learning for tourism innovation: Evidence from the Western Cape, *Tourism Geographies*, 19(3), 340–361.

Bressan, A., & Pedrini, M. (2020). Exploring sustainable-oriented innovation within micro and small tourism firms. *Tourism Planning and Development*, 17(5), 497–514.

Brier, M., Kallmuenzer, A., Clauss, T., Gast, J., Kraus, S., & Tiberius, V. (2021). The role of business model innovation in the hospitality industry during the COVID-19 crisis. *International Journal of Hospitality Management, 92*, 102723.

BVA-BDRC. (May, 2021). *ClearSight on Recovery and COVID-19.* Available at: www.bva-bdrc.com/wp-content/uploads/2021/04/ClearSight-on-Recovery-May-2021.pdf?utm_campaign=ClearSight&utm_medium=email&_hsmi=124382159&_hsenc=p2ANqtz-_IcxCSSJSx13nl7dVcBVieEPeP8hbnk-vtfTe9yak4HWTLJSZ6ffR2O3F1WIXg_K_3V8ybbdEMGbPK8kmrRrzwQgMfQA&utm_content=124382159&utm_source=hs_email [Accessed November 3, 2021].

Christensen, C., Raynor, M., & McDonald, R. (December, 2015). What is disruptive innovation? *Harvard Business Review, 2015*, 44–53.

Coles, T. E., Liasidou, S., & Shaw, G. (2008). Tourism and New Economic Geography: Issues and challenges in moving from advocacy to adoption. *Journal of Travel and Tourism Marketing, 25*(3–4), 312–324.

Coles, T. E., Warren, N., Borden, D. S., & Dinan, C. R. (2017). Business models among SMTEs: Identifying attitudes to environmental costs and their implications for sustainable tourism. *Journal of Sustainable Tourism, 25*, 471–488.

Coles, T. E. (2022a). Business models. In D. Buhalis (Ed.), *Encyclopaedia of Tourism Management and Marketing*. Edward Elgar.

Coles, T. (2022b). The sharing economy in tourism and property markets: A comment on the darker side of conceptual stretching. *Current Issues in Tourism, 25*(19), 3068–3075.

Debbage, K. (2018). Economic geographies of tourism: A critical and contested discourse. In C. Cooper, S. Volo, W. Gartner, & N. Scott (Eds.), *The SAGE Handbook of Tourism Management* (pp. 53–68). Sage.

Department of Culture, Media and Sport (DCMS). (2021). *The Tourism Recovery Plan.* Available at: https://assets.publishing.service.gov.uk/government/uploads/system/uploads/attachment_data/file/992974/Tourism_Recovery_Plan__Web_Accessible_.pdf [Accessed October 25, 2021].

Dziedzic, M., & Warnock-Smith, D. (2016). The role of secondary airports for today's low-cost carrier business models: The European case. *Research in Transportation Business & Management, 21*, 19–32.

Fayos-Solà E., & Cooper, C. (2019. Introduction: Innovation and the future of tourism. In E. Fayos-Solà, & C. Cooper (eds) *The Future of Tourism* (pp. 1–16). Springer.

Garay, L., Font, X., & Corrons, A. (2019). Sustainability-oriented innovation in tourism: an analysis based on the decomposed theory of planned behaviour. *Journal of Travel Research, 58*(4), 622–636.

Gomezelj, D. O. (2016). A systematic review of research on innovation in hospitality and tourism. *International Journal of Contemporary Hospitality Management, 28*(3), 516–558.

Government, Her Majesty's. (2021). *Working Safely during Coronavirus (COVID-19). Hotels and Guest Accommodation.* Available at: www.gov.uk/guidance/working-safely-during-covid-19/hotels-and-guest-accommodation [Accessed November 3, 2021]

Groß, S., & Schröder, A. (Eds.) (2007). *Handbook of Lost Cost Airlines: Strategies, Business Processes and Market Environment*. Erich Schmidt Verlag.

Guttentag, D. (2015). Airbnb: Disruptive innovation and the rise of an informal tourism accommodation sector. *Current Issues in Tourism, 18*(12), 1192–1217.

Hall, C. M., & Williams, A. M. (2008). *Tourism and Innovation.* 1st ed. (2nd ed. 2019). Routledge.

Hall, C. M., & Williams, A. M. (2019). *Tourism and Innovation.* 2nd edition. Routledge

Hartley, L. P. (1953). *The Go-Between.* Penguin.

Hjalager, A. M. (2002). Repairing innovation defectiveness in tourism. *Tourism Management, 23*(5), 465–474.

Hjalager, A. M. (2010). A review of innovation research in tourism. *Tourism Management, 31*(1), 1–12.

Hjalager, A. M. (2015). 100 innovations that transformed tourism. *Journal of Travel Research, 54*(1), 3–21.

Iglesias-Sánchez, P., Correia, M., & Jambrino-Maldonado, C. (2019). Challenges of open innovation in the tourism sector. *Tourism Planning & Development, 16*(1), 22–42.

Ioannides, D., & Debbage, K. (1997). Post-fordism and flexibility: The travel industry polyglot. *Tourism Management, 18*(4), 229–241.

Ioannides, D., & Gyimóthy, S. (2020). The COVID-19 crisis as an opportunity for escaping the unsustainable global tourism path. *Tourism Geographies, 22*(3), 624–632.

Lowe, M., Williams, A., Shaw, G., & Cudworth, K. (2012). Self-organizing innovation networks, mobile knowledge carriers and diasporas: Insights from a pioneering boutique hotel chain. *Journal of Economic Geography, 12*(5), 1113–1138.

Organisation for Economic Co-operation and Development (OECD). (2020). *Tourism Policy Responses to the Coronavirus (COVID-19)*. Available at: www.oecd.org/coronavirus/policy-responses/tourism-policy-responses-to-the-coronavirus-covid-19-6466aa20/ [Accessed October 25, 2021].

Organisation for Economic Co-operation and Development (OECD) and Eurostat. (2018). *Guidelines for Collecting, Reporting and Using Data on Innovation*, 4th ed. Available at: www.oecd-ilibrary.org/science-and-technology/oslo-manual-2018_9789264304604-en [Accessed November 3, 2021].

Niewiadomski, P. (2014). Towards an economic-geographical approach to the globalisation of the hotel industry. *Tourism Geographies*, *16*(1), 48–67.

Nunes, S., & Cooke, P. (2021). New global tourism innovation in a post-coronavirus era. *European Planning Studies*, *29*(1), 1–19.

Pikkemaat, B., Peters, M., & Bichler, B. (2019). Innovation research in tourism: Research streams and actions for the future. *Journal of Hospitality and Tourism Management*, *41*, 184–196.

Reinhold, S., Beritelli, P., & Grünig, R. (2019a). A business model typology for destination management organizations. *Tourism Review*, *74*(6), 1135–1152.

Reinhold, S., Zach, F., & Krizaj, D. (2017). Business models in tourism a review and research agenda. *Tourism Review*, *72*(4), 462–482.

Reinhold, S., Zach, F., & Krizaj, D. (2019b). Business models in tourism – State of the art. *Tourism Review*, *74*(6), 1120–1134.

Schumpeter, J. (1934). *The Theory of Economic Development*. Harvard University Press.

Shaw, G., & Agarwal, S. (2018). The development of service-dominant logic within tourism management. In C. Cooper, S. Volo, W. Gartner, & N. Scott (Eds.) *The SAGE Handbook of Tourism Management* (pp. 173–190). Sage.

Shaw, G., & Williams, A. M. (1994). *Critical Issues in Tourism: A Geographical Perspective*. Blackwell.

Sørensen, F. (2007). The geographies of social networks and innovation in tourism. *Tourism Geographies*, *9*(1), 22–48.

Sørensen, F., & Balsby, N. (2021). Brokers and saboteurs: Actor roles in destination innovation network development. *Tourism Planning & Development*, *18*(5), 547–572.

Tang, J., Williams, A., Makkonen, T., & Jiang, J. (2019). Are different types of interfirm linkages conducive to different types of tourism innovation? *International Journal of Tourism Research*, *21*, 901–913.

Teece, D. (2010). Business models, business strategy and innovation. *Long Range Planning*, *43*, 172–194.

Thomas, R., & Wood, E. (2014). Innovation in tourism: Reconceptualising and measuring the absorptive capacity of the hotel sector. *Tourism Management*, *45*, 39–48.

Turner, B. L. (2002). Contested identities: Human-environment geography and disciplinary implications in a restructuring academy. *Annals of the Association of American Geographers*, *92*(1), 52–74.

Velu, C., Smart, A., & Phillips, M. (2015). *The Imperative for Business Model Innovation. A Research and Practice Perspective*. Available at: www.nemode.ac.uk/wp-content/uploads/2012/12/Velu-2015-BMI-White-Paper.pdf [Accessed February 25, 2021]

Visit Britain-Visit England. (2021). *Business Recovery Stories*. Available at: www.visitbritain.org/business-advice/business-recovery-stories [Accessed November 3, 2021].

Walsh, M., Lynch, P., & Harrington, D. (2011). A capability-based framework for tourism innovativeness. *Irish Journal of Management*, *31*(1), 21–41.

Williams, A. M., & Shaw, G. (1988). Tourism: Candyfloss industry or job generator? *The Town Planning Review*, *59*(1), 81–103.

Wu, A. (2020). Improving tourism innovation performance: Linking perspectives of asset specificity, intellectual capacity and absorptive capacity. *Journal of Hospitality & Tourism Research*, *44*(6), 908–930.

35

MAKING SENSE OF TOURISM ENTREPRENEURSHIP

The Nexus between Entrepreneurs and Their Spatial Environment

Dimitri Ioannides and Jonathan Yachin

Introduction

The last decades have seen an explosion of research on tourism entrepreneurship, which reflects a growing appreciation concerning the prominence of small-scale businesses in the global tourism system (Morrison et al., 2010; Shaw, 2014, Thomas et al., 2011; Yachin, 2020). This enhanced interest demonstrates the substantial progress made since Page et al. (1999) described research on small tourism businesses as a *terra incognita*. Nevertheless, most such works are inductive and case-study oriented, leading to a recurring critique that tourism entrepreneurship research remains rich in practice but poor in theory (Fu et al., 2019; Thomas et al., 2011). Here, we argue that viewing tourism entrepreneurship as the processes whereby actors interact with their spatial environment to create value from a locality's intrinsic features lends itself to a geographical perspective.

Tourism is place-based, meaning that geographers are suitable investigators of issues like business location and the production of space (Hall & Page, 2008). Furthermore, geographers' interest in tourism entrepreneurship is rooted in the discipline's tradition of exploring entrepreneurs' role in regional development and comprehending varying patterns of business creation (Qian, 2016). Entrepreneurship is a vital driver of economic evolution, despite being a distinctly spatially uneven process (Stam, 2009). In other words, entrepreneurship is a contextual phenomenon (Welter, 2011). To uncover why some places are more entrepreneurial than others, geographers investigate, *inter alia*, dimensions such as the place's demographic characteristics, workers' educational background and how existing businesses result in further entrepreneurial activity (Mack, 2016).

A geographical perspective of tourism entrepreneurship constitutes a promising research area related to tourism's economic geography (Debbage & Ioannides, 2012; Ioannides & Debbage, 1998). In this chapter, we focus almost exclusively on small-scale businesses, since such enterprises constitute the majority of tourism businesses worldwide, and their contribution to destinations is expressed in economic, social and cultural terms (Morrison et al., 2010). Chiefly, we propose 'spatial bricolage' (Korsgaard et al., 2021) as a valuable framework for investigating tourism entrepreneurship, since this reflects the interplay between owner-managers of tourism businesses and their

DOI: 10.4324/9781003286301-42

geographical context. We flesh out essential contributions made by geographers to the tourism entrepreneurship literature, highlighting writings that promote the theoretical advancement of the field. We end by proposing future research avenues, arguing briefly that this research field would benefit by including the role of institutions and policymakers in determining destinations' future development trajectory. This is especially pertinent for the ability of businesses and tourism destinations to deal with change and unforeseen circumstances (e.g., pandemics).

The *Why, How* and *What* of Tourism Entrepreneurship

Entrepreneurship researchers focus mostly on three main questions: *why entrepreneurs act; how entrepreneurs act;* and, *what happens when entrepreneurs act?* (Stevenson & Jarillo, 1990). Most studies investigate why entrepreneurs act, particularly focusing on the notion of lifestyle entrepreneurship (Ateljevic & Doorne 2000; Carson et al., 2018). By examining the entrepreneurs' background, motivations and perception of success, scholars have promoted our understanding of tourism entrepreneurship as a means to accomplish a desired lifestyle, which typically involves combining leisure interests with work (see Thulemark & Duncan, Chapter 29 of this volume; Shaw, 2014; Yachin, 2020). While certain observers (Hjalager et al., 2018; Peters et al., 2009) criticise tourism lifestyle entrepreneurs for operating in suboptimal production levels and restricting the potential for tourism development, others demonstrate lifestyle entrepreneurs' value and contribution (Ateljevic & Doorne, 2000; Komppula, 2014). Notably, in-migrant lifestyle entrepreneurs can be particularly important for rural destination development by injecting the local economy with entrepreneurial spirit, capital, and new ways of imagining local resources as tourism experiences (Bosworth & Farrell, 2011).

The perceived contributions of tourism entrepreneurs or outcomes of tourism entrepreneurship link to investigations of *what happens when entrepreneurs act*. Such research typically relies on the discipline of economics, inspired by classical entrepreneurship scholars like Joseph Schumpeter (1883–1950) and Israel Kirzner (b. 1930). Several tourism-related studies focusing on entrepreneurial outcomes exist (Fu et al., 2019). Some seek to establish causal relationships between different attributes of the entrepreneur (e.g., background, proactiveness and access to resources) and the firm's financial performance or chances of survival (Brouder & Eriksson, 2013; Kallmuenzer et al., 2019). Others argue for the potential economic, social and environmental contribution of tourism entrepreneurs to destination development (Bosworth & Farrell, 2011; Komppula, 2014; Tervo-Kankare, 2019). Nevertheless, it is difficult to assess the contribution of small-scale tourism enterprises to destination sustainability or local wellbeing. Furthermore, considering their lifestyle aspirations, it is arguably wrong to measure firm success merely by economic indicators (Dewhurst & Horobin,1998). Therefore, we suggest that the key to understanding the underlying mechanisms of tourism entrepreneurship is to ask *how entrepreneurs act*.

Entrepreneurship is manifested in the entrepreneurs' actions and decisions (Yachin, 2020). The growing body of research providing a behind-the-scenes perspective into how entrepreneurs act truly advances our understanding of tourism entrepreneurship. Examples include adapting to risks and uncertainties (Williams et al., 2021), learning and developing adaptive capabilities (Kelly et al., 2020; Yachin, 2018); and experimentation during the firm formation and product development phases (Rodriguez-Sanchez et al., 2019). All these studies illustrate entrepreneurship as a dynamic process whereby entrepreneurs continuously react to opportunities and adapt to changes in circumstances. Furthermore, they illuminate how the lifestyle character and values associated with tourism entrepreneurs come into practice in their decision making and business practices (Kornilaki et al., 2019; Power et al., 2020). Yet another dominant trait of tourism entrepreneurship is that it is rooted in the interplay between the entrepreneur and their environment, a feature which can be best explained through the spatial bricolage concept.

Spatial Bricolage

Spatial bricolage is defined as "making do by applying combinations of the resources at hand in the immediate spatial context to new problems and opportunities" (Korsgaard et al., 2021: 160). This is a recent development of the entrepreneurial bricolage concept (Baker & Nelson, 2005), which explains the resourcing behaviour of entrepreneurs in constrained environments, as practices building on flexibility, trial-and-error and challenging conventions. Spatial bricolage involves the extensive and intentional use of local resources, suppliers and the involvement of local people in the operation (Korsgaard et al., 2021). It is enabled by, and closely related to the notion of embeddedness. The key idea is that both one's local ties and access to knowledge and resources help entrepreneurs to overcome challenges (Yachin & Ioannides, 2020). Furthermore, a premise of spatial bricolage is that the creative interpretation of elements in their geographical context, and the commodification of these through storytelling helps entrepreneurs to design attractive value propositions (Yachin, 2020).

Since spatial bricolage embodies both a strong appreciation of and dependency on the locality, its pertinence to a place-based activity like tourism is conspicuous. Indeed, several ideas embraced by the spatial bricolage concept are not new to tourism research where notions of embeddedness, commodification through storytelling and use of local elements like culture and nature for tourism purposes, resonate through the literature. Here, we stress the need to utilise spatial bricolage as a framework combining these ideas, allowing "us to describe, explain and explore the *modus operandi* of tourism entrepreneurs in ways that move us away from a persistent focus on their identity and motivations, to research focusing on their entrepreneurial behaviours" (Yachin & Ioannides, 2020: 1018). This can promote the theoretical conceptualisation of tourism entrepreneurship and establish academic links to contemporary entrepreneurship research.

Table 35.1 presents the behaviours and activities associated with entrepreneurial and spatial bricolage. Central to these is 'resource transfer', which relates to the processes by which entrepreneurs gain the right to use local resources that are beyond their control or ownership (e.g., nature, culture or heritage) for tourism purposes. Resource transfer is grounded in a commitment to the wellbeing of the local community and ecosystem and manifests itself by assuming responsibility for the used resources. It is a behaviour that further strengthens the reciprocity between the entrepreneurs and local community, and the fundamental connection between tourism businesses and the place.

Spatial bricolage directs the research focus to the nexus between entrepreneurs and their spatial environment and underlines the meaning of geographical location in tourism entrepreneurship. This calls for a geographical perspective, provoking questions concerning the *where* of tourism entrepreneurship.

The *Where* of Tourism Entrepreneurship, an Evolutionary Perspective

From a geographical perspective, several authors incorporate entrepreneurship and innovation as key factors in explaining how particular geographical spaces are commodified (Debbage, 2019; Page & Ateljevic, 2009). Further, studies have sought to unravel the dynamics of marketing alliances between companies (Milne & Pohlman, 1998) and apply industrial district thinking for

Table 35.1 Spatial bricolage behaviours and activities

Behaviours and activities	Definition
Making do	Creating something from nothing. Using cheap or free discarded or disused resources for new purposes
Bias towards action and improvisation	Working by trial and error. Initiating a range of projects and constantly responding to opportunities.
Refusal to enact limitations	Doing things differently. Consciously disregarding limitations of commonly accepted definitions of inputs, practices and standards, and those imposed by institutions or political settings.
Community involvement	Drawing on the local human capital and collaborating with local actors; contributing to the local community.
Local sourcing	Prioritising locally available physical and non-material resources.
Commodification through storytelling	The entrepreneur commodifies local history, culture, and features of the place through storytelling to enhance the product value.
Embeddedness	The nature, depth, and extent of an entrepreneur's ties into the local environment and social structure.
Resource transfer	A process by which entrepreneurs and resource-holders agree on the deployment and governance of resources; aims to tackle opportunistic behaviour and address operational circumstances, in which the entrepreneurs' value propositions are based on resources beyond their control.

Source: Adapted from Yachin & Ioannides (2020).

understanding the interrelationship of tourism-related companies in a destination (Hjalager, 2000). Important in this line of research are concepts like 'learning regions', the 'knowledge economy' and 'innovation' with discussions concerning how the knowledge produced in specific regions explains why some places are more innovative and competitive than others (Nelson & Winter, 1982; MacKinnon et al., 2002).

Shaw and Williams (2009) see 'learning regions' as significant for spreading tacit knowledge between firms given their geographical proximity to one another. They contend that knowledge within such learning regions is diffused "via inter-firm linkages and partnerships, firms and other knowledge-creating bodies such as universities and government agencies, and informally based exchanges of work-related gossip in both workplace and social settings within the region" (p. 329). Likewise, employees are an important source of knowledge since when several tourism firms are clustered, workers' mobility between employers can lead to knowledge transfer and generation of new ideas (Shaw & Williams. 2009; Sørensen, 2007).

We also must not underestimate the influence of networks. Especially for small-scale tourism businesses, networks allow the pursuit of opportunities and compensate for lack of resources and missing skills (Shaw, 2014; Yachin, 2021). Owner-managers of tourism micro-firms typically resort to utilising existing relationships for multiple purposes, meaning that social, business-related and interest-based ties fill an amalgam of functions that together comprise the ego-networks of tourism entrepreneurs (see also Coles, Chapter 34 of this volume; Yachin, 2021). An alternative perspective views tourism networks as the interaction between several firms (e.g., transportation, accommodation, tour guide) to provide a holistic tour product (van der Zee & Vanneste, 2015). Shaw and Williams (2009) list four types of tourism business networks: networks between destination-based firms and exogenous companies like tour operators; between firms offering similar products (e.g., accommodation establishments); between tourism firms and their suppliers (e.g., those providing technology support); and between distinct companies (e.g., an accommodation provider and an outdoor adventure specialist). Shaw and Williams maintain that personal relationships between actors are especially significant when it comes to the transfer of knowledge. This coincides with Yachin's (2021) observation that cognitive proximity and personal comparability are crucial in determining with whom tourism entrepreneurs associate.

Debbage (2019) cautions that the verdict is still out when it comes to establishing connections between the effectiveness of networks and the spread of tacit knowledge in boosting entrepreneurship in the tourism sector. However, he cites literature arguing, for instance, that non-localised networks can be strong. Considering the tendency of tourism entrepreneurs to rely on cognitive and geographically close alters, the role of 'brokers', the actors forming links to distant networks, is important for innovation and for mitigating risks of lock in, over-embeddedness and reliance on informal networks (Bürcher et al., 2016; Yachin, 2021).

Lock-in relates to path dependence, a key concept incorporated in evolutionary economic geography (EEG) (see also Ioannides & Brouder; Chapter 36 of this volume). Stam (2009) stresses that examining entrepreneurship through an EEG lens provides a perspective of the role of both firms and individuals in shaping regional economic development. He argues that when recent arrivals in an area embrace new ideas, existing firms must adapt to survive. Since EEG offers a dynamic approach to understanding entrepreneurship, it is an improvement over studies that traditionally have been conducted in a static manner (Qian, 2016). In tourism's case, Debbage (2018: 348) describes how among the so-called 'theoretical turns' characterising tourism research over two decades, the evolutionary turn reinforces the political economy perspective in tourism research leading to the development of enhanced understanding, among others, of

> how path dependence and new path creation are inter-linked; how knowledge transfer between tourism firms can shape regional processes from a more Darwinian perspective; and a closer examination of regional branching particularly regarding how tourism emerges at the nascent stage in different regions.
>
> (See also Brouder, 2014)

Tourism Entrepreneurship: Setting a Research Agenda

The study of tourism entrepreneurs and entrepreneurship has evolved dramatically since Williams, Shaw, and Greenwood (1989) first explained tourism entrepreneurship as a combination of consuming lifestyle and producing tourism services. Notably, the first edition of this handbook omitted a chapter dealing explicitly with tourism entrepreneurship (Wilson, 2012). However, the decade that has since passed witnessed a dramatic increase in publications, reflecting the vital roles of entrepreneurs and small-scale businesses in the global tourism system. There have been encouraging developments towards a more theoretical approach to investigating tourism entrepreneurship (Rodriguez-Sanchez et al., 2019; Power et al., 2020; Williams et al., 2021). These studies demonstrate how tourism research can benefit from implementing contemporary entrepreneurship theories in the tourism context.

Solvoll et al. (2015) lament that the tourism entrepreneurship scholarship is hidden in tourism journals and that cross-fertilisation with mainstream entrepreneurship literature is not as obvious as it should be. This, in their opinion, limits what tourism entrepreneurship research can offer the entrepreneurship field overall. These authors also stress that research on tourism entrepreneurship ignores what could be useful theoretical insights from the mainstream literature. For instance, they contend that tourism researchers can build on their knowledge about tourism firm networks by showing greater awareness of writings on networks in the service industries.

Shaw and Williams (2009) highlight the absence of a clear understanding about management practices in several sectors that constitute the tourism industry, arguing that this is a research direction worthwhile pursuing. Others have called for gender-based perspectives in tourism entrepreneurship, including investigations on how rural-based women can be empowered to develop and manage businesses, which can subsidise more traditional male-dominated farm-based activities (Pettersson & Heldt Cassel, 2014). Figueroa-Domecq et al. (2020) lament the atheoretical manner in which most writings deal with female tourism entrepreneurs. They advocate (p. 10) that by using a feminist postcolonial viewpoint: "[W]omen entrepreneurship and tourism research might rethink its prioritisation with a Global North focus… . Research needs to become attuned with the multiple, fluid and unfolding ways both gender and entrepreneurship are performed through the tourism industry in specific places."

Further, these authors suggest shifting attention on tourism entrepreneurship to regions outside the Global North and paying greater attention to women not only as providers of services (e.g., managerial roles). Greater attention to gendered activities in entrepreneurship should also be placed on urban settings (see also Honggang & Fan, Chapter 8).

Given that tourism has been rapidly transforming through new technologies, this trend has undeniably influenced entrepreneurship practices, opening up novel possibilities for academic research (Ratten, 2020). Important has been the effect of the rise of the so-called peer-to-peer platform-based accommodation sector. Debbage (2019) observes that companies like Airbnb and HomeAway may drastically reshape the geography of traditional tourism destinations. In many non-traditional tourism destinations, homeowners can double up as micro-entrepreneurs by transforming their homes or second homes into tourism business assets. Questions that arise are: [C]ould these online platforms open up opportunities for declining rural areas by enabling properties to be transformed into tourism accommodation? To what extent can such developments boost spin-off businesses and overall regional economic development? Such questions, according to Debbage, are important, given that these new technologies have further blurred the boundaries between the production and consumption of tourism, a sector that already suffers from numerous definitional ambiguities (Ioannides & Debbage, 1998).

The role of rural-based small-scale entrepreneurs as stewards of sustainability and ethical practices is another research avenue worth investigating (Debbage, 2019). Reflecting on the resilience shown by of micro-entrepreneurs in the Swedish countryside who quickly adapted their products following the outbreak of the COVID-19 pandemic, Yachin (2020) wonders whether this could signify a "paradigm shift towards a most just, sustainable and holistic approach to tourism development" (Yachin, 2020: 90, see also Ioannides & Gyimóthy, 2020).

Finally, we believe it would be useful to expand the study of tourism entrepreneurship beyond the role of small firms by investigating so-called policy as an instigator of change in tourism destinations. Van der Zee and Vanneste (2015: 51) recognise the lack of attention towards the role of organisations like DMOs or specific individuals as "network champions" who act entrepreneurially to promote an agenda such as diversifying existing tourism products. Indeed, most existing studies on policy-oriented networks focus on network structure and discuss the limited active participation of tourism firms (Petridou et al., 2019) but also the fact that varying priorities of stakeholders in a network generate a chasm between theory and practice (Beaumont & Dredge, 2010). Thus, we wonder if such discussions could benefit by linking them to the topic of so-called policy entrepreneurs. These are individuals who are either elected or appointed officials but can also be other actors, including business owners, who lead to effective change through their attributes of "ambition, social acuity, credibility, sociability, and tenacity" (Petridou & Mintrom, 2020: 13). In a destination, tourism policy entrepreneurs would, for example, be those individuals who pursue a goal of a desirable transformation towards a more sustainable tourism product. While the field of policy administration is rich with studies on such policy entrepreneurs, it is unclear if they exist in the highly fragmented tourism arena. Studies on topics that could expand the tourism entrepreneurship literature beyond the business domain could pursue questions like: Do tourism policy entrepreneurs exist and, if so, who are they? How does the spatial or political context play out in triggering policy entrepreneurship in destinations? How do policy entrepreneurs nudge others in their network to concur with their ideas?

Conclusion

The place-based attributes of the tourism sector make the study of tourism entrepreneurship a fruitful research area for tourism geographers. Indeed, in recent years several geographers have paid attention to various aspects relating to this topic. Here, we have highlighted the value of spatial bricolage, which explores the interplay of entrepreneurs and their spatial environment, as an important framework in the study of tourism entrepreneurship, especially in the context of micro-firms in rural environments. We suggest that this constitutes a fruitful avenue of further research while also highlighting several additional avenues for further exploration that can help us better comprehend, among others, the geographical dimensions of tourism firms and the role these entities play in determining the competitiveness of destinations.

We end by reflecting on whether tourism entrepreneurs can be key actors in the transformation of tourism products and, ultimately, a destination's goal of achieving a holistic approach to sustainable development. Indeed, are there particular stakeholders (either from the private sector or public organisations) within a destination who stand out as policy entrepreneurs by acting creatively to implement their agendas? This is a relevant question especially during present times, when numerous destinations worldwide have been struggling to recover from the effects of the devastating COVID-19 pandemic. It remains to be seen if eventually at least some of these places will witness a drastic transformation of their tourism sectors and to assess to what extent either public or private entrepreneurs can play a key role in this process.

References

Ateljevic, I., & Doorne, S. (2000). Staying within the fence: Lifestyle entrepreneurship in tourism. *Journal of Sustainable Tourism* 8(5), 378–392.

Baker, T., & Nelson, R. E. (2005). Creating something from nothing: Resource construction through entrepreneurial bricolage. *Administrative Science Quarterly 50*(3), 329–366.

Beaumont, N., & Dredge, D. (2010). Local tourism governance: A comparison of three network approaches. *Journal of Sustainable Tourism* 18(1), 7–28.

Bosworth, G., & Farrell, H. (2011). Tourism entrepreneurs in Northumberland. *Annals of Tourism Research* 38(4), 1474–1494.

Brouder, P. (2014). Evolutionary economic geography: A new path for tourism studies? *Tourism Geographies* 16(4), 540–545.

Brouder, P., & Eriksson, R. H. (2013). Staying power: What influences micro- firm survival in tourism? *Tourism Geographies* 15(1), 125–144.

Bürcher, S., Habersetzer, A., & Mayer, H. (2016). Entrepreneurship in peripheral regions: A relational perspective. In E. A. Mack, & H. Qian, (Eds.), *Geographies of Entrepreneurship* (pp. 143–164). Routledge.

Carson, D. A., Carson, D. B., & Eimermann, M. (2018). International winter tourism entrepreneurs in northern Sweden: Understanding migration, lifestyle, and business motivations. *Scandinavian Journal of Hospitality and Tourism* 18(2), 183–198.

Debbage, K. (2018) Economic geographies of tourism: One 'turn' leads to another. *Tourism Geographies*, 20(2), 347–353.

Debbage, K. (2019). Geographies of tourism entrepreneurship and innovation: An evolving research agenda. In D. Müller (ed.) *A Research Agenda for Tourism Geographies* (pp. 79–88). Edward Elgar Publishing.

Debbage, K., & Ioannides, D. (2012). The economy of tourism spaces: A multiplicity of 'critical turns'? In J. Wilson (ed.) *The Routledge Handbook of Tourism Geographies* (1st ed.) (pp.149–156). Routledge.

Dewhurst, P., & Horobin, H. (1998) Small business owners. In R. Thomas (Ed.) *The Management of Small Tourism and Hospitality Firms* (pp. 19–39). Cengage Learning.

Figueroa-Domecq, C., de Jong, A., & Williams, A. (2020). Gender, tourism and entrepreneurship: A critical review. *Annals of Tourism Research*, 84, https://doi.org/10.1016/j.annals.2020.102980.

Fu, H., Okumus, F., Wu, K., & Köseoglu, M. A. (2019). The entrepreneurship research in hospitality and tourism. *International Journal of Hospitality Management* 78, 1–12.

Hall, C. M., & Page, S. J. (2008). Progress in tourism management: From the geography of tourism to geographies of tourism – A review. *Tourism Management* 30(1), 3–16.

Hjalager, A. M. (2000). Tourism destinations and the concept of industrial districts. *Tourism and Hospitality Research* 2(3), 199–213.

Hjalager, A. M., Kwiatkowski, G., & Østervig Larsen, M. (2018). Innovation gaps in Scandinavian rural tourism. *Scandinavian Journal of Hospitality and Tourism* 18(1), 1–17.

Ioannides, D., & Debbage, K. (1998). *The Economic Geography of the Tourist Industry: A Supply-Side Analysis*. Routledge.

Ioannides, D., & Gyimóthy, S. (2020). The COVID-19 crisis as an opportunity for escaping the unsustainable global tourism path. *Tourism Geographies* 22(3), 624–632.

Kallmuenzer, A., Kraus, S., Peters, M., Steiner, J., & Cheng, C. F. (2019). Entrepreneurship in tourism firms: A mixed-methods analysis of performance driver configurations. *Tourism Management* 74, 319–330.

Kelly, N., Kelliher, F., Power, J., & Lynch, P. (2020). Unlocking the niche potential of senior tourism through micro-firm owner-manager adaptive capability development. *Tourism Management* 79, 104081. https://doi.org/10.1016/j.tourman.2020.104081.

Komppula, R. (2014). The role of individual entrepreneurs in the development of competitiveness for a rural tourism destination – A case study. *Tourism Management* 40, 361–371.

Kornilaki, M., Thomas, R., & Font, X. (2019). The sustainability behaviour of small firms in tourism: The role of self efficacy and contextual constraints. *Journal of Sustainable Tourism* 27(1), 97–117.

Korsgaard, S., Müller, S., & Welter, F. (2021). It's right nearby: How entrepreneurs use spatial bricolage to overcome resource constraints. *Entrepreneurship & Regional Development* 33(1–2), 147–173.

Mack, E. A. (2016). The geography of entrepreneurship. In E. A. Mack, & H. Qian, (Eds.), *Geographies of Entrepreneurship* (pp. 1–12). Routledge.

MacKinnon, D., Cumbers, A., & Chapman, K. (2002). Learning, innovation and regional development: A critical appraisal of recent debates. *Progress in Human Geography* 26(3), 293–311.

Milne, S., & Pohlman, C. (1998). Continuity and change in the hotel sector: Some evidence from Montreal. In D. Ioannides, & K. Debbage (Eds.), *The Economic Geography of the Tourist Industry: A Supply-Side Analysis* (pp. 180–196). Routledge.

Morrison, A., Carlsen, J., & Weber, P. (2010). Small tourism business research change and evolution. *International Journal of Tourism Research* 12(6), 739–749.

Nelson, R. R., & Winter, S. G. (1982). *An Evolutionary Theory of Economic Change*. The Belknap Press of Harvard University Press.

Page, S. J., & Ateljevic, J. (Eds.) (2009). *Tourism and Entrepreneurship: International Perspectives*. Butterworth-Heinemann.

Page, S. J., Forer, P., & Lawton, G. R. (1999). Small business development and tourism: Terra incognita? *Tourism Management* 20(4), 435–459.

Peters, M., Frehse, J., & Buhalis, D. (2009). The importance of lifestyle entrepreneurship: A conceptual study of the tourism industry. *PASOS Revista de Turismo y Patrimonio Cultural* 7(3), 393–405.

Petridou, E., & Mintrom, M. (2020). A research agenda for the study of policy entrepreneurs. *Policy Studies Journal* 49(4), 943–967.

Petridou, E., Olausson, P., & Ioannides, D. (2019). Nascent island tourism policy development in Greenland: A network perspective. *Island Studies Journal* 14(2), 227–244.

Pettersson, K., & Heldt Cassel, S. (2014). Women tourism entrepreneurs: Doing gender on farms in Sweden. *Gender in Management: An International Journal* 29(8), 487–504.

Power, S., Di Domenico, M., & Miller, G. (2020). Risk types and coping mechanisms for ethical tourism entrepreneurs: A new conceptual framework. *Journal of Travel Research* 59(6), 1091–1104.

Quan, H. (2016). The geography of entrepreneurship: Where are we? Where do we go? In E. A. Mack, & H. Qian (Eds.), *Geographies of Entrepreneurship* (pp. 165–174). Routledge.

Ratten, V. (2020). Tourism entrepreneurship research: A perspective article. *Tourism Review* 75(1), 122–125.

Rodriguez-Sanchez, I., Williams, A. M., & Brotons, M. (2019). The innovation journey of new-to-tourism entrepreneurs. *Current Issues in Tourism* 22(8), 877–904.

Shaw, G. (2014). Entrepreneurial cultures and small business enterprises in tourism. In A. A. Lew, C. M. Hall, & A. M. Williams (Eds). *A Companion to Tourism* (pp. 120–131). Blackwell.

Shaw, G., & Williams, A. (2009). Knowledge transfer and management in tourism organisations: An emerging research agenda. *Tourism Management* 30, 325–335.

Solvoll, S., Alsos, G. A., & Bulanova, O. (2015). Tourism entrepreneurship: Review and future directions. *Scandinavian Journal of Hospitality and Tourism* 15(1), 120–137.

Sørensen, F. (2007). The geographies of social networks and innovation in tourism. *Tourism Geographies* 9(1), 22–48.

Stam, E. (2009). Entrepreneurship, evolution and geography. *Papers on Economics and Evolution*, no. 0907. Max Planck Institute of Economics.

Stevenson, H. H., & Jarillo, J. C. (1990). A paradigm of entrepreneurship: Entrepreneurial management. *Strategic Management Journal* 11, 17–27.

Tervo-Kankare, K. (2019). Entrepreneurship in nature-based tourism under a changing climate. *Current Issues in Tourism* 22(11), 1380–1392.

Thomas, R., Shaw, G., & Page, S. J. (2011). Understanding small firms in tourism: A perspective on research trends and challenges. *Tourism Management* 32(5), 963–976.

van der Zee, E., & Vanneste, D. (2015). Tourism networks unravelled: A review of the literature on networks in tourism management studies. *Tourism Management Perspectives*, 15, 46–56.

Welter, F. (2011). Contextualizing entrepreneurship—Conceptual challenges and ways forward. *Entrepreneurship Theory and Practice* 35(1), 165–184.

Williams, A. M., Rodriguez Sanchez, I., & Škokić, V. (2021). Innovation, risk, and uncertainty: A study of tourism entrepreneurs. *Journal of Travel Research* 60(2), 293–311.

Williams, A. M., Shaw, G., & Greenwood, J. (1989). From tourist to tourism entrepreneur, from consumption to production: Evidence from Cornwall, England. *Environment and Planning A* 21(12), 1639–1653.

Wilson, J. (2012). *The Routledge Handbook of Tourism Geographies* (1st ed.). Routledge.

Yachin, J. M. (2018). The 'customer journey': Learning from customers in tourism experience encounters. *Tourism Management Perspectives, 28*, 201–210.

Yachin, J. M. (2020). *Behind the Scenes of Rural Tourism: A Study of Entrepreneurship in Micro-Firms*. Doctoral Thesis in Tourism Studies. Mid-Sweden University.

Yachin, J. M. (2021). Alters & functions: Exploring the ego-networks of tourism micro-firms. *Tourism Recreation Research* 46(3), 319–332.

Yachin, J. M., & Ioannides, D. (2020). "Making do" in rural tourism: The resourcing behavior of tourism micro-firms. *Journal of Sustainable Tourism* 28(7), 1003–1021.

36

TOURISM AND ECONOMIC GEOGRAPHY

An Evolving Agenda

Dimitri Ioannides and Patrick Brouder

Introduction

Tourism, as an applied field of study, naturally lends itself to economic geography inquiry. As a place-based, regionally specific, albeit multifarious economic sector, tourism offers rich cases for analysis. Yet tourism has retained a marginal place in the broader area of economic geography, a point that Britton (1991) highlighted more than three decades ago when he made a passionate plea to view tourism through a political economy lens (see also Bianchi, 2009; 2012). Over the years, Ioannides and Debbage (1998) repeatedly highlighted the need to bridge the gap between tourism studies and economic geography (see also Debbage & Ioannides, 2012; Ioannides & Debbage, 2014). More than a quarter of a century, later and the reality looks more like a river that has been forded only sporadically. However, some of those fords have provided solid enough footing for improved analyses of tourism and economic development and, while they must be carefully approached and negotiated, they certainly provide a basis for further investigation.

In this chapter, we provide an overview of how evolutionary economic geography (EEG) (Boschma & Frenken, 2018) can further our understanding of the dynamics of tourism development in various places and, more importantly, how this approach can help bring tourism-related research into a more central place within economic geography. Following a brief overview of the relationship between tourism studies and economic studies, we look more closely at the links between economic geographies and tourism geographies, highlighting the considerable successes that have been made in this sub-field. We then make the case for evolutionary economic geographies as a productive path for tourism inquiry and expand on the potential applications of this approach in the context of tourism. Finally, we link these research paths to a few emerging areas of research, which may benefit from embracing EEG approaches to augment their own modes of inquiry in the years ahead.

Economic Geography and Tourism: Reticent Research Relations

Tourism has persistently struggled for recognition within economic geography and regional economics, since the tourism sector is perceived as insignificant when compared with 'real' industries such as high technology manufacturing (Calero & Turner, 2020). This is even though tourism

DOI: 10.4324/9781003286301-43

is often touted as a tool for economic growth and diversification in rapidly transforming post-industrial places and remote rural peripheries. The reasons for this are, by now, well known (Ioannides & Debbage, 1998; Ioannides & Debbage, 2014). They include, among others, the tendency to regard manufacturing and producer services as key economic drivers, whereas consumer services, including tourism, receive only a cursory glance (Brouder, 2020; Calero & Turner, 2020; Debbage & Daniels, 1998).

Since tourism lacks a unitary set of standard industrial classification codes and given its partially industrialised nature (Leiper, 1990), these characteristics complicate matters from an economic geography standpoint since this sub-discipline focuses heavily on supply-side perspectives of economic sectors. Even the advent of tourism satellite accounts (Smith, 1998), an important step toward solving the conceptual quagmire of how to define tourism from a supply-side perspective, has not lived up to expectations because of, among others, the considerable costs involved in undertaking such accounting, especially at the regional level (Calero & Turner, 2020).

The absence of any significant mention of tourism in mainstream economic geography treatises has not hindered other researchers from using concepts and tools 'borrowed' from economic geography to investigate tourism as a phenomenon within economic geography. Indeed, since so much of tourism research in general has been produced by geographers it is unsurprising to see studies dealing with issues one would normally expect economic geographers to cover such as: seeking to comprehend the locational dynamics and agglomeration patterns of tourism-related sectors; or investigating tourism's role as an instrument of economic restructuring (Brouder, 2017; Halkier et al., 2019; Larsson & Lindström, 2014; Sanz-Ibáñez & Anton Clavé, 2014; 2016).

Gains that have been made in the confluence of tourism and economic development have tended to be diverse and pluralistic in nature (Hall & Page, 2009). Specifically, tourism researchers have approached the topic from their own perspectives in particular place-based contexts leading to the emergence of the "economic geographies of tourism" (Ioannides & Debbage, 2014: 115). Nevertheless, it is well worth mentioning that, while some success has occurred in developing the scholarship relating to the economic geographies of tourism, most of the writings have been by tourism-focussed researchers who target a tourism-focused readership. Few mainstream economic geographers have written about tourism (e.g., Gibson, 2008; 2009) and only a handful of flagship geography journals have published multiple tourism articles (e.g., *Progress in Human Geography* and *Geoforum*). It is even more astonishing that *Economic Geography*, the flagship journal in the field, has not published a single article on the tourism sector in the period 2000–2023.

This situation means, on the one hand, that economic geographers are rarely exposed to tourism-related articles in field-specific journals. On the other hand, however, there is a tendency in tourism journals to regard economic geography-informed studies, not only as innovative and enlightening, but also as a source of inspiration for new research directions in the economic geography of tourism. Thus, in recent years, there has been a noticeable upturn in the numbers of tourism related articles inspired by new research developments in economic geography. Most noticeably has been the appearance of the so-called evolutionary economic geographies of tourism, inspired by the advent and growing popularity of EEG since the late 1990s (Calero & Turner, 2020). Indeed, the growing number of tourism related studies that have embraced an EEG perspective (Brouder & Eriksson, 2013; Brouder et al., 2017; Carson & Carson, 2017; Ma & Hassink, 2013; Sanz-Ibáñez et al., 2017) have substantially strengthened our theoretical understanding of the evolutionary dynamics of tourism, a topic that has long preoccupied countless researchers (Butler, 1980; Butler, 2006; Christaller, 1964; Miossec, 1977; Stansfield, 1978; Wolfe, 1952). The adoption (and adaptation!) of evolutionary approaches in tourism studies opens the door for tourism scholars to not only learn more about their own chosen cases but also to flow new perspectives back into economic geography theory.

EEG itself has had a major influence in the field of economic geography in recent years (Boschma & Frenken, 2018; Boschma & Martin, 2010). This approach was heavily inspired by Thorstein Veblen (1898) and the field of evolutionary economics (Nelson & Winter, 1982), which focuses heavily on the level of the firm. One of its advantages, as Henning (2019, p. 602) maintains is that "EEG recognizes the importance of both time and history to a scientific understanding of regional development. In this sense, EEG offers a framework that not only enables one to say that time and history matter for regional development, but also how they matter." As Boschma (2022: 125) points out, understanding why regions exhibit varying levels of development has much to do with "contingent historical processes that are often path dependent as well as place dependent."

The adoption of EEG by tourism scholars has proved to be a significant step towards improving our conceptualisation of how tourism as a "dynamic system" (Calero & Turner, 2020: 16) relates to a region's overall development path. Specifically, as Calero and Turner maintain, the EEG lens allows the forces that shape the sector's evolutionary path in specific destinations to be more easily recognised. Previously, we (Ioannides & Brouder, 2017) have maintained that EEG could be the way to shift tourism research into a more recognised position within economic geography especially since it allows one to investigate not only how the tourism sector itself behaves over time but also to understand how its own evolutionary path is interlinked and interacting with other sectors. In other words, viewing tourism as only one part of a particular area's "complex economic structure [makes it] clearer to comprehend its evolutionary track over time and determine the forces that determine its pathway" (Ioannides & Brouder, 2017: 184).

Of course, tourism geographers have long been guilty of preaching to the choir by writing for audiences that are more likely to be sympathetic towards our scientific musings. This is somewhat understandable, and we can hardly be accused of being the only ones guilty of this approach. Nevertheless, this situation creates a tricky scenario for knowledge development in tourism studies. On the one hand, the economic geographies of tourism tend to get siloed, and only a limited set of tourism scholars acknowledge the value of such research. On the other hand, most economic geographers remain unaware of any potential contributions to the sub-discipline of economic geography from this silo. Yet there are certain research strata within economic geography that hold a marked position themselves, and which have emerged more recently meaning that tourism's oddity may not be as apparent as was once thought (Debbage & Daniels, 1998). The rapidly growing interest in evolutionary approaches and the range of potential applications of the concepts emerging open a door to otherwise idiosyncratic realms of application, such as tourism.

The Dynamics of Tourism's Evolution

Tourism has always been recognised as a spatially evolving economic and social phenomenon (Christaller, 1964; Gormsen, 1981). As well as examining change across space much research has focussed on change within specific places (Stansfield, 1978). Arguably, Richard Butler's 'Tourist Area Cycle of Evolution' (Butler, 1980) remains the single most influential concept in tourism studies. While plenty of criticisms of the concept have been made over the years (i.e., that it appears to be too descriptive and proscriptive), this is as much a result of the decoupling of the model from any evolutionary theory (it is interesting that the 'Cycle of Evolution' in the original concept was soon replaced by 'Life Cycle' to give the now ubiquitous term and acronym 'Tourism Area Life Cycle' (TALC). A recombinant TALC that embraces the latest ideas from evolutionary economic geographies (cf. Brouder & Eriksson, 2013) redeems its rightful place as a powerful heuristic of change in tourism places (Hall & Page, 2009).

A persistent challenge that limits the TALC's utility has been the tendency of most researchers to focus on tourism's evolution in a vacuum, without considering how this occurs relative to parallel changes that take place within the geographic area under investigation. In other words, tourism is examined without accounting for how this sector interacts with the institutional, socio-ecological and socioeconomic context in which it is situated. This situation limits the model's real-world application and can lead to a tendency for 'tourism think', for example, that tourism development (read growth!) is the ultimate aim, and that regional sustainability and resilience is simply a by-product. Thankfully, EEG offers the opportunity for tourism researchers to overcome this problem since it places tourism as one cog of a complex economic sector (Brouder, 2014). For instance, by considering path dependent structures, such as institutional inertia, it allows us to understand why in some places, tourism's ability to revive an ailing economy has not been as successful as in others. As Brouder and Fullerton (2015) have demonstrated, it is often hard for an ex-manufacturing dependent locality to readily transform into a tourism destination given that local contingencies, such as the existing skillset of a large portion of the workforce and/or the mindset of community and business leaders, might hinder such changes from happening. Carson and Carson (2017) demonstrated this problem in a case study of tourism development in central Australia where they identified that the creation of a new development trajectory in a specific destination is often influenced by various contingencies including its history, institutional context and culture.

In summary, the dynamics of tourism evolution are complex and can boast a rich research heritage within tourism studies, in general, and tourism geographies, in particular. However, as an applied field of study, tourism empirics readily highlight some deficiencies in the various theories being promulgated. While none of these deficiencies reduce the related theory to rubble, neither can they be dismissed as idiosyncrasies of an anomalous sector. Below, we highlight a selection of recombinant research routes which we believe could augment the evolving agenda of tourism and economic geography in the years to come.

Political Economy of Destination Evolution

Having said this, it is important to remind our readers of the need to incorporate a political economy perspective when considering the evolutionary dynamics of destinations. This was a point recognised long ago when, inspired by Britton (1991), Ioannides (1994) argued that it is imperative to consider the role of the state in influencing how tourism develops in various places. Carson and Carson (2017) elaborated on this discussion by stressing the need to understand how the institutional and political contexts in any locality influence the behaviour of businesses and, by default, affect a sector's evolutionary path. In other words, the value of these authors' approach in EEG is that it allows us to better understand the diverse range of forces that cause path dependency in particular industrial sectors as opposed to evolutionary economics (Nelson & Winter, 1982), which focuses exclusively on behaviours of individual firms. The call by Carson and Carson (2017) to adopt a political economy perspective towards understanding the dynamics of tourism's evolution in any destination builds on the ideas of mainstream scholars in evolutionary economic geography (e.g., Essletzbichler, 2012; MacKinnon et al., 2009; MacKinnon et al., 2019; see also Brouder, 2019; Brouder & Ioannides, 2014). This political economy approach is just one of several additional possible research directions whereby EEG can substantially reinforce the ties of economic geography to tourism studies.

Sustainability Transitions and Tourism Evolution

In the 50th anniversary special issue of *Regional Studies,* Boschma, Coenen, Frenken, & Truffer (2017) argued for new perspectives on regional diversification by bringing together EEG and transition studies. Their conceptual point of departure was an attempt to rebalance studies in EEG, which have tended to focus on related diversification as the strongest source of change in regional economies. By incorporating transition studies, they open space for further interrogation of paths to regional economic evolution. More recently, in tourism studies, Niewiadomski and Brouder (2022; 2023) have called for a recombination of sustainable tourism and EEG. Specifically, they highlight the sustainability transitions literature as a relatively untapped source of new approaches to understanding evolution in tourism. The potential for combining EEG, with its relatively neutral ideals for the future world, and sustainability transitions, with its clear focus on large-scale regime change for a better future, is an academic path that would resonate with tourism scholars and may be an epistemological confluence that will yield valuable insights.

Tourism Work and Workers in an Evolving Tourism Sector

Despite the growing academic attention regarding tourism work and workers over the last two decades (Baum, 2007; Bianchi & de Man, 2021; Ioannides et al., 2021; Ladkin, 2014; Tufts, 2004; Zampoukos & Ioannides, 2011), clearly, we have only just begun to scratch the surface when it comes to strengthening our theoretical understanding of this research theme. Specifically, when discussing the geographies of tourism, Gibson (2009) suggests that matters revolving around labour, including worker mobility, the gendered and ethnic division of labour, and the embodiment of tourism work should become mainstream in geographic research. This is becoming ever more urgent as changing technology processes further undermine the position of labour in the economic system and since, in the cases that EEG does consider labour, this takes place through the prism of inter-firm labour mobility. Thus, tourism with its labour-intensive structure offers a window into the evolution of twenty-first century work.

Gibson's assertion dovetails with the argument made by labour geographer Andrew Herod (1997) who has long argued that workers' agency must be considered when seeking to comprehend how they create "their own historical geographies under capitalism" (Herod, 1997: 16). Based on this perspective and the ideas of McKinnon et al. (2009) that it is vital to account for the role of workers in innovation processes such as through knowledge transfer, we believe that adopting an EEG lens would serve to clarify how, over time, tourism workers in any given locality interact and respond to other agents and institutional practices in shaping its development trajectory.

Policymakers' Role in Shaping Tourism's Evolutionary Path

Bohn et al. (2023) recently reflected that publicly funded projects often influence the behaviour of both private as well as public stakeholders in terms of deciding how the evolutionary pathway of a specific destination will play out. Nevertheless, their study in Arctic regions of Finland and Sweden demonstrates how the funding itself has a limited effect in transforming a destination's development trajectory. The authors explain that since the public funding only lasts for a limited period of time, it is hard to maintain momentum in the projects that depended on this financing. To be sure, financing can often foster networks of stakeholders, albeit ones that might be short lived (Shepherd & Ioannides, 2020), but what is less obvious from this research is whether in certain contexts, longer term success stories emerge due to the persistent agency of a particular influential stakeholder or groups of stakeholders. Do such stakeholders exist and, if so, to what extent do they play an active role in creating new tourism pathways in destination? Baekkelund (2021) argues that, while within EEG agency is often thought to be attributed to one actor, in fact, any changes that occur result from the collective actions of several actors and, as such, "an actor's social network is an integral part of the actor's agency" (Baekkelund, 2021: 758). To illustrate her argument, Baekkelund provides the example of how the collective actions of public and private agents in the Norwegian municipality of Odda led to its transformation from a manufacturing community into one that is increasingly dependent on tourism.

Based on Baekkelund's arguments and drawing inspiration from the literature in policy entrepreneurship (Narbutaite Aflaki et al., 2015) we wonder if so-called policy or political entrepreneurs, namely individuals from the political sphere, bureaucrats or even private citizens exist in tourism-related contexts who might take the initiative to act in a manner that can drastically influence the destination's evolutionary pathway. Is it possible, for instance, to narrow down within a specific institutional setting the identity of actors who can be key agents in escaping the shackles imposed by path dependency (cf. Gill & Williams, 2014)? These questions coincide with the argument made by MacKinnon et al. (2009) that it is imperative to understand the power influences and political structures that might lead to entirely new development trajectories. In turn, these latter points are central for enhancing our political economy perspective with regards to tourism (Britton, 1991; Brouder, 2019; Gibson, 2009).

Additional Thoughts on New Directions in the Economic Geography of Tourism

A fruitful way to further our understanding of tourism's evolutionary dynamics is to better comprehend how this sector can benefit from the transfer of knowledge from other economic sectors that, at first glance, seem to be entirely unrelated. A useful study pointing us in that direction is the one by Larsson and Lindström (2014: 1551), relating to the knowledge transfer between the leisure boatbuilding industry and the tourism sector in the Swedish island of Orust. These authors' argument is that over time, the fortunes of the two sectors have become increasingly interlinked and knowledge transfer between them can be seen "as a way to spur innovation in experience production." This leads us to suggest that the time is ripe to investigate the evolutionary dynamics of tourism in relation to those of other sectors, especially in contexts such as metropolitan regions where multiple sectors co-exist at the same time.

The approach prescribed above relates to intra-regional sectoral connections. However, as Boschma (2022) suggests in a recent paper where he examines the links between Global Production Networks and Global Value Chains to EEG, it would be very useful to also consider inter-regional linkages in such discussions. Specifically, how do leading firms, including multinational companies (but also markets), which exist outside a specific region influence regional path dependency, and to what extent can they influence new development trajectories (Niewiadomski, 2014)? Exploring questions such as this would be useful in terms of furthering our understanding of the dynamics of tourism in specific destinations.

The Telos of Tourism Evolution: It is Going Somewhere, but is it Reaching Something?

We, along with our co-editors of the volume *Tourism Destination Evolution* once asked, 'Why is tourism not an evolutionary science?' (Brouder, Anton Clave, Gill, & Ioannides, 2017). We derived this incisive title from Boschma and Frenken's (2006) paper, where the authors asked the same of economic geography. In turn, they had, of course, creatively 'borrowed' their own title from Veblen's (1898) famous titular query relating to economics.

Unfortunately, what we have not yet managed to do is answer our question! Certainly, there are clear advantages to adopting an evolutionary perspective in tourism studies. Strengthening the links between tourism and economic geography (cf. Ioannides & Debbage, 1998) is an obvious one, though on its own could be viewed as rather self-indulgent. A second advantage is the fact that, despite almost a century of studies (albeit a century of sporadic engagement) on the socio-spatial economy of tourism, we have still only scratched the surface of understanding economic change in tourism places. Happily, a new generation of tourism scholars is taking up the mantle of investigating change in tourism places at various scales from the local to the supra-national. It is our sincere hope that others will join them and immerse themselves in this rich space of inquiry.

We are also aware of the animadversions that abound when EEG and tourism are brought together. To this day, economic geographers may see tourism as frivolous, and EEG is perceived as lacking a positive, normative target (e.g., sustainability) (cf. MacKinnon et al., 2009). It is our hopeful conclusion that tourism evolution, with its recombinant epistemology borne of sustainable development and evolutionary economic geographies, remains a fertile field for meaningful inquiry in tourism.

References

Baekkelund, N.G. (2021). Change agency and reproductive agency in the course of industrial path evolution. *Regional Studies* 55(4), 757–768.

Baum, T. (2007). Human resources in tourism: Still waiting for change. *Tourism Management, 28*, 1383–1399.

Bianchi, R. (2009). The "critical turn" in tourism studies: A radical critique. *Tourism Geographies* 11, 484–504.

Bianchi, R. (2012). A radical departure: A critique of the critical turn in tourism studies. In J. Wilson (Ed.), *The Routledge Handbook of Tourism Geographies* (pp. 46–54). Routledge.

Bianchi, R., & de Man, F. (2021) Tourism, inclusive growth and decent work: A political economy critique, *Journal of Sustainable Tourism* 29(2–3), 353–371.

Bohn, D., Carson, D.A., Demiroglu, O.C., & Lundmark, L. (2023). Public funding and destination evolution in sparsely populated Arctic regions. *Tourism Geographies* 25(8), 1833–1855. DOI: 10.1080/14616688.2023.2193947

Boschma, R. (2022). Global value chains from an evolutionary economic geography perspective: A research agenda. *Area Development and Policy* 7(2), 123–146.

Boschma, R., Coenen, L., Frenken, K., & Truffer, B. (2017). Towards a theory of regional diversification: Combining insights from evolutionary economic geography and transition studies. *Regional Studies* 51(1), 31–45.

Boschma, R., & Frenken, K. (2006). Why is economic geography not an evolutionary science? Towards an evolutionary economic geography. *Journal of Economic Geography* 6, 273–302.

Boschma, R., & Frenken, K. (2018). Evolutionary economic geography. In G.L. Clark et al. (Eds.), *The New Oxford Handbook of Economic Geography* (pp. 213–229). Oxford University Press.

Boschma, R. & Martin, R. (2010). *The Handbook of Evolutionary Economic Geography*. Edward Elgar Publishing.

Britton, S. (1991). Tourism, capital and place: Towards a critical geography of tourism.' *Environment and Planning D: Society and Space* 9, 451–478.

Brouder, P. (2014). Evolutionary economic geography and tourism studies: Extant studies and future research directions. *Tourism Geographies* 16(4), 540–545.

Brouder, P. (2017). Evolutionary economic geography: Reflections from a sustainable tourism perspective. *Tourism Geographies* 19(3), 438–447.

Brouder, P. (2019). Towards a geographical political economy of tourism. In D.K. Muller (Ed.), *A Research Agenda for Tourism Geographies* (pp. 71–78). Edward Elgar Publishing.

Brouder, P. (2020). Reset redux: Possible evolutionary pathways towards the transformation of tourism in a COVID-19 world. *Tourism Geographies* 22(3), 484–490.

Brouder, P. & Eriksson, R. (2013). Tourism evolution: On the synergies of tourism studies and evolutionary economic geography. *Annals of Tourism Research* 43, 370–389.

Brouder, P. & Fullerton, C. (2015). Exploring heterogeneous tourism development paths: Cascade effect or co-evolution in Niagara? *Scandinavian Journal of Hospitality and Tourism* 15(1–2), 152–166.

Brouder, P. & Ioannides, D. (2014). Urban tourism and evolutionary economic geography: Complexity and co-evolution in contested spaces. *Urban Forum* 25(4), 419–430.

Brouder, P., Clavé, S.A., Gill, A., & Ioannides, D. (2017). *Tourism Destination Evolution*. Routledge.

Butler, R. (1980). The concept of a tourist area cycle of evolution: Implications for management of resources. *The Canadian Geographer* 24(1), 5–12.

Butler, R. (2006). *The Tourism Area Life Cycle, Vols. 1 & 2*. Channel View Publications.

Calero, C. & Turner, L.W. (2020). Regional economic development and tourism: A literature review to highlight future directions for regional development research. *Tourism Economics* 26(1), 3–26.

Carson, D.A. & Carson, D.B. (2017). Path dependence in remote area tourism development: Why institutional legacies matter. In P. Brouder, S.A. Clavé, A. Gill & D. Ioannides (Eds.), *Tourism Destination Evolution* (pp. 103–122). Routledge.

Christaller, W. (1964). Some considerations of tourism location in Europe: The peripheral regions-underdeveloped countries-recreation areas. *Regional Science Association Papers* 12(1), 95–105.

Debbage, K.G., & Daniels, P. (1998). The tourist industry and economic geography: Missed opportunities. In D. Ioannides & K. Debbage (Eds.), *The Economic Geography of the Tourist Industry: A Supply-Side Analysis* (pp. 17–30). Routledge.

Debbage, K.G., & Ioannides, D. (2012). The economy of tourism spaces: A multiplicity of "critical turns". In J. Wilson (Ed.), *The Routledge Handbook of Tourism Geographies* (pp. 149–156). Routledge.

Essletzbichler, J. (2012). Generalized Darwinism, group selection and evolutionary economic geography. *Zeitschrift für Wirtschaftsgeographie* 56(3), 129–146.

Gibson, S. (2008). Locating geographies of tourism. *Progress in Human Geography* 32(3), 407–422.

Gibson, S. (2009). Geographies of tourism: Critical research on capitalism and local livelihoods. *Progress in Human Geography* 33(4), 527–534.

Gill, A., & Williams, P. (2014). Mindful deviation in creating a governance path towards sustainability in resort destinations. *Tourism Geographies* 16(4), 546–562.

Gormsen, E. (1981). The spatio-temporal development of international tourism: Attempt at a centre-periphery model. In Baretje R. (Ed.) *La Consommation d' Espace par le Tourisme et sa Preservation* (pp. 150–170). CHET.

Halkier, H., Müller, D.K., Goncharova, N.A., Kiriyanova, L., Kolupanova, I.A., Yumatov, K.V., & Yakimova, N.S. (2019). Destination development in Western Siberia: Tourism governance and evolutionary economic geography. *Tourism Geographies* 21(2), 261–283.

Hall, C.M. & Page, S. (2009). Progress in tourism management: From the geography of tourism to geographies of tourism: A review. *Tourism Management* 30, 3–16.

Henning, M. (2019). Time should tell (more): Evolutionary economic geography and the challenge of history. *Regional Studies* 53(4), 602–613.

Herod, A. (1997). From a geography of labor to a labor geography: Labor's spatial fix and the geography of capitalism. *Antipode* 29(1), 1–31.

Ioannides, D. (1994). *The State, Transnationals, and the Dynamics of Tourism Evolution in Small Island Nations*. Unpublished PhD dissertation. Rutgers University.

Ioannides, D., & Brouder, P. (2017). Tourism and economic geography redux: Evolutionary economic geography's role in scholarship bridge construction. In P. Brouder, S.A. Clavé, A. Gill, & D. Ioannides (Eds.), *Tourism Destination Evolution* (pp. 183–190). Routledge.

Ioannides, D. & Debbage, K.G. (1998). *The Economic Geography of the Tourist Industry: A Supply-side Analysis*. Routledge.

Ioannides, D., & Debbage, K.G. (2014). Economic geographies of tourism revisited: From theory to practice. In A.A. Lew, C.M. Hall, & A.M. Williams (Eds.), *The Wiley Blackwell Companion to Tourism* (pp. 107–119). John Wiley and Sons.

Ioannides, D., Gyimóthy, S., & James, L. (2021). From liminal labor to decent work: A human-centered perspective on sustainable tourism employment. *Sustainability* 13(2), 851.

Ladkin, A. (2014). Labor mobility and labor market structures in tourism. In A.A. Lew, C.M. Hall, & A.M. Williams (Eds.), *The Wiley Blackwell Companion to Tourism* (pp. 132–142). John Wiley and Sons.

Larsson, A. & Lindström, K.N. (2014). Bridging the knowledge-gap between the old and the new: Regional marine experience production in Orust, Västra Götaland, Sweden. *European Planning Studies* 22(8), 1551–1568.

Leiper, N. (1990). Partial industrialization of tourism systems. *Annals of Tourism Research* 17, 600–605.

Ma, M., & Hassink, R. (2013). An evolutionary perspective on tourism area development. *Annals of Tourism Research* 41, 89–109.

MacKinnon, D., Cumbers, A., Pike, A., Birch, K., & McMaster, R. (2009). Evolution in economic geography: Institutions, political economy, and adaptation. *Economic Geography* 85(2), 129–150.

MacKinnon, D., Dawley, S., Pike, A., & Cumbers, A. (2019). Rethinking path creation: A geographical political economy approach. *Economic Geography* 95(2), 113–135.

Miossec, J.M. (1977). Un modele de l' espace touristique. *L'Espace Geographique* 6(1), 41–48.

Narbutaite Aflaki, I., Petridou, E., & Miles, L. (Eds.) (2015). *Entrepreneurship in the Polis: Understanding Political Entrepreneurship*. Ashgate.

Nelson, R.R. & Winter, S.G. (1982). *An Evolutionary Theory of Economic Change*. Harvard University Press.

Niewiadomski, P. (2014). Towards an economic-geographical approach to the globalisation of the hotel industry. *Tourism Geographies* 16(1), 48–67.

Niewiadomski, P. & Brouder, P. (2022). Towards an evolutionary approach to sustainability transitions in tourism. In I. Booyens & P. Brouder (Eds.), *Handbook of Innovation for Sustainable Tourism* (pp. 82–92). Edward Elgar Publishing.

Niewiadomski, P. & Brouder, P. (2023). From 'sustainable tourism' to 'sustainability transitions in tourism'? *Tourism Geographies* 26(2), 141–150.

Sanz-Ibáñez, C., & Anton Clavé, S. (2014). The evolution of destinations: Towards an evolutionary and relational economic geography approach. *Tourism Geographies* 16(4), 563–579.

Sanz-Ibáñez, C., & Clavé, S.A. (2016). Strategic coupling evolution and destination upgrading. *Annals of Tourism Research* 56, 1–15.

Sanz-Ibáñez, C., Wilson, J., & Clavé, S.A. (2017). Moments as catalysts for change in tourism destinations' evolutionary paths. In P. Brouder, S.A. Clavé, A., Gill, & Ioannides, D. (Eds.), *Tourism Destination Evolution* (pp. 81–102). Routledge.

Shepherd, J. & Ioannides, D. (2020). Useful funds, disappointing framework: Tourism stakeholder experiences of INTERREG. *Scandinavian Journal of Hospitality and Tourism* 20(5), 485–502.

Smith, S.L.J. (1998). Tourism as an industry. In D. Ioannides & K.G. Debbage (Eds.), *The Economic Geography of the Tourist Industry: A Supply-side Analysis* (pp. 31–52). Routledge.

Stansfield, C.A. (1978). Atlantic City and the resort cycle: Background to the legalization of gambling. *Annals of Tourism Research* 5(2), 238–251.

Tufts, S. (2004). Building a "competitive city": Labour and Toronto's bid to host the Olympic Games. *Geoforum* 20, 47–58.

Veblen, T. (1898). Why is economics not an evolutionary science? *Quarterly Journal of Economics* 12(4), 373–397.

Wolfe, R.I. (1952). Wasaga Beach – The divorce from the geographic environment. *The Canadian Geographer* 2, 57–66.

Zampoukos, K., & Ioannides, D. (2011). The tourism labour conundrum: Agenda for new research in the geography of hospitality workers. *Hospitality and Society* 1(1), 25–45.

PART VIII

Challenges for the Future of Tourism Geographies Education

37

TOWARD STRATEGY DEVELOPMENT FOR TOURISM GEOGRAPHIES EDUCATION USING TOWS MATRIX

Velvet Nelson

Introduction

While scholarly publications on tourism geographies have increased in recent years (Müller, 2019; Müller & Hall, Chapter 2 of this volume), the subject of tourism geographies education has received relatively little attention in the literature. To some extent, the problems and prospects discussed for the study of tourism geographies may be extrapolated to tourism geographies education; however, education plays a far more important role in the ongoing success of the field than generally acknowledged. Students in large general education classes are future tourists and potentially future decision-makers; students in advanced classes and graduate programmes will work in the industry or lead the next generation of research and problem solving. Moreover, many scholars depend on student enrollment and major numbers to support academic departments and positions, which allow them to continue investigating the important current issues in tourism geographies, from inequalities to the impacts of climate change.

This chapter conceptualises the state of tourism geographies education and proposes strategies for the future through the TOWS (threats, opportunities, weaknesses, and strengths) matrix. The TOWS matrix considers the same elements as a SWOT (strengths, weaknesses, opportunities, and threats) analysis; however, the two differ in perspective and purpose. As indicated by the order of letters in the acronym, SWOT analysis places emphasis on the internal environment (i.e., strengths and weaknesses). In contrast, TOWS analysis requires examination of the external environment first (i.e., threats and opportunities). SWOT is considered a planning tool that allows entities to identify and prioritise key points in each of the four areas. TOWS is considered an action tool. It allows entities to not only identify key points but to understand relationships between the four areas (i.e., the ways in which threats and opportunities connect with weaknesses and strengths), and to develop strategies that will minimise the threats and weaknesses while maximising the opportunities and strengths (Weihrich, 1982).

Drawing from the geography and tourism geographies literature, this chapter identifies threats, opportunities, weaknesses, and strengths shaping contemporary tourism geographies education. Next, the chapter uses the framework of the TOWS matrix to consider strategies to support tourism

DOI: 10.4324/9781003286301-45

geographies classes and programmes. Perspectives from North American universities have typically dominated the literature, limiting insight into other contexts and regions. Thus, the chapter concludes with a call for (1) further discussion of tourism geographies education at different types of higher education institutions and in diverse international contexts and (2) greater efforts to support this education to sustain tourism geographies in the future.

Threats

In higher education, tourism geographies have faced a variety of external threats. Over four decades ago, Jafari and Ritchie (1981: 25) wrote, "For a variety of reasons, both governments and universities appear unwilling to recognise tourism as an important, legitimate field of study which merits the levels of funding accorded to other professional schools and faculties." Authors have also identified threats to university geography programmes. Karan and Mather (1986) noted declining enrollments in geography at American universities, largely attributed to ineffective teaching, and the closure of a major geography department. Similarly, Tapiador and Martí-Henneberg (2007) cited problems attracting new students to geography in Spanish universities. Harvey, Forster, and Bourman (2002) raised concerns that the majority of geography departments in Australian universities had been subsumed by larger, multidisciplinary units. Mather (2007) found a similar situation among several South African universities. As recently as 2019, Kaplan (2019) wrote in an American Association of Geographers (AAG) President's Column, "the last year or so has been sobering, at least for Geography in the United States." American universities continue to face issues such as rising tuition, decreasing public confidence, and decreasing enrollment, all of which pose threats to geography programmes.

The COVID-19 pandemic exacerbated the situation for universities globally. An International Association of Universities report (Marinoni, van't Land, and Jensen, 2020) found that 80 percent of respondents from 109 countries felt that COVID would negatively impact enrollment, which would in turn have negative financial consequences. Travel restrictions during the pandemic caused a sharp decline in international student enrollments in particular, and countries' responses to the pandemic could have lasting implications on whether those students return (Fischer, 2021; Visentin and Bagshaw, 2021). This particularly poses a threat to tourism programmes that rely on these students.

In addition to these threats to tourism and geography programmes and departments, scholars have raised concerns about the longevity of tourism geographies specifically. With few formal tourism geographies programmes (Dornan & Truly, 2009), some tourism geographers have been able to find niches in general geography programmes. Although there are exceptions (see, for example, Han's (2018) discussion of tourism geographies in China), these individuals are frequently the only tourism geographer in their department. In the Canadian context, Gill (2018: 185) noted that tourism geographers "may find themselves marginalised within their departments, often operating as 'lone-wolves'." Any courses or programmes in tourism geographies, such as minors, approved in geography departments are conceptualised and operated by individuals as opposed to collaborative groups. Moreover, few departments offer advanced degrees in tourism geographies (Han et al., 2015).

Commentators have also noted that many tourism geographers are employed in tourism, recreation, hospitality, and/or business departments (Dornan & Truly, 2009; Gill, 2018; Hall, 2005; Hall & Page, 2009; Han et al., 2015). Such departments may again only employ a single tourism geographer. While these appointments can introduce tourism students to a geographic perspective, programmes are not intended to train the next generation of tourism geographers. Additionally, when standalone tourism geographers retire, there is often little interest in replacing them with other tourism geographers (Han et al., 2015). This is viewed as a long-term threat to the continuation of tourism geographies as a field of study, from fewer undergraduate courses exposing students to tourism geographies to fewer PhD graduates in the field (Gill, 2018). Graduates in tourism will lack a foundation in geographic knowledge and theory, while geography graduates with geospatial skills will lack the background to apply those skills in the context of tourism (Che, 2017).

Finally, Bao and Ma (2010) highlighted the global publishing system as an external threat to the perpetuation of Chinese tourism geographies. These authors write,

> It is encouraging to see that the number of research papers … published in Western journals has been on the rise. However, the impact of such publications on Chinese tourism geography is difficult to assess as it is a problem for the vast majority of students and scholars in China to access or read them.
>
> (Bao & Ma, 2010: 15–16)

Opportunities

In the face of these overarching threats to both geography and tourism programmes, tourism geographers remain optimistic about the potential for their field to grow and thrive. Tourism and geography, including geography education, are complementary (Che, 2003; Nelson, 2018, 2021). Tourism geography courses support geography curricula by reinforcing geographic concepts (Nelson, 2018) as well as providing the 'vehicle' through which students can examine geographic questions (Schmelzkopf, 2002). Moreover, tourism is an appealing topic for students. This can be exploited in efforts to raise the visibility of geography among both students and university administrators (Che, 2009) as well as bolster students' enrollment in undergraduate geography classes and programmes (Che, 2003; Gill, 2018; Han et al., 2015).

At a small American university without a geography department, Schmelzkopf (2002) found that it was often difficult to entice undergraduate students to take a course in geography. One of her explicit goals in creating a general education tourism geographies course was to generate student interest in geography. Such courses play a vital role as geography programmes, particularly in the US context, face the threat of elimination due to the combination of low enrollment and reduced budgets. Likewise, Hall and Page (2009) stated that tourism has been viewed as a means of expanding the relevance of geography programmes at universities in Eastern Europe. In addition, Prakapienė and Olberkytė (2013) and Honcharuk, Rozhi, Dutchak, Poplavskyi, Rybinska, and Horbatiuk (2021) made cases for integrating tourism in geography education programmes in Lithuania and Ukraine respectively. Compared to cases in which geography programmes faced declining enrollments, Han, Ng, and Guo (2015) cited the addition of tourism geographies courses in Taiwan to support growing programmes.

Tourism geographies educators have argued that undergraduate and masters-level tourism geographies programmes help meet the demand for knowledge workers in the global tourism system (Che, 2003, 2009; Dornan & Truly, 2009; Han et al., 2015; Seremet et al., 2021). With the tremendous growth of tourism prior to the COVID-19 pandemic, as well as rising awareness of the consequences of tourism for peoples and places, this demand for workers who understand and can work to solve key issues was only expected to increase. For example, Scherle and Hopfinger (2013) noted that climate change had been a key focus of German-speaking tourism geographers.

Given the economic significance of tourism in Alpine countries, the authors cited pressure from the business community to develop a problem-centred perspective.

Despite the disruptive effects of COVID-19 lockdowns on tourism geographies courses, necessary adaptations could also be viewed as opportunities to experiment and innovate. Some educators found that the flipped learning approach (Graham et al., 2017) had benefits for their courses. Students were able to independently review pre-recorded lectures of key concepts (Nelson, 2022), which helped free up in-class time for active learning exercises and intensive discussions (Schmelzkopf, 2002).

While virtual visits and navigations were generally deemed inadequate substitutes for field experiences, educators understood the value of new online activities in overcoming the spatial, temporal, and monetary challenges to student travel, particularly in international contexts (Nelson, 2022). Collaborations that were formed in the context of the pandemic, such as virtual connections between classes in other countries, can be maintained in the future, for example, in the form of international cooperative learning (ICL) as outlined by Lai and Wang (2013).

Talking about tourism during the pandemic crisis also brought new urgency to classroom discussion topics. These discussions helped bring into focus the widespread and often complicated nature of tourism's effects on real peoples and places, as was clearly seen when tourism came to an abrupt halt (Nelson, 2022). For example, the COVID-19 pandemic created an opportunity to delve into the issues of inequality that have long been embedded in tourism practices (Benjamin et al., 2020; Higgins-Desbiolles, 2020). This creates an opportunity to change the way people – from students to departmental colleagues, university administrators, and more – think about tourism, from just about 'fun' to a global phenomenon with significant real-world implications (Nelson, 2022).

Bear and Skorton (2019) cited students' increasing interest in understanding and working toward solutions for real local and global issues as a key factor in recent calls for more holistic and interdisciplinary approaches in higher education. Although there are barriers to these approaches in traditional institutions, there is growing recognition that the integration of knowledge and ways of thinking from different disciplines presents opportunities to tackle complex problems (Hannon et al., 2018; Jacob, 2015). Both geography and tourism are well positioned to capitalise on this trend.

Weaknesses

Despite a century-long history of geographical research on tourism (Dornan and Truly, 2009), tourism geographers have complained that the subject is marginalised within academic geography (Gibson, 2008; Gill, 2018; Hall, 2005; 2013; Han, 2018; Han et al., 2015; Hughes, 1998; Müller, 2014; Saarinen, 2019; Scherle & Hopfinger, 2013). These scholars contest the perception that they study 'fun'. Saarinen (2019: 33) wrote, "This kind of argument should have evaporated into thin air, but nevertheless, despite the massive scale, importance, and obvious or potential consequences of tourism for people and the environment, it is not seen as a serious or relevant enough topic."

This lack of tangible support (i.e., positions and funding) and intangible support (i.e., derision) within geography can constitute a push factor, impelling tourism geographers to leave geography programmes for tourism and recreation departments or to simply pursue other areas of specialisation within geography. Those who remain in geography programmes may be able to research topics in tourism geographies, but their principal teaching responsibilities will be in other areas – if they teach a tourism geographies course at all. Thus, as Jafari and Ritchie (1981) noted years ago, tourism education continues to exist largely independent of tourism research as opposed to benefiting from a close relationship.

In the absence of formal tourism geographies programmes, tenure and promotion systems may not give appropriate credit to publications on aspects of tourism geographies education as compared to thematic research, thus discouraging this kind of work. Accordingly, the academic literature on tourism geographies education remains thin and selective. Han, Ng, and Guo (2015) found that the literature primarily focuses on undergraduate tourism geographies education in geography departments in North America and New Zealand. These countries, along with the UK, also constitute the predominant sources for leading tourism geographies publications (Hall, 2013), but the study of tourism geographies has been rapidly globalising (Müller, 2019; Wilson & Anton Clavé, 2013).

The twentieth anniversary *Tourism Geographies* commentaries on tourism geographies in the authors' countries (Lew, 2018) provided a small glimpse into the state of tourism geographies education around the world. For example, Han (2018: 190) wrote, "In China … tourism geography, as a research and teaching area, operates in an entirely different academic landscape, and one that is both prosperous and fruitful for tourism geographers." In Japan, Funck (2018) noted that the study of tourism geographies is primarily concentrated in a few graduate schools in the Tokyo area. The existing literature offers little insight into the tourism geographies education that occurs in tourism, recreation, hospitality, and other related programmes. In addition, scant attention has been given to tourism geographies education in graduate programmes, community/junior colleges (Han et al., 2015), or even secondary schools. An understanding of the latter programmes in particular could support initiatives to increase enrollment in undergraduate geography programmes.

Strengths

There are no shortages of arguments for tourism geographies education. The relationship between geography and tourism is well-established (Jafari & Ritchie 1981). Indeed, tourism geographers have often argued that geography should be at the heart of tourism studies (Gill, 2018). Dornan and Truly (2009) maintained that geography education provides insights into tourism that other disciplines do not. Ferreira (2018) contended that a geographic background supports a holistic approach to understanding highly complex problems, and Müller (2018: 173) noted that "there is already a tradition within tourism geographies to address complex problems that require insights from different bodies of knowledge." Geography is inherently interdisciplinary, and tourism education requires such an approach (Jafari & Ritchie, 1981; Schmelzkopf, 2002; Wilson, 2012).

The trend of moving away from core geographic ideas, such as place and space, as organising concepts in tourism geographies research (Müller, 2019) can also be seen in tourism geographies education. In the education literature, powerful knowledge refers to the knowledge that is produced within disciplinary communities and taught to students to help them better understand and explain complex phenomena (Maude, 2017). Applied in the context of geography education, geography provides students with the tools to understand important current and future issues and work toward solutions for these issues (Maude, 2016). Likewise, this problem-based learning approach has the potential to guide tourism geographies education as it brings significant issues into the classroom, including climate change mitigation and adaptation, water consumption and scarcity, social inequality and injustice, geopolitical instability and conflict, AI and the digital transformation, and more.

Tourism geographies educators have emphasised the importance of field trips, field courses, and/or study abroad to give students first-hand experience with the phenomena studied (Nelson, 2018). Schmelzkopf (2002) described an end-of-term field experience in which students visited a local destination (e.g., a declining coastal resort town), talked with stakeholders, and weighed the issues facing that destination. She felt this experience played an important role as students applied what they learned in the class in the form of social practice. Similarly, De Bres and Coomansingh (2006) discussed an assignment-based field experience at a local destination (e.g., a small town with heritage resources). Students contacted the stakeholders for a selected attraction and prepared a questionnaire that would be distributed to their classmates after the site visit. Results of the questionnaires were compiled and reported to attraction stakeholders. The authors concluded that students gained a clearer understanding of the challenges faced by the destination. Che (2009) summarised several international tourism geographies study abroad courses and highlighted the potential to raise awareness about thematic issues in real-world contexts.

Che (2009) also discussed service-learning projects as a component of tourism geographies education. Service-learning refers to experiential learning based on partnerships between universities and communities. In higher education, this approach helps meet diverse student learning needs, foster a sense of civic responsibility, and prepare students for employment (Dorsey, 2001). Specifically, Che (2009) noted that projects encouraged collaboration among diverse student groups, connected students with local communities, supported community organisations, and developed students' project management skills.

During the COVID-19 pandemic, educators expressed concerns about the continuation of these types of learning experiences in their tourism geographies courses. Nelson (2022) noted that educators incorporated a variety of virtual alternatives. Reported in her study, a European tourism geographies educator was able to arrange virtual site visits and talks with stakeholders in lieu of the field trip she had planned. This had value in allowing students to see sites and hear the perspectives of different stakeholders. However, the educator was eager to return to a field trip model as soon as it could be done safely. In Turkey, Seremet, Haigh, and Cihangir (2021) described using game theory in a tourism geographies course to practise real-world problem solving in lieu of research-based projects. In South Korea, Choi (2022) was able to get permission to conduct site rural visits for a course service-learning project but was required to take students in small groups for a more limited time.

Strategies

The TOWS Matrix (see Figure 37.1) is designed to match external opportunities and threats with internal strengths and weaknesses to facilitate the formulation of strategic alternatives. First, the Weaknesses and Threats (WT), or mini-mini strategy is oriented toward minimising both weaknesses and threats. Second, the Weaknesses and Opportunities (WO), or mini-maxi strategy attempts to minimise weaknesses while maximising opportunities. Third, the Strength and Threats (ST) strategy or maxi-mini aims to maximise strengths while minimising weaknesses. Finally, the Strength and Opportunities (SO) strategy or maxi-maxi maximises both strengths and opportunities (Weihrich, 1982).

Facing both external threats and internal weaknesses can be a daunting prospect. Tourism geographers are often faced with a lack of external support from governments or university administrators as well as a lack of internal support from departmental colleagues. A WT strategy should consider broader educational efforts and community engagement. Educators can raise awareness about key issues in tourism and tourism geographies research externally through public lectures, news media, and social media as well as internally through colloquia, visual informational displays, and social media. For example, articles on the recently launched Tourism

Internal factors External factors	Strengths (S) • Problem-solving approaches through the powerful geography approach • Field experiences and service-learning projects in tourism geographies courses	Weaknesses (W) • Lack of departmental support for tourism geographies • Separation of tourism geographies teaching and research
Opportunities (O) • Tourism is an appealing topic that can generate interest in geography • Tourism geographies can meet the demand for knowledge workers and problem-solvers	SO: maxi-maxi strategy • Support interdisciplinary programs that draw on the appeal of tourism and the problem-solving approaches in geography • E.g., sustainable tourism, global tourism, cultural heritage tourism	WO: mini-maxi strategy • Scholarly publications and conference presentations on tourism geographies • E.g., share ideas and inspiration for tourism geographies classes and/or programs
Threats (T) • Lack of administrative support for both geography and tourism programs • Declines in student enrollment (i.e., universities, international students, geography programs)	ST: maxi-mini strategy • Increased emphasis on course-based service-learning projects to highlight real-world problem solving • E.g., social media documentation of students working to address tourism-related issues in their communities	WT: mini-mini strategy • Education and community engagement to highlight the value or tourism geographies • E.g., *Tourism Geographic* (TGx) articles to communicate topics in tourism geographies to wider audiences

Figure 37.1   Tourism Geographies Education TOWS Matrix.
Source: Compiled by author.

Geographic provide examples of scholarly work that is written to be engaging and accessible to general audiences.

The general lack of knowledge about tourism geographies education presents a significant weakness that could prevent tourism geographers from capitalising on opportunities. A WO strategy should prioritise an increase in scholarly publications and conference presentations on tourism geographies education. This will offer new insight into tourism geographies education at all levels of higher education (e.g., community or junior colleges as well as undergraduate and graduate programmes) and in diverse international contexts. Sharing successes can provide ideas and inspiration particularly to standalone tourism geographers responsible for developing classes or programmes on their own. These discussions also have the potential to lead to collaboration between tourism geographies educators.

The powerful geography approach strengthens tourism geographies education by working toward solutions to real world issues. An ST strategy would emphasise this approach to highlight the significance of tourism, and the relevance of tourism geographies, and to combat the threat to tourism geographies programmes. Such a strategy would be well supported by service learning and community engagement projects. Through these projects, students are able to apply the tools and concepts obtained in their coursework to addressing real world issues in their communities. These projects support the development of the communication, decision-making, problem-solving, and leadership skills that contribute to career success. In addition, projects foster community support for tourism geographies programmes.

Despite the various threats to and weaknesses within tourism geographies education, tourism geographers remain optimistic about the field. A key SO strategy would focus on the interdisciplinary nature of both geography and tourism to capitalise on the trend toward interdisciplinarity in higher education. As Bear and Skorton (2019: online) hypothesised:

If this movement toward greater integration in higher education persists, perhaps in the future students will no longer major in a specific science, engineering, art, or humanities discipline, but rather will focus their studies on addressing the many real-world challenges they are likely to encounter in the context of their careers, lives, and civic engagement.

Thus, a strategy might involve programmes that focus on addressing specific issues in tourism such as tourism impacts and sustainable tourism development. Additionally, tourism geographers can support interdisciplinary programmes that span diverse departments, from geography to business administration, to reduce the programmes' vulnerability to university cutbacks.

Conclusion

As Weihrich (1982) noted in his proposal of the TOWS Matrix, external and internal environments are dynamic, and factors will change. External threats and internal weaknesses have pushed tourism geographies educators to evolve over time (Che, 2003; Dornan and Truly, 2009; Lai and Wang, 2013). For our field to remain viable, we will need to continue to evolve. The TOWS analysis discussed here considers potential strategies to minimise threats and weaknesses while maximising opportunities and strengths. However, this is simply a starting point. There are still many gaps in the knowledge about tourism geographies education.

Thus, this chapter concludes with a call for greater discussion of tourism geographies education in the literature and at conferences. This should acknowledge the challenges that academic tourism geographers face, which can range from a lack of support for tourism geographies within geography and/or tourism programmes to generally declining enrollments that threaten these programmes. At the same time, tourism geographers need to share their successes, to show how they were able to mobilise the strengths of our field and capitalise on opportunities to provide ideas and inspiration for others. If tourism geographers are able to strengthen courses, programmes, and positions, the field as a whole will benefit.

References

Bao, Jigang & Laurence J.C. Ma. (2010). Tourism Geography in China, 1978–2008: Whence, What and Whither? *Progress in Human Geography* 35(1), 3–20.

Bear, Ashley & David Skorton. (2019). The World Needs Students with Interdisciplinary Education. *Issues in Science and Technology* 35(2), 60–62. https://issues.org/the-world-needs-students-with-interdisciplinary-education/.

Benjamin, Stefanie, Alana Dillette & Derek H. Alderman. (2020). 'We Can't Return to Normal': Committing to Tourism Equity in the Post-Pandemic Age. *Tourism Geographies* 22(3), 476–483.

Che, Deborah. (2003). Tourism Geography Education: Continuities and Changes. *Papers of the Applied Geography Conferences* 26, 205–211.

Che, Deborah. (2009). Teaching Tourism Geography. *Tourism Geographies* 11(1), 120–123.

Che, Deborah. (2017). Tourism Geography and its Central Role in a Globalized World. *Tourism Geographies* 20(1), 164–165.

Choi, Suh-Hee. (2022). Service-Learning During the Pandemic Through a Tourism Geography Course. *Journal of Teaching in Travel & Tourism* 22(3), 220–228.

De Bres, Karen & Johnny Coomansingh. (2006). A Student Run Field Exercise in Applied Tourism Geography. *Journal of Geography* 105, 67–72.

Dornan, D'Arcy & David Truly. (2009). Tourism Geography Education: Opportunities, Obstacles and the Production of Tourism Geographers. *Tourism Geographies* 11(1), 73–94.

Dorsey, Bryan. (2001). Linking Theories of Service-Learning and Undergraduate Geography Education. *Journal of Geography* 100, 124–132.

Ferreira, Sanette. (2018). Tourism Through the Lens of a Human Geographer: A View From the South. *Tourism Geographies* 20(1), 178–181.

Fischer, Karin. (2021). Is This the End of the Romance Between Chinese Students and American Colleges? *The Chronicle of Higher Education.* www.chronicle.com/article/is-this-the-end-of-the-romance-between-chinese-students-and-u-s-colleges (accessed February 2024).

Funck, Carolin. (2018). 'Cool Japan' – A Hot Research Topic: Tourism Geography in Japan. *Tourism Geographies* 20(1), 187–189.

Gibson, Chris. (2008). Locating Geographies of Tourism. *Progress in Human Geography* 32(3) 407–422.

Gill, Alison. (2018). Reflections on Institutional and Paradigmatic Changes in Tourism Geography: A Canadian Perspective. *Tourism Geographies* 20(1), 185–186.

Graham, Marnie, Jessica McLean, Alexander Read, Sandie Suchet-Pearson & Venessa Viner. (2017). Flipping and Still Learning: Experiences of a Flipped Classroom Approach for a Third-Year Undergraduate Human Geography Course. *Journal of Geography in Higher Education* 41(3), 403–417.

Hall, C. Michael. (2005). Reconsidering the Geography of Tourism and Contemporary Mobility. *Geographical Research* 43(2), 125–139.

Hall, C. Michael. (2013). Framing Tourism Geography: Notes from the Underground. *Annals of Tourism Research* 43, 601–623.

Hall, C.M. & S.J. Page. (2009). Progress in Tourism Management: From the Geography of Tourism to Geographies of Tourism – A Review. *Tourism Management* 30, 3–16.

Han, Guosheng. (2018). Tourism Geographies in China: Comparisons and Reflections. *Tourism Geographies* 20(1), 190–192.

Han, Guosheng, Ping Ng, & Yingjie Guo. (2015). The State of Tourism Geography Education in Taiwan: A Content Analysis. *Tourism Geographies* 17(2), 279–299.

Hannon, John, Colin Hocking, Katherine Legge & Alison Lugg. (2018). Sustaining Interdisciplinary Education: Developing Boundary Crossing Governance. *Higher Education & Research Development* 37(7), 1424–1438.

Harvey, N., C. Forster & R.P. Bourman. (2002). Geography and Environmental Studies in Australia: Symbiosis for Survival in the 21st Century? *Australian Geographical Studies* 40(1), 21–32.

Higgins-Desbiolles, Freya. (2020). Socialising Tourism for Social and Ecological Justice After COVID-19. *Tourism Geographies* 22(3), 610–623.

Honcharuk, Vitalii, Inna Rozhi, Olena Dutchak, Myhailo Poplavskyi, Yuliia Rybinska & Nataliia Horbatiuk. (2021). Training of Future Geography Teachers to Local Lore and Tourist Work on the Basis of Competence Approach. *Revista Româneasca pentru Educatie Multidimensionala* 13(3), 429–447.

Hughes, George .(1998). Tourism and the Semiological Realization of Space. In George Ringer (Ed.) *Destinations: Cultural Landscape of Tourism* (pp. 17–32). Routledge.

Jacob, W. James. (2015). Interdisciplinary Trends in Higher Education. *Palgrave Communications* 1, 15001. https://doi.org/10.1057/palcomms.2015.1.

Jafari, Jafar & J.R. Brent Ritchie. (1981). Toward a Framework for Tourism Education: Problems and Prospects. *Annals of Tourism Research* 8(1), 13–34.

Kaplan, David H. (2019, July). Should We Be Worried? Or How to Maintain and Expand the Number of Geographers in Our Schools. *AAG Newsletter*.

Karan, Pradyumna P. & Cotton Mather. (1986). The Trouble with College Geography. *Journal of Geography* 85(3), 95–97.

Lai, Kun & Suosheng Wang. (2013). International Cooperative Learning and Its Applicability to Teaching Tourism Geography: A Comparative Study of Chinese and American Undergraduates. *Journal of Teaching in Travel and Tourism* 13(1), 75–99.

Lew, Alan A. (2018). Tourism Geographies Today. *Tourism Geographies* 20(1), 163.

Marinoni, Giorgio, Hilligje van't Land & Trine Jensen. (2020). The Impact of COVID-19 on Higher Education Around the World. *International Association of Universities*. www.iau-aiu.net/IMG/pdf/iau_covid19_ and_he_survey_report_final_may_2020.pdf (accessed February 2024).

Mather, Charles. (2007). Between the 'Local' and the 'Global': South African Geography after Apartheid. *Journal of Geography in Higher Education* 31(1), 143–159.

Maude, Alaric. (2016). What Might Powerful Geographical Knowledge Look Like? *Geography* 101, 70–76.

Maude, Alaric. (2017). Applying the Concept of Powerful Knowledge to School Geography. In Clare Brooks, Graham Butt & Mary Fargher (Eds.), *The Power of Geographical Thinking* (pp. 27–40). Springer.

Müller, Dieter K. (2014). 'Tourism Geographies are Moving Out' – A Comment on the Current State of Institutional Geographies of Tourism Geographies. *Geographia Polonica* 3(87), 299–312.

Müller, Dieter K. (2018). Time to Reconsider Tourism Geographies? *Tourism Geographies* 20(1), 172–174.

Müller, Dieter K. (2019). Tourism Geographies: A Bibliometric Review. In Dieter K. Müller (Ed.) *A Research Agenda for Tourism Geographies* (pp. 7–22). Edward Elgar Publishing.

Nelson, Velvet. (2018). Geography and Tourism in Undergraduate Education. *Tourism Geographies* 20(5), 919–920.

Nelson, Velvet. (2021). *An Introduction to the Geography of Tourism, 3rd Edition.* Rowman and Littlefield Publishers,

Nelson, Velvet. (2022). COVID-19 and Tourism Geography at a Crossroads: Challenges, Opportunities and Moving Forward. In Stanley D. Brunn & Donna Gilbreath (Eds.), *COVID-19 and An Emerging World of Ad Hoc Geographies* (pp. 1913–1925). Springer.

Prakapienė, Dalia & Loreta Olberkytė. (2013). Using Educational Tourism in Geographical Education. *Review of International Geographical Education Online* 3(2), 138–151.

Saarinen, Jarkko. (2019). Not a Serious Subject?! Academic Relevancy and Critical Tourism Geographies. In Ieter K. Müller (Ed.), *A Research Agenda for Tourism Geographies* (pp. 33–41). Edward Elgar Publishing.

Scherle, Nicolai & Hans Hopfinger. (2013). German Perspectives on Tourism Geography. In Julie Wilson & Salvador Anton Clavé (Eds.), *Geographies of Tourism (Tourism Social Science Series, Vol. 19)* (pp. 69–89). Emerald Group Publishing Limited.

Schmelzkopf, Karen. (2002). Interdisciplinarity, Participatory Learning and the Geography of Tourism. *Journal of Geography in Higher Education* 26(3), 181–195.

Seremet, Mehmet, Martin Haigh & Emine Cihangir. (2021). Fostering Constructive Thinking about the 'Wicked Problems' of Team-Work and Decision-Making in Tourism and Geography. *Journal of Geography in Higher Education* 45(4), 517–537.

Tapiador, Francisco J. & Jordi Martí-Henneberg. (2007). Best of Times, Worst of Times: A Tale of Two (Spanish) Geographies. *Journal of Geography in Higher Education* 31(1), 81–96.

Visentin, Lisa & Eryk Bagshaw. (2021). Universities Brace for Chinese Student Hit After COVID. *The Age.* www.theage.com.au/politics/federal/universities-brace-for-chinese-student-hit-after-covid-20210226-p5764m.html (accessed February 2024).

Weihrich, Heinz. (1982). The TOWS Matrix – A Tool for Situational Analysis. *Long Range Planning* 15(2), 54–66.

Wilson, Julie (Ed.) (2012). *The Routledge Handbook of Tourism Geographies* (1st ed.). Routledge.

Wilson, Julie & Anton Clavé, Salvador (Eds.) (2013). *Geographies of Tourism: European Research Perspectives.* Emerald Group Publishing.

PART IX

Conclusions

38

TOURISM GEOGRAPHIES FOR THE 2020S

Dieter K. Müller and Julie Wilson

Introduction

While the COVID-19 pandemic hitting the world during 2020 and 2021 was anticipated as a game changer by some, as the *Tourism Geographies* special issue on the topic illustrated (Lew et al., 2021), indeed there is little to indicate that this has been the case. This is in line with other commentators' perceptions that tourism can be a part of a solution to pertinent challenges related to employment and injustice (Butcher, 2023). Hence, many of the rooted questions that tourism geographers have put forward have not lost their relevance – issues of local and regional development and various other kinds of impacts on host communities and environments. Still, one may wonder whether anything at all has changed?

Tourism geographies are a vibrant field of research, as this volume vividly demonstrates. As a scientific practice, they are also a successful global undertaking and it seems that tourism geographies are not moulded to the same extent as other scientific fields by the hegemonic Anglo–American structures (Müller, 2021). Even though no compilation is perfect, this volume gathers authors from Europe, the Americas, Asia, Africa and Oceania, and the comprehensive global reach of tourism geographies is seen at conferences and in other publications as well (Müller, 2019).

Indeed, several of the topics raised in the previous volume (Wilson, 2012) remain firmly on the research agenda, while others have emerged during the past decade. This new Handbook is published in a time of change, challenging our imaginaries of what tourism will look like in the future. Certainly, during the production of this book, several crises occurred, which challenged our expectations and thinking of how tourism would and should develop. However, we learnt that development is difficult to foresee, and that tourism indeed is a constant force with the potential to survive a pandemic.

It is thus a risky undertaking to project a future for tourism as a phenomenon, but also for the practice of tourism research. Still, the 2020s seem to continue amidst troubled times – indeed, in an era of polycrises (see Wilson & Müller, Chapter 1) – and there is currently little to indicate a future with fewer challenges. Hence, over the course of the following sections, a couple of challenges are listed that we argue have the potential to change the future course of tourism in the short to mid term.

DOI: 10.4324/9781003286301-47

Tourism and Climate Change

For a while now, climate change has been identified as a major threat to the environment and to humanity (see also Huijbens, this volume; Knowles & Scott, this volume). Recent extreme weather events and undesired heat records, almost on an annual basis, underline the urgency of the problem and the need to mitigate and adapt. Not least air travel has been spotted as a major contributor of greenhouse gas emissions and thus, global tourism is a field where significant change ought to be achieved (Tourism Panel on Climate Change, 2023). In parallel, alternatives such as proximity tourism have been promoted to find ways towards a more sustainable tourism consumption (Rantala et al., 2020; Romagosa, 2020).

However, even though awareness of the climate crisis and its impacts on the planet may be on the rise, actions to cope with them seem to remain limited. Certainly, while many destinations work actively to mitigate or at least minimise negative environmental impacts of tourism, few have however chosen an exit strategy that discontinues efforts to continue developing tourism. In many destinations, while the sector can play an important role for employment and welfare and other less environmentally harmful activities, allowing for the same level of economic and social wellbeing is simply not in sight. This path dependency makes it difficult to imagine radically alternative futures.

Besides efforts on the local or regional destination level, attempts to tackle the more structural dimensions of tourism on the national and international scales remain scarce. At least, the United Nations' Climate Change Conference COP28 agreed to discontinue dependency on fossil fuels (UNFCCC, 2024). However, this is a reaction to countries not delivering a reduction of emissions as planned and, thus, further negative deviations from an anticipated declining curve of emissions can be expected.

On a national scale, especially in the Global North, governments' willingness to regulate transportation is low. Transportation networks are central to economic development and, thus, governments are interested in sustaining connectivity. At the same time, political leadership does not always seem to be willing to take the necessary action to handle the climate crisis. This can be because of pressure triggered by material phenomenology the rise of populist parties doubting climate change or unwilling to bear its cost, or because leadership is itself part of such movements (see even Duval & Macilree, Chapter 32, and Papatheochari et al., Chapter 22, this volume). Often, responsibility for action is thus forwarded to individuals and households who are expected to change behaviour for the sake of the environment. While some protagonists try to demonstrate alternative tourism futures, little suggests that a majority will join them in their undertaking. Instead, the surprisingly quick recovery of travel after the pandemic indicates quite the opposite.

Regardless of the success of mitigation and adaptation efforts, it is already clear that extreme weather events will continue to influence tourism as they already do today. This will likely change the geographies of tourism, too. A lack of snow, floodings, and extremely hot summers will erode the resource base for some destinations, while others will benefit from such change and the new mobilities thereby generated (for example 'coolcations'). Climate change will thus remain on the agenda for tourism geographers and indeed, it can be expected that geographers' positioning at the intersection of the natural and the social sciences provides an ideal point of departure for enabling contributions of relevant knowledge, for understanding and tackling future challenges. This is already demonstrated today by research on resilience and regenerative approaches in tourism (see Ólafsdóttir & Sæþórsdóttir, Chapter 20; Bellato et al., Chapter 23), although this stretches to political, economic and social change as well (Cheer & Lew 2017; Hall et al., 2017).

(Geo)political Turmoil

Ultimately, global tourism benefits from peace and stability, implying open borders and freedom to travel. While this has been a privilege of citizens of the Global North and of well-off elites of the Global South, globalisation has indeed implied that growing numbers of people in countries like China and India had the opportunity to take part in tourism since the late twentieth century (UNWTO, 2024). This does not imply just access for everybody. However, it indicates that tourism indeed plays a core role in many households' lives, not only for spending leisure in pleasant environments, but also for sustaining global family networks or work relations.

This globalisation of tourism has recently been threatened, not only by the COVID-19 pandemic with its mobility restrictions, but also by warfare, such as the Russian invasion of Ukraine, which has caused travel restrictions and closed borders. Multiple other armed conflicts, piracy and terror attacks continue to shock the world. Moreover, these are accompanied by sometimes chauvinist, nationalist, populist and even racist governments and societies, not very open to international tourism. Recent developments seem thus to be in contrast to those of the late twentieth century and, indeed, the temporary de-globalisation during the pandemic that was seen as an opportunity to rethink (Niewiadomski, 2020), might just still turn into a more permanent phase. However, this time, it might just be marred by xenophobic sentiments or geopolitical unrest.

However, tourism is not just an innocent victim of global change driven by extraneous forces. At a more local level, tourism is indeed also considered as a force challenging the social and sometimes economic order. Not least in the context of overtourism, tourism has been seen as a unsustainable force creating unacceptable pressure on the environment, housing markets and the social fabric of neighbourhoods and communities (Milano et al., 2019; see also Papatheochari et al., Chapter 22, Russo & Ivars, Chapter 26, and Buhr & Cocola-Gant, Chapter 31 of this volume). While the notion of overtourism has also contributed to debates on degrowing tourism (Blázquez et al., Chapter 5 of this volume, Fletcher et al., 2020; Hall et al., 2020), a more recent approach is presented by Bellato and colleagues who argue for a regenerative tourism contributing to destinations rather than just consuming them (Chapter 32, this volume).

Obviously, tourism geographers are already engaged in analysing the (geo)political dimensions of tourism in various ways. However, the scope of political and social unrest is growing and, thus, tourism geographers may have reason to intensify their efforts to address this topic in the future.

Digitalisation and Artificial Intelligence

Technological advancement has for a long time affected how tourism is performed and organised (see Roelofsen, Chapter 25, Russo & Ivars, Chapter 26, Dudley, Chapter 27, and van der Zee, Chapter 28 of this volume). Of course, the future will see even more changes triggered by innovation. During the last decennium, platform economies have altered the tourism industry and changed the way people and companies organise travel, accommodation and various related services. Today, artificial intelligence is considered a decisive innovation that will undoubtedly affect tourism. Indeed, already by 2023, the number of comments on the potential impact of AI applications justified a meta-analysis of this discourse (García-Madurga & Grilló-Méndez, 2023). Forecasting, operational efficiency improvements, the enhancement of customer experiences and sustainability have been identified as fields where AI is expected to make a change.

The COVID-19 pandemic arguably demonstrated how technological change influences mobility as well. Lockdowns and various forms of mobility restrictions during the pandemic forced people to work from home. However, while this has not been welcomed by everyone and everywhere, the pandemic provided a showcase that works in many sectors, indeed, can be organised in spatially flexible ways. This has triggered new ways of organising life in space and time, and it challenges traditional dichotomies between 'home' and 'away' or 'work' and 'leisure'. In this volume, Hannonen discusses this topic in relation to digital nomads who utilise a newly gained geographical independence and allocate their work to places (often) in tourist destinations offering good Internet connections and preconditions to achieving a desired lifestyle. Destinations and countries reacted quickly and tried to improve their 'stickiness' by providing a set of special visa regulations and offers, such as the availability of coworking spaces. This is expected to attract and retain groups that are considered creative and economically powerful. While digital nomadism may currently be considered somewhat avant-garde, and an option only for a limited amount of people during certain periods of the lifecourse, second-homes may cater for potentially larger shares of society to achieve similar outcomes (Hall & Müller, 2018; Pitkänen et al., 2020).

In any case, digitalisation and other such current developments indicate that notions of tourists being people on vacations have increasingly become outdated and obsolete. Instead, the blurring of categorical boundaries as indicated above reaffirms the call for a mobilities turn (Cresswell, 2006; Hannam et al., 2006) and indeed, for a thorough rethinking of our understanding and theoretical conceptualisation of tourism geographies as well.

Moreover, as Roelofsen reminds us in Chapter 25, digitalisation should not skew geographers' perspectives from the material conditions in one place that enable others to yield the benefits of such processes elsewhere. The provision of energy and various raw materials needed to facilitate a digital development are examples of such materialities. Such a notion also underlines the spatial and functional relationships between tourism and other sectors of society that from a superficial perspective are only loosely connected. Inherently, this calls for geographers with an ability to integrate knowledge to push the boundaries of their interests and move beyond the pure study of an industry.

This is not an AI-generated text …

AI technologies, from algorithmic management to automated decision-making systems and everything in between, already have the potential to exacerbate existing inequalities and injustices within tourism. However, as Malik (2024) reminds us, the concern lies not only in a dystopian future where machines dominate humans, but in the current reality where AI can serve as a tool for consolidating power and authority, often at the expense of the most vulnerable – scenarios that are easy to imagine as reproducible in tourism contexts.

However, besides changing practices of tourism consumption and production, AI can also be expected to change the practice of doing tourism research. This book is based on the expertise of tourism geographers, which has been assembled through years of scientific work including reading, thinking, empirical field studies and debate. These assets were used to provide reviews of what has been done within the field and for identifying research gaps and providing outlooks concerning what could and should be done in the future. Optimists regarding the potential of technological advancements may see that in a not-too-distant future this role has been taken on by generative AI tools, which will be able to review the existing literature quicker, more comprehensively and unconstrained by language barriers.

This calls for a reconsideration of what research is, and should be, about. Already today the constantly growing number of publications, fueled not least by journals with questionable peer review practices, imply that gaining a comprehensive knowledge of a field increasingly becomes a utopia. Utilising generative AI tools for summarising this literature will thus most likely become a self-evident part of future scientific work. Whether AI tools will be able to assess the literature more qualitatively and to identify new questions remains to be seen. In an ideal situation, the new preconditions will trigger a renewed focus on creativity and innovative blue skies thinking among researchers. Filling current gaps in tourism knowledge, rather than reproducing already existing knowledge one more time, will hopefully take a more dominant role in academic work. The legacy of geographical research combining an interest for real-world problems and 'being' in the landscape are otherwise assets that should be valuable for maintaining a successful path into the future.

Hence, the future will no doubt see new handbooks on tourism geographies. Maybe AI tools will be used to generate parts of them, but essentially an ever-changing spectacle such as tourism will also in the future require fresh examination and creative thinking to make sense out of an increasingly complex phenomenon. Some of the perspectives of this volume will hold into the future, but they will certainly be accompanied by hitherto unthought-of approaches and ideas. Hence, tourism geographies will remain a vibrant and complex field of investigation.

References

Butcher, J. (2023). COVID-19, tourism and the advocacy of degrowth. *Tourism Recreation Research* 48(5), 633–642.

Cheer, J.M., & Lew, A.A. (Eds.) (2017). *Tourism, Resilience and Sustainability: Adapting to Social, Political and Economic Change*. Routledge.

Cresswell, T. (2006). *On the Move: Mobility in the Modern Western World*. Routledge.

Fletcher, R., Mas, I.M., Romero, A.B., & Blázquez Salom, M. (Eds.) (2020). *Tourism and Degrowth: Towards a Truly Sustainable Tourism*. Routledge.

García-Madurga, M.Á., & Grilló-Méndez, A.J. (2023). Artificial intelligence in the tourism industry: An overview of reviews. *Administrative Sciences* 13(8), 172.

Hall, C.M., & Müller, D.K. (Eds.) (2018). *The Routledge Handbook of Second Home Tourism and Mobilites*. Routledge.

Hall, C.M., Lundmark, L., & Zhang, J.J. (Eds.) (2020). *Degrowth and Tourism: New Perspectives on Tourism Entrepreneurship, Destinations and Policy*. Routledge.

Hall, C.M., Prayag, G., & Amore, A. (2017). *Tourism and Resilience: Individual, Organisational and Destination Perspectives*. Channel View Publications.

Hannam, K., Sheller, M., & Urry, J. (2006). Mobilities, immobilities and moorings. *Mobilities,* 1(1), 1–22.

Lew, A., Cheer, J., Brouder, P., Teoh, S., Balslev Clausen, H., Hall, C.M., Haywood, M., …, & Salazar, N. (Guest Eds.) (2021). Visions of travel and tourism after the global COVID-19 transformation of 2020. *Tourism Geographies, 22*(3).

Malik, K. (2024). Elon Musk v OpenAI: Tech giants are inciting existential fears to evade scrutiny. *The Observer Artificial intelligence (AI),* 10 March 2024.

Milano, C., Cheer, J.M., & Novelli, M. (Eds.) (2019). *Overtourism: Excesses, Discontents and Measures in Travel and Tourism.* CABI.

Müller, D.K. (2019). Tourism geographies: A bibliometric review. In D.K. Müller (Ed.), *A Research Agenda for Tourism Geographies* (pp. 7–22). Edward Elgar.

Müller, M. (2021). Worlding geography: From linguistic privilege to decolonial anywheres. *Progress in Human Geography, 45*(6), 1440–1466.

Niewiadomski, P. (2020). COVID-19: From temporary de-globalisation to a re-discovery of tourism? *Tourism Geographies* 22(3), 651–656.

Pitkänen, K., Hannonen, O., Toso, S., Gallent, N., Hamiduddin, I., Halseth, G., Hall, C.M., Müller, D.K., Treivish, A., & Nevedova, T. (2020). Second homes during Corona: Safe or unsafe haven and for whom? Reflections from researchers around the world. *Finnish Journal of Tourism Research* 16(2), 20–39.

Rantala, O., Salmela, T., Valtonen, A., & Höckert, E. (2020). Envisioning tourism and proximity after the anthropocene. *Sustainability* 12(10), 3948.

Romagosa, F. (2020). The COVID-19 crisis: Opportunities for sustainable and proximity tourism. *Tourism Geographies* 22(3), 690–694.

Tourism Panel on Climate Change. (2023). *Tourism and Climate Change Stocktake 2023*. Available at: https://tpcc.info/

UNFCCC. (2024). *UN Climate Change Conference – United Arab Emirates*. United Nations Framework Convention on Climate Change. Available at: https://unfccc.int/cop28 [accessed February 2024].

UNWTO. (2024). *UN Tourism Barometer*. UN World Tourism Organization. Available at: www.unwto.org/un-tourism-world-tourism-barometer-data [accessed February 2024].

Wilson, J. (Ed.) (2012). *The Routledge Handbook of Tourism Geographies*. Routledge.

INDEX

gentrification processes 16–17, 27–31, 99, 142, 149, 154, 173, 182, 286, 291, 296, 313, 317–19, 336, 344–50
geoarbitrage 336, 349
geo-ethics 207
geoforce of humanity, the 111–12
Geoforum (journal) 15, 399
geo-heritage 179
geographical knowledge 12, 273
geographic information systems (GIS) 368, 372–3
geographies of marginalisation 32
geography: as a mother/parent discipline 10–16, 21; mainstream 10, 16, 164, 399–402
geography departments 4, 10–12, 15, 21, 410, 413
geohumanities 163
Geojournal (journal) 15
geopolitical instability/turmoil 224, 413, 423
geopolitics 4–5, 31, 73–7, 120, 274, 302
Georgescu-Roegen, Nicholas 51
geosocial relations 112
geothermal energy production 111
geotourism 133
Giddens, Anthony 367–8
global diasporas 180
global environmental–economic paradigm, the 42
globalisation 16, 28, 53, 134, 139, 146–7, 160, 183–5, 262, 332, 362, 423; de-globalisation 362, 423; economic and financial 53; temporary de-globalisation (during COVID-19) 4, 423
global-local nexus 264
'global mobility regime' 169
Global North 26–30, 52, 98, 139, 143, 146, 154, 277, 344, 394, 422–3
global positioning system (GPS) 19, 101, 302–3, 368, 372–3
global production networks 403
global publishing system 411
Global South 26–30, 41, 52, 98, 101, 143, 225–7, 261, 264, 277, 301, 344, 423
Global Tourism Crisis Committee 228
Global Travel Association Coalition (GTAC) 220
global value chains 403
global warming 194–5, 220–1
good living 51
governance 13, 16, 19, 31, 51, 70, 75–7, 132–3, 194, 195, 204, 215, 226–7, 230, 236, 239–43, 262, 265–6, 278, 288–92, 295, 302, 392; governmentality 301–2
GPS tracking systems 302
Great Acceleration, the 50, 201, 204
green transportation 241
growth paradigm 208, 246, 251, 304

Hägerstrand, Torsten 367–72
haptic sensualities 116
Harvey, David 170, 203

heritage 12–14, 19, 31, 131–3, 138–43, 150–1, 172, 179–86, 196–7, 225, 237–8, 288, 293, 391, 401, 414–15
heritage tourism 75, 151, 179, 181, 184–6, 415
heteropatriarchal discourses 117, 278
higher education 10, 410–15
high speed rail (HSR) 361
historical geographies 120, 179–80, 184–6, 402
'hockey stick' shape of exponential increase, the 201
holistays 327
holistic reductionism 214–16
'hollowing out' (of communities) 384
holocene, the 201
homonormalisation 97, 100
H1N1 pandemic (2009) 31
'hope-as-utopia' 43
'hopeful tourism' 41, 246
hotspots, destinations as 100, 334, 338–9
housing 52, 97, 195–6, 285, 295, 311–19, 344–50, 384, 423
hubbing (in air transport) 358
human flourishing 252
human mobilities *see* mobilities
human wellbeing 197, 214
human–wildlife interactions 327
hybridisation (of the platform economy) 76, 319
hypermobility 26, 31

Ice Age, the 201–2, 221
identity 12–14, 19, 28, 66, 72, 75, 91, 96–100, 106, 128–31, 152, 160, 172–3, 215, 261–3, 294, 335, 371, 379, 384, 391, 403
image 14, 70, 88, 108–10, 116–18, 128–30, 142, 171–2, 205–7, 221, 225, 237–9, 277–8, 312–14, 319, 335, 338, 348
(im)mobility/(im)mobilities 26, 61, 67, 325–9
indigenous perspectives 41–2, 246–57; indigenous peoples 143, 202, 247–50, 254–6, 261; indigenous practitioners 120, 143, 248–52
industrial brownfields 182
industrialisation 168, 303
industrial revolution, the 202, 250
inequalities 28, 42–4, 50–3, 74, 83–6, 89–91, 119, 143, 161, 196, 202, 246, 261–5, 275, 279–80, 285–7, 339, 344, 386, 409, 412–13, 424; affective inequality 119
information and communication technologies (ICTs) 285, 312
innovation 19, 83, 150–2, 161, 227, 252, 274, 278–80, 286–91, 350, 379–86, 391–3, 402–3, 423; business innovation 379–86; disruptive innovation 309–13; innovation as a fuzzy concept 380; neologisms in innovation 380; open innovation 291, 383; social innovation 286, 290–1